T0360808

P-adic Analytic Functions

P-adic Analytic Functions

Alain Escassut
Université Blaise Pascal, France

NEW JERSEY • LONDON • SINGAPORE • BEIJING • SHANGHAI • HONG KONG • TAIPEI • CHENNAI • TOKYO

Published by

World Scientific Publishing Co. Pte. Ltd.

5 Toh Tuck Link, Singapore 596224

USA office: 27 Warren Street, Suite 401-402, Hackensack, NJ 07601

UK office: 57 Shelton Street, Covent Garden, London WC2H 9HE

Library of Congress Cataloging-in-Publication Data
Names: Escassut, Alain, author.
Title: *P*-adic analytic functions / Alain Escassut.
Description: New Jersey : World Scientific, [2021] | Includes bibliographical references and index.
Identifiers: LCCN 2020046177 | ISBN 9789811226212 (hardcover) |
ISBN 9789811226229 (ebook for institutions) | ISBN 9789811226236 (ebook for individuals)
Subjects: LCSH: p-adic analysis. | Analytic functions. | Nevanlinna theory.
Classification: LCC QA241 .E835 2021 | DDC 512.7/4--dc23
LC record available at https://lccn.loc.gov/2020046177

British Library Cataloguing-in-Publication Data
A catalogue record for this book is available from the British Library.

For any available supplementary material, please visit
https://www.worldscientific.com/worldscibooks/10.1142/11990#t=suppl

Desk Editors: Britta Ramaraj/Michael Beale

Typeset by Stallion Press
Email: enquiries@stallionpress.com

Printed in Singapore

Contents

Introduction

The theory of analytic and meromorphic functions is well known on the field $\mathbb{C}$. Consider now an algebraically closed ultrametric field that is complete with respect to its ultrametric absolute value, such as the field $\mathbb{C}_p$ which is the completion of the algebraic closure of $\mathbb{Q}_p$ whose absolute value is the p-adic absolute value. It is possible to make a theory of analytic and meromorphic functions and this was done in the 20-th century, with new results recently obtained. However, understanding the behaviour of an analytic function in a domain of such a field $\mathbb{K}$ requires to know all particular properties of $\mathbb{K}$ which are very different from those of the field $\mathbb{C}$. Particularly $\mathbb{C}_p$ is not spherically complete, which means that certain decreasing sequences of disks may have an empty intersection, though the field is complete...All constructions of fields are made in the first Part A in order to work in a field the properties of which are clearly known. Thus, the book is aimed at being autocontained in order to provide readers all basic properties without looking at several other books.

Analytic elements where defined by Marc Krasner in order to make analytic functions. Due to the absence of connected sets in an ultrametric field, we defined infraconnected sets which make the biggest class of sets where analytic elements have a coherent behaviour. Here we do not recall the theory of analytic functions on quasi-connaected subsets, made by Marc Krasner, nor the more general theory of analytic functions on infraconnected subsets (without T-filters) made by Philippe Robba but we will restrict ourselves to the properties of analytic and meromorphic functions on classical sets such as the whole field $\mathbb{K}$, or a disk, or an annulus and the complement of a disk and we will study many properties of these functions, such as the growth order for analytic functions and overall the Nevanlinna theory for meromorphic functions, in the whole field $\mathbb{K}$ made by Abdelbaki Boutabaa and this one inside a disk and finally in the complement of an open disk. We can then obtain many applications on value sharing, parametrization and small functions.

All properties of analytic functions or meromorphic functions are based on the properties of analytic elements inside disks and annuli, with sometimes the use of the famous Mittag-Leffler Theorem for analytic elements on an infraconnected set, due to Marc Krasner and also the factorization of analytic elements on an infraconnected set due to Elhanan Motzkin.

Problems linked to exponentials are well known in complex analysis. Similar problems may be considered in an ultrametric field: Hermite-Lindemann's Theorem and transcendence of one among a few eponentials, for instance. Most of proofs require specific ultrametric methodes. Here we give an original proof of Hermite-Lindemann's Theorem in an ultrametric field which applies not only to $\mathbb{C}_p$ but also the Levi-Civita field.

Given an open disk of center 0 and diameter R and a sequence $(a_n, q_n)_{n\in\mathbb{N}}$ with $\lim\limits_{n\to+\infty} |a_n| = R$, $|a_n| < |a_{n+1}| < R$ and $q_n \in \mathbb{N}$, the problem to construct an anaytic function admitting each a_n as a zero of order q_n was solved by Michel Lazard in a spherically complete field. The construction is a very big work which is recalled here.

The order of growth for entire functions is recalled with all relations to the type and the cotype of growth. Next, a study is made on the order of growth and the type and the cotype of growth for an analytic function inside an open disk.

The p-adic Nevanlinna theory is constructed, in the whole field and inside an open disk, but also in the complement of a hole, with the help of Motzkin's factors.

Branched values are studied in connection with the order of growth for the numerator and the denominator.

The zeros of meromorphic functions and their derivatives are thoroughly examined and are used to study the Hayman conjecture, in the p-adic context. The case of $f'f^2$ was only solved in 2013 in a p-adic field and requires here many intermediate results, some of them due to Jean-Paul Bézsivin.

Concerning small functions in complex analysis, a famous theorem due to K. Yamanoi is well known and unfortunately we don't have an equivalent in p-adic analysis. However, we present here a kind of theorem providing us with an inequality that is not as good as Yamanoi's inequality in complex analysis but lets us obtain some new results in problems on sharing small functions. For instance, two meromorphic functions sharing 7 small functions (ignoring multiplicity) are equal; that applies to analytic functions: analytic functions sharing 3 small functions (ignoring multiplicity) are equal.

As previously remarked, the situation in fields with residue characteristic zero particularly involves Levi-Civita fields and this gets an increasing importance. When results require a specific statement, that is mentioned, particularly in Chapters B.16 and B.17.

A. Ultrametric fields

A.1. Basic definitions and properties of ultrametric fields

In this chapter, we will recall basic definitions and properties on ultrametric fields: ultrametric absolute values, valuation rings, and residue fields. We must define holes of a subset and infraconnected subsets that are essential for the behavior of analytic functions (certain authors improperly call such sets "connected sets," which makes no sense in topology since there are no connected sets except singletons in an ultrametric field). A major interest of the class of infraconnected sets is that it is the biggest class of sets in an ultrametric complete algebraically closed field where the famous Krasner-Mittag-Leffler theorem applies.

Definitions and notations: Throughout the book, we denote by $\mathbb{N}$ the set of integers ≥ 0, by $\mathbb{Z}$ the ring of relative integers, by $\mathbb{Q}$ the field of rational numbers, by $\mathbb{R}$ the field of real numbers, and by $\mathbb{C}$ the field of complex numbers.

Given a topological space T and a subset S of T, we denote by $\overline{S}$ its closure (also called adherence) and by $\overset{\circ}{S}$ its interior (also called opening).

Let E be a field provided with an absolute value $|\ .\ |$ and let log be a real logarithm function of basis $\theta > 1$. We call *valuation associated to that absolute value* the mapping v from E to $\mathbb{R}$ defined as $v(x) = -\log|x|$ and here, we set $\Psi(x) = \log|x|$ and $|E| = \{|x| \mid x \in E\}$. If a set F contains the zero of a ring, we denote by F^* the set $F \setminus \{0\}$. An absolute value is said to be *trivial* if $|x| = 1\ \forall x \in E \setminus \{0\}$.

Throughout the book, we will denote by $\mathbb{L}$ a field complete with respect to a non-trivial ultrametric absolute value and by $\mathbb{K}$ an algebraically closed field complete with respect to a non-trivial ultrametric absolute value. We will denote by $|\ .\ |_\infty$ the Archimedean absolute value defined on $\mathbb{R}$.

Lemma A.1.1: *Let $\mathbb{E}$ be a field provided with an ultrametric absolute value $|\ .\ |$. The completion of $\mathbb{E}$ with respect to that absolute value is provided with an ultrametric absolute value, which continues that of $\mathbb{E}$. The set $\{|x| \mid x \in \mathbb{E}^*\}$ is a subgroup of the multiplicative group $\mathbb{R}_+^*$.*

Definitions and notations: Given a field $\mathbb{E}$ provided with an ultrametric absolute value $|\ .\ |$, the multiplicative group $\{|x| \mid x \in \mathbb{E}^*\}$ is called *the value group of* $\mathbb{E}$ and the additive group $\{v(x) \mid x \in \mathbb{E}\}$ is called *valuation group of* $\mathbb{E}$.

Similarly, the set $\{\Psi(x) \mid x \in \mathbb{E}^*\}$ is a subgroup of $\mathbb{R}$ called *valuation group of* $\mathbb{E}$.

The field $\mathbb{E}$ is said to have *discrete valuation* or *discrete absolute value* if its valuation group is a discrete subgroup of $\mathbb{R}$ and hence is isomorphic to $\mathbb{Z}$. Else, the valuation group is dense in $\mathbb{R}$ and $\mathbb{E}$ is said to have *dense valuation* or a *dense absolute value.*

Lemma A.1.2 is classical and proven in the same way no matter what the absolute value of $\mathbb{E}$.

Lemma A.1.2: *Let $\mathbb{E}$ be a field provided with two absolute values whose associated valuations are v and w, respectively. They are equivalent if and only if there exists $r > 0$ such that $w(x) = rv(x)$ whenever $x \in \mathbb{E}$.*

Proof. If such an r exists, the two absolute values are seen to be equivalent. Reciprocally, we assume them to be equivalent and take $a \in \mathbb{E}$ such that $v(a) \geq 0$. It is seen that $w(a) \geq 0$. On the other hand, for all $x \in \mathbb{E}$ and for all $m, n \in \mathbb{N}$, we have $v\left(\dfrac{x^m}{a^n}\right) > 0$ if and only if $w\left(\dfrac{x^m}{a^n}\right) > 0$. Therefore, we see that $\dfrac{v(x)}{v(a)} > \dfrac{n}{m}$ is equivalent to $\dfrac{w(x)}{w(a)} > \dfrac{n}{m}$. Then, since $\mathbb{Q}$ is dense in $\mathbb{R}$, we have $\dfrac{v(x)}{v(a)} = \dfrac{w(x)}{w(a)}$ whenever $x \in \mathbb{E}$ and therefore $\dfrac{w(x)}{v(x)} = \dfrac{w(a)}{v(a)}$. □

Definitions and notations: The set of the $x \in E$ such that $|x| \leq 1$ will be denoted by U_E and the set of the $x \in E$ such that $|x| < 1$ will be denoted by M_E.

Then Lemma A.1.3 is immediate:

Lemma A.1.3: *$U_{\mathbb{E}}$ is a local subring of E whose maximal ideal is $M_{\mathbb{E}}$.*

Definitions and notations: Henceforth, U_E is called *the valuation ring of E*. The maximal ideal M_E of U_E is called the *valuation ideal* and the field $\mathcal{E} = \dfrac{U_E}{M_E}$ is called *the residue class field of* $\mathbb{E}$. For any $a \in \mathbb{E}$, the residue class of a will be denoted by $\overline{a}$.
The characteristic of $\mathcal{E}$ is named *the residue characteristic of* $\mathbb{E}$ and will be denoted by p.

Lemma A.1.4: *Let F be a subfield of $\mathbb{E}$ and $\mathcal{E}$ (resp. $\mathcal{F}$) be the residue class field of $\mathbb{E}$ (resp. of F). Then $\mathcal{F}$ is a subfield of $\mathcal{E}$. If $\mathbb{E}$ is algebraically closed and if its valuation is not trivial, it is dense.*

Proof. The first statement is immediate. Next, given $\alpha \in \mathbb{E}$ such that $0 < |\alpha| < 1$ and $\beta \in \mathbb{E}$ such that $\beta^q = \alpha^s$, we have $v(\beta) = \dfrac{q}{s}v(\alpha)$ whenever $s \in \mathbb{N}^*$ and $q \in \mathbb{Z}$. □

Lemma A.1.5: *Let V be a $\mathbb{L}$-vector space of finite dimension provided with two norms. Then the two norms are equivalent.*

Proof. Let $\| \,.\, \|$ and $\| \,.\, \|'$ be the two norms on V. We proceed by induction on the dimension of V and assume the equivalence true for subspaces of dimension $n < q$. Let V have dimension q. Let $e_1, ..., e_q$ be a base of V. Let us suppose that the two norms are not equivalent on V. Then there exists a sequence $(u_n)_{n\in\mathbb{N}}$ of the form $u_n = \sum_{j=1}^{q} a_{j,n}e_j$, with $\|u_n\| \geq 1$, such that $\lim_{n\to\infty} \|u_n\|' = 0$. Let S be the subspace of V generated by $\{e_1, ..., e_{q-1}\}$. For every $n \in \mathbb{N}$, we put $v_n = \sum_{j=1}^{q-1} a_{j,n}e_j$.

First, we suppose that

(1) $\lim_{n\to\infty} |a_{q,n}| = 0$.

Since $\lim_{n\to\infty} \|u_n\|' = 0$, we have $\lim_{n\to\infty} \|v_n\|' = 0$. By hypothesis, the restrictions of the two norms to S are equivalent, hence we have $\lim_{n\to\infty} \|v_n\| = 0$. But since $\|u_n\| \geq 1$ for all $n \in \mathbb{N}$, this contradicts (1).

Now, since (1) is not true, there exists a subsequence of the sequence $(|a_{q,n}|)_{n\in\mathbb{N}}$ that admits a strictly positive lower bound and therefore, without loss of generality, we can clearly assume that there exists $r > 0$ such that $|a_{q,n}| \geq r$ for all $n \in \mathbb{N}$. Let $(x_n)_{n\in\mathbb{N}}$ be the sequence defined as $x_n = \dfrac{u_n}{a_{q,n}}$. It is seen that

(2) $\lim_{n\to\infty} \|x_n\|' = 0$.

The two norms $\| \,.\, \|$ and $\| \,.\, \|'$ are equivalent on S and they both are equivalent to the product norm $\| \,.\, \|''$ defined as $\|\sum_{j=1}^{q-1} b_je_j\|'' = \max_{1\leq j\leq q-1} |b_j|$. Since $\mathbb{L}$ is complete, S is complete with respect to $\| \,.\, \|''$. Hence, S is closed in V with respect to the two norms $\| \,.\, \|$ and $\| \,.\, \|'$. Hence by (2), e_q belongs to S, which is absurd and finishes the proof. □

Theorem A.1.6: *Let F be an algebraic extension of $\mathbb{L}$, provided with two absolute values extending the one $\mathbb{L}$. These absolute values are equal.*

Proof. Let v, w be the valuations associated to these absolute values. Let $a \in F$. By Lemma A.1.5, the two absolute values are equivalent on $\mathbb{L}[a]$. Hence, by Lemma A.1.2, there exists $r > 0$ such that $w(x) = rv(x)$ whenever $x \in \mathbb{L}[a]$. But since $v(x) = w(x)$ whenever $x \in \mathbb{L}$ and since there exists $u \in \mathbb{L}$ such that $v(u) \neq 0$, we have $r = 1$. □

Lemma A.1.7: *Let A be a $\mathbb{L}$-algebra. Let ϕ be a semi-norm of $\mathbb{L}$-algebra satisfying $\phi(x^n) = (\phi(x))^n$ $\forall x \in A$. Then ϕ is ultrametric.*

Proof. Let $a, b \in A$ satisfy $\phi(a) \geq \varphi(b)$. We just have to show that $\phi(a+b) \leq \phi(a)$. Obviously we have $\phi((a+b)^n) = \phi(\sum_{k=0}^{n} C_n^k a^k b^{n-k})$. For each $k = 0, ..., n$ we have $\phi(C_n^k a^k b^{n-k}) = |C_n^k| \phi(a^k b^{n-k}) \leq \phi(a)^k \phi(b)^{n-k} \leq \phi(a)^n$, hence $\phi((a+b)^n) \leq (n+1)\phi(a)^n$ and therefore $\phi(a+b) \leq \sqrt[n]{n+1}\ \phi(a)$ for all $n \in \mathbb{N}^*$. Finally, we obtain $\phi(a+b) \leq \phi(a)$. □

The most classical example of an ultrametric complete algebraically closed field is the field $\mathbb{C}_p$ that will be described later.

Definitions and notations: Consider the field $\mathbb{E}$ provided with an ultrametric absolute value. Let $a \in \mathbb{E}$ and let $r \in \mathbb{R}_+$. We denote by $d(a,r)$ the disk $\{x \in \mathbb{E} |\ |x-a| \ \leq r\}$, by $d(a,r^-)$ the disk $\{x \in \mathbb{E} |\ |x-a| \ < r\}$, and we call *circle of center* a, *of radius* r the set $C(a,r) = d(a,r) \setminus d(a,r^-)$.

Given r_1 and r_2 such that $0 < r_1 < r_2$, we denote by $\Gamma(a, r_1, r_2)$ the annulus $\{x \in E |\ r_1 < |x-a| \ < r_2\}$ and by $\Delta(a, r_1, r_2)$ the annulus $\{x \in E |\ r_1 \ \leq |x-a| \ \leq r_2\}$.

We know that if $b \in d(a,r)$, then $d(b,r) = d(a,r)$. In the same way, if $b \in d(a,r^-)$, then $d(b,r^-) = d(a,r^-)$. Moreover, given two disks T and T' such that $T \cap T' \neq \emptyset$, then either $T \subset T'$ or $T' \subset T$.

We denote by δ the distance defined on $\mathbb{E}$ by $\delta(a,b) = |a-b|$. Given $a \in \mathbb{E}$ and a subset D of $\mathbb{E}$, we set $\delta(a, D) = \inf\{|x-a| \mid x \in D\}$ and given two subsets D, F of $\mathbb{E}$, we set $\delta(D, F) = \inf\{|x-y| \mid x \in D,\ y \in F\}$.

We set $\mathrm{diam}(D) = \sup\{|x-y| \mid |x \in D, y \in D\}$ and $\mathrm{diam}(D)$ is named the *diameter* of D.

Similarly, we set $\mathrm{codiam}(D) = \sup\{|x-y|\ |x \in D, y \notin D\}$ and $\mathrm{codiam}(D)$ is named the *codiameter of* D.

Of course, the following three statements are seen to be equivalent:

(i) $d(a,r) = d(a,r^-)$
(ii) $C(a,r) = \emptyset$
(iii) $r \notin \mid E \mid$

Further, the disks $d(b,r^-)$ included in $C(a,r)$ (resp. in $d(a,r)$) are the disks $d(b,r^-)$ such that $b \in C(a,r)$ (resp. in $d(a,r)$). They are called *the classes* of $C(a,r)$ (resp. of $d(a,r)$).

Henceforth, D *will denote a subset of the field* $\mathbb{L}$.

The closure of D (also called *adherence* of D) is denoted by $\overline{D}$ and the interior of D (also called *opening* of D) is denoted by $\overset{\circ}{D}$.

Given a point $a \in L$, we put $\delta(a, D) = \inf\{|x-a|\ |x \in D\}$. Then $\delta(a, D)$ is named *the distance of* a *to* D.

Given two subsets $D,\ D'$ of L, we put $\delta(D,D')=\inf\{|x-y|\ |x\in D,\ u\in D'\}$. Then $\delta(D,D')$ is called *the distance between D and D'*.

We will denote by $\widehat{\mathbb{L}}$ an extension of $\mathbb{L}$ provided with an absolute value that extends that of $\mathbb{L}$. Given $a\in\widehat{\mathbb{L}},\ r>0,\ \widehat{d}(a,r)$ (resp. $\widehat{d}(a,r^-)$) will denote the disk $\{x\in\widehat{\mathbb{L}}|\ \ |x-a|\le r\}$ (resp. $\{x\in\widehat{\mathbb{L}}|\ \ |x-a|<r\}$).

Let D be a subset of $\mathbb{L}$, of diameter $R\in\mathbb{R}$ (resp. $+\infty$), whose holes form a family $\big(d(a_i,r_i^-)\big)_{i\in\mathbb{N}}$. Let $a\in D$. We will denote by $\widehat{D}$ the set $\widehat{d}(a,R)\setminus \Big(\big(\bigcup_{i\in I}\widehat{d}(a_i,r_i^-)\big)\bigcup(\overline{D}\setminus D)\Big)$ (resp. $\widehat{L}\setminus\Big(\bigcup_{i\in I}\widehat{d}(a_i,r_i^-)\Big)$).

Lemma A.1.8: *Let $d(a,r),\ d(b,s)$ be disks such that $d(a,r)\cap d(b,s)\neq\emptyset$ with $r\le s$. Then $d(a,r)\subset d(b,s)$.*

Let us also notice the basic Lemma A.1.9.

Lemma A.1.9: *Suppose that the residue class field $\mathcal{E}$ of the field $\mathbb{E}$ is finite, of cardinal q. Then for every disk $d(a,r)$ with $a\in\mathbb{E}$ and $r\in|\mathbb{E}|$, admits only q classes.*

Lemma A.1.10: *$\widetilde{D}\setminus\overline{D}$ admits a unique partition of the form $(T_i)_{i\in I}$, whereas each T_i is a disk of the form $d(a_i,r_i^-)$ with $r_i=\delta(a_i,D)$.*

Proof. For every $a\in\widetilde{D}\setminus\overline{D}$, let $r(a)=\delta(a,D)$. Let α and β be two points in $\widetilde{D}\setminus\overline{D}$ such that $|\ \beta-\alpha\ |<r(\alpha)$. It is easily seen that for every $x\in D$, we have $|x-\beta|=|x-\alpha\ |$, and then the family of the disks $T(\alpha)=d(\alpha,r(\alpha)^-)$ $(\alpha\in\widetilde{D}\setminus\overline{D})$ makes a partition of $\widetilde{D}\setminus D$ because given α and $\beta\in\widetilde{D}\setminus\overline{D}$, either $|\alpha-\beta|<r(\alpha)$ and then $T(\alpha)=T(\beta)$, or $|\alpha-\beta|\ge r(\alpha)$ and then $|\alpha-\beta|\ge r(\beta)$. Hence, $T(\alpha)\cap T(\beta)=\emptyset$. □

Definitions and notations: Such disks $d(a_i,r_i^-)$ are called *the holes* of D. If D is bounded of diameter R, we denote by $\widetilde{D}$ the disk $d(a,R)$ for any $a\in D$. If D is not bounded, we put $\widetilde{D}=\mathbb{L}$.

Example 1: The holes of a disk $d(a,r^-)$, with $r\in|\ \mathbb{L}\ |$, are the classes of $C(a,r)$.
Example 2: The only one hole of $\mathbb{L}\setminus d(0,1^-)$ is $d(0,1^-)$.
Example 3: The holes of $\mathbb{L}\setminus d(0,1)$ are the disks $d(a,1^-)$ with $a\in d(0,1)$.

Definitions and notations: D is said to be *infraconnected* [44, 50, 58] if for every $a\in D$, the mapping I_a from D to $\mathbb{R}_+$ defined by $I_a(x)=|x-a|$ has an image whose closure in $\mathbb{R}_+$ is an interval. In other words, D is not infraconnected if and only if there exist a and $b\in D$ and an annulus $\Gamma(a,r_1,r_2)$ with $0<r_1<r_2<|a-b|$ such that $\Gamma(a,r_1,r_2)\cap D=\emptyset$.

Lemma A.1.11 is obvious:

Lemma A.1.11: *If D is infraconnected of diameter $R \in \mathbb{R}$ (resp. $+\infty$), then $\overline{I_a(D)} = [0, R]$ (resp. $\overline{I_a(D)} = [0, +\infty[$).*

Lemma A.1.12 gives a point of view from a hole of D.

Lemma A.1.12: *Let D be infraconnected and let α belong to a hole T of diameter ρ. The closure of the set $\{|x - \alpha| \; |x \in D\}$ is an interval whose lower bound is ρ.*

Proof. We just have to show that for every r and r' such that $\rho < r < r' < \text{diam}(D)$, there exists $\beta \in D$ such that $r < |\beta - \alpha| < r'$. By definition of the holes, there exists $b \in D$ such that $|\alpha - b| < r$ and then, since D is infraconnected, there exists $\beta \in D$ such that $r < |b - \beta| < r'$. But it is seen that $|\beta - \alpha| = |b - \beta|$. □

Given two infraconnected sets A and B, we may prove $A \cup B$ to be infraconnected in Theorems A.1.13 and A.1.15.

Theorem A.1.13: *Let A and B be two infraconnected sets such that $A \cap B \neq \emptyset$. Then $A \cup B$ is infraconnected.*

Proof. If A and B are not bounded, the statement is obvious because for every $a \in A$, $\overline{I_a(A)} = \mathbb{R}_+$ and for every $a \in B$, $\overline{I_a(B)} = \mathbb{R}_+$. Now we may assume A to be bounded, of diameter R, whereas B has diameter $R' \geq R$ (resp. is not bounded). Then $A \cup B$ has diameter R' (resp. is not bounded). Let $c \in A \cap B$, let $a \in A \cup B$, and let us show that $\overline{I_a(A \cup B)} = [0, R']$ (resp. $[0, +\infty[$).

For convenience, we first assume B to be bounded. Since $c \in A \cap B$, we see that $|x - a| \leq \max(|x - c|, |c - a|) \leq R'$ whenever $x \in A \cup B$. Hence, $\overline{I_a(A \cup B)} \subset [0, R']$. Hence, we just have to show that $\overline{I_a(A \cup B)} \supset [0, R']$. Obviously, $\overline{I_a(A \cup B)} = \overline{I_a(A)} \cup \overline{I_a(B)} = [0, R] \cup \overline{I_a(B)}$. Hence, we have to show that $\overline{I_a(B)} \supset [R, R']$. But when $x \in B$ with $|x - a| > R$, we see that $|x - a| = |x - c|$ (because $|c - a| \leq R$). Hence, $I_a(B) \cap]R, R'] = I_c(B) \cap]R, R']$ and finally $\overline{I_a(B)} \supset [R, R']$ because $\overline{I_c(B)} \supset [R, R']$.

When B is not bounded, in the same way it is seen that $\overline{I_a(A \cup B)} = [0, +\infty[$. This finishes showing that $A \cup B$ is infraconnected. □

Corollary A.1.14: *The relation $\mathcal{R}$ defined by $x\mathcal{R}y$, if there exists an infraconnected subset of D that contains x and y, is an equivalence relation.*

Proof. $\mathcal{R}$ is obviously reflexive and symmetric. It is transitive by Theorem A.1.13. □

Definitions and notations: The equivalence classes with respect to this relation are called *the infraconnected componants.*

Examples: (1) $d(0, 1^-) \cup d(1, 1^-)$ is infraconnected. Its holes are the disks $d(\alpha, 1^-)$ with $|\alpha| = |\alpha - 1| = 1$.

(2) Let $r \in]0,1[$ and $D = d(0,1^-) \cup d(1,r)$. Then D is not infraconnected, its infraconnected components are $d(0,1^-)$ and $d(1,r)$. The holes of D are the disks $d(\alpha,1^-)$ with $|\alpha| = |\alpha - 1| = 1$ and the disks $d(\alpha, |\alpha-1|^-)$ with $r < |\alpha - 1| < 1$.

Theorem A.1.15: *Let A and B be infraconnected sets such that $\widetilde{A} = \widetilde{B}$. Then $A \cup B$ is infraconnected.*

Proof. Obviously, $\widetilde{A \cup B} = \widetilde{A}$. If A is bounded, let $\widetilde{A} = d(\alpha, R)$ and otherwise let $\widetilde{A} = L$. First let us assume A to be bounded. For $a \in A$, the set $\{|x-a| \mid x \in A\}$ is dense in $[0,R]$. Hence, so is the set $\{|x - a \mid \mid x \in A \cup B\}$. In the same way, B plays the same role. Hence, this still holds for $a \in B$. Finally, if A is not bounded we just replace $[0,R]$ by $[0,+\infty[$. That finishes proving Theorem A.1.15. □

Definitions and notations: An infraconnected subset D of L is said to be *affinoid* if it is of the form $d(a,R) \setminus \bigcup_{k=1}^{q} d(b_k, r_k^-)$ with R and $r_k \in |\mathbb{L}|\ \forall k$. A subset D of L is said to be *affinoid* if it is a finite union of infraconnected affinoid subsets.

Proposition A.1.16: *Let D_1, D_2 be two infraconnected affinoid subsets of $\mathbb{L}$ such that $D_1 \cap D_2 \neq \emptyset$ and set $D = D_1 \cup D_2$ and $E = D_1 \cap D_2$. Then both D and E are infraconnected affinoid. Moreover, $\widetilde{D}$ is either $\widetilde{D_1}$ or $\widetilde{D_2}$ and each hole of D is either a hole of D_1 or a hole of D_2.*

Proof. By Theorem A.1.13 D is infraconnected. Consider now D_1 of the form $d(a,r) \setminus (\cup_{i=1}^{m} d(a_i, r_i^-))$ and D_2 of the form $d(b,s) \setminus (\cup_{i=1}^{n} d(b_i, s_i^-))$. Suppose for instance $r \leq s$ and let $c \in D_1 \cap D_2$. Then we can check that

$$E = d(c,r) \setminus \Big(\big(\cup_{i=1}^{d} (a_i, r_i^-)\big) \cup \big(\cup_{i=1}^{n} d(b_i, s_i^-)\big)\Big),$$

which is an infraconnected affinoid again. Since $D_1 \cap D_2 \neq \emptyset$, we have $\widetilde{D_1} \cap \widetilde{D_2} \neq \emptyset$, hence $\widetilde{D}$ is either $\widetilde{D_1}$ or $\widetilde{D_2}$, hence $\operatorname{diam}(D) \in |\mathbb{L}|$. Next, since the holes of both sets are in finite number, each hole of D is either a hole of D_1 or a hole of D_2, so each hole of D has a diameter in $|\mathbb{L}|$ and of course they are in finite number. □

Definitions and notations: We will call *empty annulus of D* an annulus $\Gamma(a, r_1, r_2)$ such that

(i) $r_1 = \sup\{|x-a| \quad |x \in D, |x-a| \leq r_2\}$
(ii) $r_2 = \inf\{\ |x-a| \quad |x \in D, |x-a| \geq r_1\}$

The set $d(a,r_1) \cap D$ will be denoted by $\mathcal{I}_D(\Gamma(a,r_1,r_2))$, whereas the set $(\mathbb{L} \setminus d(a, r_2^-)) \cap D$ will be denoted by $\mathcal{E}_D(\Gamma(a,r_1,r_2))$. When there is no risk of confusion about the set D, we will just write $\mathcal{I}(\Gamma(a,r_1,r_2))$, (resp. $\mathcal{E}(\Gamma(a,r_1,r_2))$), instead of $\mathcal{I}_D(\Gamma(a,r_1,r_2))$, (resp. $\mathcal{E}_D(\Gamma(a,r_1,r_2))$).

Remark 1: By definition, D is not infraconnected if and only if it admits an empty annulus.

Remark 2: By definition $\{\mathcal{I}(\Gamma(a,r_1,r_2)),\mathcal{E}(\Gamma(a,r_1,r_2))\}$ is a partition of D.

Examples: Let $r\in]0,1[$, $D=d(0,r)\cup d(1,1^-)$, and $D'=d(0,r^-)\cup d(1,r)$. Then $\Gamma(0,r,1)$ is an empty annulus of D and also of D'. In the same way $\Gamma(1,r,1)$ is also an empty annulus of D'.

Definitions and notations: Let $\mathcal{X}(D)$ be the set of the empty annuli of D. Given Λ_1 and $\Lambda_2\in\mathcal{X}(D)$, it is easily seen that $\mathcal{I}(\Lambda_1)\subset\mathcal{I}(\Lambda_2)$ is equivalent to $\mathcal{E}(\Lambda_1)\supset\mathcal{E}(\Lambda_2)$. We will denote by $\le$ the relation defined on $\mathcal{X}(D)$ by $\Lambda_1\le\Lambda_2$ if $\mathcal{I}(\Lambda_1)\subset\mathcal{I}(\Lambda_2)$ and we set $\Lambda_1<\Lambda_2$ if $\Lambda_1\le\Lambda_2$ and $\Lambda_1\ne\Lambda_2$.

Lemmas A.1.17 and A.1.18 are easily seen.

Lemma A.1.17: *The relation $\le$ is a relation of order on $\mathcal{X}(D)$. Let Λ_1 and Λ_2 be two empty annuli of D. The following assertions are equivalent:*

(i) Λ_1 *and* Λ_2 *are not comparable with respect to the order* $\le$

(ii) $\mathcal{I}(\Lambda_1)\subset\mathcal{E}(\Lambda_2)$

(iii) $\mathcal{I}(\Lambda_2)\subset\mathcal{E}(\Lambda_1)$

(iv) $\mathcal{I}(\Lambda_1)\cap\mathcal{I}(\Lambda_2)=\emptyset$

Lemma A.1.18: *Let $\Lambda\in\mathcal{X}(D)$ and let $x\in\mathcal{I}(\Lambda)$ (resp. $x\in\mathcal{E}(\Lambda)$). The infraconnected component of x is included in $\mathcal{I}(\Lambda)$ (resp. in $\mathcal{E}(\Lambda)$). If $\Lambda'\in\mathcal{X}(D)$ is such that $\Lambda<\Lambda'$, then $\mathcal{I}(\Lambda')\cap\mathcal{E}(\Lambda)\ne\emptyset$.*

Lemma A.1.19 is a direct consequence of Lemmas A.1.17 and A.1.18.

Lemma A.1.19: *Let Θ be an empty annulus of D. The family of the empty annuli $\Lambda\ge S$ is totally ordered.*

Proof. Let Λ_1 and $\Lambda_2\in\mathcal{X}(D)$ satisfy $\Lambda_1\ge\Theta,\Lambda_2\ge\Theta$. Then, $\mathcal{I}(\Lambda_1)\cap\mathcal{I}(\Lambda_2)\supset\mathcal{I}(\Theta)\ne\emptyset$, hence $\mathcal{I}(\Lambda_1)$ is not included in $\mathcal{E}(\Lambda_2)$, hence Λ_1 and Λ_2 are comparable. □

Lemma A.1.20: *Let Θ be a minimal element of $\mathcal{X}(D)$ for the order $\le$. Then $\mathcal{I}(\Theta)$ is an infraconnected component of D.*

Proof. Suppose that $\mathcal{I}(\Theta)$ is not infraconnected. By definition, $\mathcal{I}(\Theta)$ is of the form $d(a,R)\cap D$, hence there exists an empty annulus $\Lambda=\Gamma(\alpha,r_1,r_2)$ of $\mathcal{I}(\Theta)$ with $\alpha\in d(a,R), r_1<r_2\le R$ and some $\beta\in\mathcal{I}(\Theta)$ such that $r_2\le|\alpha-\beta|\le R$. Since $\Lambda\subset d(a,R)$, we see that $\Lambda\cap D=\emptyset$, hence Λ is an empty annulus of D and therefore $\Lambda<\Theta$. This ends the proof of Lemma A.1.20. □

Theorem A.1.21: *D has finitely many infraconnected components if and only if it has finitely many empty annuli. Moreover, if so does D then one of the infraconnected components is $A_0=\bigcap\limits_{\Theta\in\mathcal{X}(D)}\mathcal{E}(\Theta)$, whereas the others are of the form*

$$A_i=\mathcal{I}(\Lambda_i)\bigcap\Big(\bigcap_{\Theta<\Lambda_i}\mathcal{E}(\Theta)\Big),\ \textit{with}\ \Lambda_i\in\mathcal{X}(D).$$

Proof. We will first assume $\mathcal{X}(D)$ to be finite and we will prove that the infraconnected components are in the form A_i, above, so that there will be finitely many ones.

Let $\Lambda_1, ..., \Lambda_n$ be these empty annuli of D and for every $i = 0, ..., n$, let A_i be the subsets of D defined from $\Lambda_1, ..., \Lambda_n$ as above. For every $x \in D$, for every $i = 1, ..., n$, either $x \in \mathcal{I}(\Lambda_i)$ or $x \in \mathcal{E}(\Lambda_i)$. Hence, it is easily seen that x belongs to one of the A_i, hence $D = \bigcup_{i=0}^{n} A_i$. We check that $A_i \cap A_j = \emptyset$ whenever $i \neq j$. First we assume $i = 0, j > 0$. Hence, $A_0 \subset \mathcal{E}(\Lambda_j)$ while $A_j \subset \mathcal{I}(\Lambda_j)$, hence $A_0 \cap A_j = \emptyset$. Now we suppose $i > 0, j > 0$. If $\Lambda_i < \Lambda_j$, then $a_j \subset \mathcal{E}(\Lambda_i)$ while $A_i \subset \mathcal{I}(\Lambda_i)$ and then $A_i \cap A_j = \emptyset$. Hence, we may assume that Λ_1 and Λ_2 are not comparable and then by Lemma A.1.17 we have $\mathcal{I}(\Lambda_i) \cap \mathcal{I}(\Lambda_j) = \emptyset$, hence $A_i \cap A_j = \emptyset$. Consequently, the family $(A_i)_{0 \leq i \leq n}$ makes a partition of D.

Now we will show that each A_i is infraconnected. Suppose that a certain A_h is not infraconnected for some $h > 0$ (resp. $h = 0$). Then it admits an empty annulus $\Lambda = \Gamma(a, r_1, r_2)$. First we notice that if $h = 0$, then $\Lambda_h > \Lambda$ because both a, b are centers of Λ_h. Now, if $h = 0$ (resp. $h > 0$), let $\Theta \in \mathcal{X}(D)$ (resp. let $\Theta \in \mathcal{X}(D)$ be such that $\Theta < \Lambda_h$). Since both a, b belong to $\mathcal{E}(\Theta)$, it is seen that all Λ is included in $\mathcal{E}(\Theta)$ and therefore is included in A_h. This contradicts the hypothesis and finishes proving that A_h is infraconnected.

Next we check that each A_j is maximal in the set of the infraconnected subsets of D. Indeed, let B be a subset of D that strictly contains a certain A_h and let $a \in B \setminus A_h$. If $h = 0$, there exists $\Theta \in \mathcal{X}(D)$ such that $a \in \mathcal{I}(\Theta)$, but $A_h \subset \mathcal{E}(\Theta)$ and therefore Θ is included in an empty annulus of B. If $h > 0$, either a belongs to $\mathcal{E}(\Lambda_h)$, whereas $A_h \subset \mathcal{I}(\Lambda_h)$ and then Λ_h is included in an empty annulus of B, or there exists $\Theta \in \mathcal{X}(D)$ satisfying $\Theta < \Lambda_h$ and $a \in \mathcal{I}(\Theta)$, but then $A_h \subset \mathcal{I}(\Theta)$ and therefore Θ is included in an empty annulus of B. Thus, in each case B is not infraconnected and this finishes showing that each A_i is maximal in the set of the infraconnected subsets of D. As a consequence, the infraconnected components of D are the A_i.

Now conversely, we assume D to have infinitely many empty annuli. First let us suppose that D has a sequence of empty annuli $(\Lambda_n)_{n \in \mathbb{N}}$ such that $\Lambda_n < \Lambda_{n+1}$ (resp. $\Lambda_n > \Lambda_{n+1}$) for all $n \in \mathbb{N}$. By Lemma A.1.15, for every $n \in \mathbb{N}$ there exists $x_n \in \mathcal{E}(\Lambda_n) \cap \mathcal{I}(\Lambda_{n+1})$ (resp. $x_n \in \mathcal{I}(\Lambda_n) \cap \mathcal{E}(\Lambda_{n+1})$) and then the infraconnected component X_n of x_n satisfies $X_n \subset \mathcal{E}(\Lambda_n) \cap \mathcal{I}(\Lambda_{n+1})$ (resp. $X_n \subset \mathcal{I}(\Lambda_n) \cap \mathcal{E}(\Lambda_{n+1})$). Hence, $X_n \cap X_m = \emptyset$ for all $n \neq m$, hence D has infinitely many infraconnected components.

Finally, we may assume that every totally ordered set of empty annuli is finite. Hence, there exists a sequence of empty annuli Λ_n that are minimal elements for the order $\leq$ on $\mathcal{X}(D)$ and then $\mathcal{I}(\Lambda_n) \cap \mathcal{I}(\Lambda_m) = \emptyset$ whenever $n \neq m$. By Lemma A.1.19, $\mathcal{I}(\Lambda_n)$ is an infraconnected component D_n of D such that $D_n \cap D_m = \emptyset$ whenever $n \neq m$. This finishes proving that D has infinitely many infraconnected components and this ends the proof. □

A.2. Monotonous and circular filters

Monotonous and circular filters are essential on an ultrametric field, mainly because for any rational function, its absolute value admits a limit along each circular filter [50, 58, 61] and circular filters are the least thin filters having this property. Most of properties of analytic functions of all kinds derive from that property of circular filters. Certain authors call "generic disk" a notion that is not clearly defined but actually represents a circular filter... We will see that, given a bounded sequence, there exists a subsequence thinner than a circular filter.

For certain problems, we can reduce ourselves to consider monotonous filters instead of circular filters. Monotonous filters are linked to sequences (a_n) such that $|a_{n+1} - a_n|$ is strictly monotonous. Moreover, decreasing filters let us define spherically complete fields.

Definitions and notations: Let J be set. A filter $\mathcal{F}$ on J is said to be *thinner* than a filter $\mathcal{G}$ if every element of $\mathcal{G}$ belongs to $\mathcal{F}$. In such a case, $\mathcal{G}$ is said to be *less thin than* $\mathcal{F}$. Two filters $\mathcal{F}$, $\mathcal{G}$ are said *to be secant* if for all $A \in \mathcal{F}$, $B \in \mathcal{G}$ we have $A \cap B \neq \emptyset$.

A filter $\mathcal{F}$ is said to be *secant* to a subset $B \subset J$ if $\{F \cap B \mid F \in \mathcal{F}\}$ is a filter.

A sequence $(u_n)_{n\in\mathbb{N}}$ in J is said to be *thinner* than a filter $\mathcal{G}$ if so is the filter defined by the sets $A_q = \{u_n | n \geq q\}$ $(q \in \mathbb{N})$. In such a case, $\mathcal{G}$ is said to be *less thin than* the sequence $(u_n)_{n\in\mathbb{N}}$.

A sequence $(u_n)_{n\in\mathbb{N}}$ in $\mathbb{L}$ will be said to be *an increasing distances sequence* (resp. *a decreasing distances sequence*) if the sequence $|u_{n+1} - u_n|$ is strictly increasing (resp. decreasing) and has a limit $\ell \in \mathbb{R}_+^*$.

The sequence $(u_n)_{n\in\mathbb{N}}$ will be said to be *a monotonous distances sequence* if it is either an increasing distances sequence or a decreasing distances sequence.

A sequence $(u_n)_{n\in\mathbb{N}}$ in $\mathbb{L}$ will be said to be *an equal distances sequence* if $|u_n - u_m| = |u_m - u_q|$ whenever $n, m, q \in \mathbb{N}$ such that $n \neq m \neq q$.

Theorem A.2.1: *Let $\mathbb{E}$ be a field provided with an ultrametric absolute value. Let $(u_n)_{n\in\mathbb{N}}$ be a bounded sequence in $\mathbb{E}$. Either we may extract a Cauchy subsequence or we may extract a monotonous distances subsequence or we may extract an equal distances subsequence from the sequence $(u_n)_{n\in\mathbb{N}}$. Further, if the absolute value of $\mathbb{E}$ is discrete, there is no monotonous distances sequence in $\mathbb{E}$. And if the residue class field of $\mathbb{E}$ is finite, there is no equal distances sequence in $\mathbb{E}$.*

Proof. Suppose Theorem A.2.1 to be false. For every $q \in \mathbb{N}$, the set of the circles $C(u_q, r)$ that contain some u_n is then finite.

Suppose that we have already defined integers n_q for $q \leq t$ satisfying

(1) $|u_{n_q} - u_{n_{q-1}}| < |u_{n_{q-1}} - u_{n_{q-2}}|$ for $2 \leq q \leq t$

and such that $d(u_{n_q}, |u_{n_q} - u_{n_{q-1}}|^-)$ contains infinitely many terms of the sequence (u_n). For every $q = 2, ..., t$, let $r_q = |u_{n_q} - n_{n_{q-1}}|$. Obviously, at least one of the circles $C(u_{n_t}, r)$, with $r < r_t$ contains infinitely many terms of the sequence $(u_n)_{n\in\mathbb{N}}$. Let $C(u_{n_t}, r_{t+1})$ be such a circle. It is seen that at least one class Λ of this circle contains infinitely many terms of the sequence because otherwise we would have a sequence of classes (Λ_j) each one containing at least one term $u_{\tau(j)}$ and then they should satisfy $|u_{\tau(j)} - u_{\tau(i)}| = r_{t+1}$ whenever $i \neq j$. Hence, the sequence $(u_n)_{n\in\mathbb{N}}$ should admit an equal distances subsequence. Then we may pick up one term $u_{n_{t+1}}$ in Λ and we have constructed the finite subsequence up to the rank $t+1$, satisfying the properties mentioned above. In the same way we may initiate the induction by defining n_2 from arbitrary n_0, n_1. The sequence $(u_{n_t})_{t\in\mathbb{N}}$ is then defined for every $t \in \mathbb{N}$ and satisfies (1) for $t > 1$. Let $\ell = \lim_{t\to\infty} |u_{n_t} - u_{n_{t+1}}|$. If $\ell = 0$, the subsequence $(u_{n_t})_{t\in\mathbb{N}}$ is a Cauchy subsequence. If $\ell > 0$, this is a decreasing distances subsequence. Thus, we have proven that we can extract a sequence that is either a convergent sequence, or a monotonous distances sequence, or an equal distances sequence.

Now, suppose that the absolute value is discrete and suppose that we have extracted a monotonous distances sequences $(b_m)_{m\in\mathbb{N}}$ form the sequence (u_n). Then the strictly monotonous sequence $|b_{m+1} - b_m|$ must tend to 0, a contradiction. Finally, suppose that the residue class field $\mathcal{E}$ of $\mathbb{E}$ is finite and suppose that we have extracted an equal distances sequences $(b_m)_{m\in\mathbb{N}}$. So, $|b_0 - b_m| = |b_0 - b_n| = |b_m - b_n|\ \forall m \neq n,\ m \neq 0, n \neq 0$. Let q be the cardinal of $\mathcal{E}$. Then, by Lemma A.1.9 the set of terms b_m is at most q, a contradiction. □

Henceforth, throughout the chapter, the field $\mathbb{L}$ is supposed to have a dense valuation and D is an infraconnected subset of $\mathbb{L}$.

Definitions and notations: Let $a \in \widetilde{D}$ and $R \in \mathbb{R}_+^*$ be such that $\Gamma(a, r, R) \cap D \neq \emptyset$ whenever $r \in]0, R[$ (resp. $\Gamma(a, R, r) \cap D \neq \emptyset$ whenever $r > R$). We call *an increasing* (resp. *a decreasing*) *filter of center a and diameter R, on D* the filter $\mathcal{F}$ on D that admits for basis the family of sets $\Gamma(a, r, R) \cap D$ (resp. $\Gamma(a, R, r) \cap D$). For every sequence $(r_n)_{n\in\mathbb{N}}$ such that $r_n < r_{n+1}$ (resp. $r_n > r_{n+1}$) and $\lim_{n\to\infty} r_n = R$, it is seen that the sequence $\Gamma(a, r_n, R) \cap D$ (resp. $\Gamma(a, R, r_n) \cap D$) is a basis of $\mathcal{F}$ and such a basis will be called *a canonical basis.* We call *a decreasing filter with no center of canonical basis $(D_n)_{n\in\mathbb{N}}$ and diameter $R > 0$, on D* a filter $\mathcal{F}$ on D that admits for basis a sequence $(D_n)_n \in \mathbb{N}$ in the form $D_n = d(a_n, r_n) \cap D$ with $D_{n+1} \subset D_n$, $r_{n+1} < r_n$, $\lim_{n\to\infty} r_n = R$, and $\bigcap_{n\in\mathbb{N}} d(a_n, r_n) = \emptyset$.

Given an increasing (resp. a decreasing) filter $\mathcal{F}$ on D of center a and diameter r, we will denote by $\mathcal{B}_D(\mathcal{F})$ the set $\{x \in D|\ \ |x - a| \geq r\}$ (resp. the set $\{x \in D|\ |x - a| \leq r\}$ and by $\mathcal{C}_D(\mathcal{F})$ the set $\{x \in D|\ |x - a| < r\}$ (resp. the set $\{x \in D|\ |x - a| > r\}$. When there is no risk of confusion we will only write $\mathcal{B}(\mathcal{F})$ instead of $\mathcal{B}_D(\mathcal{F})$ and $\mathcal{C}(\mathcal{F})$ instead of $\mathcal{C}_D(\mathcal{F})$. Next, $\mathcal{C}_D(\mathcal{F})$ will be named *the body of* $\mathcal{F}$ and $\mathcal{B}_D(\mathcal{F})$ will be named *the beach of* $\mathcal{F}$.

We call *a monotonous filter on D* a filter is either an increasing filter or a decreasing filter (with or without a center). Given a monotonous filter $\mathcal{F}$, we will denote by $\text{diam}(\mathcal{F})$ its diameter.

The field $\mathbb{L}$ is said to be *spherically complete* if every decreasing filter on $\mathbb{L}$ has a center in $\mathbb{L}$. The field $\mathbb{C}_p$, e.g., is not spherically complete (see Chapter A.5). However, every algebraically closed complete ultrametric field admits a spherically complete algebraically closed extension and this will be recalled in Chapter A.7.

Lemma A.2.2: *Let $(a_n)_n \in \mathbb{N}$ be an increasing distances (resp. a decreasing distances) sequence in D. There exists a unique increasing (resp. decreasing) filter $\mathcal{F}$ on D such that the sequence $(a_n)_n \in \mathbb{N}$ is thinner than $\mathcal{F}$.*

Proof. Let $r_n = |a_{n+1} - a_n|$ and let $R = \lim\limits_{n\to\infty} r_n$.

We first suppose $(a_n)_{n\in\mathbb{N}}$ to be an increasing distances sequence. The increasing filter $\mathcal{F}$ of center a_0, of diameter R is obviously less thin than the sequence $(a_n)_{n\in\mathbb{N}}$. We will show that $\mathcal{F}$ is unique. Let $\mathcal{G}$ be an increasing filter of center a, of diameter R', less thin than the sequence $(a_n)_{n\in\mathbb{N}}$. For every $r < R'$, there exists $q \in \mathbb{N}$ such that $a_n \in \Gamma(a, r, R')$ whenever $n \geq q$. If $a \in d(a_0, R^-)$, this clearly requires that $R = R'$ and then $\mathcal{G} = \mathcal{F}$. Let us suppose that $a \notin d(a_0, R^-)$. Then we have $|a - a_n| = |a_n - a_m| = C$ whenever $n \neq m$ so $R' > R$ and then $\Gamma(a, r, R')$ does not contain the a_n whenever $r > R$. Finally, $\mathcal{G} = \mathcal{F}$.

We now suppose the sequence $(a_n)_{n\in\mathbb{N}}$ to be a decreasing distances sequence with a point a such that $|a - a_n| = |a_{n+1} - a_n|$ whenever $n \in \mathbb{N}$. Then the decreasing filter of center a, of diameter R is a decreasing filter less thin than the sequence $(a_n)_{n\in\mathbb{N}}$. We will show it to be the only decreasing filter less thin than the sequence $(a_n)_{n\in\mathbb{N}}$. Indeed given a decreasing filter $\mathcal{G}$ less thin than the sequence $(a_n)_{n\in\mathbb{N}}$, it must have a center because if it had no center, the sequence $d(a_{n+1}, |a_{n+1} - a_n|)$ would be one of its canonical basis, but by definition it has an intersection that contains a. Then, symmetrical to the case when $\mathcal{F}$ is increasing, it is easily seen that $\mathcal{F}$ is unique.

Now we suppose that the sequence $(a_n)_{n\in\mathbb{N}}$ is a decreasing distances sequence and that there does not exist $a \in L$ such that $|a - a_n| = |a_{n+1} - a_n|$ whenever $n \in \mathbb{N}$. We put $|a_{n+1} - a_n| = r_n$. Hence, the sequence of disks $d(a_{n+1}, r_n)$ has empty intersection and then the filter $\mathcal{F}$, a basis of which is the sequence $(D_n)_{n\in\mathbb{N}}$ with $D_n = d(a_{n+1}, r_n) \cap D$, is a decreasing filter with no center, of diameter R. There is no decreasing filter with center $a \in \mathbb{L}$, less thin than the sequence (a_n) because we should have $|a - a_n| = r_n$ whenever $n \in \mathbb{N}$. Hence, it just remains to show that $\mathcal{F}$ is the only decreasing filter with no center less thin than the sequence (a_n). Let us suppose that there exists another decreasing filter $\mathcal{G}$ of diameter R' with no center, of canonical basis $(D'_m)_{m\in\mathbb{N}}$ less thin than the sequence (a_n). If $R' > R$, since every D'_m contains points a_n, it is seen that all the a_n lie in $D \cap d(a_0, R) \subset D'_m$ whenever $m \in \mathbb{N}$ and this contradicts that $\mathcal{G}$ has no center. Hence, we have $R' \leq R$.

But symmetrically we have $R \leq R'$. Hence, $R = R'$. We will show that $\mathcal{G} = \mathcal{F}$. For every $m \in \mathbb{N}$, let ρ_m be the diameter of D'_m and let $a_q \in D'_m$ be such that $r_q \leq \rho_m$. Clearly, $a_n \in D'_m$ whenever $n \geq q$, hence $D_n \subset D'_m$ whenever $n > q$. In the same way, let $n \in \mathbb{N}$ and $t \in \mathbb{N}$ be such that $\rho_m < r_n$ whenever $m \geq t$. Then, D'_m contains some a_s that belongs to $d(a_{n+1}, r_n) \cap D = D_n$, hence $D'_m \subset D_n$ whenever $m \geq t$. That finishes showing that $\mathcal{G} = \mathcal{F}$ and that ends the proof of Lemma A.2.2. □

Lemma A.2.3: *Let $\mathcal{F}$ be an increasing filter (resp. a decreasing filter) on $\mathbb{L}$, of center $a \in \widetilde{D}$ and diameter $R \leq \operatorname{diam}(D)$ (resp. $R < \operatorname{diam}(D)$) such that a does not belong to a hole of diameter $\rho \geq R$ (resp. $\rho > R$). Then $\mathcal{F}$ is secant with D and induces on D an increasing filter (resp. a decreasing filter) of center a and diameter R, on D.*

Proof. We just have to check that $\Gamma(a, r, R) \cap D \neq \emptyset$ whenever $r \in]0, R[$ (resp. $\Gamma(a, R, r) \cap D \neq \emptyset$ whenever $r > R$) and this is obvious when $a \in \overline{D}$ because D is infraconnected and $R \leq \operatorname{diam}(D)$ (resp. $R < \operatorname{diam}(D)$). Now let us assume a to belong to a hole T of diameter $\rho < R$ (resp. $\rho \leq R$). Since $|\mathbb{L}|$ is dense in $[0, +\infty[$, for every $r < R$ (resp. $r > R$), D has points α such that $r < |a - \alpha| < R$ (resp. $R < |a - \alpha| < r$) and this ends the proof. □

Definitions and notations: Let $\mathcal{F}$ be an increasing (resp. a decreasing) filter of center a and diameter R on D. $\mathcal{F}$ is said *to be pierced* if for every $r \in]0, R[$, (resp. $r > R$), $\Gamma(a, r, R)$ (resp. $\Gamma(a, R, r)$) contains some hole T_m of D.

A decreasing filter with no center $\mathcal{F}$ of canonical basis $(D_m)_{m \in \mathbb{N}}$ on D is said *to be pierced* if for every $m \in \mathbb{N}$, $\widetilde{D}_m \setminus \widetilde{D}_{m+1}$ contains some hole T_m of D.

Remarks: The definition of a pierced filter with no center also applies to a decreasing filter with a center and then is equivalent to that given just above for such a filter.

If $\mathcal{F}$ is an increasing (resp. a decreasing) filter of center a, of diameter R, $\mathcal{F}$ is pierced if and only if there exists a sequence of holes $(T_n)_{n \in \mathbb{N}}$ of D such that $\delta(a, T_n) < \delta(a, T_{t+1})$, (resp. $\delta(a, T_n) > \delta(a, T_{n+1})$), $\lim\limits_{n \to \infty} \delta(a, T_n) = R$.

Given a Cauchy filter $\mathcal{F}$ on D, of limit a in L, we will call *a canonical basis of* $\mathcal{F}$ a sequence D_m in the form $d(a, r_m) \cap D$ with $0 < r_m < r_{m+1}$ and $\lim\limits_{m \to \infty} r_m = 0$. The filter $\mathcal{F}$ is said *to be pierced* if for every $m \in \mathbb{N}$, $\widetilde{D}_m$ contains some hole of D.

Let $a \in \widetilde{D}$. Let $(T_{m,i})_{\substack{1 \leq i \leq s(m) \\ m \in \mathbb{N}}}$ be a sequence of holes of D, which satisfies

$\delta(a, T_{m,i}) = d_m \quad (1 \leq i \leq h_m)$, $d_m < d_{m+1}$ (resp. $d_m > d_{m+1}$), $\lim\limits_{m \to \infty} d_m = S > 0$.

The sequence $(T_{m,i})_{\substack{1 \leq i \leq s(m) \\ m \in \mathbb{N}}}$ is called *an increasing (resp. a decreasing) distances holes sequence that runs the increasing (resp. decreasing) filter of center a, of diameter R.*

Now let $(T_{m,i})_{\substack{1 \leq i \leq s(m) \\ m \in \mathbb{N}}}$ be a sequence of holes of D that satisfies

$\delta(a_m, T_{m,i}) = d_m$ $(1 \leq i \leq s(m))$, $d_m > d_{m+1}$, $\lim\limits_{m\to\infty} d_m = R > 0$, where the filter $\mathcal{F}$ of basis $D_m = d(a_m, d_m) \cap D$ is a decreasing filter with no center. The sequence $(T_{m,i})_{\substack{1\leq i\leq s(m)\\ m\in\mathbb{N}}}$ is called *a decreasing distances holes sequence that runs* $\mathcal{F}$.

Summarizing these definitions, an increasing (resp. decreasing) distances holes sequence that runs an increasing (resp. decreasing) filter $\mathcal{F}$ will be just named *an increasing (resp. decreasing) distances holes sequence* and the filter $\mathcal{F}$ will be named *the increasing (resp. decreasing) filter associated to the sequence* $(T_{m,i})_{\substack{1\leq i\leq s(m)\\ m\in\mathbb{N}}}$. The diameter of $\mathcal{F}$ will be called *the diameter of the sequence* $(T_{m,i})_{\substack{1\leq i\leq s(m)\\ m\in\mathbb{N}}}$. If $\mathcal{F}$ has a center a, a will be named *the center of the sequence* $(T_{m,i})_{\substack{1\leq i\leq s(m)\\ m\in\mathbb{N}}}$. If $\mathcal{F}$ has no center, the sequence $(T_{m,i})$ will be called *a decreasing distances holes sequence with no center.*

Finally, an increasing (resp. decreasing) distances holes sequence will be called *a monotonous distances holes sequence* and the sequence $(d_m)_{m\in\mathbb{N}}$ is called *the monotony* of the monotonous distances holes sequence.

Let $(T_{m,i})_{\substack{1\leq i\leq s(m)\\ m\in\mathbb{N}}}$ be a monotonous distances holes sequences and for every $(m,i)_{\substack{1\leq i\leq s(m)\\ m\in\mathbb{N}}}$ let $\rho_{m,i} = \text{diam}(T_{m,i})$. The number $\inf\limits_{\substack{1\leq i\leq s(m)\\ m\in\mathbb{N}}} \rho_{m,i}$ will be called *piercing of the sequence* $(T_{m,i})_{\substack{1\leq i\leq s(m)\\ m\in\mathbb{N}}}$.

If a monotonous holes sequence has a piercing $\rho > 0$, it will be said to be *well pierced.* If a monotonous filter $\mathcal{F}$ is run by a well-pierced monotonous holes sequence, $\mathcal{F}$ will be said to be well pierced.

In each case, the sequence of circles $C(a, d_m)$ when $\mathcal{F}$ has center a (resp. $C(a_{m+1}, d_m)$ when $\mathcal{F}$ has no center) will be said *to run the filter* $\mathcal{F}$ and *to carry* the monotonous distances holes sequence $(T_{m,i})_{\substack{1\leq i\leq s(m)\\ m\in\mathbb{N}}}$.

A monotonous distances holes sequences $(T_{m,i})_{\substack{1\leq i\leq s(m)\\ m\in\mathbb{N}}}$ will be said to be *simple* if $s(m) = 1$ for all $m \in \mathbb{N}$.

Next, a sequence of holes $(T_m)_{m\in\mathbb{N}}$ of D will be called *a Cauchy sequence of holes of limit* $a \in \mathbb{L}$ if $\lim\limits_{m\to\infty} \delta(a, T_m) = 0$. Such a sequence will be said *to run* the Cauchy filter of basis $\{d(a,r) \cap D | r > 0\}$.

Definitions and notations: In all the Propositions, Theorems, Corollaries and Lemmas, A.2.4, A.2.5, A.2.6, A.2.7, A.2.8, A.2.9, and A.2.10, γ is the Moebius function $b + \dfrac{1}{x-a}$ with $a, b \in \mathbb{L}$.

Proposition A.2.4: *Let $\alpha \in D, r > 0$ be such that $|a - \alpha| < t$. Then* $\gamma(C(\alpha, r)) = C\left(b, \frac{1}{r}\right)$.

Proof. We may assume $b = 0$ and then the proof is immediate. □

Corollary A.2.5: *Let $\alpha \in \mathbb{L}, r_1, r_2 \in]0, +\infty[$ with $|a-\alpha| < r_1 < r_2$. Then* $\gamma(\Gamma(\alpha, r_1, r_2)) = \Gamma\left(b, \frac{1}{r_2}, \frac{1}{r_1}\right)$.

Corollary A.2.6: *Let $\mathcal{F}$ be the increasing (resp. decreasing) filter of center α and diameter $R > |a-\alpha|$, on $\mathbb{L}\setminus\{a\}$. Then $\gamma(\mathcal{F})$ is the decreasing (resp. increasing) filter of center b and diameter $\frac{1}{R}$.*

Lemma A.2.7: *Let $\alpha \in \mathbb{L}$ be such that $|\alpha - a| \neq r$. Then*

$$\gamma(C(\alpha, r)) = C\Big(\gamma(\alpha), \frac{r}{|a-\alpha|^2}\Big).$$

Proof. When x belongs to $C(\alpha, r)$, we have

$$\gamma(x) - \gamma(\alpha) = |\frac{\alpha - x}{(x-\alpha)(a-\alpha)}| = \frac{r}{|a-\alpha|^2}.$$

Hence, $\gamma(C(\alpha, r)) \subset C\Big(\gamma(\alpha), \frac{r}{|a-\alpha|^2}\Big)$. Now let $\xi(u) = \gamma^{-1}(u) = a + \frac{1}{u-b}$. We see that $C\Big((\gamma(\alpha), \frac{r}{|a-\alpha|^2}\Big) \subset C(\alpha, r)$. Since γ and ξ are injective we see that γ must be a surjection onto $C\Big(\gamma(\alpha), \frac{r}{|a-\alpha|^2}\Big)$. □

Corollary A.2.8: *Let $\alpha \in \mathbb{L}$ and $r, r' \in]0, +\infty[$ be such that $0 < r < r' < |a-\alpha|$. Then we have* $\gamma\Big(\Gamma(\alpha, r, r')\Big) = \Gamma\Big(\gamma(\alpha), \frac{r}{|a-\alpha|^2}, \frac{r'}{|a-\alpha|^2}\Big)$,
$\gamma(d(\alpha, r)) = d\Big(\gamma(\alpha), \frac{r}{|a-\alpha|^2}\Big)$, $\gamma(d(\alpha, r^-)) = d\Big(\gamma(\alpha), \Big(\frac{r}{|a-\alpha|^2}\Big)^-\Big)$.

Corollary A.2.9: *Let $\mathcal{F}$ be the increasing (resp. decreasing) filter of center α and diameter R on $\mathbb{L} \setminus \{a\}$ with $|a-\alpha| > R$. Then $\gamma(\mathcal{F})$ is an increasing (resp. a decraesing) filter of center $\gamma(\alpha)$, of diameter $\frac{R}{|a-\alpha|^2}$ on $\mathbb{L} \setminus \{b\}$.*

Corollary A.2.10: *Let $\mathcal{F}$ be a decreasing filter with no center, of basis $(D_n)_{n\in\mathbb{N}}$ on $\mathbb{L} \setminus \{a\}$ such that $a \notin D_0$. Then $\gamma(\mathcal{F})$ is a decreasing filter with no center, of canonical basis $\Big(\gamma(D_n)\Big)_{n\in\mathbb{N}}$ on $\mathbb{L} \setminus \{b\}$.*

Theorem A.2.11: *We suppose $a \in D$. Let $D' = \gamma(D)$. Let $\mathcal{F}$ be a filter on D which is either a monotonous filter or a Cauchy filter. Then $\mathcal{F}$ is pierced if and only if $\gamma(\mathcal{F})$ is a pierced filter on D'.*

Proof. For example we suppose first $\mathcal{F}$ to be a monotonous filter. By definition, $\mathcal{F}$ is the intersection with D of a monotonous filter $\mathcal{G}$ of $\mathbb{L} \setminus \{a\}$. Hence, $\mathcal{F}$ is pierced if and only if $\mathcal{G}$ is secant with $(\mathbb{L} \setminus \{a\}) \setminus \overline{D}$. Since γ is bicontinuous in $\mathbb{L} \setminus \{a\}$ we see that $\gamma(\mathcal{G})$ is secant with $(\mathbb{L} \setminus \{b\}) \setminus \overline{D'}$ if and only if $\mathcal{G}$ is secant with $(\mathbb{L} \setminus \{a\}) \setminus \overline{D}$ because $\gamma(\overline{D}) \setminus \{a\} = \overline{D'} \setminus \{b\}$. Hence, the conclusion is clear. In the same way, if $\mathcal{F}$ is a Cauchy filter of limit $\alpha \in \mathbb{L}$, we consider the filter $\mathcal{G}'$ of the neighborhoods of $\gamma(\alpha)$ in $L \setminus \{b\}$ and we see that $\mathcal{G}$ is secant with $(L \setminus \{a\}) \setminus \overline{D}$ if and only if $\mathcal{G}'$ is secant with $(\mathbb{L} \setminus \{b\}) \setminus \overline{D'}$. □

We are now going to define circular filters, which roughly characterize the absolute values on $\mathbb{L}(x)$ when $\mathbb{L}$ is algebraically closed.

Definitions and notations: Let $a \in \mathbb{L}$ and let $R \in]0, +\infty[$. We call *circular filter of center a and diameter R on L* the filter $\mathcal{F}$, which admits as a generating system the family of sets $\Gamma(\alpha, r', r'') \cap D$ with $\alpha \in d(a, R), r' < R < r''$, i.e., $\mathcal{F}$ is the filter that admits for basis the family of sets of the form $\left(\bigcap_{i=1}^{q} \Gamma(\alpha_i, r'_i, r''_i)\right)$ with $\alpha_i \in d(a, R), r'_i < R < r''_i \ \ (1 \leq i \leq q \ , \ q \in \mathbb{N})$.

For reasons that will appear when characterizing the absolute values of $L(x)$ when $\mathbb{L}$ is algebraically closed, a decreasing filter with no center on $\mathbb{L}$, of canonical basis $(D_n)_{n \in \mathbb{N}}$ will also be called *circular filter on $\mathbb{L}$ with no center, of canonical basis $(D_n)_{n \in \mathbb{N}}$*.

Finally, the filter of the neighborhoods of a point $a \in \mathbb{L}$ will be called *circular filter of the neighborhoods of a on $\mathbb{L}$*. It will also be named *circular filter of center a and diameter* 0. A circular filter on L will be said to be *large* if it has a diameter different from 0.

A circular filter on $\mathbb{L}$ secant with D will be called *circular filter* on D. Given a circular filter $\mathcal{F}$ on $\mathbb{L}$, its diameter will be denoted by $\text{diam}(\mathcal{F})$ and we will call *$\mathcal{F}$-affinoid* any infraconnected affinoid subset of L lying in $\mathcal{F}$.

Lemma A.2.12 describes circular filters on an infraconnected subset of $\mathbb{L}$.

Lemma A.2.12: *Let $a \in \widetilde{D}$, let ρ be the distance from a to D and let R be such that $\rho \leq R \leq \text{diam}(D)$. For $j = 1, ..., q$ let $\alpha_j \in d(a, R)$ and let $r'_j, r''_j \in \mathbb{R}_+$ be such that $r'_j < R < r''_j$. Then,* $\bigcap_{j=0}^{q} (\Gamma(\alpha_j, r'_j, r''_j) \cap D) \neq \emptyset$.

Proof. If $\rho < R$ we put $r' = \max_{1 \leq j \leq q} r'_j$ and we see that $\Gamma(a, r', R) \cap D$ is not empty (because D is infraconnected) and is included in every set $\Gamma(\alpha_j, r'_j, r''_j) \cap D$. If $R < \text{diam}(D)$ we put $r'' = \min_{1 \leq j \leq q} r''_j$ and in the same way, $\Gamma(a, R, r'') \cap D$ is not empty and is included in every set $\Gamma(\alpha_j, r'_j, r''_j) \cap D$.

Now if $\rho = R = \text{diam}(D)$, let $b \in D$ and let $r' = \max_{1 \leq j \leq q} r'_j$. Then $\Gamma(b, r', R) \cap D$ is not empty and is included in every set $\Gamma(\alpha_j, r'_j, r''_j) \cap D$. □

Corollary A.2.13: *Let $a \in \widetilde{D}$, let ρ be the distance from a to D, and let R be such that $\rho \leq R \leq \text{diam}(D)$. The circular filter on $\mathbb{L}$ of center a and diameter R is secant with D.*

Proposition A.2.14 is immediate according to definitions:

Proposition A.2.14: *Let $\mathcal{F}$ be an increasing filter (resp. a decreasing filter) of center a and diameter R on D. Then the circular filter of center a and diameter R on L is secant with D and is the only circular filter on D less thin than $\mathcal{F}$.*

Conversely, let $\mathcal{F}$ be a circular filter of center a and diameter R, on D secant with $d(a, R^-)$ (resp. $\mathbb{L} \setminus d(a, R)$). Then the increasing filter (resp. decreasing filter) of center a and diameter R on $\mathbb{L}$ is secant with D and thinner than $\mathcal{F}$.

Lemma A.2.15: *Let $\mathcal{F}$ be a circular filter. Then $\mathcal{F}$ admits a basis consisting of the family of all $\mathcal{F}$-affinoids. If $\mathcal{F}$ does not admit a countable basis, it has a center and its diameter belongs to $|\mathbb{L}|$. If $\mathcal{F}$ has no center and is secant with an infraconnected affinoid subset B of $\mathbb{L}$, then B lies in $\mathcal{F}$. If $\mathcal{F}$ has center a and diameter r, then an infraconnected affinoid set B lies in $\mathcal{F}$ if and only if it satisfies $E \cap (\mathbb{L} \setminus d(a, r)) \neq \emptyset$, $B \cap d(b, r^-) \neq \emptyset \ \forall b \in d(a, r)$.*

Proof. By definition, a circular filter with no center has a countable basis and of course so does a Cauchy circular filter. In both cases, it admits a basis consisting of a family of disks that are $\mathcal{F}$-affinoid sets.

Now, consider a circular filter of center a and diameter $r > 0$. Then, $\mathcal{F}$ admits for basis the family of sets of the form $d(a, r + \frac{1}{n}) \setminus (\bigcup_{i=1}^{q} d(a_i, (r - \frac{1}{n})^-)$ where a_i is the center of $\mathcal{F}$ satisfying $|a_i - a_j| = r$. In particular, if $r \notin |L|$ we have $q = 1$ and we obtain a basis of the form $\Gamma(a, r - \frac{1}{n}, r + \frac{1}{n})$, which is countable.

Now, suppose that $\mathcal{F}$ is secant with an infraconnected affinoid subset B of $\mathbb{L}$. Suppose first that $\mathcal{F}$ has no center. Let $(A_n)_{n \in \mathbb{N}}$ be a canonical basis of $\mathcal{F}$. Since each A_n admits common points with B, each is included in $\widetilde{B}$ and therefore it is included in B if and only if it contains no hole of B. But since $\mathcal{F}$ has no center, $\bigcap_{n=0}^{\infty} A_n = \emptyset$, hence there exists $q \in \mathbb{N}$ such that $A_n \subset B \ \forall n \geq q$ and therefore $B \in \mathcal{F}$.

Now suppose that $\mathcal{F}$ has center a and diameter r. If $B \in \mathcal{F}$, it obviously satisfies $B \cap (\mathbb{L} \setminus d(a, r)) \neq \emptyset$, $B \cap d(b, r^-) \neq \emptyset \ \forall b \in d(a, r)$. Since B has finitely many holes, on one hand there exists $s > r$ such that $\Gamma(a, r, s) \subset E$ and on the other hand, all classes of $d(a, r)$ are included in B, except finitely many: $d(b_j, r^-)$, $1 \leq j \leq n$. And for each $j = 1, ..., n$, there exists $r_j < r$ such that $\Gamma(b_j, r_j, r) \subset B$. Finally, B contains the set $d(a, s) \setminus \bigcup_{j=1}^{n} d(b_j, r_j)$, which obviously lies in $\mathcal{F}$. □

Corollary A.2.16: *Let $\mathcal{F}$ be a circular filter of diameter r. For every $s \in]0, r[$, the family of $\mathcal{F}$-affinoid of codiameter $\rho > s$ is a basis of $\mathcal{F}$. If two disks $d(a, r)$ and*

$d(b,s)$ have no common points and if $\mathcal{F}$ is secant with $d(a,r)$, it is not secant with $d(b,s)$.

Proposition A.2.17: *Let $a \in \widetilde{D}$, let S be the closure of $\{|x-a| \; |x \in D\}$ in $\mathbb{R}$. For every $r \in S$, the circular filter $\mathcal{F}$ of center a and diameter r is secant with D.*

Proof. Let $a \in \widetilde{D}$. We first suppose that $C(a,r) \cap D = \emptyset$. Then either there exists a sequence $(x_n)_{n\in\mathbb{N}}$ in D such that $r < |x_{n+1}-a| < |x_n - a|$, $\lim_{n\to\infty} |x_n - a| = r$, or there exists a sequence $(x_n)_{n\in\mathbb{N}}$ in D such that $|x_n - a| < |x_{n+1}-a| < r$, $\lim_{n\to\infty} |x_n - a| = r$. In both cases, the circular filter $\mathcal{F}$ of center a and diameter r is clearly secant with D.

Now we may suppose that $C(a,r) \cap D \neq \emptyset$. Let $b \in C(a,r) \cap D$. We see that b is also a center of $\mathcal{F}$. Since D is infraconnected and since $|a-b| = r \leq \mathrm{diam}(D)$ there does exist a sequence $(x_n)_{n\in\mathbb{N}}$ in D such that $\lim_{n\to\infty} |x_n - b| = r$. Hence, $\mathcal{F}$ (that has center b and diameter r) is secant with D. □

Proposition A.2.18: *Let $(a_n)_{n\in\mathbb{N}}$ be a sequence in D that is either a monotonous distances sequence or a constant distances sequence. Then there exists a unique circular filter on D less thin than the sequence (a_n).*

Proof. First we suppose that the sequence $(a_n)_{n\in\mathbb{N}}$ is an increasing (resp. a decreasing) distances sequence. By Lemma A.2.2 there exists a unique increasing (resp. decreasing) filter $\mathcal{F}$ on D less thin than the sequence $(a_n)_{n\in\mathbb{N}}$. If $\mathcal{F}$ has center a and diameter R, by Proposition A.2.14, $\mathcal{F}$ is less thin than the circular filter of center a, of diameter R on D. If $\mathcal{F}$ is decreasing with no center, $\mathcal{F}$ is a circular filter.

Now we suppose that $(a_n)_{n\in\mathbb{N}}$ is a constant distances sequence. We put $R = |a_n - a_m|$ for $n \neq m$ and $a = a_0$. The circular filter $\mathcal{F}$ of center a of diameter R on $\mathbb{L}$ is clearly secant with D because each set $\Lambda_n = \Gamma(a_n, r', r'')$ with $r' < R < r''$ belongs to a generating system of $\mathcal{F}$ and contains a_m for every $m > n$, hence its intersection with D is a circular filter $\mathcal{C}$ on D less thin than the sequence $(a_n)_{n\in\mathbb{N}}$. That ends the proof. □

Corollary A.2.19: *Let $\mathcal{F}$ and $\mathcal{G}$ be two circular filters that are secant. Then they are equal.*

Proof. We can find a monotonous sequence thinner than $\mathcal{F}$. Then the sequence is thinner than $\mathcal{G}$ and hence $\mathcal{G} = \mathcal{F}$. □

Definitions and notations: Let $\mathbb{L}'$ be an extension of L provided with an absolute value that extends the one of $\mathbb{L}$. Let D be a set in $\mathbb{L}$, let $\mathcal{F}$ be a monotonous filter on D, and let D' be a set in $\mathbb{L}'$ that contains D. Let $(a_n)_{n\in\mathbb{N}}$ be a monotonous distances sequence that runs $\mathcal{F}$. In $\widehat{D}$, there is a unique monotonous filter less thin than the sequence $(a_n)_{n\in\mathbb{N}}$. This filter will be denoted by $\widehat{\mathcal{F}}$.

In the same way, let $\mathcal{G}$ be a circular filter of center a and diameter r on D. We will denote by $\widehat{\mathcal{G}}$ the filter of center a and diameter r on D'. Finally, let $\mathcal{G}$ be a circular filter with no center. Then it is a decreasing filter, hence we have already previously defined $\widehat{\mathcal{G}}$.

Corollary A.2.20: *Let $(a_n)_{n\in\mathbb{N}}$ be a bounded sequence in $\mathbb{L}$. Then there exists a subsequence $(a_{n_t})_{t\in\mathbb{N}}$ and a unique circular filter $\mathcal{F}$ on $\mathbb{L}$ less thin than the subsequence $(a_{n_t})_{t\in\mathbb{N}}$.*

Proof. Since the sequence $(a_n)_{n\in\mathbb{N}}$ is bounded, by Theorem A.2.1 we can extract either a monotonous distances subsequence or a constant distances subsequence, or a converging subsequence. In all cases, once such a subsequence is chosen, there exists a unique circular filter $\mathcal{F}$ on $\mathbb{L}$ less thin than the subsequence. $\square$

Theorem A.2.21: *Let $(a_n)_{n\in\mathbb{N}}$, $(b_n)_{n\in\mathbb{N}}$ be two sequences such that $|a_n - b_n| \leq t < r$ $\forall n \in \mathbb{N}$. Suppose that the sequence $(a_n)_{n\in\mathbb{N}}$ is thinner than a circular filter $\mathcal{F}$ of diameter r. Then the sequence $(b_n)_{n\in\mathbb{N}}$ also is thinner than $\mathcal{F}$.*

Proof. By Corollary A.2.16, $\mathcal{F}$ admits a basis consisting of $\mathcal{F}$-affinoids S of codiameter $\rho > t$. Consider such a $\mathcal{F}$-affinoid S. Then if a_n belongs to S, so does b_n. Now, when n is big enough, all a_n belong to S and hence so do all b_n. And since $\mathcal{F}$ admits a basis of $\mathcal{F}$-affinoids with a codiameter $s > t$, we see that the sequence $(b_n)_{n\in\mathbb{N}}$ is thinner than $\mathcal{F}$. $\square$

A.3. Ultrametric absolute values for rational functions

Definitions and notations: As mentioned in Chapter A.1, log denotes a real logarithm function of basis $\theta > 1$ (eventually we can take for θ an integer p that is the residue characteristic of $\mathbb{K}$). When a function f from an interval I to $\mathbb{R}$ admits a right-side (resp. a left-side) derivative at a point $a \in I$, we will denote it by $f'^r(a)$ (resp. $f'^l(a)$). If the variable is μ, we will also denote it by $\dfrac{d_r f}{d\mu}$ (resp. $\dfrac{d_l f}{d\mu}$).

Moreover, throughout the chapter, we will denote by $\mathbb{L}$ a field provided with an ultrametric absolute value $|\ .\ |$.

The set of circular filters on $\mathbb{K}$ secant with a subset D of $\mathbb{K}$ will be denoted by $\Phi(D)$ and the subset of large circular filters on $\mathbb{K}$ secant with D will be denoted by $\Phi^\circ(D)$. We will show the absolute values on the field $\mathbb{K}(x)$ of rational functions to be characterized by the circular filters on $\mathbb{K}$. Actually, the most important property of such absolute values comes from the fact that the logarithm of an absolute value is a piecewise affine function of the logarithm of the absolute value of the variable. And next, a valuation function is then defined for any $h \in \mathbb{K}(x)$ in the following way: let $r \in]0, +\infty[$ be such that $\mu = -\log r$ and let $\mathcal{F}$ be the circular filter of center 0 and diameter r. Following classical notations [2, 50, 72], one sets

$$v(h, \mu) = -\log(\lim_{\mathcal{F}} |h(x)|).$$

This function $v(h, \mu)$, called *the valuation function of* h, is convenient mainly because it is piecewise affine. However, in order to avoid many changes of sign, here we will consider $\Psi(h, \log r) = \log(\lim_{\mathcal{F}} |h(x)|)$ and we will show that if $\frac{d_r\Psi(h,\mu)}{d\mu} \neq \frac{d_l\Psi(h,\mu)}{d\mu}$, then $\frac{d_r\Psi(h,\mu)}{d\mu} - \frac{d_l\Psi(h,\mu)}{d\mu}$ is equal to the difference between the number of zeros and the number of poles of h (taking multiplicity into account) on the circle $C(0, r)$ such that $\log r = \mu$. This translates properties of $|h(x)|$ into terms of piecewise affine functions.

However, this kind of definition presents the inconvenience of changing the sense of monotony for both $|x|$ and $|h(x)|$. Moreover, its sign is opposite to this of the counting function of zeros for entire functions in the Nevanlinna theory. This is why, here we will adopt another set of notation and put $\Psi(x) = \log |x| \ \forall x \in \mathbb{K}$. First, we have to state several basic properties that work not only in an algebraically closed field such as $\mathbb{K}$ but more generally in a field $\mathbb{E}$ that is just provided with an ultrametric absolute value.

Let $\mathbb{L}[x_1, ..., x_q]$ be an algebra of polynomials in q indeterminates, with coefficients in $\mathbb{L}$. For each

$P(x_1, ..., x_q) = \sum_{j_1+,...,+j_q \leq t} a_{i_1,...,i_q} x_1^{j_1}, ..., x_q^{j_q}$, we set

$$\overline{P(x_1, ..., x_q)} := \sum_{j_1+,...,+j_q \leq t} \overline{a_{i_1,...,i_q}} x_1^{j_1}, ..., x_q^{j_q}.$$

On $\mathbb{L}[x_1, ..., x_q]$, we put $\|P\| := \sup_{j_1+,...,+j_q \leq t} |a_{i_1,...,i_q}|_0$.

However, when there is no risk of confusion, we will just write $\| \, . \, \|$ instead of $\| \, . \, \|_0$.

Lemma A.3.1: *$\| \, . \, \|$ is a multiplicative norm of $\mathbb{L}$-algebra.*

Proof. Let $B = \mathbb{L}[x_1, ..., x_q]$. Clearly, $\| \, . \, \|$ is an ultrametric norm of $\mathbb{L}$-vector space on B. We check that $\|PQ\| = \|P\|\|Q\|$ whenever $P, Q \in B$. Both $\|P\|$, $\|Q\|$ belong to $|\mathbb{L}|$. Hence, without loss of generality, we may clearly assume $\|P\| = \|Q\| = 1$. Thus, we have $\overline{P} = \overline{Q} = \overline{1}$. Let $\mathcal{L}$ be the residue field of $\mathbb{L}$. Since $\mathcal{L}[x_1, ..., x_q]$ is a ring without divisors of zeros, we have $\overline{PQ} \neq \overline{0}$ and therefore $\|PQ\| = 1$. This ends the proof. □

Definitions and notations: The norm $\| \, . \, \|$ on $\mathbb{L}[x_1, ..., x_q]_0$ is called *the Gauss norm*. Given a polynomial $P(x) = \sum_{j=1}^{q} a_j x^j \in \mathbb{L}[x]$, for any $r > 0$ we set

$|P|(r) = \max_{0 \leq j \leq q} (|a_j| r^j)$.

By ultrametricity, Lemma A.3.2 is then immediate:

Lemma A.3.2: *Let $P(x) \in \mathbb{L}[x]$. For all $x \in \mathbb{L}$, one has $|P(x)| \leq |P|(|x|)$.*

Lemma A.3.3: *Suppose $\mathbb{L}$ is algebraically closed. Let $P(x) = \sum_{j=0}^{n} a_j x^j \in \mathbb{L}[x] \setminus \{0\}$ and let $r \in \mathbb{R}_+$. Then $|P(x)|$ admits a limit equal to $|P|(r)$ when $|x|$ approaches r but remains different from r. Let $x \in d(0,r)$. Then $|P(x)| \leq |P|(r)$. If P has no zero in the class of x in $d(0,r)$, then $|P(x)| = |P|(r)$. If P has at least one zero in that class, then $|P(x)| < |P|(r)$.*

Proof. Let $r' < r$ and $r'' > r$ be such that P has no zero in $\Gamma(0,r',r) \cup \Gamma(0,r,r'')$. We may obviously assume P to be monic. Let $P(x) = \prod_{i=1}^{n}(x - \alpha_i)$ be the factorization of P in an algebraic closure of $\mathbb{L}$, with

$|\alpha_i| < r'$ for $i \leq h$
$|\alpha_i| > r''$ for $i \geq \ell$
$|\alpha_i| = r$ for $i = h, ..., \ell$.

Now let $x \in \Gamma(0,r',r)$. Clearly, $|x - \alpha_i| = |x|$ whenever $i \leq h$, whereas $|x - \alpha_i| = |\alpha_i|$ whenever $i > h$, hence $|P(x)| = |x|^h \prod_{i=h+1}^{n} |\alpha_i|$, hence $\lim_{|x| \to r^-} |P(x)| = r^h \prod_{i=h+1}^{n} |\alpha_i|$.

Symmetrically, we show that $\lim_{|x| \to r^+} |P(x)| = r^{\ell-1} \prod_{i=\ell}^{n} |\alpha_i| = r^h \prod_{i=h+1}^{n} |\alpha_i|$. But the terms $|a_j x^j|$ are all different for every $|x|$ except for finitely many values so that there exist $\rho' \in [r', r[$ and $\rho'' \in]r, r'']$ such that the $|a_j x^j|$ are all different when $x \in \Gamma(0,\rho',r) \bigcup \Gamma(0,r,\rho'')$. Then we have $|P(x)| = \max_{0 \leq j \leq n} |a_j||x|^j$ and hence $\lim_{\substack{|x| \to r \\ |x| \neq r}} |P(x)| = \max_{0 \leq j \leq n} |a_j| r^j$.

Now let $x \in d(0,r)$ and let us assume P to have no zero in the class Λ of x in $d(0,r)$. This means that $|x - \alpha_i| = r$ whenever $i = 0, ..., \ell - 1$ and $|x - \alpha_i| = |\alpha_i|$ whenever $i \geq \ell$. Thus, $|P(x)| = r^{\ell-1} \prod_{i=\ell}^{n} |\alpha_i|$. If P has at least one zero α_{h+1} in the class of x we see that $|x - \alpha_{h+1}| < r$, whereas $|x - \alpha_i| \leq r$ for every $i = h+2, ..., \ell - 1$, hence $|P(x)| < r^{\ell-1} \prod_{i=\ell}^{n} |\alpha_i|$ and finally $|P(x)| < |P|(r)$. □

Theorem A.3.4: *Let $P \in \mathbb{L}[X]$ and let U be the unit disk $\{x \in \ \mid |x| \leq 1\}$ of $\mathbb{K}$. Then $\|P\| = \sup_{x \in U}(|P(x)|)$.*

Proof. On one hand, $|P(x)| \leq \|P\| \ \forall x \in U$. On the other hand, by Theorem A.3.3, $\lim_{\substack{|x| \to 1,\ x \in U, \\ |x| \neq 1}} |P(x)| = \|P\|$. Consequently, the equality holds. □

Corollary A.3.5: *Let $P \in \mathbb{L}[X]$ and let $t \in \mathbb{L}$ be such that $|t| \leq 1$. Let $Q(X) = P(X+t)$. Then $\|P\| = \|Q\|$. If $\|P\| \leq 1$, then P is 1-Lipschitzian.*

Theorem A.3.6: *For every $r > 0$ the mapping from $\mathbb{L}[x]$ to $\mathbb{R}_+$ defined by $P \to |P|(r)$ is an absolute value on $\mathbb{L}[x]$ such that $|P(a)| \leq |P|(|a|)\ \forall a \in \mathbb{L}$.*

Proof. Due to the definition of $|P|(r)$, it is easily checked that $|P|(r) = 0$ if and only if $P = 0$ and that

$$|P+Q|(r) \leq \max(|P|(r), |Q|(r)).$$

We can also check that $|P(a)| \leq |P|(|a|)\ \forall a \in \mathbb{L}$. Now, set $P(x) = \sum_{j=1}^{m} a_j x^j$, $Q(x) = \sum_{j=1}^{n} b_j x^j$ and let $P(x)Q(x) = \sum_{j=1}^{m+n} c_j x^j$. Let s (resp. t) be the biggest of the integers such that $|P|(r) = |a_s|r^s$ (resp. $|Q|(r) = |b_t|r^t$). Then $|P|(r)|Q|(r) = |a_s b_t|r^{s+t}$. On one hand, we can check that, obviously, $|c_j|r^j \leq |a_s b_t|r^{s+t}\ \forall j = 0, ..., m+n$, hence $|PQ|(r) \leq |P|(r)|Q|(r)$. On the other hand, since $|a_j|r^j < |a_s|r^s\ \forall j > s$ and $|b_j|r^j < |b_t|r^t\ \forall j > t$, we have $|c_{s+t}|r^{s+t} = |a_s b_t|r^{s+t}$, which proves that $|PQ|(r) \geq |P|(r)|Q|(r)$ and hence ends the proof. □

Now, Lemma A.3.7 shows that we can change the origin, inside the disk $d(0,r)$:

Lemma A.3.7: Suppose E is algebraically closed. *Let $r \in \mathbb{R}_+$ and let $a \in \mathbb{L}$ be such that $|a| \leq r$. Then $|P(x)|$ has a limit when $|x-a|$ approaches r but remains different from r. Further, that limit does not depend on $a \in d(0,r)$ and it belongs to $|\mathbb{K}|$ if and only if so does r.*

Proof. We set $x = a + u$ and $P_a(u) = P(a+u)$. For every $P \in \mathbb{L}[x]$ we have $\lim_{\substack{|u|\to r \\ |u|\neq r}} |P_a(u)| = |P_a|(r)$. In particular, $|P_a|(r) = \lim_{\substack{|u|\to r^+ \\ |u|<r}} |P_a(u)|$. But for every $\rho > r$, $C(0,\rho) = C(a,\rho)$, hence $\lim_{|x|\to r^+} |P(x)| = \lim_{|u|\to r^+} |P(a+u)| = \lim_{|u|\to r^+} |P_a(u)|$. That ends the proof of Lemma A.3.7. □

Theorem A.3.8: *Suppose $\mathbb{L}$ is algebraically closed. Let $P(x) = \sum_{j=0}^{q} a_j x^j \in \mathbb{E}[x]$ be a monic polynomial such that $a_j \in d(0,1)$ whenever $j = 0, ..., q$. Then the q zeros of P belong to $d(0,1)$.*

Proof. Let ψ be the absolute value defined on $E[x]$ by $\psi(P) = \lim_{|u|\to 1, |u|\neq 1} |P(u)|$ and let $P(x) = \prod_{j=1}^{q} (x - c_j)$. By Lemma A.3.3, for each $j = 1, ..., q$, it is seen that $\psi(x - c_j) \geq 1$, whereas $\psi(P) = 1$. Hence, $\psi(x - c_j) = 1$ for every $j = 1, ..., q$ and therefore by Lemma A.3.3 again, we have $c_j \leq 1$ whenever $j = 1, ..., q$. □

Definitions and notations: These ultrametric absolute values defined on $\mathbb{L}[x]$ are immediately extended to rational functions by $\left|\frac{P}{Q}\right|(r) := \frac{|P|(r)}{|Q|(r)}$.

Then Lemma A.3.9 is immediate:

Lemma A.3.9: *Suppose $\mathbb{L}$ is algebraically closed. Let $h \in \mathbb{L}(x)$ and let $r \in \mathbb{R}_+$. For every $a \in d(0,r)$ we have $\lim_{\substack{|x-a|\to r\\ |x-a|\neq 0}} |h(x)| = |h|(r)$. Let $x \in C(0,r)$. If h has no zero and no pole in the class of x in $C(0,r)$ then $|h(x)| = |h|(r)$. Further, $|h|(r)$ belongs to $|\mathbb{K}|$ if and only if so does r.*

Circular filters characterize the multiplicative norms defined on $\mathbb{K}(x)$ [50, 52, 61, 62].

Theorem A.3.10 (B. Guennebaud): *For every circular filter $\mathcal{F}$ on $\mathbb{K}$, for every rational function $P(x) \in \mathbb{K}[x], |P(x)|$ has a limit $\varphi_{\mathcal{F}}(P)$ along the filter $\mathcal{F}$. The mapping $\mathcal{F} \to \varphi_{\mathcal{F}}$ from $\Phi(\mathbb{K})$ into the set of the multiplicative semi-norms on $\mathbb{K}[x]$ is a bijection. Moreover, for every large circular filter on $\mathbb{K}$, $\varphi_{\mathcal{F}}$ has continuation to $\mathbb{K}(x)$ and the mapping $\mathcal{F} \to \varphi_{\mathcal{F}}$ from $\Phi^\circ(\mathbb{K})$ into the set of the multiplicative norms on $\mathbb{K}(x)$ is a bijection.*

If $\mathcal{F}$ has center 0 and diameter r, then $\varphi_{\mathcal{F}}(h) = |h|(r)$.

Proof. We first suppose that $\mathcal{F}$ has center $a \in \mathbb{K}$ and diameter $r > 0$. With no loss of generality we may obviously assume $a = 0$ by means of the change of variable $x = a + u$. Then by Lemma A.3.3, $|h(x)| = |h|(r)$ holds in every class of $C(0,r)$ but finitely many ones $\Lambda_1, ..., \Lambda_q$. For every $j = 1, ..., q$, we take $\alpha_j \in \Lambda_j$ and set $\alpha_0 = 0$. Let ϵ be > 0. By Lemma A.3.3 there exist $\rho' \in]0, r[$ and $\rho'' > r$ such that $|\ |h(x)| - |h|(r)\ |_\infty \leq \epsilon$ for every $x \in \bigcap_{j=0}^{q} \Gamma(\alpha_j, \rho', \rho'')$, so $\lim_{\mathcal{F}} |h(x)| = |h|(r)$.

Now we suppose that $\mathcal{F}$ has no center in $\mathbb{K}$. It admits a canonical basis $(D_n)_{n\in\mathbb{N}}$ and then given $h \in \mathbb{K}(x)$, there exists $q \in \mathbb{N}$ such that h has neither any zero nor any pole inside D_q. Hence, $|h(x)|$ is equal to a constant l in D_q and therefore we have $\lim_{\mathcal{F}} |h(x)| = l$. By the same kind of reasoning as in Theorem A.3.6, it is easily seen that $\varphi_{\mathcal{F}}$ is an absolute value on $\mathbb{K}(x)$.

Now, we will check that the mapping $\mathcal{F} \to \varphi_{\mathcal{F}}$ is injective. Indeed let $\mathcal{F}_1, \mathcal{F}_2$ be two different circular filters and let r_1 (resp. $r_2 \geq r_1$) be the diameter of $\mathcal{F}_1$ (resp. $\mathcal{F}_2$). We first suppose that we may find disks $\Lambda_1 = d(a_1, \rho_1)$ and $\Lambda_2 = d(a_2, \rho_2)$ such that $\Lambda_1 \cap \Lambda_2 = \emptyset$ and $\mathcal{F}_1$ (resp. $\mathcal{F}_2$) is secant with Λ_1 (resp. Λ_2). Then it is seen that $|a_1 - a_2| > \rho_1 \geq r_1$. Hence, $\varphi_{\mathcal{F}_1}(x - a_1) \leq r_1$, whereas $\varphi_{\mathcal{F}_2}(x - a_1) = |a_1 - a_2| > r_1$ and therefore $\varphi_{\mathcal{F}_1} \neq \varphi_{\mathcal{F}_2}$.

We now suppose that we cannot find disks Λ_1, Λ_2 defined as above. Since $r_1 \leq r_2$, any disk Λ that belongs to $\mathcal{F}_1$ is included in any disk that belongs to $\mathcal{F}_2$ and therefore any point of Λ_1 is a center of $\mathcal{F}_2$. Thus, $\mathcal{F}_2$ admits a center $a \in \Lambda$ and then, $\mathcal{F}_1$ is secant with $d(a, r_2)$. Hence, we have $r_1 < r_2$ because otherwise $\mathcal{F}_1$ would be equal to $\mathcal{F}_2$. In particular, $\mathcal{F}_2$ is secant with one class $d(\alpha, r_2^-)$ of $d(a, r_2)$. Then we have $\varphi_{\mathcal{F}_1}(x - \alpha) \leq r_1$, whereas $\varphi_{\mathcal{F}_2}(x - \alpha) = r_2$. This finishes showing that the mapping $\mathcal{F} \to \varphi_{\mathcal{F}}$ is injective.

Now we will show that this mapping defined on $\Phi^\circ(\mathbb{K})$ is also surjective onto the set of multiplicative norms, i.e., the absolute values on $\mathbb{K}[x]$ continuing these of $\mathbb{K}$. Indeed, let ψ be such an absolute value on $\mathbb{K}[x]$ and let $r = \inf_{\lambda \in \mathbb{K}} \psi(x - \lambda)$.

We first suppose that there exists $a \in \mathbb{K}$ such that $\psi(x - a) = r$. Since ψ is an absolute value, we check that $r > 0$ because if $r = 0$, we have $\psi(h) = h(a)$ for every $h \in \mathbb{K}[x]$ and then ψ is not an absolute value. Hence, we can assume $r > 0$. Let $\mathcal{F}$ be the circular filter of center a, of diameter r. By Lemma A.1.7, we know that ψ is ultrametric and then for every $b \in \mathbb{K}$, we have $\psi(x - b) \leq \max(\psi(x - a), |a - b|) = \max(|a - b|, r)$. But by definition we have $\psi(x - b) \geq r$, hence (1) $r \leq \psi(x - b) \leq \max(r, |a - b|)$.

If $|a - b| > r$, then both $\psi(x - b), \varphi_{\mathcal{F}}(x - b)$ are equal to $|a - b|$. If $|a - b| \leq r$, then b is another center of $\mathcal{F}$ and we have $\varphi_{\mathcal{F}}(x - a) = \varphi_{\mathcal{F}}(x - b) = r$. But by (1) we see that $\psi(x - b) = r$. So we have shown that $\varphi_{\mathcal{F}}(x - b) = \psi(x - b)$ for all $b \in \mathbb{K}$ and since $\mathbb{K}$ is algebraically closed, this finishes proving that $\psi = \varphi_{\mathcal{F}}$.

We now suppose that there does not exist $a \in \mathbb{K}$ such that $r = \psi(x - a)$. There exists $\alpha_n \in \mathbb{K}$ such that $r < \psi(x - \alpha_n) < r + \frac{1}{n}$. Let $\rho_n = \psi(x - \alpha_n)$. For $b \in \mathbb{K} \setminus d(\alpha_n, \rho_n)$ clearly we have

(2) $\psi(x - b) = |b - \alpha_n| > \rho_n$

because $\psi(x - \alpha_n) < |\alpha_n - b|$. Further, if $\mathcal{F}_n$ is the circular filter of center α_n and diameter ρ_n, we have

(3) $\psi(x - b) = |b - \alpha_n| = \varphi_{\mathcal{F}_n}(x - b)$.

However, there exists $\alpha_{n+1} \in \mathbb{K}$ such that $r < \psi(x - \alpha_{n+1}) < \min\left(\rho_n, r + \frac{1}{n+1}\right)$. Hence, by (2) we see that $\alpha_{n+1} \in d(\alpha_n, \rho_n)$. That way, we may define a decreasing sequence of disks $D_n = d(\alpha_n, \rho_n)$ such that $r < \rho_n < r + \frac{1}{n}$ and $\psi(x - \alpha_n) = \rho_n$. Let $D'_n = D_n \cap D$. Then the decreasing filter $\mathcal{F}$ of basis $(D'_n)_{n \in \mathbb{N}}$ satisfies $\lim_{\mathcal{F}} \psi(x - \alpha_n) = r$. It is easily seen that $\mathcal{F}$ has no center because if α is a center of $\mathcal{F}$, then $\psi(x - \alpha) \leq \max(|\psi - \alpha_n|, |\alpha_n - \alpha|)$, hence $\psi(x - \alpha) = r$. We will show that $\psi = \varphi_{\mathcal{F}}$. Let $b \in \mathbb{K}$ and $q \in \mathbb{N}$ be such that $b \notin D_q$. Then by (3), for $n \geq q$ we have $\psi(x - b) = \varphi_{\mathcal{F}_n}(x - b)$. On the other hand, it is easily seen that $\varphi_{\mathcal{F}_n}(x - b) = \varphi_{\mathcal{F}}(x - b)$. Thus, $\psi(x - b) = \varphi_{\mathcal{F}}(x - b)$ whenever $b \in \mathbb{K}$ and then $\psi = \varphi_{\mathcal{F}}$.

Finally, let ψ be a multiplicative semi-norm that is not a norm: there exists a polynomial P such that $\psi(P) = 0$ and hence, there exists $a \in \mathbb{K}$ such that

$\psi(x-a)=0$. Let $b \in \mathbb{K}$. Then $\psi(x-b) = \psi((x-a)+(a-b)) \leq \max\psi(x-a), \psi(a-b))$. But since $\psi(t) = |t|\ \forall t \in \mathbb{K}$, we have $\psi(x-b) = |a-b|$, hence putting $h(x) = x-b$, we have $\psi(h) = |h(a)|$, which shows that the equality $\psi(h) = |h(a)|$ holds for every polynomial of degree 1 and therefore it holds in all $\mathbb{K}[x]$. This ends the proof of Theorem A.3.10. □

Theorem A.3.11: *Let $\mathcal{H}$ be a filter on $\mathbb{K}$ and let $\mathcal{F}$, $\mathcal{G} \in \Phi(\mathbb{K})$ be less thin than $\mathcal{H}$. Then $\mathcal{F} = \mathcal{G}$.*

Proof. Since $\mathcal{H}$ is thinner than $\mathcal{F}$ and $\mathcal{G}$ we have $\lim_{\mathcal{H}} |P(x)| = \varphi_{\mathcal{F}}(P) = \varphi_{\mathcal{G}}(P)\ \forall P \in \mathbb{K}[x]$, hence $\varphi_{\mathcal{F}} = \varphi_{\mathcal{G}}$. But by Theorem A.3.10, the mapping that associates to each circular filter $\mathcal{F}$ the multlplicative semi-norm $\varphi_{\mathcal{F}}$ is injective and hence $\mathcal{F} = \mathcal{G}$. □

Definitions and notations: When $\mathcal{F}$ is the circular filter of center a, of diameter r, we will also denote by $\varphi_{a,r}$ the absolute value $\varphi_{\mathcal{F}}$. Hence, by definition we have $\varphi_{a,r}(h) = \lim_{\substack{|x-a|\to r \\ |x-a|\neq a}} |h(x)|$. In particular, we notice that $\varphi_{0,r}(h) = |h|(r)$.

Finally, we will denote by φ_a the multiplicative semi-norm defined on rational functions with no pole at a as $\varphi_a(h) = |h(a)|$.

Now, let us go back to the field L. For $\mu \in \mathbb{R}$, we set $\Psi(h,\mu) = \log(|h|(\theta^{\mu}))$ for simplicity, we set $\Psi(h) = \Psi(h,0)$. Thus, comparatively to the valuation function $v(h,\mu)$ defined and used in previous works [2, 58], we have $\Psi(h,\mu) = -v(h,-\mu)$. The advantage of the function Ψ is to respect the sense of variation of $|h|(r)$.

The translation of statements A.3.1, A.3.2, A.3.3, A.3.6, A.3.7, A.3.8, and A.3.9 into terms of valuation allows us to obtain the Lemmas A.3.12 and A.3.13.

Lemma A.3.12: *Let $P(x) = \sum_{j=0}^{n} a_j x^j \in \mathbb{K}[x] \setminus \{0\}$. For every $\mu \in \mathbb{R}$, we have $\Psi(P,\mu) = \max_{0\leq j\leq n} \Psi(a_j) + j\mu$. Moreover, $\Psi(P(a)) \leq \Psi(P, \Psi(a))\ \forall a \in L$.*

Suppose $P \in \mathbb{K}[x]$. The equality $\Psi(P(x)) = \Psi(P,\Psi(x))$ holds if and only if P has no zero α such that $\Psi(x-\alpha) < \Psi(x)$.

Lemma A.3.13: *Let $h \in \mathbb{K}(x) \setminus \{0\}$. We have $\Psi(h(x)) = \Psi(h,\Psi(x))$ for every $x \in \mathbb{K}$ such that h has no zero α satisfying $\Psi(x-\alpha) < \Psi(x)$ and no pole β satisfying $\Psi(x-\beta) < \Psi(x)$.*

Lemma A.3.14: *Let $h_1, h_2 \in \mathbb{K}(x) \setminus \{0\}$. Then we have*

$$\Psi(h_1+h_2,\mu) \leq \max\bigl(\Psi(h_1,\mu), \Psi(h_2,\mu)\bigr).$$

When $\Psi(h_1,\mu) > \Psi(h_2,\mu)$, then we have $\Psi(h_1+h_2,\mu) = \Psi(h_1,\mu)$. Moreover, $\Psi(h_1 \,.\, h_2,\mu) = \Psi(h_1,\mu) + \Psi(h_2,\mu)$.

Definitions and notations: In order to perform easily any change of origin, for every $a \in \mathbb{K}$ and $h \in \mathbb{K}(x) \setminus \{0\}$ we put $\Psi_a(h,\mu) = \Psi(h_a,\mu)$ with $h_a(u) = h(a+u)$. Thus, if $\mathcal{F}$ denotes the circular filter of center a and diameter $\theta^{-\mu}$, then $\Psi_a(h,\mu) = \log(\varphi_{\mathcal{F}}(h))$.

We now consider again a polynomial $P(x) = \sum_{j=0}^{n} a_j x^j \neq 0$. We denote by $\nu^+(P,\mu)$ (resp. $\nu^-(P,\mu)$) the biggest (resp. the smallest) index j such that $\Psi(a_j) + j\mu = \Psi(P,\mu)$.

Lemma A.3.15 is a consequence of Lemma A.3.9:

Lemma A.3.15: *Let $h \in \mathbb{K}(x) \setminus \{0\}$ and let $a, b \in \mathbb{K}$. For every $\mu \geq \Psi(a-b)$, we have $\Psi_a(h,\mu) = \Psi_b(h,\mu)$.*

Theorem A.3.16: *Let $P(x) = \sum_{j=0}^{n} a_j x^j \in \mathbb{K}[x]$. For every $\mu \in \mathbb{R}, \nu^+(P,\mu) - \nu^-(P,\mu)$ is equal to the number of zeros admitted by P in the circle $C(0,\theta^{\mu})$ in $\mathbb{K}$. The function $\nu^+(P,.)$ (resp. $\nu^-(P,.)$) is increasing and continuous on the right (resp. on the left). Moreover, given $Q \in \mathbb{K}[x]$, then ν^+, ν^- satisfy $\nu^+(PQ,\mu) = \nu^+(P,\mu) + \nu^+(Q,\mu)$, $\nu^-(PQ,\mu) = \nu^-(P,\mu) + \nu^-(Q,\mu)$. Further, if $\nu^+(P,\mu) = \nu^-(P,\mu)$, then both are constant in a neighborhood of μ.*

The function $\Psi(P,.)$ is continuous, piecewise affine, increasing, convex, and has a right-side derivative (resp. a left-side derivative) equal to $\nu^+(P,\mu)$ (resp. $\nu^-(P,\mu)$).

Proof. It is easily seen that the equality
(1) $\nu^+(P,\mu) = \nu^-(P,\mu)$
holds for every μ but finitely many values, at most n. It is also clear that the functions $\nu^+(P,.)$ and $\nu^-(P,.)$ are increasing. By continuity, we see that the function $\nu^+(P,\mu)$ is continuous on the right at each point, whereas $\nu^-(P,\mu)$ is continuous on the left at each point. Finally, if (1) holds in an interval $]\mu',\mu''[$, then the functions $\nu^+(P,.)$ and $\nu^-(P,.)$ are constant and equal. Consider an interval $]\mu',\mu''[$ such that $\nu^+(P,\mu) = \nu^-(P,\mu)$ for all $\mu \in]\mu',\mu''[$ and let $j = \nu^+(P,\mu)$ whenever $\mu \in]\mu',\mu''[$. Then $\Psi(P,\mu) = \Psi(a_j) + j\mu$ so that the function $\Psi(P,.)$ is in the form $A + i\mu$ in this interval.

Now let μ be such that $\nu^+(P,\mu) < \nu^-(P,\mu)$. We see that $\Psi(P,.)$ is still continuous at μ and has a left-side derivative equal to $\nu^-(P,\mu)$ and a right-side derivative equal to $\nu^+(P,\mu)$. Finally, the function $\Psi(P,.)$ is continuous, piecewise affine, convex, and largely increasing.

If P and $Q \in \mathbb{K}(x) \setminus \{0\}$, then $\nu^+(PQ,\mu)$ is the right-side derivative of the function $\Psi(PQ,.)$. But $\Psi(PQ,.) = \Psi(P,.) + \Psi(Q,.)$, hence its right-side derivative at μ is just $\nu^+(P,\mu) + \nu^+(Q,\mu)$. In the same way, we have $\nu^-(PQ,\mu) = \nu^-(P,\mu) + \nu^-(Q,\mu)$ by considering left-side derivatives.

Then, to prove that $\nu^+(P,\mu)-\nu^-(P,\mu)$ is the number of zeros of P in $C(0,\theta^\mu)$, it is sufficient to show this when P is a binomial $x-a$. But then, this is obvious because $\nu^+(P,\mu)=\nu^-(P,\mu)=0$ whenever $\mu<\Psi(\alpha), \nu^+(P,\mu)=\nu^-(P,\mu)=1$ whenever $\mu>\Psi(\alpha)$, whereas $\nu^+(P,\Psi(\alpha))=1, \nu^-(P,\Psi(\alpha))=0$. So all the statements of Theorem A.3.16 have been proven. □

Applying Lemma A.3.12 and Theorem A.3.16 to the numerator and the denominator of a rational function, we obtain Corollary A.3.17.

Corollary A.3.17: *Let* $P(x)=\sum_{j=0}^{n} a_j x^j \in \mathbb{K}[x]$. *For every* $\mu \in \mathbb{R}, \nu^+(P,\log(r))$ *is equal to the number of zeros admitted by* P *in* $d(0,r)$.

Corollary A.3.18: *Let* $h \in \mathbb{K}(x)\setminus\{0\}$. *The function in* μ $\Psi(h,\mu)$ *is continuous and piecewise affine.*

If μ *is such that* $d(0,\theta^\mu)$ *contains* s *zeros and* t *poles of* h *(taking multiplicity into account), but neither any zero nor any pole in* $C(0,\theta^\mu)$, *then* $\Psi(h,.)$ *has a derivative equal to* $s-t$ *at* μ.

If μ *is such that* $C(0,\theta^\mu)$ *contains* s *zeros and* t *poles of* h *(taking multiplicity into account), then we have* $\frac{d_l\Psi}{d\mu}(h,\mu)-\frac{d_r\Psi}{d\mu}(h,\mu)=s-t$. *Further, if the function* $\Psi(f,\mu)$ *is not derivable at* μ, *then* μ *lies in* $\Psi(\mathbb{K})$.

Corollary A.3.19: *Let* $h \in \mathbb{K}(x)\setminus\{0\}$ *have no pole (resp. no zero) in an annulus* $\Gamma(0,r',r'')$. *Then,* $\Psi(h,\mu)$ *is convex (resp. concave) in* $[\log r', \log r'']$.

Corollary A.3.20: *Let* $h \in \mathbb{K}(x)\setminus\{0\}$ *have* s *zeros and* t *poles in* $d(0,r')$ *and have neither any zero nor any pole in an annulus* $\Gamma(0,r',r'')$. *Then in* $\Gamma(0,r',r'')$, $\Psi(h,\log|x|)$ *is of the form* $A+(s-t)\log|x|$. *Moreover,* $\nu^+(h,\mu)-\nu^-(h,\mu)=s-t$ $\forall\mu\in]\log r',\log r''[$ *and* $\nu^+(h,\mu)$ *(resp.* $\nu^-(h,\mu)$*) is continuous on the right (resp. on the left). Finally, given* $g\in\mathbb{K}(x)$, *we have* $\nu^+(gh,\mu)=\nu^+(g,\mu)+\nu^+(h,\mu)$, $\nu^-(gh,\mu)=\nu^-(g,\mu)+\nu^-(h,\mu)$.

Theorem A.3.21: *Let* $\mathcal{G}\in\Phi(\mathbb{K})$. *Let* $f\in\mathbb{K}(x)$ *and take* $\epsilon>0$. *There exists a* $\mathcal{G}$*-affinoid* E *such that* $|\,|f(x)|-\varphi_{\mathcal{G}}(f)|_\infty\le\epsilon$, $\forall x\in E$.

Proof. Let $r=\operatorname{diam}(\mathcal{G})$ and let $l>r$. If $\mathcal{G}$ has no center, there exists a disk $d(a,l)\in\mathcal{G}$, with $r\in|\mathbb{K}|$, containing neither zeros nor poles of f, therefore, by Lemma A.3.9 $|f(x)|$ is a constant equal to $\varphi_{\mathcal{F}}(f)$ in $d(a,r)$, so our claim is obvious. Now, suppose that $\mathcal{G}=\mathcal{F}_{a,r}$. Let $\Lambda_1,\dots,\Lambda_q$ be the classes of $d(a,r)$ containing at least one zero or one pole of f. By Lemma A.3.9, $|f(x)|$ is a constant equal to $\varphi_{\mathcal{G}}(f)$ in $d(a,r)\setminus\left(\bigcup_{j=1}^q\Lambda_j\right)$. Consider a class $\Lambda_j=d(a_j,r^-)$ and let s_j (resp. t_j) be the number of zeros (resp. poles) of f inside Λ_j and let s_0 (resp. t_0) be the number of zeros (resp. poles) of f in all $d(a,r)$. Let $\rho\in]0,r[\cap|\mathbb{K}|$ be such that

$|(\frac{r}{\rho})^{s_j-t_j}-1|\varphi_{\mathcal{G}}(f)\leq\epsilon\ \forall j=0,...,q$. Let $l=\frac{r^2}{\rho}$ and $E=d(a,l)\setminus\bigcup_{j=1}^{q}d(a_j,\rho^-)$. By Corollary A.3.19, we can check that the inequality $|\ |f(x)|-\varphi_{\mathcal{G}}(f)|_\infty\leq\epsilon$ holds in all E. Since $\rho<r$, E is an infraconnected affinoid set that belongs to $\mathcal{G}$. Moreover, by definition, $l>r$. □

Theorem A.3.22: *Let $\mathcal{F}$ be a filter on $\mathbb{K}$ such that for every $h\in\mathbb{K}(x)$, $|h(x)|$ admits a limit along $\mathcal{F}$. Then there exists a circular filter $\mathcal{H}$ less thin than $\mathcal{F}$.*

Proof. For every $h\in\mathbb{K}(x)$, set $\phi(h)=\lim_{\mathcal{F}}|h(x)|$. Then ϕ belongs to $Mult(|K(x))$ and hence by Theorem A.3.10, there exists a unique circular filter $\mathcal{H}$ such that $\phi=\varphi_{\mathcal{H}}$. Suppose that $\mathcal{F}$ is not thinner than $\mathcal{H}$. There exists a subset B of $\mathbb{K}$ such that $\mathcal{F}$ is secant with B but $\mathcal{H}$ is not. Since $\mathcal{H}$ admits a basis consisting of affinoid subsets, there exists a $\mathcal{H}$-affinoid D such that $D\cap B=\emptyset$. Since D is affinoid, it is of the form $d(a,R)\setminus\Big(\bigcup_{k=1}^{q}d(a_i,r_i^-)\Big)$ and $\mathcal{H}$ also admits a $\mathcal{H}$-affinoid E of the form $d(a,S)\setminus\Big(\bigcup_{k=1}^{q}d(a_i,s_i^-)\Big)$ with $S<R$ and $r_i<s_i\ \forall i=1,...,q$. Let $b\in\Gamma(a,S,R)$ and $h(x)=\frac{\prod_{i=1}^{q}(x-a_i)^{m_i}}{(x-b)^n}$. Then with integers m_i and n big enough, we can get

$$\inf\{|h(x)|\mid x\in E\}>\sup\{|h(x)|\mid x\in B\}.$$

A contradiction to the hypothesis: $|h(x)|$ admits a limit along $\mathcal{F}$. □

A.4. Hensel Lemma

The Hensel Lemma is a classical tool for studying the factorization of analytic functions on a circle [2, 50, 58] and is indispensable in Chapter A.5. It is a strong result that roughly says: "In a complete field $\mathbb{L}$, if $\overline{P}$ splits in the form $\gamma\eta$ with $(\gamma,\eta)=\overline{1}$, then P also splits in $\mathbb{L}[x]$ in the form gh with $\overline{g}=\gamma$, $\overline{h}=h$, $\deg(\overline{g})=\deg(\gamma)$." The proof is not very easy and requires a serious preparation. Here, we will roughly follow the same process as in [2, 50, 58] and more precisely in [58], with a few corrections.

Definitions and notations: Given a field E and $g,h\in\mathbb{L}[x]$, here (g,h) will denote the monic greatest common divisor of g and h. Given $Q(x)=\sum_{j=0}^{n}b_jx^j\in U_{\mathbb{L}}[x]$, as in Chapter A.3 we denote by $\overline{Q}$ the polynomial $\sum_{j=0}^{q}\overline{a_j}\ x^j\in\mathcal{L}[x]$. In this chapter, $P(x)=\sum_{j=0}^{q}a_jx^j\in L[x]$ will denote a polynomial of degree q.

Lemma A.4.1. is immediate:

Lemma A.4.1: *The quotient ring* $\dfrac{U_{\mathbb{L}}[x]}{M_{\mathbb{L}}[x]}$ *is isomorphic to* $\mathcal{L}[x]$.

Lemma A.4.2: *For all* $\alpha \in \mathbb{L}$, *we have* $\Psi(P(\alpha)) \leq \Psi(P,0) + \max(0, q\Psi(\alpha))$.

Proof. By Lemma A.3.12 we have $\Psi(P(\alpha)) \leq \max_{0\leq j\leq q}(\Psi(a_j) + j\Psi(\alpha)) \leq \max_{0\leq j\leq q} \Psi(a_j) + \max_{0\leq j\leq q} j\Psi(\alpha)$. But $\max_{0\leq j\leq q} \Psi(a_j) = \Psi(P,0)$ and $\max_{0\leq j\leq q} j\Psi(\alpha) = \max(0, q\Psi(\alpha))$. □

Definitions and notations: A polynomial $\sum_{j=0}^{q} a_j x^j \in L[x]$ will be said to be *quasi-monic* if $|a_q| = 1$.

Lemma A.4.3: *Let* $F, D \in U_{\mathbb{L}}[x]$ *with* D *quasi-monic. Let* $Q, R \in U_{\mathbb{L}}[x]$ *satisfy* $F = DQ+R$ *and* $\deg(R) < \deg(D)$. *Then we have* $\Psi(Q,0) \leq \Psi(F,0)$ *and* $\Psi(R,0) \leq \Psi(F,0)$.

Proof. We can clearly assume $F \neq 0$. Then, by multiplying F by a suitable constant λ, we can also assume $\Psi(F,0) = 0$. Since D is quasi-monic, the Euclidean division of F by D is clearly possible in $U_{\mathbb{L}}[x]$ and therefore Q is the quotient, R is the rest of this division, due to the fact that $\deg(R) < \deg(D)$. So we have $\Psi(Q,0) \leq 0$, $\Psi(R,0) \leq 0$ because both Q, R belong to $U_{\mathbb{L}}[x]$. □

Corollary A.4.4: *Let* $F, D \in \mathbb{K}[x]$ *with* D *having all its zeros in* $d(0,1)$. *Let* $Q, R \in U_{\mathbb{K}}[x]$ *satisfy* $F = DQ + R$ *and* $\deg(R) < \deg(D)$. *Then we have* $\Psi(Q,0) \leq \Psi(F,0) - \Psi(D,0)$ *and* $\Psi(R,0) \leq \Psi(F,0)$. *Moreover, if* F *has all its zeros in* $d(0,1)$, *then* $\Psi(Q,0) = \Psi(F,0) - \Psi(D,0)$.

Proof. The first statement is just an application of Lemma A.4.3. Next, if F has all its zeros in $d(0,1)$, we can assume that both F, D are monic and satisfy $\|F\| = \|D\| = 1$. Consequently, Q also must be monic and hence $\|Q\| = 1$, which ends the proof. □

Lemma A.4.5: *Let* $g, h \in U_{\mathbb{L}}[x]$ *be quasi-monic, such that* $(\overline{g}, \overline{h}) = \overline{1}$ *and* $\deg(P) < \deg(g)+\deg(h)$. *There exist* $V, W \in \mathbb{L}[x]$ *satisfying* $\Psi(Vg+Wh-P,0) < \Psi(P,0)$, $\Psi(V,0) \leq \Psi(P,0)$, $\Psi(W,0) \leq \Psi(P,0)$, $\deg(V) < \deg(h)$, $\deg(W) < \deg(g)$.

Proof. Since $(\overline{g}, \overline{h}) = \overline{1}$, by Bezout's theorem there exists υ and $\tau \in \mathcal{L}[x]$ such that $\upsilon\overline{g} + \tau\overline{h} = \overline{1}$, $\deg(\upsilon) < \deg(\overline{h})$, and $\deg(\tau) < \deg(\overline{g})$. Let $S, T \in U\mathbb{L}[x]$ satisfy $\overline{S} = \upsilon$, $\overline{T} = \tau, \deg\,(S) = \deg\,(\upsilon), \textit{and} \deg\,(T) = \deg\,(\tau)$. Thus, we have $\overline{Sg + Th - 1} = \overline{0}$, i.e.,

(1) $\Psi(Sg + Th - 1) < 0$.

We now consider the Euclidean division of SP by h and TP by g, respectively. We obtain $SP = S_0h + V$and $TP = T_0g + W$. By Lemma A.4.2, it is seen that $\max(\Psi(V,0), \Psi(S_0,0)) \leq \Psi(SP,0) \leq \Psi(P,0)$. Moreover, by hypothesis we have

(1) $\deg(V) < \deg(h)$ and
(2) $\deg(W) < \deg(g)$.

Let $M = Sg + Th - 1$. Then we have $MP = (S_0 + T_0)gh + Vg + Wh - P$. Since $\deg(P) < \deg(g) + \deg(h)$, by (1) and (2) we see that $\deg(Vg + Wh - P) < \deg(g) + \deg(h)$ and therefore $Vg + Wh - P$ is just the remainder of the Euclidean division of BP by gh. But then, by Lemma A.4.3, we have $\Psi(Vg + Wh - P, 0) \leq \Psi(MP, 0) = \Psi(M, 0) + \Psi(P, 0)$, and therefore by (1) and by definition of M it is seen that $\Psi(M, 0) < 0$. This finishes proving that $\Psi(Vg + Wh - P, 0) < \Psi(P, 0)$ and this ends the proof of Lemma A.4.5. □

Definitions and notations: Let $g, h \in U_{\mathbb{L}}[x]$, be monic and satisfy $(\overline{g}, \overline{h}) = \overline{1}$. We will denote by $B(f,g)$ the set of constants $c \in \mathbb{R}_+$ such that, for every polynomial $Q \in \mathbb{L}[x]$ satisfying $\deg(Q) < \deg(g) + \deg(h)$, there exist $V, W \in \mathbb{L}[x]$ satisfying $\Psi(Vg + Wh - Q, 0) \leq \Psi(Q, 0) + c$, $\Psi(V, 0) \leq \Psi(Q, 0)$, $\Psi(W, 0) \leq \Psi(Q, 0)$, $\deg(V) < \deg(h)$, *and* $\deg(W) < \deg(g)$.

Lemma A.4.6: *Let $g, h \in U_{\mathbb{L}}[x]$, be quasi-monic and satisfy $(\overline{g}, \overline{h}) = \overline{1}$ and let $d = \deg(g) + \deg(h)$. Then $B(f,g)$ is a not an empty interval whose lower bound is 0. Moreover, given $\lambda \in B(f,g)$ and monic polynomials s, $t \in U_{\mathbb{L}}[x]$ such that $\Psi(g - s, 0) \leq \lambda$, $\Psi(h - t, 0) \leq \lambda$, then $B(s,t) = B(g,h)$.*

Proof. Let $d = \deg(g) + \deg(h)$. We can apply Lemma A.4.5 to each polynomial $Q_n = x^n$ for every $n = 0, \dots, d-1$. Thus, we have polynomials V_n, W_n satisfying $\Psi(V_ng + W_nh - x^n, 0) < 0$, $\Psi(V_n, 0) \leq 0$, $\Psi(W_n, 0) \leq 0$, $\deg(V_n) < \deg(h)$, and $\deg(W_n) < \deg(g)$. We put $\lambda_n = \Psi(V_ng + W_nh - x^n, 0)$, $(0 \leq n \leq d-1)$. Now let $Q = \sum_{n=0}^{d-1} a_nx^n$, let $V = \sum_{n=0}^{d-1} a_nV_n$, $W_n = \sum_{n=0}^{d-1} a_nW_n$, and let $\lambda(g,h) = \max_{0 \leq n \leq d-1} \lambda_n$. Clearly, we have $\Psi(Vg + Wh - Q, 0) \leq \max_{0 \leq n \leq d-1}(\Psi(a_n) + \lambda_n) \leq \max_{0 \leq n \leq d-1} \Psi(a_n) + \max_{0 \leq n \leq d-1} \lambda_n = \Psi(Q, 0) + \lambda(g,h)$.

But trivially

$$\Psi(V, 0) \leq \max_{0 \leq n \leq d-1} \Psi(a_n),\ \Psi(W, 0) \leq \Psi(Q, 0),\ \deg(V) \leq \max_{0 \leq n \leq d-1}(\deg(V_n))$$

$$< \deg(h),\ \deg(W) \leq \max_{0 \leq n \leq d-1}(\deg(W_n)) < \deg(h).$$

So, $\lambda(g,h)$ lies in $B(f,g)$ and hence it is obviously seen that $B(f,g)$ is a not an empty interval and that its lower bound is 0.

Now, let $c \in B(f,g)$ and let $s,\ t \in U_{\mathbb{L}}[x]$ be monic and satisfy $\Psi(g-s,0) < c$, $\Psi(h-t,0) < c$. Since $\Psi(V,0) \leq \Psi(Q,0)$, $\Psi(W,0) \leq \Psi(Q,0)$, it is easily seen that $\Psi(V(g-s)+W(h-t),0) < c+\Psi(Q,0)$, and therefore $\Psi(Vs+Wt-Q,0) < c+\Psi(Q,0)$. This shows that $\lambda(s,t) \leq c$ and therefore $B(f,g) \subset B(s,t)$. But similarly we have $B(s,t) \subset B(g,h)$ and this ends the proof of Lemma A.4.6. □

Lemma A.4.7: *Let $Q \in L[x]$ and let $g,h \in U_{\mathbb{L}}[x]$ be quasi-monic and satisfy $(\overline{g},\overline{h}) = \overline{1}$. Let $c \in B(g,h)$. There exist monic polynomials V, $W \in L[x]$ satisfying*
$\Psi(Vg+Wh-Q,0)\ \leq c+\Psi(Q,0)$,
$\deg(W)\ < \deg(g)$, $\deg(V) \leq \max(\deg(h), \deg(Q)-\deg(g))$,
$\Psi(V,0) \leq \Psi(Q,0)$, and $\Psi(W,0) \leq \Psi(Q,0)$.

Proof. We consider the Euclidean division of Q by gh : $Q = \ell gh + Q_1$. Hence, $\deg(Q_1)\ < \deg(g)+\deg(h)$. By Lemma A.4.3, we have

(1) $\Psi(Q_1,0) \leq \Psi(Q,0) and$
(2) $\Psi(\ell,0) \leq \Psi(Q,0)$.
By Lemma A.4.6, there exist $V_1, W_1 \in \mathbb{L}[x]$ satisfying
(3) $\Psi(V_1g+W_1h-Q_1,0)\ \leq \Psi(Q_1,0)+c$,
(4) $\Psi(V_1,0)\ \leq \Psi(Q_1,0)$,
(5) $\Psi(W_1,0)\ \leq \Psi(Q_1,0)$,
(6) $\deg(V_1)\ < \deg(h)$, and
(7) $\deg(W_1)\ < \deg(g)$.

Now we put $V = V_1+\ell h$, $W = W_1$. So we have $Vg+Wh-Q = V_1+\ell gh+W_1h-\ell gh-Q_1$ and therefore by (3) we obtain $\Psi(Vg+Wh-Q,0)\ \leq \Psi(Q_1,0)+c$. Hence, by (1) we obtain
(8) $v(Vg+Wh-Q)\ \geq\ v(Q)+c$.
Now, by (1) and (5) it is seen that
(9) $\Psi(W,0)\ \leq \Psi(Q,0)$.
We will check
(10) $\Psi(V,0)\ \leq \Psi(Q,0)$.
Indeed, we have $\Psi(h,0) = 0$, hence by (2) we see
(11) $\Psi(\ell h,0)\ \leq\ \Psi(Q,0)$.
But by (4) we have $\Psi(V_1,0)\ \ \leq\ \Psi(Q,0)$ and therefore by (11) we obtain (10). Finally, by definition we have $\deg(\ell) = \deg(Q)-\deg(gh)$ and therefore
(12) $\deg(V)\ \leq \max(\deg(V_1), \deg(\ell h))\ \leq\ \max(\deg(h), \deg(Q)-\deg(g))$.
Thanks to (7), (8), (9), (10), and (12). Lemma A.4.7 is now proven. □

Theorem A.4.8 (Hensel Lemma): *L is suppposed to be complete. Let $P \in U_{\mathbb{L}}[x]$ be such that $\overline{P}$ splits in $\mathcal{L}[x]$ in the form $\gamma\eta$ with γ, η relatively prime. There exists $g,h \in U_{\mathbb{L}}[x]$ such that $P = gh, \overline{g} = \gamma, \overline{h} = \eta, \deg(g) = \deg(\gamma)$.*

Proof. We can obviously take quasi-monic polynomials $g_0, h_0 \in U_{\mathbb{L}}[x]$ such that $\overline{g_0} = \gamma$, $\overline{h_0} = \eta$. We put $\xi = \Psi(P-g_0h_0,0)$ and take $\zeta \in B(g_0,h_0)$ satisfying $\zeta \leq \xi$.

We will construct sequences $(g_n)_{n\in\mathbb{N}}, (h_n)_{n\in\mathbb{N}}$ in $\mathbb{L}[x]$ satisfying for all $n \geq 0$:

$i_n)$ $\Psi(P - g_n h_n, 0) \leq (n+1)\zeta$,
$ii_n)$ $\Psi(g_n - g_{n-1}, 0) \leq n\zeta$, $\Psi(h_n - h_{n-1}, 0) \leq n\zeta$,
$iii_n)$ $\deg(h_n) \leq \deg(P) - \deg(g_0)$, $\deg(g_n) = \deg(g_0)$,
$iv_n)$ $\overline{g_n} = \gamma$, $\overline{h_n} = \eta$, and
$v_n)$ $\zeta \in B(g_n, h_n)$.

First we put $P_1 = P - g_0 h_0$. We notice that $\deg(P_1) = \deg(P)$. We now apply Lemma A.4.7 to the case when $(Q, g, h) = (P_1, g_0, h_0)$: there exist $V_1, W_1 \in \mathbb{L}[x]$ satisfying (1) $\deg(W_1) < \deg(g_0)$, (2) $\deg(V_1) < \deg(P) - \deg(g_0)$, (3) $\Psi(V_1, 0) \leq \zeta$, (4) $\Psi(W_1, 0) \leq \zeta$, and (5) $\Psi(V_1 g_0 + W_1 h_0 - P_1, 0) \leq \zeta + \Psi(P_1, 0)$. Next we put $g_1 = g_0 + W_1$, $h_1 = h_0 + V_1$. We check that $i_1), ii_1), iii_1)$, and $iv_1)$ are satisfied. Moreover, by (3) and (4) and by Lemma A.4.6, ζ lies in $B(g_1, h_1)$, hence v_1 is satisfied.

Now we suppose we have already constructed the pairs (g_m, h_m) satisfying $i_m), ii_m), iii_m), iv_m)$, and $v_m)$ for every $m = 0, \ldots, n$. Then we put $P_{n+1} = P - g_n h_n$.

We can apply Lemma A.4.7 to the case when (Q, g, h) is equal to (P_{n+1}, g_n, h_n). So, we can obtain $V_{n+1}, W_{n+1} \in L[x]$ satisfying (6) $\Psi(W_{n+1} g_n + V_{n+1} h_n - P_{n+1}, 0) \leq \zeta + \Psi(P_{n+1}, 0)$, (7) $\deg(W_{n+1}) < \deg(g_n)$, (8) $\deg(V_{n+1}) \leq \max(\deg(h_n), \deg(P_{n+1}) - \deg(g_n))$, and (9) $\Psi(V_{n+1}, 0) \leq \Psi(P_{n+1}, 0)$, $\Psi(W_{n+1}, 0) \leq \Psi(P_{n+1}, 0)$. By (6) and by (v_n) we obtain (10) $\Psi(W_{n+1} g_n + V_{n+1} h_n - P_{n+1}, 0) \leq (n+2)\zeta$. Now we put $g_{n+1} = g_n + W_{n+1}$, $h_{n+1} = h_n + W_{n+1}$. We check that
$P - g_{n+1} h_{n+1} = (P_{n+1} - h_n W_{n+1} - g_n V_{n+1}) - V_{n+1} W_{n+1} =$
$= P_{n+1} - h W_{n+1} - g_{n+1} V_{n+1} + (h_{n+1} - h_n) W_{n+1} + (g_{n+1} - g_n) V_{n+1} + V_{n+1} W_{n+1}$.
By $ii_m)$, true for $m \leq n$, we notice that
(11) $\Psi(g_n - g_{n+1}, 0) \leq (n+1)\zeta$, $\Psi(h_n - h_{n+1}, 0) \leq (n+1)\zeta$,
and then, by (9) and (10), we obtain
$i_{n+1})$ $\Psi(P - g_{n+1} h_{n+1}, 0) \leq (n+2)\zeta$.

Relation $i_{n+1})$ is true by definition and $iii_{n+1})$ and $iv_{n+1})$ are easily checked. By (11) and by Lemma A.4.6, Relation $v_{n+1})$ is also clear.

Therefore, the sequences $(g_n)_{n\in\mathbb{N}}$, $(h_n)_{n\in\mathbb{N}}$ satisfying $i_n)$, $ii_n)$, $iii_n)$, $iv_n)$, and $v_n)$ are now constructed. Since $\mathbb{L}$ is complete, the vector space $\mathbb{L}_q[x]$ of polynomial of degree $m \leq q$ is obviously complete with respect to the Gauss norm $\|\cdot\|$, which is characterized by $\log \|Q\| = \Psi(Q, 0)$.

Then by Relation $ii_n)$, the sequences $(g_n)_{n\in\mathbb{N}}, (h_n)_{n\in\mathbb{N}}$ converge in $\mathbb{L}_q[x]$. We put $g = \lim_{n\to\infty} g_n$, $h = \lim_{n\to\infty} h_n$. By $iii_n)$, we have $\deg(g) = \deg(g_0) = \deg(\gamma)$. By $iv_n)$, we have $\overline{g} = \gamma$, $\overline{h} = \eta$ and finally by $i_n)$, we have $\Psi(P - gh, 0) = +\infty$, hence $P = gh$. That ends the proof of Theorem A.4.8. □

A.5. Extensions of ultrametric fields: The field $\mathbb{C}_p$

All considerations on analytic and meromorphic functions require to consider a complete ultrametric algebraically closed field $\mathbb{K}$. Here, we will construct the field $\mathbb{C}_p$ and study finite extensions of $\mathbb{Q}_p$. And we show that $\mathbb{C}_p$ is not spherically complete.

Definitions and notations: As in Chapters A.2, A.3 and A.4, $\mathbb{L}$ denotes a complete ultrametric field whose absolute value is not trivial and whose residue class field is $\mathcal{L}$. We will denote by F an algebraically closed ultrametric field whose absolute value is not trivial.

Let $\mathbb{E}$ be a field, let $\mathbb{B}$ be a finite algebraic extension of $\mathbb{E}$, and let $q = [\mathbb{B} : \mathbb{E}]$. We will denote by $\mathcal{N}$ the algebraic norm of $\mathbb{B}$ over $\mathbb{E}$. Given $a \in \mathbb{B}$, we will denote by $\mathrm{irr(a, \mathbb{E})}$ the minimal polynomial of a over $\mathbb{E}$.

Lemma A.5.1 is classical in algebra [70].

Lemma A.5.1: *Let $q = [\mathbb{B} : \mathbb{E}]$ and let $\mathcal{N}$ be the norm of $\mathbb{B}$ over $\mathbb{E}$. Let $a \in \mathbb{B}$, let $P_a = \mathrm{irr(a, \mathbb{E})}$, and let $d = \deg(P_a)$. Then $\mathcal{N}$ satisfies $\mathcal{N}(a) = \big((-1)^d P_a(0)\big)^{\frac{q}{d}}$ and $\mathcal{N}(ab) = \mathcal{N}(a)\mathcal{N}(b)$, $\forall b \in \mathbb{B}$.*

Theorem A.5.2: *Let $\mathbb{E}$ be an algebraic extension of $\mathbb{L}$, let $a \in \mathbb{E}$, and let $P = \mathrm{irr(a, \mathbb{L})}$. Then a is an integral over $U_{\mathbb{L}}$ if and only if $|P(0)| \leq 1$. Moreover, if $|P(0)| \leq 1$, then $\|P\| = 1$. Finally, if $|P(0)| = 1$, then $\mathrm{irr(\overline{a}, \mathcal{L})} = \overline{\mathrm{P}}$.*

Proof. First we assume a to be integral over $U_{\mathbb{L}}$. Then there exists a monic polynomial $Q \in U_{\mathbb{L}}[x]$ such that $Q(a) = 0$. Therefore, P divides Q in $\mathbb{L}[x]$. Let $Q(x) = P(x)T(x)$. Since both P, Q are monic, so is T. Therefore, $\Psi(P(,0) \geq 0$, $\Psi(T(,0) \geq 0$. But since $\Psi(Q, 0) = 0$ and since $\Psi(Q, 0) = \Psi(P, 0) + \Psi(T, 0)$, P, T must satisfy $\Psi(P, 0) = \Psi(T, 0) = 0$ and therefore $|P(0)| \leq 1$.

Now we assume $|P(0)| \leq 1$. Suppose $\Psi(P, 0) > 0$. There exists $b \in M_{\mathbb{L}}$ such that $\Psi(bP, 0) = 0$ and then $|bP(0)| < 1$, hence $\overline{0}$ is a zero of $\overline{bP}$. Further, we notice (1) $\deg(\overline{bP}) < \deg(P)$.

Let $\overline{bP} = x^d\phi$ with $\phi(\overline{0}) \neq \overline{0}$. Then x^d and ϕ are relatively prime in $\mathcal{L}[x]$. Therefore, by Theorem A.4.8 there exist g, $h \in U_{\mathbb{L}}[x]$ such that $\overline{g} = x^d$, $\overline{h} = \phi$, $\deg(g) = d$, and $P = gh$. But since P is irreducible in $\mathbb{L}[x]$ and since $d > 0$, h must be a constant and therefore $\deg(P) = d$ a contradiction to (1). Consequently, $\Psi(P, 0) = 0$. Now, suppose $|P(0)| = 1$. Since P is irreducible in $\mathbb{L}[X]$, by Theorem A.4.8 so is $\overline{P}$ in $\mathcal{L}[X]$, hence $\mathrm{irr(\overline{a}, \mathcal{L})} = \overline{\mathrm{P}}$. □

Corollary A.5.3: *Let $\mathbb{E}$ be an algebraic extension of $\mathbb{L}$ equipped with the unique extension of the absolute value of $\mathbb{L}$. Let $a \in \mathbb{E}$ be such that $|a| = 1$, of degree l over $\mathbb{L}$. Then the residue class $\overline{a}$ of a in the residue class field of $\mathbb{E}$ is algebraic, of degree l over $\mathcal{L}$.*

Proof. Let $P(x) = \sum_{j=0}^{m} a_j x^j = \mathrm{irr}(\mathrm{a}, \mathbb{L})$. Since $|a| = 1$ we have $|a_0| = 1$, hence by Theorem A.5.2 we have $\|P\| = 1$ and $\overline{a}$ satisfies $\mathrm{irr}(\overline{\mathrm{a}}, \mathcal{L}) = \overline{\mathrm{P}}$, which ends the proof. □

Theorem A.5.4: *Let $\mathbb{E}$ be an algebraic extension of $\mathbb{L}$. There exists a unique absolute value φ on $\mathbb{E}$ that extends the one of $\mathbb{L}$. Further, this absolute value is ultrametric and defined as follows: given $a \in \mathbb{E}$, $Q = \mathrm{irr}(\mathrm{a}, \mathbb{L})$ and $t = \deg(Q)$, then $\varphi(a) = \sqrt[t]{|Q(0)|}$.*

Proof. We first notice that $\varphi(a) = |a|$ whenever $a \in \mathbb{L}$. We will show that φ is an ultrametric absolute value on $\mathbb{E}$. Clearly, we have $\varphi(a) \neq 0$ whenever $a \in \mathbb{E}$. By Lemma A.5.1, it is easily seen that we have $\varphi(ab) = \varphi(a)\varphi(b)$ whenever $a,\ b \in \mathbb{E}$ and therefore $\varphi(a)^{-1} = \varphi(a^{-1})$. So it remains to show the ultrametric inequality. For this, we will show $\varphi(1+z) \leq 1$ for every $z \in U_{\mathbb{E}}$. For convenience, we put again $P_z = \mathrm{irr}(\mathrm{z}, \mathrm{E})$ whenever $z \in \mathbb{E}$. Let $z \in U_{\mathbb{E}}$. So we have $|P_z(0)| \leq 1$ and then by Theorem A.5.2, z is integral over $U_{\mathbb{L}}$, hence so is $1 + z$. Hence, by Theorem A.5.2, we have $|P_{1+z}(0)| \leq 1$ and therefore $\varphi(1 + z) \leq 1$. Now, the ultrametric inequality will be easily derived. Let $a, b \in \mathbb{E}$ satisfy $0 < |a| \leq |b|$. We have $\varphi(a + b) = \varphi\big(b(1 + \frac{a}{b})\big) = \varphi(b)\varphi(1 + \frac{a}{b})$. But $\varphi(1 + \frac{a}{b}) \leq 1$, hence finally $\varphi(a + b) \leq \varphi(b)$. Thus, we have now proven φ to be an ultrametric absolute value that extends that of $\mathbb{L}$. Then by Theorem A.1.6, this absolute value on $\mathbb{E}$ is unique, which ends the proof. □

Corollary A.5.5: *Let Ω be an algebraic closure of $\mathbb{L}$. There exists a unique absolute value φ on Ω that extends the one of $\mathbb{L}$. Further, this absolute value is ultrametric and defined as follows: given $a \in \Omega$, $Q = \mathrm{irr}(\mathrm{a}, \mathbb{L})$ and $t = \deg(Q)$, then $\varphi(a) = \sqrt[t]{|Q(0)|}$.*

Corollary A.5.6: *Let $P(x) \in \mathbb{L}[x]$ be irreducible over $\mathbb{L}$, let Ω be an algebraic closure of $\mathbb{L}$ provided with the absolute value extending that of $\mathbb{L}$ and let $b_1, ..., b_q$ be the zeros of P in Ω. Then $|b_i| = |b_j|\ \forall i, j \leq q$.*

Corollary A.5.7: *Let Ω be an algebraic closure of $\mathbb{L}$ provided with the unique absolute value $|\ .\ |$ that extends the one of $\mathbb{L}$. Then U_Ω is equal to the integral closure of $U_{\mathbb{L}}$. Moreover, $|\Omega| = \{\sqrt[n]{r} \mid r \in |\mathbb{L}|,\ n \in \mathbb{N}^*\}$.*

Corollary A.5.8: *Suppose that the value group of $\mathbb{L}$ is $\mathbb{Z}$. Let $\mathbb{E}$ be a finite algebraic extension of $\mathbb{L}$ of degree t provided with the unique absolute value $|\ .\ |$ that extends the one of $\mathbb{L}$. There exists a rational r of the form $\dfrac{s}{t}$ such that the value group of $\mathbb{E}$ is $r\mathbb{Z}$.*

Lemma A.5.9: *Let $\mathbb{B}$ be an algebraic extension of $\mathbb{E}$ provided with an absolute value extending that of $\mathbb{E}$. Then the residue class field of $\mathbb{B}$ is algebraic over the residue class field of $\mathbb{E}$.*

Proof. Let $t \in U_{\mathbb{B}}$ and let $\widehat{\mathbb{E}}$ be the completion of $\mathbb{E}$ with respect to the absolute value of $\mathbb{E}$. Since t is algebraic over $\mathbb{E}$, so much the more it is algebraic over $\widehat{\mathbb{E}}$. Then by Corollary A.5.5, the residue class $\bar{t}$ of t is algebraic over the residue class field of $\widehat{\mathbb{E}}$. But $\widehat{\mathbb{E}}$ obviously has the same residue class field as $\mathbb{E}$. □

Corollary A.5.10: *Let $\mathbb{B}$ be an algebraic extension of $\mathbb{L}$ provided with the unique absolute value that extends the one of $\mathbb{L}$. Then the residue class field of $\mathbb{B}$ is an algebraic extension of the residue class field of $\mathbb{L}$. Moreover, if $\mathbb{B}$ is finite over $\mathbb{L}$, then residue class field of $\mathbb{B}$ is finite over the residue class field of $\mathbb{L}$.*

Proof. Suppose that $\mathbb{B}$ is of the form $\mathbb{L}[u]$. Without loss of generality, we may assume that $|u| = 1$. Then the residue class field of $\mathbb{B}$ is $\mathcal{L}[\overline{u}]$. Next, we can generalize by induction. □

Theorem A.5.11: *Let Ω be an algebraic closure of $\mathbb{L}$ provided with the unique absolute value extending the one of $\mathbb{L}$. Then the residue class field of Ω is an algebraic closure of $\mathcal{L}$.*

Proof. Let $\mathcal{T}$ be the residue class field of Ω. Let $u \in U_\Omega$ and let $P = \mathrm{irr}(u, \mathbb{L})$. By Corollary A.5.7, P belongs to $U_{\mathbb{L}}[x]$ and obviously satisfies $\overline{P}(\overline{u}) = \overline{0}$, hence $\overline{u}$ is algebraic over $\mathcal{L}$. So $\mathcal{T}$ is an algebraic extension of $\mathcal{L}$. Now let $q \in \mathcal{L}[x]$, and let $Q \in U_{\mathbb{L}}[x]$ be a polynomial such that $\overline{Q} = q$. Then Q factorizes in $\Omega[x]$ in the form $\prod_{j=1}^{s}(x - a_j)$ with $|a_j| \leq 1$ $\forall j = 1, ..., s$, hence $\overline{a}_j$ belongs to $\mathcal{T}$ and $q(x) = \overline{a}_j = \prod_{j=1}^{s}(x - \overline{a}_j)$. So, $\mathcal{T}$ contains the algebraic closure of $\mathcal{L}$. And since it is an algebraic extension of $\mathcal{L}$, then it is the algebraic closure of $\mathcal{L}$. □

Corollary A.5.12: *The residue class field $\mathcal{K}$ of $\mathbb{K}$ is algebraically closed.*

Lemma A.5.13: *Let $P(x) = \sum_{j=0}^{t} a_j x^j$, $Q(x) = \sum_{j=0}^{t} b_j x^j$ be monic, belong to $\mathbb{B}[x]$ and satisfy $\|P\|\|Q\| = 1$. For each zero α of P, Q admits at least one zero β such that $|\alpha - \beta|^t \leq \max_{0 \leq j \leq t} |a_j - b_j|$.*

Proof. Let $s = \max_{0 \leq j \leq t} |a_j - b_j|$ and let α be a zero of P. By Theorem A.3.6 we have $|(P - Q)(x)| \leq s$ whenever $x \in d(0, 1)$, hence in particular $|Q(\alpha)| \leq s$. Let $\beta_1, ..., \beta_t$ be the zeros of Q (taking multiplicities into account). So we have $\prod_{j=1}^{t} |\beta_j - \alpha| \leq s$ and then that at least one of the β_j satisfies $|\beta_j - \alpha| \leq s^{\frac{1}{t}}$. □

Theorem A.5.14: *Let $\mathbb{B}$ be an algebraically closed extension of $\mathbb{L}$ provided with the unique absolute value that extends the one of $\mathbb{L}$. The completion of $\mathbb{B}$ also is algebraically closed.*

Proof. Let $\widetilde{\mathbb{B}}$ be the completion of $\mathbb{B}$ and let $P(x) = \sum_{j=0}^{t} a_j x^j \in \widetilde{\mathbb{B}}[x]$ be monic. Let Ω be an algebraic closure of $\widetilde{\mathbb{B}}$, provided with the unique absolute value extending that of $\widetilde{\mathbb{B}}$ and let $\alpha_1, ..., \alpha_t$ be the zeros of P in Ω. Up to a change of variable, we may assume that $|\alpha_j| \leq 1 \ \forall j = 1, ..., t$. Let $\varepsilon \in]0, 1[$ and let $Q(x) = \sum_{j=0}^{t} b_j x^j \in \mathbb{B}[x]$ be such that $\max_{0 \leq j \leq t} |a_j - b_j| \leq \varepsilon^t$. For each $j = 1, ..., t$, by Lemma A.5.13 Q admits a zero β such that $|\alpha_j - \beta|^t \leq \varepsilon$. Since $Q \in \mathbb{B}[x]$, obviously β belongs to $\mathbb{B}$ and therefore we see that α_j belongs to $\widetilde{\mathbb{B}}$. That ends the proof. □

Theorem A.5.15 is due to Marc Krasner [69, 74].

Theorem A.5.15 (M. Krasner): *Let $\mathbb{L}$ have characteristic zero. Let Ω be an algebraic closure of $\mathbb{L}$ provided with the unique absolute value extending the one of $\mathbb{L}$. Let $a \in \Omega$, let $a_2, ..., a_n$ be the conjugates of a in Ω, and let $b \in \Omega$ satisfy $|b - a| < |b - a_j|$ for every $j = 2, ..., n$. Then we have $\mathbb{L}[a] \subset \mathbb{L}[b]$.*

Proof. Let $a_1 = a$ and let $P(x) = \mathrm{irr}(a, \mathbb{L})$. In $\Omega[x]$, the polynomial P splits in the form $\prod_{j=1}^{n}(x - a_j)$. Let $Q(x) = \mathrm{irr}(a, \mathbb{L}[b])$. Then Q divides P. Let $t = \deg(Q)$ and suppose the a_j ranged in such a way that $Q(x) = \prod_{j=1}^{t}(x - a_j)$. Let $R(y) = Q(b+y)$. Then $R(y)$ is seen to be irreducible in $\mathbb{L}[b][y]$ like $Q(x)$ in $\mathbb{L}[b][x]$. Moreover, the zeros of R are just the $a_j - b$, with $1 \leq j \leq t$. Thus, we have $R = \mathrm{irr}(a - b, \mathbb{L}[b])$. But since $\mathbb{L}$ is complete, by Corollary A.5.5 we have $|a_j - b| = \sqrt[q]{|R(0)|}$ for every $j = 1, ..., t$. In particular, for $j = 2, ..., t$, we have $|a_j - b| = |a - b|$ and this contradicts the hypothesis. Finally, we have $t = 1$ and therefore a lies in $\mathbb{L}[b]$. □

Corollary A.5.16: *Let $\mathbb{L}$ have characteristic zero. Let Ω be an algebraic closure of $\mathbb{L}$ provided with the unique absolute value extending the one of $\mathbb{L}$. Let $a \in \Omega$, let $a_2, ...a_n$ be the conjugates of a in Ω, and let $b \in \Omega$ satisfy $|b - a| < |b - a_j|$ for every $j = 2, ...n$ and $[\mathbb{L}[b] : \mathbb{L}] \leq n$. Then we have $\mathbb{L}[a] = \mathbb{L}[b]$.*

We can now recall the construction of p-adic fields.

Notations and definitions: Let p be a prime number. On $\mathbb{Z}$, the p-adic absolute value is defined as follows: given $n \in \mathbb{Z}^*$, it factorizes in a unique way in the form $p^s q$, with $q \in \mathbb{Z}^*$, prime to p. So, here we take $\theta = p$ and set $|n|_p = p^{-s}$.

Lemma A.5.17 is immediate.

Lemma A.5.17: *$|\ .\ |_p$ is an ultrametric absolute value on $\mathbb{Z}$ that has continuation to $\mathbb{Q}$ and defines an ultrametric absolute value on $\mathbb{Q}$ and $\mathbb{N}$ is dense in $\mathbb{Z}$. Then $|n| \geq \frac{1}{n}\ \forall n \in \mathbb{N}^*$.*

$$U_{\mathbb{Q}} = \{p^n\Big(\frac{a}{b}\Big) \mid n \in \mathbb{N},\ a \in \mathbb{Z}, b \in \mathbb{Z}^*,\ \gcd(a,p) = \gcd(b,p) = 1\},$$

$$M_{\mathbb{Q}} = \{p^n\Big(\frac{a}{b}\Big) \mid n \in \mathbb{N}^*,\ a \in \mathbb{Z}, b \in \mathbb{Z}^*,\ \gcd(a,p) = \gcd(b,p) = 1\},$$

and the residue characteristic of $\mathbb{Q}$ is p. The residue class field of $\mathbb{Q}$ is the field of p elements $\mathbb{F}_p$. The valuation group of $\mathbb{Q}$ is isomorphic to the additive group $\mathbb{Z}$.

Remarks and notations: Now $\mathbb{Q}$ admits a completion with respect to the p-adic absolute value and its completion is denoted by $\mathbb{Q}_p$. The closure of $\mathbb{Z}$ in $\mathbb{Q}_p$ is denoted by $\mathbb{Z}_p$.

On $\mathbb{Q}_p$, we extend the valuation and the absolute value $|\ .\ |_p$ defined on $\mathbb{Q}$ and we set again $\Psi_p(x) = -v_p(x)$.

An algebraic closure Ω_p of $\mathbb{Q}_p$ is equipped with the unique extension of the p-adic absolute value defined on $\mathbb{Q}_p$ and we will again denote it by $|\ .\ |_p$. The valuation group of $\mathbb{Q}_p$ is obviously equal to the one of $\mathbb{Q}$. Next, the valuation group of Ω_p is easily seen to be isomorphic to $(\mathbb{Q}, +)$. In Chapter A.8, we will see that Ω_p is not complete.

By Theorem A.5.14, Ω_p has a completion denoted by $\mathbb{C}_p$ that is algebraically closed. The valuation group of $\mathbb{C}_p$ is then isomorphic to $(\mathbb{Q}, +)$ like the one of Ω_p. Moreover by Theorem A.5.11, the residue class field of Ω_p is an algebraic closure of $\mathbb{F}_p$ and the one of $\mathbb{C}_p$ is seen to be the same. The absolute value $|\ .\ |_p$ defined on Ω_p has a natural extension to $\mathbb{C}_p$ and the associated valuation will be denoted by v_p again, and we set again $\Psi_p(x) = -v_p(x)\ \forall x \in \mathbb{C}_p$. However, when there is no risk of confusion, we will just write Ψ instead of Ψ_p.

Theorem A.5.18: *Let a be integral over $\mathbb{Z}$ and let $a_2, ..., a_q$ be the conjugates of a over $\mathbb{Q}$. Then $|a| \leq 1$ and $a_j| \leq 1\ \forall j = 2, ..., q$.*

Proof. Since a is integral over $\mathbb{Z}$, it is integral over $\mathbb{Z}_p$. Let $P(X) = irr(a, \mathbb{Q})$ and let $B(X) = irr(a, \mathbb{Q}_p)$. Let $a_1, ..., a_h$ be the conjugates of a over $\mathbb{Q}_p$ (with $a_1 = a$). Then $P(0) = \prod_{j=1}^{q} a_j$. Then B divides P in $\mathbb{Q}_p[X]$. Moreover, $\|P\| = 1$ and $\|B\| = 1$. Next, $B(0) = \prod_{j=1}^{h} a_j$. By Corollary A.5.6, we have $|a_1| =, ..., |a_h|$, whereas $|B(0)| \leq 1$, hence $|a_j| \leq 1$, i.e., $|a| \leq 1$. Next, what is true for a_1 also holds for every $a_j, ..., j = 2, ..., q$, hence $|a_j|/leq1$. □

In the future, we will use Lemma A.5.19.

Lemma A.5.19: *Let $a \in \mathbb{C}_p$ be algebraic over $\mathbb{Q}_p$, such that $\log_p(|a|)$ is of the form $\frac{\lambda}{t}$, with $\lambda \in \mathbb{Z}$ and t in $\mathbb{N}^*$. Take m, $n \in \mathbb{N}$ and $b \in \mathbb{C}_p$ such that $\log_p(|b|)$ is of the form $\frac{u}{w}$ with $u \in \mathbb{N}$ and $w \in \mathbb{N}$ prime, prime with u and such that $w > \max(m, n, t)$.*

Let f, $g \in \mathbb{Q}_p[a]$ be such that $|fb^m|_p = |gb^n|_p$. Then $m = n$.

Proof. We notice that for every $x \in \mathbb{Q}_p[a]$, $\Psi_p(|x|)$ is of the form $\frac{\ell}{t}$ with $\ell \in \mathbb{Z}$. Consequently, $\Psi_p(|f|)$ is of the form $\frac{h}{t}$ and $\log_p(|g|)$ is of the form $\frac{k}{t}$ with h and $k \in \mathbb{Z}$. Consequently, $\Psi_p(|fb^m|) = \frac{h}{t} + \frac{mu}{w}$ and $\Psi_p(|gb^n|) = \frac{k}{t} + \frac{nu}{w}$ and therefore, due to the equality $|fb^m|_p = |gb^n|_p$, we have $(h - k)w = ut(n - m)$. But since $w > t$, it is prime with ut, hence it must divide $n - m$, which is impossible because $\max(m, n) < w$, except if $m = n$. □

Lemma A.5.20: *Let $|\ .\ |$ be an ultrametric absolute value on $\mathbb{Q}$. If this absolute value is trivial, the residue characteristic is zero. If the absolute value is not trivial, there exists a prime number q such that $|\ .\ |$ is equivalent to $|\ .\ |_q$.*

Proof. If this absolute value is trivial, it is clear that the residue characteristic is zero. So we suppose that the absolute value is not trivial. For every $n \in \mathbb{N}^*$, we have $|n| \leq 1$. Since $|\ .\ |$ is not trivial, there certainly exists $s \in \mathbb{N}^*$ such that $|s| < 1$. Let q be the smallest $s \in \mathbb{N}^*$ such that $|s| < 1$. It is easily checked that q is prime. Since $M_{\mathbb{Q}}$ is a principal ideal of $\mathbb{Z}$, we have $M_{\mathbb{Q}} = q\mathbb{Z}$.

Let $t = |\frac{1}{q}|$. It is easily checked that given $m \in \mathbb{Z}^*$, of the form $q^s n$, with $n \in \mathbb{Z}^*$, prime to q, we have $|m| = t^{-s}$. Let w be the valuation associated to this absolute value. Then w is clearly proportional to v_q and by Lemma A.1.2, is equivalent to v_q. This ends the proof. □

Lemma A.5.21 is easily seen.

Lemma A.5.21: *$\mathbb{N}$ is dense in $\mathbb{Z}_p$, the invertible elements in $\mathbb{Z}_p$ are the ones whose absolute value is 1, $\mathbb{Z}_p$ is compact, equal to $U_{\mathbb{Q}_p}$, $p\mathbb{Z}_p$ is equal to $M_{\mathbb{Q}_p}$. $\mathbb{Q}_p$ is locally compact. The residue class field of $\mathbb{Q}_p$ is equal to the field of p elements $\mathbb{F}_p$. Finally, U_p is the union of p disks $d\left(u, \frac{1}{p}\right)$.*

Proof. All statements are immediate except the compacity of $\mathbb{Z}_p$. Consider a sequence (a_n) in $\mathbb{Z}_p$. Since it is bounded, by Theorem A.2.1 we can extract either a monotonous distances sequence or an equal distances sequence, or a converging sequence. But since each circle $d(a, r)$ with $r \in |\mathbb{Q}_p|$ only has p classes, there are no equal distances sequence in $\mathbb{Q}_p$. And since the absolute value is discrete there is no monotonous distances sequence in $\mathbb{Q}_p$. Hence, we can extract a converging sequence from the sequence (a_n). □

Corollary A.5.22: *For each $m \in \mathbb{N}$, U_p is the union of p^m distinct disks $d\left(u, \frac{1}{p^m}\right)$. Let E be a finite algebraic extension of $\mathbb{Q}_p$. There exists a constant $B > 0$ such that for all $m \in \mathbb{N}$, the number of distinct $d\left(u, \frac{1}{p^m}\right)$ is inferior or equal to Bp^m.*

By definition and construction of $\mathbb{C}_p$, we have this corollary:

Corollary A.5.23: *The field of algebraic numbers is dense in $\mathbb{C}_p$. Hence, $\mathbb{C}_p$ contains a dense countable subset.*

In Theorem A.5.24, we will follow the method of [84].

Theorem A.5.24: *$\mathbb{C}_p$ is not spherically complete.*

Proof. Let $(r_n)_{n\in\mathbb{N}}$ be a real sequence such that $0 < r < r_{n+1} < r_n < 1 \ \forall n \in \mathbb{N}$. Let $\mathcal{S}$ be the set of sequences of the set $\{0, 1\}$. Suppose that for each $n = 0, ..., q$ we have defined 2^n disks $d(a_{n,k}, r_n), k = 0, 1$, such that for each $n = 1, ..., q$ and for each $k = 0, 1$, the disks $d(a_{n,0}, r_n)$ and $d(a_{n,1}, r_n)$ are included in some $d(a_{n-1,k}, r_{n-1})$ (with $k = 0$ or $k = 1$) and have an empty intersection. It is then immediate to define in each disk $d(a_{q,k} r_q)$ two disks $d(a_{q+1,0}, r_{q+1})$ and $d(a_{q+1,1}, r_{q+1})$ having an empty intersection. So, the family is defined for every $n \in \mathbb{N}$.

Now, let $(u_n) \in \mathcal{S}$ and let (V_n) be a decreasing sequence of disks defined as follows: suppose (V_n) is defined for $n \leq q$. If $u_{q+1} = 0$, we set $V_{q+1} = d(a_{q+1,0}, r_{q+1})$ and if $u_{q+1} = 1$, we set $V_{q+1} = d(a_{q+1,1}, r_{q+1})$. In that mapping, which associates to each sequence (u_n), the decreasing sequence of disks (V_n) is clearly injective. Now, consider two such distinct sequences (u_n) and (u'_n). Let q be the smallest integer such that $u_q \neq u'_q$. The distance from V_q to V'_q is at least r_q. Consequently, for every $n \geq q$, the distance between V_n and V'_n is at least $r_q \geq r$

Suppose now that $\mathbb{C}_p$ is spherically complete. For each sequence (u_n), the intersection of the decreasing sequence of disks (V_n) contains a point $\alpha((u_n))$ and hence, by the last conclusion, if (u_n) and (u'_n) are two different sequences, we have $|\alpha((u_n)) - \alpha((u'_n))| \geq r$. But we know that the set of sequences (u_n) is not countable and hence the set of the $\alpha((u_n))$, $((u_n) \in \mathcal{S})$ is not countable. Consequently, $\mathbb{C}_p$ contains an uncountable subset Σ such that $|x - y| \geq r \ \forall x \neq y, \ x, \ y \in \Sigma$. This contradicts the fact that $\mathbb{C}_p$ contains a dense countable subset. □

A.6. Normal extensions of $\mathbb{Q}_p$ inside $\mathbb{C}_p$

Definitions and notations: Recall that $\mathbb{L}$ is a complete field with respect to an ultrametric absolute value. For every $s \in \mathbb{N}^*$, we put $u_s = \dfrac{1}{p^{s-1}(p-1)}$ and $r_s = p^{-u_s}$. We will study the p^s-th roots of 1 and we will show that they lie in circles of center 1 and radius r_s. We will examine normal extensions of $\mathbb{Q}_p$ and totally ramified extensions and show the role of Eisenstein polynomials.

Remark: In [50], due to a misprint, u_s is defined as $\dfrac{1}{p^s(p-1)}$ instead of $\dfrac{1}{p^{s-1}(p-1)}$.

We will need certain technical lemmas.

Lemma A.6.1: *Let $s \in \mathbb{N}^*$. For every $n \in \mathbb{N}^*$ such that $n < p^s$, we have* $\left|\binom{p^s}{n}\right|_p = \dfrac{1}{p^s|n|_p}$.

Proof. We notice that for any $n \in \mathbb{N}^*$ such that $n < p^s$, n is a multiple of p^h for some $h < s$, if and only if so is $p^s - n$. Now, let B be the bijection from $\{1, ..., n-1\}$ onto $\{(p^s - n+1), .., p^s - 1\}$, defined as $B(j) = p^s - j$. Thus, for every $j = 1, ..., n-1$, we have $|j|_p = |B(j)|_p$. Now, obviously $\left|\binom{p^s}{n}\right|_p = \left|\dfrac{p^s}{n}\right|_p \cdot \left|\prod_{j=1}^{n-1} \dfrac{B(j)}{j}\right|_p$. But as we just saw, each factor $\left|\dfrac{B(j)}{j}\right|_p$ is equal to 1 and therefore the conclusion is clear. □

Definitions and notations: Let $s \in \mathbb{N}$. We will denote by W_s the group of the p^s-th roots of 1 in $\mathbb{C}_p$, i.e., the $\zeta \in \mathbb{C}_p$ such that $\zeta^{p^s} = 1$ and we will denote by B_s the set $W_s \setminus W_{s-1}$ and we set $W = \bigcup_{s\in\mathbb{N}} W_s$.

F_s will denote the polynomial $\sum_{j=0}^{p-1} x^{jp^{s-1}}$ and we put $G_s(x) = F_s(1 + x)$.

Definitions and notations: A monic polynomial $P(x) = \sum_{j=0}^{q} a_j x^j \in \mathbb{L}[x]$ will be called *an Eisenstein polynomial* if it satisfies $a_j \in M_{\mathbb{L}}$ whenever $j = 0, \dots, q-1$ and $a_0 \notin (M_{\mathbb{L}})^2$.

Theorem A.6.2 (Eisenstein): *Let $\mathbb{L}$ have a discrete valuation. Let $P(x) = \sum_{j=0}^{q} a_j x^j \in \mathbb{L}[x]$ be an Eisenstein polynomial. Then P is irreducible in $\mathbb{L}[x]$.*

Proof. We suppose P not irreducible. Then P splits in $\mathbb{L}[x]$ in the form $S(x)T(x)$ with $S(x) = \sum_{j=0}^{m} \alpha_j x^j$, $T(x) = \sum_{j=0}^{n} \beta_j x^j$, and $\alpha_m = \beta_n = 1$. Since S, T are monic we have $\|S\| \geq 1$, $\|T\| \geq 1$, and since $\|S\|\|T\| = \|ST\| = \|P\| = 1$, we have $\|S\| = \|T\| = 1$. Hence, both S, T belong to $U_{\mathbb{L}}[x]$. First we notice that if α_0 belongs to $M_{\mathbb{L}}$, then β_0 does not, because $a_0 \notin (M_{\mathbb{L}})^2$. Hence, we may assume $\alpha_0 \in M_{\mathbb{L}}$ and $\beta_0 \notin M_{\mathbb{L}}$. Further, we have $\alpha_j \in M_{\mathbb{L}}$ for every $j = 0, \dots, m-1$.

Indeed, let ℓ be the smallest of the integers h such that $|\alpha_h| = 1$. Then we have $|a_l| = |\beta_0\alpha_\ell + \sum_{j=1}^{\ell} \beta_j\alpha_{\ell-j}| = 1$ because $|\beta_0\alpha_\ell| = 1$ and $\sum_{j=1}^{\ell} \beta_j\alpha_{\ell-j} \in M_{\mathbb{L}}$. Consequently, $\ell = q$, therefore P is irreducible. □

Lemmas below will be useful in the sequel.

Lemma A.6.3: *Let $\mathcal{G}$ be a subgroup of the multiplicative group $(\mathbb{K}^*, \cdot)$ included in $C(0,1)$ and let $u \in \mathcal{G}$. The bijection γ from $\mathcal{G}$ onto $\mathcal{G}$ defined as $\gamma(x) = ux$ is isometric.*

Lemma A.6.4: *Let $(j,n) \in \mathbb{N} \times \mathbb{N}^*$ be such that $j < n$. Then we have*

$$\binom{n}{j+1} = \sum_{h=j}^{n-1} \binom{h}{j}.$$

Lemma A.6.5: *For every $s \in \mathbb{N}$, G_s is an Eisenstein polynomial.*

Proof. First we suppose $s = 1$. We have

$$G_1(x) = \sum_{h=0}^{p-1} \Big(\sum_{j=0}^{h} \binom{h}{j} x^j \Big) = \sum_{j=0}^{p-1} \Big(\sum_{h=j}^{p-1} \binom{h}{j} \Big) x^j.$$

Hence, by Lemma A.6.4, we have $G_1(x) = \sum_{j=0}^{p-1} \binom{p}{j+1} x^j$. Moreover, by Lemma A.6.1, we have $\left|\binom{p}{j+1}\right|_p = \frac{1}{p}$ for every $j = 0, \dots, p-2$ and therefore G_1 is an Eisenstein polynomial.

Now we consider the general case $s \in \mathbb{N}^*$. First we put $T_s(x) = (1+x)^{p^s}$. By Lemma A.6.1, it is seen that $T_1(x)$ is of the form $1 + x^p + \gamma_1(x)$ with $\gamma_1(x) \in px\mathbb{Z}_p[x]$ and $\deg(\gamma_1) = p - 1$. Then by an immediate induction, we see that $T_s(x)$ is of the form $1 + x^{p^s} + \gamma_s(x)$ with $\gamma_s(x) \in px\mathbb{Z}_p[x]$ and $\deg(\gamma_s) = p^s - 1$.

As a consequence, it is easily seen that G_s is an Eisenstein polynomial if and only if so is the polynomial $g_s(x) = \sum_{j=0}^{p-1} (1 + x^{p^s})^j$. But we have $g_s(x) = G_1(x^{p^s})$. Since G_1 is an Eisenstein polynomial, so is g_s. This ends the proof. □

Theorem A.6.6: *For each $s \in \mathbb{N}^*$, B_s consists of $p^s - p^{s-1}$ roots of 1 of order p^s that lie in $C(1, r_s)$. For every $t \in B_s, \mathrm{irr}(t, \mathbb{Q}_p)$ is equal to F_s and $B_s \cap d(t, r_s^-)$ is equal to tW_{s-1}.*

Proof. Let $t \in B_s$ and let $F(x) = \mathrm{irr}(t, \mathbb{Q}_p)$. Then F divides $x^{p^s} - 1$ and has a degree $d > p^{s-1}$. Since $x^{p^s} - 1 = (x^{p^{s-1}} - 1)F_s$, F divides F_s. But by Lemma

A.6.5 and Lemma A.6.1, G_s is irreducible over $\mathbb{Q}_p$, hence so is F_s and therefore $F = F_s$. Then by Corollary A.5.5, we have $\Psi_p(t-1) = \dfrac{\Psi_p(G_s(0))}{p^s - p^{s-1}} = -u_s$. Hence, B_s is included in $C(1, r_s)$ and obviously consists of $p^s - p^{s-1}$ different points in this circle. Let ϕ be the mapping defined in W_{s-1} as $\phi(\xi) = t\xi$. Since t is of order p^s and any element of W_{s-1} is of order $< p^s$, one sees that ϕ is an injection from W_{s-1} into B_s and then, by Lemma A.6.3, we have $\phi(W_{s-1}) \subset B_s \cap d(t, r_s^{\ -})$. Conversely, given $u \in B_s \cap d(t, r_s^{\ -})$, then $t^{-1}u$ lies in W_{s-1}, which shows that ϕ is a bijection from W_{s-1} onto $B_s \cap d(t, r_s^{\ -})$. □

Corollary A.6.7: *For every* $\xi,\ \zeta \in W_1$ *of* $\xi \neq \zeta$, *we have* $\Psi_p(\xi - \zeta) = -\frac{1}{p-1}$.

Theorem A.6.8: *Let* $n \in \mathbb{N}^*$ *and let* ζ *be a* n*-th root of 1 of order* n. *Then* ζ *belongs to* $d(1, 1^-)$ *if and only if* n *is of the form* p^s $(s \in \mathbb{N})$.

Proof. By Theorem A.6.6 we know that if n is of the form p^s, then ζ belongs to $d(1, 1^-)$. Now we suppose $\zeta \in d(1, 1^-)$ and put $n = q\, p^s$ with q prime to p. Let $\xi = \zeta^{(p^s)}$. It is seen that ξ also belongs to $d(1, 1^-)$, because $|\xi - 1|_p = |\zeta - 1|_p \left| \sum_{j=0}^{p^s-1} \zeta^j \right|_p$. Let $P(x) = x^q - 1$. If $\xi \neq 1$, then $\overline{P}(x)$ admits 1 as a zero of order $t \geq 2$. But $\overline{P}'(\overline{1}) = \overline{q} \neq \overline{0}$, hence $\xi = 1$. Therefore, we have $q = 1$ and this ends the proof. □

Definitions and notations: Let $\mathbb{E}$ be a finite extension of $\mathbb{Q}_p$. Recall that the residue class field of $\mathbb{Q}_p$ is $\mathbb{F}_p$. Let $\mathcal{E}$ be the residue class field of $\mathbb{E}$. Since $\mathbb{E}$ is finite over $\mathbb{Q}_p$, it is locally compact, hence $\sup\{|x| \ |x \in \mathbb{E}\ |x| < 1\} < 1$. So, we can choose an element $s \in \mathbb{E}$ such that $|s| < 1$ and such that $|s| = \sup\{|x| \ |x \in \mathbb{E}\ |x| < 1\}$. Such an element s is called *a uniformizer* of $\mathbb{E}$. Since s is algebraic over $\mathbb{Q}_p$, $|p|$ is of the form $|s|^e$ with $e \in \mathbb{N}^*$. The number e is called *ramification index of* $\mathbb{E}$.

Next, by Corollary A.5.3 we know that if $a \in \mathbb{E}$ is algebraic over $\mathbb{Q}_p$ of degree q and such that $|a| = 1$, then its residue class $\overline{a}$ is algebraic over $\mathbb{F}_p$, of degree $\leq q$. Consequently, if $\mathbb{E}$ is finite over $\mathbb{Q}_p$, then $\mathcal{E}$ is finite over $\mathbb{F}_p$. The number $[\mathcal{E} : \mathbb{F}_p]$ is called residual degree of $\mathbb{E}$ and will be denoted by f.

The extension $\mathbb{E}$ is said to be *ramified* if $e > 1$ and *unramified* if $e = 1$.

Lemma A.6.9 is just a remark:

Lemma A.6.9: *Let* T *be an unramified extension of* $\mathbb{Q}_p$ *and let* $\mathbb{E}$ *be an extension of* $\mathbb{Q}_p$ *such that* $\mathbb{E} \subset T$. *Then* $\mathbb{E}$ *is unramified.*

Theorem A.6.10: *Let* $n \in \mathbb{N}^*$ *be prime to* p *and let* u, w *be two distinct roots of* 1 *in* Ω_p *of order* n. *Then in* Ω_p *we have* $|u - w| = 1$.

Proof. Let $h = \dfrac{u}{w}$. Then $h^n = 1$. Suppose $|h - 1| < 1$. By Theorem A.6.8, n is of the form p^s, a contradiction. □

A.7. Spherically complete extensions

Several problems on p-adic analytic functions require one to consider an ultrametric algebraically closed extension of $\mathbb{K}$, which is spherically complete, in order to give every circular filter a center. Others require to have a complete algebraically closed extension that admits a non-countable residue class field. Proving the existence of a spherically complete algebraically closed extension of the ground field $\mathbb{K}$ isn't easy, most of the ways involving basic considerations in logic. Here, we will follow the method proposed by Bertin Diarra in [58], that is only based on the notion of ultraproducts [40, 41].

Notations and definitions: Here we denote by $(\mathbb{E}_i)_{i\in I}$ an infinite family of field extensions of $\mathbb{L}$, provided each with an ultrametric absolute value $|\ .\ |_i$ extending that of $\mathbb{L}$. Next, $\mathcal{U}$ will denote an ultrafilter on I. We remember that $\mathcal{U}$ is said to be *principal* if there exists $\alpha \in I$ such that $\mathcal{U}$ is the set of the subsets of I that contain α. Then $\mathcal{U}$ is said to be *incomplete* if there exists a decreasing sequence $(X_n)_{n\in\mathbb{N}}$ of elements of $\mathcal{U}$, such that $\bigcap_{n\in\mathbb{N}} X_n = \emptyset$. Since I is infinite, there obviously exist incomplete ultrafilters on I. In particular, any incomplete ultrafilter is not principal. $\mathcal{R}$ will denote the subring of $\prod_{i\in I} \mathbb{E}_i$ that consists of the set $(a_i)_{i\in I} \in \prod_{i\in I} \mathbb{E}_i$ such that $\sup_{i\in I} |a_i|_i < +\infty$. Of course, $\mathcal{R}$ is a $\mathbb{L}$-algebra.

We will denote by φ the mapping from $\mathcal{R}$ into $\mathbb{R}_+$ defined as $\varphi((a_i)_{i\in I}) = \lim_{\mathcal{U}} |a_i|_i$. Then φ is seen to be a multiplicative semi-norm of the $\mathbb{L}$-algebra $\mathcal{R}$.

We put $\mathcal{J} = Ker(\varphi)$ and $\mathcal{S} = \dfrac{\mathcal{R}}{\mathcal{J}}$ and we denote by ψ the canonical surjection from $\mathcal{R}$ onto $\mathcal{S}$. Then $\mathcal{S}$ is obviously provided with an absolute value $|\ .\ |$ defined as $|\psi(a)| = \varphi(a),\ (a \in \mathcal{R})$.

On the other hand, $\mathcal{R}$ is seen to be provided with a norm of $\mathbb{L}$-algebra $\|\ .\ \|$ defined as $\|(a_i)_{i\in I}\| = \sup_{i\in I} |a_i|$. Next we denote by $|||\ .\ |||$ the semi-norm quotient of the norm of L-algebra by the ideal $\mathcal{J}$, defined on $\mathcal{R}$ as $|||a||| = \inf_{t\in\mathcal{J}} \|a - t\|$.

Theorem A.7.1 (B. Diarra): *$\mathcal{S}$ is a field extension of $\mathbb{L}$ and its absolute value $|\ .\ |$ extends the one of $\mathbb{L}$. Moreover, if $\mathcal{U}$ is non-principal and if each $\mathbb{E}_i$ has a dense valuation group, then $\mathcal{S}$ has a valuation group equal to $(\mathbb{R}, +)$. Further, if each $\mathbb{E}_i$ is algebraically closed, then so is $\mathcal{S}$.*

Proof. Let $\alpha \in \mathcal{S}\setminus\{0\}$ and let $a = (a_i)_{i\in I} \in \mathcal{R}$ be such that $\psi(a) = \alpha$. By definition we have $\lim_{\mathcal{U}} |a_i|_i \neq 0$. Hence, there exists $J \in \mathcal{U}$ such that for every $i \in J$ we have $\dfrac{\varphi(a)}{2} < |a_i|_i < \dfrac{3\varphi(a)}{2}$, hence $a_i \neq 0$, and therefore

(1) $\dfrac{2}{3\varphi(a)} < |a_i^{-1}|_i < \dfrac{2}{\varphi(a)}$ whenever $i \in J$.

Now let $b = (b_i)_{i\in I} \in \prod_{i\in I} \mathbb{E}_i$ be defined as $b_i = a_i^{-1}$ whenever $i \in J$ and $b_i = 1$ whenever $i \in I \setminus J$. By (1) it is seen that b does belong to $\mathcal{R}$. But now, $ab - 1$ is an element $(c_i)_{i\in I}$ of $\mathcal{R}$ that satisfies $c_i = 0$ whenever $i \in J$, hence $\lim_{\mathcal{U}} |c_i| = 0$. Therefore, $ab - 1$ belongs to $\mathcal{J}$ and finally, in $\mathcal{S}$ we have $\psi(a)\psi(b) = 1$. This shows $\mathcal{S}$ to be a field.

Next we suppose that for each $i \in I$ the valuation group of $\mathbb{E}_i$ is dense. Let $r \in]0, +\infty[$. We can obviously find a family $(\varepsilon_i)_{i\in I}$ in $]0, +\infty[$ such that $\lim_{\mathcal{U}} \varepsilon_i = 0$. For every $i \in I$, let $a_i \in \mathbb{E}_i$ satisfy $r - \varepsilon_i < |a_i| < r$ and let $a = (a_i)_{i\in I}$. Of course, a belongs to $\mathcal{R}$ and satisfies $\lim_{\mathcal{U}} |a_i|_i = r$, hence r belongs to $|\mathcal{S}|$. This shows that the valuation group of $\mathcal{S}$ is equal to $\mathbb{R}$.

Finally, we suppose that each field $\mathbb{E}_i$ is algebraically closed. Let

$$P(x) = \sum_{n=0}^{q} \lambda_n x^n \in \mathcal{S}[x],$$

with $\lambda_q = 1$ and $q > 0$. We will show that P admits at least one zero in $\mathcal{S}$. For every $n = 0, 1, ..., q-1$, let $(a_{i,n})_{i\in I} \in \mathcal{R}$ satisfy $\psi((a_{i,n})_{i\in I}) = \lambda_n$ and let $a_{i,q} = 1$ whenever $i \in I$. So, we have $\psi((a_{i,n})_{i\in I}) = \lambda_n$ whenever $n = 0, ..., q$. For every $i \in I$, we put $T_i(x) = \sum_{n=0}^{q} a_{i,n}x^n$. Since $\mathbb{E}_i$ is algebraically closed and since $a_{i,q} = 1$ for every $i \in I$, at least one of the zeros α_i of T_i in $\mathbb{E}_i$ satisfies $|\alpha_i|^q \le |a_{i,0}|$. But by hypothesis $(a_{i,0})_{i\in I}$ belongs to $\mathcal{R}$, hence so does $(\alpha_i)_{i\in I}$. Hence, we can put $\alpha = \psi((\alpha_i)_{i\in I})$ and then we have $P(\alpha) = 0$. This finishes showing that $\mathcal{S}$ is algebraically closed and this ends the proof of Theorem A.7.1. □

Lemma A.7.2 (B. Diarra): *Let $a = (a_i)_{i\in I} \in \mathcal{R}$. Then we have $|||a||| = \varphi(a)$.*

Proof. Let $J \in \mathcal{U}$ and let $e = (e_i)_{i\in I} \in \mathcal{R}$ be defined as $e_i = 0$ whenever $i \in J$ and $e_i = 1$ whenever $i \notin J$. For convenience, we put $b = ae$. Clearly b belongs to $\mathcal{J}$, hence we have $|||a||| = \inf_{t\in\mathcal{J}} \|a - t\| \le \|a - b\|$. But now, we check that $\|a - b\| = \sup_{i\in J} |a_i|_i$. Further, this is true for every $J \in \mathcal{U}$. Hence, we obtain $|||a||| \le \inf_{J\in\mathcal{U}} \left(\sup_{i\in J} |a_i|_i\right) = \lim_{\mathcal{U}} |a_i|_i = \varphi(a)$. On the other hand, for all $t \in \mathcal{J}$, we have $\varphi(a - t) = \varphi(a) \le \|a - t\|$, hence $\varphi(a) \le |||a|||$. This ends the proof of Lemma A.7.2. □

Theorem A.7.3 (B. Diarra): *If $\mathcal{U}$ is incomplete, then $\mathcal{S}$ is spherically complete.*

Proof. Let $(\alpha_n)_{n\in\mathbb{N}}$ be a decreasing distances sequence in $\mathcal{S}$ and for every $n \in \mathbb{N}$, let $a_n \in \mathcal{R}$ be such that $\psi(a_n) = \alpha_n$. By induction, we can easily construct another sequence $(b_n)_{n\in\mathbb{N}}$ in $\mathcal{R}$ satisfying

$(\mathcal{V}_n)$ $\psi(b_n) = \alpha_n$ for every $n \in \mathbb{N}$.
$(\mathcal{W}_n)$ $\|b_n - b_{n-1}\| < r_{n-2}$ whenever $n > 1$.

Indeed, let $q \in \mathbb{N}^*$ suppose that we have defined $b_0, ..., b_q$ satisfying $(\mathcal{V}_n)$ for every $n = 0, ..., q$. Of course, we have
(1) $\varphi(b_q - a_{q+1}) = \varphi(a_q - a_{q+1}) = r_q < r_{q-1}$.
By Lemma A.7.2, we have $\varphi(b_q - a_{q+1}) = |||b_q - a_{q+1}|||$, hence by (1) there exists $c \in \mathcal{J}$ such that $\|b_q - a_{q+1} - c\| < r_{q-1}$. So, we put $b_{q+1} = a_{q+1} + c$ and then $(\mathcal{V}_{q+1})$, $(\mathcal{W}_{q+1})$ are satisfied. In order to begin the induction, we put $b_0 = a_0$, $b_1 = a_1$, and then we can define the sequence $(b_n)_{n\in\mathbb{N}}$ for every $n \in \mathbb{N}$, satisfying $(\mathcal{V}_n)$ and $(\mathcal{W}_n)$.
Now for each $n \in \mathbb{N}$, we put $b_n = (b_{i,n})_{i\in I}$. Since $\mathcal{U}$ is incomplete, we can take a decreasing sequence $(X_n)_{n\in\mathbb{N}}$ of elements of $\mathcal{U}$ such that $\bigcap_{n\in\mathbb{N}} X_n = \emptyset$. We put $I_0 = I \setminus X_0$ and for each $n \in \mathbb{N}^*$, $I_{n+1} = X_n \setminus X_{n+1}$. Thus, the family $(I_n)_{n\in\mathbb{N}}$ makes a partition of I. Further, for each $q \in \mathbb{N}$, $(I_n)_{n\geq q}$ makes a partition of X_q. Hence, we can define a surjective mapping g from I onto $\mathbb{N}$ as $g(i) = n$ whenever $i \in I_n$. Now for every $i \in I$, we put $h_i = b_{i,g(i)}$. By $(\mathcal{W}_n)$, we have $\|b_n\| \leq r_0$ whenever $n \in \mathbb{N}$, hence $|h_i|_i \leq r_0$ for each $i \in I$ and therefore $(h_i)_{i\in I}$ belongs to $\mathcal{R}$. We put $h = (h_i)_{i\in I}$ and $w = \psi(h)$. We will show
(2) $|w - \alpha_n| \leq r_{n-1}$ whenever $n \in \mathbb{N}^*$.

Let $n \in \mathbb{N}^*$ be fixed. It is seen that for every $m > n$, we have $\|b_n - b_m\| < r_{n-1}$, hence for every $i \in I$ we have $|b_{i,n} - b_{i,m}|_i < r_{n-1}$. Moreover, since $(I_m)_{m>n}$ makes a partition of X_{n+1}, for every $i \in X_{n+1}$ there exists $m > n$ such that $i \in I_m$ and then we have $|b_{i,n} - h_i|_i = |b_{i,n} - b_{i,g(i)}|_i = |b_{i,n} - b_{i,m}|_i < r_{n-1}$ whenever $i \in X_{n+1}$. But X_{n+1} belongs to $\mathcal{U}$ and therefore in $\mathcal{S}$ we have $|b_n - h| = \lim_{\mathcal{U}} |b_{i,n} - h_i|_i \leq r_{n-1}$. This is true for every $n \in \mathbb{N}^*$ and finally this shows (2). Hence, w belongs to $\bigcap_{n\in\mathbb{N}} d_{\mathcal{S}}(\alpha_{n+1}, r_n)$ and this finishes proving that $\mathcal{S}$ is spherically complete. $\square$

Theorem A.7.4: *$\mathbb{K}$ admits a spherically complete algebraically closed extension whose residue class field is not countable and whose valuation group is equal to $\mathbb{R}$.*

Proof. First we will construct a complete algebraically closed extension of $\mathbb{K}$ whose residue class field is not countable. Let T be a transcendental extension of the form $\mathbb{K}((x_j)_{j\in\mathbb{R}})$ provided with the absolute value $|\ .\ |$ defined on $\mathbb{K}[(x_j)_{j\in\mathbb{R}}]$ by

$$\Big|\sum_{j_1,...,j_q\leq N} a_{j_1,...,j_q} x_{j_1}^{t_1},...,x_{j_q}^{t_q}\Big| = \max_{j_1,...,j_q\leq N} |a_{j_1,...,j_q}|.$$

It is seen that $|x_j - x_h| = 1$ whenever $j, h \in \mathbb{R}$ such that $j \neq h$ and therefore the residue class field of T is not countable. Let T' be the completion of T and let $\mathbb{E}$ be an algebraic closure of T', provided with the unique absolute value that extends the one of T'. Let $\mathbb{E}'$ be the completion of $\mathbb{E}$. By Theorem A.5.14, $\mathbb{E}'$ is algebraically closed. Obviously, its residue class field contains the one of T and therefore is not countable.

Now, we can construct $\mathcal{S}$ by taking $I = \mathbb{N}$ and $\mathbb{E}_i = \mathbb{E}'$ for every $i \in \mathbb{N}$. Since $\mathbb{E}'$ is algebraically closed, by Theorem A.7.1 so is $\mathcal{S}$. Moreover, the valuation group of $\mathbb{E}'$ obviously is dense and therefore, by Theorem A.7.1, $\mathcal{S}$ has a valuation group equal to $\mathbb{R}$. Finally, by Theorem A.7.3, $\mathcal{S}$ is spherically complete. That ends the proof of Theorem A.7.4. □

Thanks to Theorem A.7.3, we can generalize Corollary A.4.4.

Theorem A.7.5: *Let $P, F \in \mathbb{K}[x]$ with F having all its zeros in $d(0,r)$. Let $Q, R \in \mathbb{K}[x]$ satisfy $P = FQ+R$ and $\deg(R) < \deg(F)$. Then we have $\Psi(Q, \log r) \leq \Psi(P, \log r) - \Psi(F, \log r)$ and $\Psi(R, \log r) \leq \Psi(P, \log r)$. Moreover, if P has all its zeros in $d(0,r)$, then $\Psi(Q, \log r) = \Psi(P, \log r) - \Psi(F, \log r)$.*

Proof. Without loss of generality, we can assume that the valuation group of $\mathbb{K}$ is $\mathbb{R}$. Consequently, up to a change of variable, we can suppose that $r = 1$. We can also assume that F is monic. Now, since F has all zeros in $d(0,r)$, this means that $\Psi(F, 0) = 0$ and hence Theorem A.7.5 is reduced to Corollary A.4.4. □

Definitions and notations: Henceforth, $\widehat{\mathbb{K}}$ will denote an algebraically closed spherically complete extension of $\mathbb{K}$.

For every disk $d(a, r^-)$ (resp. $d(a,r)$) in $\mathbb{K}$, we will denote by $\widehat{d}(a, r^-)$ (resp. $\widehat{d}(a,r)$) the disk of same center and diameter in $\widehat{\mathbb{K}}$. Similarly, we will denote by $\widehat{C}(a,r)$ the circle $\{x \in \widehat{\mathbb{K}} \mid |x-a| = r\}$.

Remarks: There exists another way to construct a spherically complete extension due to Irving Kaplansky [67].

Definitions and notations: Let $\mathbb{E}$ be an extension of $\mathbb{L}$ provided with an ultrametric absolute value that extends that of $\mathbb{L}$. The extension $\mathbb{E}$ is said to be *immediate* if its residue class field is identical to that of $\mathbb{L}$ and its value group also is identical to that of $\mathbb{L}$.

Theorem A.7.6 is due to I. Kaplansky [67].

Theorem A.7.6: *$\mathbb{L}$ admits an immediate extension that is maximal with respect to the inclusion.*

An immediate extension of $\mathbb{L}$ is spherically complete if and only if it is maximal with respect to the inclusion.

A maximal immediate extension of $\mathbb{L}$ is unique up to an $\mathbb{L}$-isomorphism.

The proof of this theorem represents a very big work. In what follows, we will not need Theorem A.7.6.

A.8. Transcendence order and transcendence type

In $\mathbb{C}_p$ we can define a notion of transcendence order stating that if a is transcendental over $\mathbb{Q}_p$ and has a transcendence order $\leq t$ and if b is trenscendental over $\mathbb{Q}_p$ but algebraic over $\mathbb{Q}_p[a]$, then b also has a transcendence order $\leq t$. We will show the existence of numbers of order less than $1+\epsilon$ for every $\epsilon > 0$.

Definitions and notations: Let $\tau \in]0, +\infty[$. Let F be a transcendental extension of $\mathbb{Q}_p$ provided with an absolute value $|\ .\ |$ extending that of $\mathbb{Q}_p$. An element $a \in F$ will be said *to have transcendence order* $\leq \tau$ or *order* $\leq \tau$ in brief, if there exists a constant $C_a \in]0, +\infty[$ such that every polynomial $P \in \mathbb{Q}_p[x]$ satisfies $\log_p(|P(a)|) \geq \log(\|P\|) - C_a(\deg(P))^\tau$. Moreover, a will be said *to have weak transcendence order* $\leq \tau^+$ or *weak order* $\leq \tau^+$ in brief if a has transcendence order $\leq \tau + \epsilon$ for every $\epsilon > 0$.

Definitions and notations: We will denote by $\mathcal{S}(\tau)$ the set of numbers $x \in \mathbb{C}_p$ having transcendence order $\leq \tau$ and by $\mathcal{S}(\tau^+)$ the set of numbers $x \in \mathbb{C}_p$ having weak transcendence order $\leq \tau^+$.

Finally, we will say that a number $x \in \mathbb{C}_p$ is *of infinite order* if it does not belong to $\mathcal{S}(\tau)$ for all $\tau \in \mathbb{R}_+^*$.

Remarks: By definition, an element $a \in \mathbb{C}_p$ having transcendence order $\leq \tau$ or weak transcendence order $\leq \tau^+$ is transcendental over $\mathbb{Q}_p$.

Theorem A.8.1: *Let $\tau \in]0, +\infty[$. If $\mathcal{S}(\tau) \neq \emptyset$ then $\tau \geq 1$.*

Proof. Let $a \in \mathbb{C}_p$, $a \neq 0$, be transcendental over $\mathbb{Q}_p$ and have transcendence order $\leq \tau$. We can find $b \in \Omega_p$, $(b \neq 0)$ such that $|a-b|_p < 1$. Consider the minimal polynomial Q of b over $\mathbb{Q}_p$. Let $b_2, ..., b_q$ be the conjugates of b over $\mathbb{Q}_p$ and set $b_1 = b$. We notice that by Corollary A.5.6, all conjugates b_j of b over $\mathbb{Q}_p$ satisfy $|b_j|_p = |b|_p$.

Suppose first that $|a|_p \leq 1$. Since $|b_j|_p = |b|_p = |a|_p \leq 1$, all coefficients of Q belong to $\mathbb{Z}_p$. Obviously, Q is monic, hence $\|Q\| = 1$. By hypothesis, there exists $C_a \in]0, +\infty[$ such that $\Psi_p(P(a)) \geq \log_p(\|P\|) - C_a(\deg(P))^\tau\ \forall P \in \mathbb{Q}_p[x]$. Consequently, $-n\Psi_p(Q(a)) = -\Psi_p((Q(a))^n) \leq C_a(n\deg(Q))^\tau\ \forall n \in \mathbb{N}^*$. Since $Q(b) = 0$ and since, by Corollary A.3.5, Q is 1-Lipschitzian in U, we have $-\Psi_p(Q(a)) > 0$ and therefore, if $\tau < 1$, the inequality $-n\Psi_p(Q(a)) \leq C_a(n\deg(Q))^\tau\ \forall n \in \mathbb{N}^*$ is impossible when n tends to $+\infty$.

Suppose now $|a|_p > 1$. Set $Q(X) = \sum_{k=0}^{q} c_k X^k$. Since the b_j satisfy $|b_j|_p = |a|_p$, $(1 \leq j \leq q)$, we have $|c_k|_p \leq (|a|_p)^{q-k}$ and particularly $|c_0|_p = \prod_{j=1}^q |b_j|_p = (|a|_p)^q$. Consequently, $\|Q\| = (|a|_p)^q$ and therefore, considering the sequence $(Q^n)_{n\in\mathbb{N}}$, for every $n \in \mathbb{N}^*$ we have,

(1) $-n\Psi_p(Q(a)) \leq -nq\Psi_p(a) + C_a(nq)^{\tau}$.

On the other hand, $Q(a) = Q(a) - Q(b) = (a - b)\sum_{k=1}^{q} c_k \sum_{j=0}^{k} a_j b^{k-j-1}$ and hence $|Q(a)|_p \leq |a - b|_p(|a|_p)^{q-1}$. Consequently, we obtain $-n\Psi_p(a - b) - n(q - 1)\Psi_p(a) \leq \Psi_p(Q(a)$ and hence by (1), $-n\Psi_p(a - b) - n(q - 1)\Psi_p(a) \leq -nq\Psi_p(a) + C_a(nq)^{\tau}$. Finally, $n(\Psi_p(a) - \Psi_p(a - b)) \leq C_a(nq)^{\tau}$. Since $|a|_p > 1$ and $|a - b|_p < 1$, this inequality is impossible again when n tends to $+\infty$, which ends the proof. □

Theorem A.8.2: *There exists $b \in \mathbb{C}_p$ transcendental over $\mathbb{Q}_p$, of order $\leq 1 + \epsilon$, for every $\epsilon > 0$.*

Proof. Consider first a strictly decreasing sequence $(\epsilon_n)_{n\in}$ such that $\lim_{n\to+\infty} \epsilon_n = 0$ and $\lim_{n\to+\infty} \epsilon_n \log(n) = +\infty$.

We can always divide any polynomial $P \in \mathbb{Q}_p[x]$ by some $\lambda \in \mathbb{Q}_p$ such that $|\lambda|_p = \|P\|$ and hence we go back to the hypothesis $\|P\| = 1$. So, if we can find some $b \in \mathbb{C}_p$ and, for every $\omega > 0$, a constant $C(\omega) > 0$ and show that for every $P \in \mathbb{Q}[X]$ such that $\|P\| = 1$, we have $-\log(|P(b)|_p) \leq C(\omega)(\deg(P))^{1+\omega}$, Theorem A.8.2 will be proven.

By induction, we can define a strictly increasing sequence $(r_n)_{n\in\mathbb{N}}$ of $\mathbb{Q}$ and a sequence $(a_n)_{n\in\mathbb{N}}$ of $\mathbb{C}_p$ with $r_n = \frac{u_n}{v_n}$, irreducible and $(v_n)_{n\in\mathbb{N}}$ a strictly increasing sequence of prime numbers satisfying further the following properties:

(i) $\lim_{n\to+\infty} r_n = +\infty$,
(ii) for every $n \in \mathbb{N}$, $n^{\epsilon_n} < r_n < (n + 1)^{\epsilon_n}$,
(iii) $v_n > \prod_{j=1}^{n-1} v_j$, and
(iv) $(a_n)^{v_n} = p^{u_n}$

By construction, the sequence $(|a_n|_p)_{n\in\mathbb{N}}$ is strictly decreasing and tends to 0 and all terms belong to U. Set $b = \sum_{n=1}^{\infty} a_n$. Now, let us fix $\varepsilon > 0$. We will show that b is transcendental over $\mathbb{Q}_p$ and has a transcendence order $\leq 1 + \varepsilon$.

Since the sequence (ϵ_n) tends to 0, we can find an integer $t(\varepsilon)$ such that $\epsilon_m < \varepsilon\ \forall m \geq t(\varepsilon)$. Thus, as a first step, let us take $q \geq t(\varepsilon)$ and let us find a constant $C(\varepsilon) > 0$, not depending on b, such that for every $P \in \mathbb{Q}[X]$ satisfying $\|P\| = 1$ and $\deg(P) = q$, we have $-\log_p(|P(b)|_p) \leq C(\varepsilon)q^{1+\varepsilon}$.

For each $n \in \mathbb{N}^*$, set $b_n = \sum_{m=1}^{n} a_m$. Since the sequence $(|a_m|_p)_{m\in\mathbb{N}}$ is strictly decreasing, we have $|b - b_n|_p = |a_{n+1}|_p$ and since P is obviously 1-Lipschitzian in the disk U, we have $|P(b) - P(b_n)|_p \leq |a_{n+1}|_p$, hence

(1) $\log_p(|P(b) - P(b_n)|_p) \leq \log_p(|a_{n+1}|_p) = -r_{n+1}$.

Now, since the sequence $\epsilon_n \log_p(n)$ tends to $+\infty$, we can choose $n(q)$ such that $(n(q)+1)^{\epsilon_{n(q)+1}} > (q+1)^{1+\varepsilon}$. Then by (1) we have,

(2) $\log_p(|P(b) - P(b_{n(q)})|_p) < \log_p(|a_{n(q)+1}|_p) = -(r_{n(q)+1}) < -(n(q)+1)^{\epsilon_{n(q)+1}} < -(q+1)^{1+\varepsilon}$.

We will show inequality (3).

(3) $-\log_p(|P(b_{n(q)})|_p) \leq (q+1)^{1+\varepsilon}$.

Thus, suppose (3) is wrong. Set $h_q = \sum_{m=q}^{n(q)} a_m$. Then $b_{n(q)} = b_{q-1} + h_q$. Now, developing P at the point b_{q-1}, we have

(4)

$$\log_p(|P(b_{n(q)})|_p) = \log_p\left(\Big|\sum_{m=0}^{q} \frac{P^{(m)}(b_{q-1})}{m!}(h_q)^m\Big|_p\right) < -(q+1)^{1+\varepsilon}.$$

Consider now the sum $\sum_{m=0}^{q} \frac{P^{(m)}(b_{q-1})}{m!}(h_q)^m$. Since the sequence $|a_m|_p$ is strictly decreasing, we have $|h_q|_p = |a_q|_p$, hence $\log_p(|h_q|_p) = -r_q$. We notice that $\mathbb{Q}_p[a_1, ..., a_{q_1}]$ is an algebraic extension of $\mathbb{Q}_p$ of degree at most $\prod_{j=1}^{q-1} v_j$. Consequently, by Corollary A.5.8, the extension $\mathbb{Q}_p[a_1, ..., a_{q_1}]$ has a value group of the form $\frac{s}{t}\mathbb{Z}$ with $t \leq \prod_{j=1}^{q-1} v_j$. On the other hand, due to the hypothesis $r_q = \frac{u_q}{v_q}$, it appears that v_q is a prime integer, prime to u_q and bigger than q and than $\prod_{j=1}^{q-1} v_j$. Consequently, we can apply Lemma A.5.19 with h_q in the role of c and b_{q-1} in the role of a. Therefore, for each $m = 0, ..., q-1$, all the $\left|\frac{P^{(m)}(b_{q-1})}{m!}(h_q)^m\right|_p$ are pairwise distinct. Consequently, we have

(5) $$\left|\sum_{m=0}^{q} \frac{P^{(m)}(b_{q-1})}{m!}(h_q)^m\right|_p = \max_{1\leq m\leq q}\left|\frac{P^{(m)}(b_{q-1})}{m!}(h_q)^m\right|_p.$$

Next, since $-\Psi_p(h_q) = r_q < (q+1)^{\omega}$, for each integer $m = 1, ..., q$, we have $\Psi_p(h_q)^m) = -mr_q > -m(q+1)^{\omega} \geq -q(q+1)^{\omega}$, hence

(6) $\Psi_p((h_q)^m) \geq -q(q+1)^{\omega} > -(q+1)^{1+\omega}\ \forall m \leq q$.

Consequently, by (4), (5), and (6), the polynomial $Q(X) = \sum_{m=0}^{q} \frac{P^{(m)}(b_{q-1})}{m!}(X)^m$ has all coefficients in $d(0,1^-)$ and hence we have $\|Q\| < 1$. But since $|b_{q-1}|_p < 1$, by Corollary A.3.5, we have $\|P\| = \|Q\| < 1$, a contradiction to the hypothesis $\|P\| = 1$. Therefore, Relation (3) is proven for every polynomial $P \in \mathbb{Q}_p[X]$ of degree $q \geq t(\omega)$, such that $\|P\| = 1$. Consequently, by (3) we obviously have a constant $C > 0$, not depending on b, such that $-\Psi_p(P(b)) \leq C(\deg(P))^{1+\omega}$ for every $P \in \mathbb{Q}_p[X]$ such that $\deg(P) \geq t(\omega)$ and $\|P\| = 1$.

Particularly b is transcendental over $\mathbb{Q}_p$ because if it was algebraic, the degrees of polynomial $P \in \mathbb{Q}_p[X]$ such that $P(b) = 0$ wouldn't be bounded. Finally, it is easily seen that there exists a constant $m > 0$ such that $|Q(b)|_p \geq m$ for every polynomial $Q \in \mathbb{Q}_p[X]$ of degree $q \leq t(\omega)$ and $\|Q\| = 1$. Therefore, b is clearly of order $\leq 1 + \omega$. □

Corollary A.8.3: *Ω_p is not complete.*

The transcendence type is defined in $\mathbb{C}_p$ in the same way as in $\mathbb{C}$ [91].

Definitions and notations: Given a complex number z, we denote by $|z|_\infty$ its modulus. Throughout this chapter, a number $a \in \mathbb{C}_p$ will just be said to be *algebraic* (resp. *transcendental*) if it is algebraic (resp. transcendental) over $\mathbb{Q}$. When a is algebraic or transcendental over $\mathbb{Q}_p$ we will precise this. Throughout the chapter, we will denote by Ω the field of algebraic numbers and by $\mathbf{A}$ the ring of algebraic integers.

Let $a \in \Omega$. We call *denominator* of a any strictly positive integer n such that na and we denote by $\text{den}(a)$ the smallest denominator of a. Let $a_2, ..., a_n$ be the conjugates of a over $\mathbb{Q}$ in $\mathbb{C}$ and put $a_1 = a$. For convenience, we will use the logarithm of base p denoted by $\log_p$. We set $\overline{|a|} = \max_{j=1,...,n}(|a_j|_\infty)$ and $s(a) = \max(\log_p(\overline{|a|}, \log(\text{den}(a)))$.

The following relations are classical and immediate:

Lemma A.8.4: *Let $a, b \in \Omega$ and let $m \in \mathbb{N}$. Then* $\text{den}(ab) \leq \text{den}(a)\text{den}(b)$, $\text{den}(a+b) \leq \text{den}(a)\text{den}(b)$, $\text{den}(ma) \leq m\text{den}(a), \text{den}(a^m) \leq (\text{den}(a))^m$, *and* $\overline{|ab|} \leq \overline{|a|}.\overline{|b|}$, $\overline{|a+b|} \leq \overline{|a|} + \overline{|b|}$, $\overline{|ma|} \leq m\overline{|a|}$, $\overline{|a^m|} = (\overline{|a|})^m$.

Let $P(X_1, ..., X_n) = \sum_{i_1,...,i_n} a_{i_1,...,i_n}(X_1)^{i_1}, ..., (X_n)^{i_n} \in \mathbb{C}[X_1, ..., X_n]$. We put $H(P) = \max_{i_1,...,i_n} |a_{i_1,...,i_n}|_\infty$ and $t(P) = \max(\log_p(H(P), \deg(P) + 1)$.

A number $a \in \mathbb{C}_p$ will be said *to have transcendence type less than* α if there exists a constant $C_a > 0$ such that, for every $Q \in \mathbb{Z}[X]$, we have either $Q(a) = 0$ or

$-\Psi_p(Q(a)) \leq C_a(t(Q))^\alpha$. We denote by $\mathcal{T}(\alpha)$ the set of numbers $a \in \mathbb{C}_p$ having a transcendence type less than or equal to α.

If a number $a \in \mathbb{C}_p$ does not belong to $\mathcal{T}(\alpha)$ for all $\alpha > 0$, we will say that a is of *infinite type*.

By Lemma A.5.17, Lemmas A.8.5, A.8.6, and A.8.7 are immediate.

Lemma A.8.5: *Let $P \in \mathbb{Z}[X]$. Then $-\Psi_p(P,0) \leq \log_p(H(P))$.*

Lemma A.8.6: *Let $P \in \mathbb{Z}[X]$ be of degree k and let $a \in \Omega$. Then*

$$\overline{|P(a)|} \leq H(P)(k+1)\big(\max(\overline{|a|},1)\big)^k.$$

Lemma A.8.7: *Let $\alpha_1, \alpha_2 \in]0,+\infty[$ satisfy $\alpha_1 \leq \alpha_2$. Then $\mathcal{T}(\alpha_1) \subset \mathcal{T}(\alpha_2)$.*

There exists a link between transcendence order over $\mathbb{Q}_p$ and transcendence type over $\mathbb{Q}$ [57].

Theorem A.8.8: *Let $\alpha \in [1,+\infty[$. Then $\mathcal{S}(\alpha) \subset \mathcal{T}(\alpha)$.*

Proof. Let $a \in \mathcal{S}(\alpha)$. By hypothesis there exists $C > 0$ such that

$$-\Psi_p(Q(a)) \leq -\Psi_p(Q,0) + C\Big(\deg(Q)\Big)^\alpha \ \forall Q \in \mathbb{Q}_p[X].$$

Hence, by Lemma A.8.5, we have $-\Psi_p(Q(a)) \leq \log_p(H(Q)) + C\Big(\deg(Q)\Big)^\alpha \ \forall Q \in \mathbb{Q}[X]$. Then, taking $C \geq 1$, we can derive

$$-\Psi_p(Q(a)) \leq C\Big(\log_p(H(Q)) + \big(\deg(Q)\big)^\alpha\Big) \leq 2C(t(Q))^\alpha \ \forall Q \in \mathbb{Q}[X],$$

which proves that $a \in \mathcal{T}(\alpha)$. □

By Theorem A.8.2, we can now state Corollary A.8.9 [45].

Corollary A.8.9: *There exists $b \in \mathbb{C}_p$, transcendental over $\mathbb{Q}_p$, such that $b \in \mathcal{T}(1+\epsilon)$ for every $\epsilon > 0$.*

Proof. Indeed, in Theorem A.8.2, we saw that there exists $b \in \mathbb{C}_p$ that belongs to $\mathcal{S}(1+\epsilon)$ for all $\epsilon > 0$. □

By Lemma A.5.17, we can immediately derive the following inequality:

Theorem A.8.10: *Let $a \in \Omega^*$ be integral of degree q, over $\mathbb{Z}$. Then $|a|_p \geq \dfrac{1}{(\overline{|a|})^q}$.*

Proof. Let $Q(X) = irr(a, \mathbb{Q})$ and let $a_1, ..., a_q$ be the conjugates of a over $\mathbb{Z}$, with $a_1 = a$. Then $\prod_{j=1}^{q} a_j$ belongs to $\mathbb{Z}^*$, hence by Lemma A.5.18 we have $|a_j| \leq 1\ \forall j = 2, ..., q$.

$$|\prod_{j=1}^{q} a_j|_p \geq \frac{1}{|\prod_{j=1}^{q} a_j|_\infty}. \tag{1}$$

Consequently,

$$|a|_p \geq |\prod_{j=1}^{q} a_j|_p \geq \frac{1}{\prod_{j=1}^{q} |a_j|_\infty}. \tag{2}$$

Now, $|\prod_{j=1}^{q} a_j|_\infty = \prod_{j=1}^{q} |a_j|_\infty \leq (\overline{|a|})^q$. Thus, by (2), we obtain

$$|a|_p \geq \frac{1}{(\overline{|a|})^q}.$$

□

Corollary A.8.11: *Let $a \in \Omega^*$ be of degree q over $\mathbb{Q}$ and let $t = \mathrm{den}(a)$. Then* $|a|_p \geq \frac{1}{t^q (\overline{|a|})^q}$.

Corollary A.8.12: *Let $a \in \Omega^*$ be of degree q over $\mathbb{Q}$ and let $t = \mathrm{den}(a)$. Then* $\log(|a|_p) \geq -2q(s(a))$.

B. Analytic elements and analytic functions

B.1. Algebras $R(D)$

The idea of considering rational functions with no pole inside a domain D, in order to define analytic functions in D, is due to Marc Krasner [68]. The behavior of rational functions in $\mathbb{K}$ is determined by circular filters, which characterize all multiplicative norms on rational functions. We will make a general study of the set of multiplicative semi-norms of a normed algebra, which is locally compact with respect to the topology of pointwise convergence. Results are first obtained by B. Guennebaud and G. Garandel [61, 62]. Henceforth, the idea of considering the topologic space of multiplicative semi-norms continuous with respect to the topology of a normed algebra, was used for many works on Banach algebra [50, 58, 61].

Definitions and notations: In this chapter, we denote by D an infinite subset of $\mathbb{K}$ and then $\mathbb{K}^D$ is provided with the topology $\mathcal{U}_D$ of uniform convergence on D.

We denote by $R(D)$ the $\mathbb{K}$-algebra of rational functions $h(x) \in \mathbb{K}(x)$ with no pole in D. Since D is infinite, $R(D)$ is clearly a $\mathbb{K}$-subalgebra of $\mathbb{K}^D$ and is provided with the topology induced by $\mathcal{U}_D$, which makes it a topological subgroup of $\mathbb{K}^D$. Algebraically, $R(D)$ is a $\mathbb{K}$-subalgebra of $\mathbb{K}(x)$ and more precisely, is of the form $S(D)^{-1}\mathbb{K}[x]$ with $S(D)$ the multiplicative set of polynomials whose zeros do not belong to D.

We denote by $R_b(D)$ the $\mathbb{K}$-subalgebra of $R(D)$ consisting of the $f \in R(D)$, which are bounded in D. Finally, if D is not bounded, we denote by $R_0(D)$ the $\mathbb{K}$-subalgebra of $R(D)$ that consists of the $f \in R(D)$ such that $\lim_{|x|\to+\infty,\ x\in D} f(x) = 0$.

For every $f \in \mathbb{K}^D$ we set $\|f\|_D = \sup_{x\in D}|f(x)| \in [0, +\infty]$.

Recall that an algebra-semi-norm ψ of a $\mathbb{K}$-algebra A is said to be *semi-multiplicative* or *power multiplicative* if it satisfies $\psi(x^n) = (\psi(x))^n\ \forall x \in A$ and is said to be *multiplicative* if $\psi(xy) = \psi(x)\psi(y)\ \forall x,\ y \in A$.

Lemma B.1.1 is then immediate.

Lemma B.1.1: *$R(D)$ is a principal ideal ring. Every ideal is of the form $P(x)R(D)$ with P a polynomial whose zeros belong to D.*

Lemma B.1.2 is an immediate application of general properties of the supremum, once the set $[0,+\infty]$ is provided with the classical extensions of the addition and the multiplication:

$$a+(+\infty)=+\infty \quad \text{for every } a\in[0,+\infty]$$
$$a.(+\infty)=+\infty \quad \text{for every } a\in]0,+\infty]$$

Lemma B.1.2: *For every $g,h\in R(D)$ we have*
(i) $\|h\|_D=0$ *if and only if* $h=0$,
(ii) $\|\lambda h\|_D=|\lambda|\ \|h\|_D$ *for every* $\lambda\in\mathbb{K}^*$,
(iii) $\|h+g\|_D\le\max(\|h\|_D,\|g\|_D)$,
(iv) *If* $(\|f\|_D,\|g\|_D)$ *is different from* $(0,+\infty)$ *and from* $(+\infty,0)$, *then* $\|hg\|_D\le\|h\|_D.\|g\|_D$,
(v) $\|h^n\|_D=(\|h\|_D)^n$ *whenever* $n\in\mathbb{N}$.

Theorem B.1.3: *$R_b(D)=R(D)$ if and only if D is closed and bounded. Moreover, if D is closed and bounded, then $\|\ .\ \|_D$ is a semi-multiplicative ultrametric norm of $\mathbb{K}$-algebra.*

Proof. We first suppose D to be bounded. By Lemma B.1.2 we just have to show that $\|h\|_D<+\infty$ for every $h\in R(D)$ in order to show that $\|\ .\ \|_D$ is a norm of $\mathbb{K}$-algebra such that $\|h^n\|_D=\|h\|_D^n$. Since D is bounded, obviously every polynomial P satisfies $\|P\|_D<+\infty$, hence by Lemma B.1.2 (iv) we just have to check that $\left\|\frac{1}{Q}\right\|_D<+\infty$. To show this, it is sufficient to prove that $\left\|\frac{1}{x-a}\right\|_D<+\infty$ for every $a\in\mathbb{K}\setminus D$. Since D is closed, the distance r from a to D is not zero, hence $\left\|\frac{1}{x-a}\right\|_D\le\frac{1}{r}$.

Now if D is not bounded, obviously $\|x\|_D=+\infty$. If D is not closed, there exists at least one point $a\in\mathbb{K}\setminus D$ with a sequence $(a_n)_{n\in\mathbb{N}}$ in D, which converges to a, hence $\left\|\frac{1}{x-a}\right\|_D=+\infty$. That ends the proof of Theorem B.1.3. □

Theorem B.1.4 (G. Garandel, B. Guennebaud) [61, 62]: *Let $\mathcal{F}$ be a large circular filter on $\mathbb{K}$ of diameter $s>0$. The following three assertions are equivalent:*
(i) $\varphi_{\mathcal{F}}(h)\le\|h\|_D$ *whenever* $h\in R(D)$.
(ii) $\varphi_{\mathcal{F}}$ *is a continuous ultrametric multiplicative norm on $R(D)$ with respect to the topology of uniform convergence.*
(iii) $\mathcal{F}$ *is secant with D.*

Proof. First (i) and (ii) are obviously equivalent. Next, (iii) clearly implies (i) because if $\mathcal{F}$ is secant with D then $\lim_{\mathcal{F}}|h(x)|=\lim_{\mathcal{F}\cap D}|h(x)|\le\|h\|_D$.

Hence, we just have to show that (i) implies (iii). For this, we assume (iii) to be false and will prove that (i) is false. We first assume $\mathcal{F}$ to have center a.

There exist annuli $\Gamma(a_i, r'_i, r''_i)$ $(1 \leq i \leq q)$ with $|a_i - a_j| = s$ whenever $i \neq j$ and $r'_i < s < r''_i$, such that the set $B = \bigcap_{i=1}^{q} \Gamma(a_i, r'_i, r''_i)$ belongs to $\mathcal{F}$ and satisfies $B \cap D = \emptyset$. We put $r' = \max_{1\leq i \leq q} r'_i$ and $r'' = \min_{1\leq i\leq q} r''_i$. Let $\rho' \in]r', s[$ and $\rho'' \in]s, r''[$, and for every $i = 1, \dots, q$ set $b_i \in \Gamma(a_i, \rho', s)$ and set $b \in \Gamma(a, s, \rho'')$.

We put $h(x) = (\prod_{i=1}^{q} (\frac{x - a_i}{x - b_i}))(\frac{\lambda}{x - b})$ with $|\lambda| = b$. We first notice that $\varphi_{\mathcal{F}}(h) = 1$ because $\varphi_{\mathcal{F}}(\frac{x - a_i}{x - b_i}) = 1$ whenever $i = 1, \dots, q$ and $\varphi_{\mathcal{F}}(\frac{\lambda}{x - b}) = 1$. Next, it is easily seen that $\|h\|_D \leq \max(\frac{r'}{\rho'}, \frac{\rho''}{r''})$. Indeed $|h(x)| = |\frac{x}{b_i}| \leq \frac{r'}{\rho'}$ when $|x - a_i| \leq r'$, and $|h(x)| = |\frac{\lambda}{x - b}| \leq \frac{\rho''}{r''}$ when $|x - a| \geq r''$. Hence, we have $\|h\|_D < 1$ while $\varphi_{\mathcal{F}}(h) = 1$ and that contradicts the assertion (i).

We now suppose that $\mathcal{F}$ has no center. Let $d(a, r)$ belong to $\mathcal{F}$ such that $d(a, r) \cap D = \emptyset$ and let $\rho \in]s, r[$. There still exists a disk $d(\alpha, \rho) \in \mathcal{F}$ such that $d(\alpha, \rho) \subset d(a, r)$. Let us take $b \in \Gamma(\alpha, \rho, r)$ and $\lambda \in \mathbb{K}$ such that $|\lambda| = |b|$. We just put $h(x) = \frac{\lambda}{x - b}$ and we have $\varphi_{\mathcal{F}}(h) = 1$ because $|h(x)| = 1$ whenever $x \in d(\alpha, \rho)$, while $|h(x)| \leq \frac{|b - \alpha|}{r} < 1$ whenever $|x - \alpha| \geq r$, hence finally $\|h\|_D < 1$. That ends the proof of Theorem B.1.4. □

In order to describe properties of the multiplicative semi-norms on $R(D)$ and next on analytic elements, we must recall a classical result on continuous multiplicative semi-norms on a normed $\mathbb{K}$-algebra A.

Definitions and notations: We denote by *Mult*(A) the set of $\mathbb{K}$-algebra multiplicative semi-norms of a $\mathbb{K}$-algebra A. Given $\psi \in Mult(A)$, we denote by $Ker(\psi)$ the set of $x \in A$ such that $\psi(x) = 0$ and $Ker(\psi)$ is called *the kernel of* ψ.

Suppose now A is a normed $\mathbb{K}$-alebra whose norm is denoted by $\| \, . \, \|$. We denote by *Mult*$(A, \| \, . \, \|)$ the set of $\mathbb{K}$-algebra multiplicative semi-norms of A, which are continuous functions on A with respect to the norm $\| \, . \, \|$ of A. Similarly, we denote by $Mult_m(A, \| \, . \, \|)$ the set of $\mathbb{K}$-algebra multiplicative semi-norms of A whose kernel is a maximal ideal, which are continuous functions on A and by $Mult_1(A, \| \, . \, \|)$ the set of $\mathbb{K}$-algebra multiplicative semi-norms of A whose kernel is a maximal ideal of codimension 1, which are continuous functions on A.

Lemma B.1.5: *Let A be a $\mathbb{K}$-algebra provided with a $\mathbb{K}$-algebra norm $\| \, . \, \|$ and let $\varphi \in Mult(A)$. Then φ belongs to $Mult(A, \| \, . \, \|)$ if and only if $\varphi(x) \leq \| \, x \, \|$ whenever $x \in A$. Moreover, if A has a unity u and if φ is not identically* 0*, then $\varphi(\lambda u) = |\lambda|$ whenever $\lambda \in \mathbb{K}$. Further if φ belongs to Mult(A), $Ker(\varphi)$ is a prime ideal and if $Mult(A, \| \, . \, \|)$, then $Ker(\varphi)$ is a closed prime ideal.*

Proof. Suppose that for some $x \in A$ we have $\varphi(x) > \| x \|$. Since the valuation group of $\mathbb{K}$ is dense, it contains a subgroup of the form $a\mathbb{Z}$, with $a > 0$. Let $q \in \mathbb{N}$ be such that $q(\log(\varphi(x)) - \log(\|x\|)) > a$. Then there clearly exists $\lambda \in \mathbb{K}$ satisfying $\| x \|^q < |\lambda| < \varphi(x)^q$. So much the more, we have $\| x^q \| < |\lambda| < \varphi(x^q)$. Let $t = x^q$. Then
$\lim_{n\to\infty}\left(\frac{t}{\lambda}\right)^n = 0$ but $\lim_{n\to\infty}\varphi\left(\left(\frac{t}{\lambda}\right)^n\right) = +\infty$, and then φ is not continuous.

Now let u be a unity in A. Either $\varphi(u) = 0$ and then $\varphi(x) = 0$ whenever $x \in A$, or $\varphi(u) = 1$ and then we have $\varphi(\lambda u) = |\lambda|\varphi(u) = |\lambda|$ whenever $\lambda \in L$. The last statement is immediate. This ends the proof of Lemma B.1.5. □

Theorem B.1.6 is well known and may be found in [61] and in Theorems 6.9 and 6.19 of [51].

Theorem B.1.6: *Let A be a $\mathbb{K}$-algebra provided with a $\mathbb{K}$-algebra norm $\| \, . \, \|$. For every $x \in A$ the sequence $\left(\| x^n \|^{\frac{1}{n}}\right)_{n\in\mathbb{N}}$ has a limit denoted by $\| x \|_{si}$, satisfying $\|x\|_{si} \leq \|x\| \; \forall x \in A$ and $\|x\|_{si} = \sup\{\phi(x) \mid \phi \in Mult(A, \|. \, .\|)$. Moreover, $\|f^n\|_{si} = (\|f\|_{si})^n \; \forall f \in A, \forall n \in \mathbb{N}$.*

Theorem B.1.7: *Let A be a $\mathbb{K}$-algebra provided with a $\mathbb{K}$-algebra norm $\| \, . \, \|$. Then Mult(A,$\| \, . \, \|$) is compact with respect to the topology of pointwise convergence.*

Proof. Let B be the unit ball of A. By Lemma B.1.5 each $\varphi \in Mult(A, \| \, . \, \|)$ has a restriction $\widehat{\varphi}$ to B, which satisfies $\widehat{\varphi}(B) \subset [0, 1]$. Hence, *Mult*($B, \| \, . \, \|$) is a closed subset of $[0, 1]^B$ provided with the topology of pointwise convergence on B. But by Tykhonov's Theorem, [0,1] is compact for this topology and then so is *Mult*($B, \| \, . \, \|$). Moreover the mapping $\varphi \to \widehat{\varphi}$ from *Mult*($A, \| \, . \, \|$) into *Mult*($B, \| \, . \, \|$) is a bijection. Indeed it is clearly injective and it is surjective because given $\psi \in Mult(B, \| \, . \, \|)$, we may extend ψ to A by putting $\overline{\psi}(x) = |\lambda|\psi(\frac{x}{\lambda})$ with $\lambda \in \mathbb{K}$, $|\lambda| \geq \|x\|$. Finally this bijection is bicontinuous with respect to the pointwise convergence on both *Mult*($A, \| \, . \, \|$) and *Mult*($B, \| \, . \, \|$), and this ends the proof of Theorem B.1.7. □

Theorem B.1.8 was given in several works. This proof is mainly given in [49].

Theorem B.1.8: *Let A be a field extension of $\mathbb{K}$ provided with a non-zero semi-norm of $\mathbb{K}$-algebra $\| \, . \, \|$. Then $\| \, . \, \|$ is a norm of $\mathbb{K}$-algebra, and there exists an absolute value φ on A extending that of $\mathbb{K}$, such that $\varphi(x) \leq \|x\|$ whenever $x \in A$.*

Proof. Let $SM(A, \| \, . \, \|)$ denote the set of continuous semi-norms ϕ of A satisfying $\phi(f^n) = (\phi(f))^n \; \forall f \in A, \; \forall n \in \mathbb{N}$. It is seen that $\| \, . \, \|$ is a norm because $Ker\| \, . \, \| = \{0\}$. In the same way, so is the spectral semi-norm $\| \, . \, \|_{si}$ associated to $\| \, . \, \|$. Now $SM(A, \| \, . \, \|_{si})$ is easily checked to be inductive with respect to the order $\geq$, i.e., given a totally ordered subset W of $SM(A, \| \, . \, \|_{si})$, the mapping ψ defined in A by $\psi(x) = \inf\{\theta(x) | \theta \in W\}$ belongs to $SM(A, \| \, . \, \|_{si})$. Then by

Zorn's Lemma, $SM(A, \| \, . \, \|_{si})$ admits a minimal element φ. As we just saw, φ is a norm of $\mathbb{K}$-algebra and we have $\varphi(x) \leq \|x\|_{si}$ whenever $x \in A$. We will prove that $\varphi(ab) = \varphi(a)\varphi(b)$ whenever a, $b \in A$. Let $a \in A \setminus \{0\}$. For every $x \in A$, we put $u_n(x) = \dfrac{\varphi(a^n x)}{\varphi(a)^n}$. The sequence $(u_n(x))_{n \in \mathbb{N}}$ is seen to be decreasing. We put $\sigma(x) = \lim_{n \to +\infty} \dfrac{\varphi(a^n x)}{\varphi(a)^n}$ whenever $x \in A$.

First we will check that σ is a norm of $\mathbb{K}$-algebra. Obviously, it is seen that for every $n \in \mathbb{N}$, u_n is a norm of $\mathbb{K}$-vector space, hence so is σ. Next, we have $u_n(x)u_n(y) = \dfrac{\phi(a^n x)\phi(a^n y)}{\phi(a^n)\phi(a^n)}, = \dfrac{\phi(a^n x)\phi(a^n y)}{\phi(a^{2n})} \geq \dfrac{\phi(a^{2n} xy)}{\phi(a^{2n})} \geq \sigma(xy)$, whenever x, $y \in A$, hence $\sigma(x)\sigma(y) \geq \sigma(xy)$. So, σ is a norm of $\mathbb{K}$-algebra. Now, we check that σ is semi-multiplicative, because:

$$\lim_{n \to +\infty} \left(\frac{\varphi(a^n x^q)}{\varphi(a^n)}\right) = \lim_{n \to +\infty} \left(\frac{\varphi(a^{qn} x^q)}{\varphi(a^{qn})}\right) = \lim_{n \to +\infty} \left(\frac{\varphi(a^n x)}{\varphi(a^n)}\right)^q = \sigma(x)^q.$$

Then, since σ satisfies (1) $\sigma(x) \leq \varphi(x) \leq \|x\|_{si}$ whenever $x \in A$, it clearly belongs to $SM(A, \| \, . \, \|_{si})$. But since φ is minimal in $SM(A, \| \, . \, \|_{si})$, actually φ is equal to σ.

Now, as the sequence $(u_n)_{n \in \mathbb{N}}$ is decreasing, we have $\sigma(x) \leq \dfrac{\varphi(ax)}{\varphi(a)} \leq \varphi(x)$, and finally by (1), $\varphi(ax) = \varphi(a)\varphi(x)$ whenever $a, x \in A$. This ends the proof of Theorem B.1.8. □

Theorem B.1.9 is immediate.

Theorem B.1.9: *Let A be a Banach $\mathbb{K}$-algebra. For every maximal ideal $\mathcal{M}$ of A, there exists $\varphi \in Mult(A, \| \, . \, \|)$ such that $Ker(\varphi) = \mathcal{M}$. If $\mathcal{M}$ is of codimension 1, the maping τ from A onto $\mathbb{K}$ admitting $\mathcal{M}$ for kernel satisfies $|\tau(f)| \leq \|f\| \ \forall f \in A$.*

Proof. Let $\mathcal{M}$ be a maximal ideal of A and let $\mathbb{B}$ be the field $\dfrac{A}{\mathcal{M}}$. Since A is complete, $\mathcal{M}$ is closed, therefore, $\mathbb{B}$ is provided with the quotient norm. By Theorem B.1.9, $\mathbb{B}$ admits an absolute value $| \, . \, |$, which extends that of $\mathbb{K}$. Let ψ be the canonical surjection from A to $\mathbb{B}$. On A we put $\varphi(x) = |\psi(x)|$. Then φ is an element of *Mult*$(A, \| \, . \, \|)$ such that $Ker(\varphi) = \mathcal{M}$.

Suppose now $\mathcal{M}$ is of codimension 1 and suppose $|\tau(f)| > \|f\|$. Then the series $\sum_{n=0}^{+\infty} \left(\dfrac{f}{\tau(f)}\right)^n$ converges and shows that $f - \tau(f)$ is invertible in A, a contradiction since $\tau(f - \tau(f)) = 0$. □

Corollary B.1.10: *Let A be a Banach $\mathbb{K}$-algebra. Every $\mathbb{K}$-algebra homomorphism from A to $\mathbb{K}$ is continuous.*

The characterization of the continuous multiplicative norms of $(R(D), \| \, . \, \|_D)$ by means of the large circular filters secant with D suggests us extending this characterization to the multiplicative semi-norms of $R(D)$.

Theorem B.1.11 [61, 62]: *Let D be a closed bounded subset of $\mathbb{K}$. The mapping Ξ from $\Phi(D)$ into Mult($R(D)$) defined as $\Xi(\mathcal{F}) = \varphi_{\mathcal{F}}$ is a bijection from $\Phi(D)$ onto Mult($R(D), \mathcal{U}_D$). Moreover, $\varphi_{\mathcal{F}}$ is an absolute value if and only if $\mathcal{F}$ does not converge in D. Further Mult($R(D, \| \cdot \|_D)$) is provided with the topology of pointwise convergence for which it is compact.*

Proof. On the one hand, by Theorem A.3.10 and Theorem B.1.4, it is clearly seen that the mapping defined on $\Phi^\circ(D)$ by $\mathcal{F} \longrightarrow \varphi_{\mathcal{F}}$ is a bijection from this set onto the set of continuous multiplicative norms on $R(D)$.

On the other hand, every $a \in D$ defines a multiplicative semi-norm ψ by $\psi(h) = |h(a)|$, the kernel of which is the maximal ideal $(x - a)R(D)$. Thus we have a mapping from the set of convergent circular filters on D into the set of multiplicative semi-norms, which are not norms: this mapping is obviously injective.

Finally let ψ be a multiplicative semi-norm whose kernel is not zero. Then $Ker(\psi)$ is a prime ideal, hence a maximal ideal of $R(D)$ and, therefore, it is of the form $(x-a)R(D)$ with $a \in D$. Then $\psi(x-a) = 0$, hence $\psi(x-b) = |a-b|$ whenever $b \in \mathbb{K}$ and, therefore, ψ is of the form $\varphi_{\mathcal{F}_a}$ with $\mathcal{F}_a$ the filter of the neighborhoods of a in D. Thus Ξ is a bijection from $\Phi(D)$ onto *Mult*($R(D), \mathcal{U}_D$) and $\varphi_{\mathcal{F}}$ is a norm if and only if $\mathcal{F}$ is not convergent. Finally, by Theorem B.1.7 *Mult*($R(D, \| \cdot \|_D)$) is compact with respect to the topology of pointwise convergence. That ends the proof of Theorem B.1.11. □

Corollary B.1.12 [61, 62]: *Mult($\mathbb{K}[X]$) is provided with the topology of pointwise convergence for which it is locally compact.*

B.2. Analytic elements

Due to the fact that any disk $d(a, r)$ is exactly the same as $d(b, r)$ for every $b \in d(a, r)$, it is easily seen that a power series $\sum_{n=0}^{\infty} a_n (x - a)^n$ that admits the disk $d(a, r)$ for disk of convergence, may not be extended outside its convergence disk as it is done in complex analysis, by means of a change of origin.

However by Runge's Theorem we remember that a holomorphic function in a compact subset D of $\mathbb{C}$ is equal to the limit of a sequence of rational functions with respect to the uniform convergence on D. This is why Marc Krasner introduced analytic elements on a subset D of $\mathbb{K}$ directly by considering limits of sequences of rational functions with respect to the uniform convergence on D [68].

Actually Marc Krasner constructed a theory of analytic functions f defined on a quasi-connected set D equal to the union of a chained family of quasi-connected sets $(D_i)_{i \in I}$ such that the restriction of f to each D_i is an analytic element on D_i. (This construction was widened to the analytic infraconnected sets by Philippe Robba [74, 83]).

Another theory was defined by John Tate, consisting (in one variable) of using infraconnected affinoid sets. Here we will only describe some basic properties of analytic elements on infraconnected sets in order to apply them to power series and various Laurent series that are used for studying meromorphic functions. A comparison between Krasner's theory and Tate's theory was made in [44] and Krasner–Tate's algebras were examined again in [50]. Here we aim at studying meromorphic functions in the field $\mathbb{K}$ and applications to problems of value distribution. This is why we will not repeat the study of Krasner–Tate algebras.

We will examine algebras of analytic elements, particularly Banach algebras of bounded analytic elements. We will see the characterization of sets D such that the space of analytic elements on D is a $\mathbb{K}$-algebra. We will examine some basic properties of analytic elements such as poles when the set D is not closed and we will see that analytic elements on a closed bounded set D are uniformly continuous. If D has finitely many infraconnected components, for each one, its characteristic function is an analytic element on D.

Definitions and notations: Let D be an infinite subset of $\mathbb{K}$. We will denote by $H(D)$ the completion of $R(D)$ for the topology $\mathcal{U}_D$ of uniform convergence on D. The elements of $H(D)$ are called the *analytic elements on* D [68].

The set $H(D)$ is then provided with the topology of uniform convergence on D for which it is complete and every $f \in H(D)$ defines a function on D, which is the uniform limit (on D) of a sequence $(h_n)_{n\in\mathbb{N}}$ in $R(D)$. Thus, given two infinite sets D, D' such that $D \subset D'$, the restriction to D of elements of $H(D')$ enables us to consider that $H(D')$ is included in $H(D)$.

Next, $H(D)$ is a $\mathbb{K}$-vectorial space and a complete topological group with respect to the topology $\mathcal{U}_D$. The question whether the product of two analytic elements on D is an analytic element on D will be studied later. However it is easily seen that given $f \in H(D)$, the function f^n also belongs to $H(D)$.

Lemma B.2.1: *For every $f, g \in H(D)$ we have*

(i) $\|f\|_D = 0$ *if and only if* $f = 0$,

(ii) $\|\lambda f\|_D = |\lambda|\ \|f\|_D$ *whenever* $\lambda \in \mathbb{K}^*$,

(iii) $\|f + g\|_D \leq \max(\|f\|_D, \|g\|_D)$,

(iv) *If* $(\|f\|_D, \|g\|_D)$ *is different from* $(0, +\infty)$ *and from* $(+\infty, 0)$ *then the function* fg *satisfies* $\|fg\|_D \leq \|f\|_D\ \|g\|_D$,

(v) $\|f^n\|_D = \|f\|_D^n$ *whenever* $n \in \mathbb{N}^*$.

Definitions and notations: We will denote by $H_b(D)$ the set of elements $f \in H(D)$ bounded on D. Then $H_b(D)$ is clearly a $\mathbb{K}$-vectorial subspace of $H(D)$ and is closed in $H(D)$. Moreover $\|\ .\ \|_D$ is a norm on $H_b(D)$ that makes it a Banach $\mathbb{K}$-algebra. If D is unbounded, we will denote by $H_0(D)$ the set of the $f \in H(D)$ such that $\lim\limits_{\substack{|x|\to+\infty \\ x\in D}} f(x) = 0$.

Theorem B.2.2 is an immediate consequence of Theorem B.1.3.

Theorem B.2.2: *$H_b(D)$ is a Banach $\mathbb{K}$-subalgebra of $\mathbb{K}^D$. The following three conditions are equivalent:*

(i) *$H_b(D) = H(D)$,*

(ii) *$H(D)$ is topological $\mathbb{K}$-vector space,*

(iii) *$(H(D), \| \,.\, \|_D)$ is a Banach $\mathbb{K}$-algebra,*

(iv) *D is closed and bounded.*

If these conditions are satisfied, then $\| \,.\, \|_D$ is a semi-multiplicative norm.

Definitions and notations: Let $f \in H(D)$ have no zero in D. The element f will be said to be *invertible in $H(D)$* if the function $\frac{1}{f}$ (also denoted by f^{-1}) belongs to $H(D)$. This definition holds even if $H(D)$ is not a ring.

Lemma B.2.3 is classical.

Lemma B.2.3: *Let $f \in H(D)$ be such that $\inf_{x\in D}|f(x)| > 0$. Then f^{-1} belongs to $H(D)$. Moreover, if D is closed and bounded, f^{-1} belongs to $H(D)$ if and only if $\inf_{x\in D}|f(x)| > 0$.*

Let $g \in H(D)$ satisfy $|g(x)| = 1$ for all $x \in D$ and $\|f-1\|_D > \|g-1\|_D$. Then we have $\|fg-1\|_D = \|f-1\|_D$.

Proof. We suppose $\inf_{x\in D}|f(x)| = \lambda > 0$. Let $(h_n)_{n\in\mathbb{N}}$ be a sequence in $R(D)$ such that $\lim_{n\to\infty}\|h_n - f\|_D = 0$. For big enough n, we have $|h_n(x)| \geq \lambda$ whenever $x \in D$, hence $\left|\frac{1}{h_n(x)} - \frac{1}{f(x)}\right| = \left|\frac{f(x)-h_n(x)}{h_n(x)f(x)}\right| \leq \frac{\|f-h_n\|_D}{\lambda^2}$, hence the sequence $\frac{1}{h_n}$ converges to $\frac{1}{f}$.

Conversely if D is closed and bounded, and if $\frac{1}{f} \in H(D)$, $\frac{1}{f}$ has to be bounded by some $M \in \mathbb{R}_+$, hence $|f(x)| \geq \frac{1}{M}$ whenever $x \in D$.

Now let $g \in H(D)$ satisfy $|g(x)| = 1$ for all $x \in D$ and $\|f-1\|_D > \|g-1\|_D$. For every $x \in D$, we have $|f(x)-1||g(x)| > |g(x)-1|$ and, therefore, $|f(x)g(x)-1| = |f(x)-1||g(x)| = |f(x)-1|$. This finishes proving Lemma B.2.3. □

Theorem B.2.4: *Let $f \in H(D)$ and let $h \in R_b(D)$. Then fh belongs to $H(D)$.*

Proof. Let ε be > 0 and let $g \in R(D)$ satisfy $\|f-g\|_D < \varepsilon$. Then we have $\|hf - hg\|_D < \varepsilon\|h\|_D$ and this clearly shows that $fh \in H(D)$. □

When D is not closed or is not bounded, we will show how to split an element $f \in H(D)$.

Theorem B.2.5: *The vector space $H(D)$ is equal to the direct sum $R_0(\mathbb{K} \setminus (\overline{D} \setminus D)) \oplus H(\overline{D})$. Moreover if D is not bounded, then $H(\overline{D})$ is equal to the direct sum $\mathbb{K}[x] \oplus H_0(\overline{D})$.*

Proof. Let $(f_n)_{n\in\mathbb{N}}$ be a sequence in $R(D)$ such that $\lim_{n\to\infty} \|f_n - f\|_D = 0$. In particular there exists $N \in \mathbb{N}$ such that $f_n - f_N$ is bounded when $n \geq N$. We put $g_n := f_n - f_N$. The sequence g_n converges in $H(D)$ to $f - f_N$. Let $g := f - f_N$. On the other hand, since each $f_n - f_N$ belongs to $R(D)$ and is bounded in D, $f_n - f_N$ belongs to $R(\overline{D})$. Obviously $\|f_n - f_N\|_D = \|f_n - f_N\|_{\overline{D}}$, hence finally g belongs to $H_b(\overline{D})$. Now we may obviously split f_N in the form $E(x) + h_1(x) + h_2(x)$ with $E(x) \in \mathbb{K}[x], h_1 \in R_0(\overline{D}), h_2 \in R_0(\mathbb{K}\setminus(\overline{D}\setminus D))$. We put $f^* = h_2$ and $\overline{f} = E+h_1+g$. We have clearly split f in the form $f^* + \overline{f}$ with $f^* \in R_0(\mathbb{K} \setminus (\overline{D} \setminus D))$, $\overline{f} \in H(\overline{D})$. Hence, we have proven that $H(D) = R_0(\mathbb{K} \setminus (\overline{D} \setminus D)) + H(\overline{D})$.

This sum is easily seen to be direct. Indeed, suppose that we have $h \in R_0(\mathbb{K} \setminus (\overline{D} \setminus D))$ and $g \in H(\overline{D})$ such that $h + g = 0$, with $h \neq 0$. Then h has a pole $\alpha \in \overline{D} \setminus D$ and it may be written in the form $\sum_{i=1}^{q} \frac{\lambda_i}{(x-\alpha)^i} + h_\alpha$ with $\lambda_i \neq 0$ $(1 \leq i \leq q)$ and $h_\alpha \in R_0(\mathbb{K} \setminus (\overline{D} \setminus (D \cup \{\alpha\})))$. But g is obviously bounded around α, hence h has to be bounded when x approaches α, hence finally α does not exist. This shows that the sum is direct.

Now, we suppose D is unbounded. First we will prove that every element $f \in H_b(D)$ admits a limit when $|x|$ tends to $+\infty$. Let $\epsilon \in \mathbb{R}_+^*$ and let $h \in R(D)$ satisfy $\|f - h\|_D < \epsilon$. Since f is bounded in D, so is h. But then h is of the form $\frac{P}{Q}$, with $\deg(P) = \deg(Q)$ and, therefore, h has a limit λ when $|x|$ tends to $+\infty$. Let $\epsilon > 0$ be such that $|h(x) - \lambda| < \epsilon$ whenever $x \in D \setminus d(0,r)$. Clearly we have $|f(x) - \lambda| < \epsilon$ whenever $x \in D \setminus d(0,r)$. This proves that f does converge along the filter $\mathcal{F}$ that admits as a basis the family of sets $D \setminus d(0,r)$ $(r > 0)$.

Let $f \in H(\overline{D})$ be unbounded. Let $(x_n)_{n\in\mathbb{N}}$ be a sequence such that $\lim_{n\to\infty} |f(x_n)| = +\infty$. Suppose the sequence $(x_n)_{n\in\mathbb{N}}$ does not tend to $+\infty$. Then there exists a bounded subsequence $(x_{n_q})_{q\in\mathbb{N}}$ such that $\lim_{q\to\infty} |f(x_{n_q})| = +\infty$, but this is impossible due to Theorem B.2.2, because such a sequence lies in a closed bounded set D' included in D. Now there exists $h \in R(\overline{D})$ such that $f - h$ is bounded and, therefore, we have $\lim_{n\to\infty} |h(x_n)| = +\infty$. Let $h(x) = P(x) + u(x)$ with $P \in \mathbb{K}[x]$ and $u \in R_0(\overline{D})$. Since $f - h$ is bounded, clearly $f - P$ belongs to $H_b(\overline{D})$, hence we have proven that $H(\overline{D}) = \mathbb{K}[x] + H_b(\overline{D})$. Moreover, since u has a limit when $|x|$ tends to $+\infty$ in D, we have $H_b(\overline{D}) = H_0(\overline{D}) + \mathbb{K}$ and, therefore, $H(\overline{D}) = \mathbb{K}[x] + H_0(\overline{D})$. Finally, by considering elements when $|x|$ tends to $+\infty$, this sum is easily seen to be direct and this ends the proof. □

Definitions and notations: Let $\alpha \in \overline{D} \setminus D$ and $f \in H(D)$, and let $f = f^* + \overline{f}$, with $f^* \in R_0(\mathbb{K} \setminus (\overline{D} \setminus D))$ and $\overline{f} \in H(\overline{D})$. Let $\alpha \in \overline{D} \setminus D$ be a pole of f^* and let $f^*(x) = \sum_{j=1}^{q} \frac{\lambda_j}{(x-\alpha)^j} + u(x)$, with $u \in R_0(\mathbb{K} \setminus (\overline{D} \setminus (D \cup \{\alpha\})))$. The pole α of order q of f^* will be called a *pole of order* q *of* f and λ_1 will be called *the residue of* f *at* α and will be denoted by $\text{res}(f, \alpha)$.

Let $a_i, 1 \leq i \leq n$ be the poles of f and for each i let q_i be the order of a_i. The polynomial $\prod_{i=1}^{n}(x - a_i)^{q_i}$ will be named *the polynomial of poles of f in $\overline{D} \setminus D$.*

Corollary B.2.6: *Let $\alpha \in \overline{D} \setminus D$ and $f \in H(D)$ be such that $|f(x)|$ is bounded in $d(\alpha, r) \cap D$ with $r > 0$. Then $f \in H(D \cup \{\alpha\})$.*

Proof. Indeed, as $\overline{f}$ is obviously bounded in $D \cap d(\alpha, r)$, so is f^* and, therefore, f^* has no pole at α. □

Corollary B.2.7: *$H(D) = H_b(\overline{D}) + R(\mathbb{K} \setminus (\overline{D} \setminus D))$, $H_b(D) = H_b(\overline{D}) \subset H(\overline{D})$, and $H_b(D) = H_b(\overline{D})$. If D is bounded, then $H_b(D) = H(\overline{D})$. If D is not bounded, for every unbounded $f \in H(D)$ there exists a unique $q \in \mathbb{N}^*$ such that $x^{-q}f(x)$ has a finite non-zero limit when $|x|$ tends to $+\infty$, $x \in D$. Let $d(a, r^-)$ be a hole of D. If f belongs to $H(\overline{D})$ and if $x^{-q}f(x)$ has a finite non-zero limit, then $\dfrac{f(x)}{(x-a)^q}$ belongs to $H_b(D)$.*

Corollary B.2.8: *If $\overline{D} = \mathbb{K}$ then $H(D) = R(D)$.*

Corollary B.2.9 comes from the definition of the poles and from Theorem B.2.5.

Corollary B.2.9: *Let $f \in H(D)$ and let $\alpha \in \overline{D} \setminus D$. Then α is a pole of order $n > 0$ for f if and only if $(x - \alpha)^n f(x)$ has a finite non-zero limit at α. If there exists no $r \in \mathbb{R}_+^*$ such that $|f(x)|$ is bounded in $d(\alpha, r) \setminus \{\alpha\}$ then α is a pole of order $n \geq 1$ for f and $(x - \alpha)^n f(x)$ has a finite non-zero limit at α.*

Theorem B.2.10: *Let $f \in H(D)$ and let $\alpha \in \overline{D} \setminus D$. Either f belongs to $H(D \cup \{\alpha\})$ or α is a pole for f.*

Proof. If f does not belong to $H(D \cup \{\alpha\})$, by Corollary B.2.6, f is unbounded in any disk $d(\alpha, r)$ whenever $r > 0$. Hence, by means of the notation of Theorem B.2.5, α is clearly a pole of f^* and, therefore, is a pole of f. □

We must notice Theorem B.2.11:

Theorem B.2.11: *Let D be closed and bounded, and let $f \in H(D)$. Then f is uniformly continuous in D.*

Proof. The claim is immediate when f is a polynomial. Suppose now that $f(x) = \dfrac{P(x)}{Q(x)} \in R(D)$. Since D is closed and bounded, there exists $m > 0$ such that $|Q(x)| \geq m \ \forall x \in D$. Consequently, $\dfrac{P(x)}{Q(x)}$ is also uniformly continuous. Then, when $f \in H(D)$, since f is a uniform limit of rational functions, f is also uniformly continuous. □

Definitions and notations: We will denote by Alg the family of sets E such that $H(E)$ is a $\mathbb{K}$-subalgebra of $\mathbb{K}^E$.

Theorem B.2.12: *Let $f \in H(D)$. There exists $W \in R_b(D)$, whose zeros lie in $\overline{D} \setminus D$ and $h \in H(\overline{D})$ such that $f = \dfrac{h}{W}$. Further, if D is bounded or if $D \in$ Alg then there exists $g \in H(\overline{D})$ such that $f = \dfrac{g}{Q}$ with Q the polynomial of poles of f in $\overline{D} \setminus D$.*

Proof. We may summarize Theorem B.2.5 in this way: f is of the form $\widetilde{f}(x) + \widehat{f}(x)$ with $\widetilde{f} \in R(\mathbb{K} \setminus (\overline{D} \setminus D))$ and $\widehat{f} \in H_b(\overline{D})$. Indeed if D is bounded we just take $\widehat{f} = \overline{f}$ and if D is not bounded, $\widehat{f}$ is the one defined in Theorem B.2.5. Thus $\widetilde{f}(x)$ can be written in the form $\dfrac{P(x)}{Q(x)}$ with $Q(x) = \prod_{i=1}^{n}(x - a_i)^{q_i}$, i.e., the polynomial of the poles of f in $\overline{D} \setminus D$, and $P(x) \in \mathbb{K}[x]$. Let $q = \sum_{i=1}^{n} q_i$. Theorem B.2.12 is obviously trivial if D has no hole, hence we may assume D to have at least one hole $T = d(a, r^-)$. Let $W(x) = \dfrac{Q(x)}{(x-a)^q}$. We know that $W \in R_b(\overline{D})$, hence $W\widehat{f} \in H_b(\overline{D})$. On the other hand, we see that $W\widetilde{f} \in R(\overline{D})$, hence $Wf \in H(\overline{D})$. We just put $h = Wf$ and have the factorization $f = \dfrac{h}{W}$.

If D is bounded we see that both Q, h are bounded in D, hence $Q\widehat{f}$ belongs to $H_b(\overline{D})$ and then Qf belongs to $H(\overline{D})$. In the same way if $H(D)$ is supposed to be a ring, then $Q\widehat{f}$ belongs to $H(\overline{D})$ and then $Qf = P + Q\widehat{f}$ belongs to $H(\overline{D})$. This ends the proof. □

Corollary B.2.13: *Let $S(D)$ be the set of polynomials whose zeros belong to $\overline{D} \setminus D$. If $D \in$ Alg, then $H(D) = S(D)^{-1}H(\overline{D})$.*

Theorem B.2.14: *Let D be closed. Let $g \in H(D)$ and let $P \in \mathbb{K}[x]$ be such that Pg belongs to $H(D)$. For every $Q \in \mathbb{K}[x]$ such that $\deg(Q) \leq \deg(P)$, Qg also belongs to $H(D)$.*

Proof. Theorem B.2.14 is clearly trivial when D belongs to Alg. Now, suppose that D does not belong to Alg, hence D is unbounded. If D has no hole, then $D = \mathbb{K}$, hence by Corollary B.2.8 $H(D)$ is equal to $\mathbb{K}[x]$. Thus, we may assume that D admits at least one hole and then, without loss of generality, we can assume that this hole is $d(0, r^-)$. Let $q = \deg(P)$. Let $P_q = P$ and let $P_{q-1} = \dfrac{P(x) - P(0)}{x}$. Then P_{q-1} be a polynomial of degree $q - 1$. We see that $(P(x) - P(a))g(x)$ belongs to $H(D)$. But $\left\|\dfrac{1}{x}\right\|_D$ is bounded and then by Theorem B.2.4, $P_{q-1}(x)g(x)$ belongs to $H(D)$. Hence, by induction, it is seen that for each $j = 1, \ldots, q$ there exists a polynomial P_j of degree j such that $P_j g$ belongs to $H(D)$ and this clearly completes the proof. □

Now when D is not infraconnected we have to notice an easy result on characteristic functions that shows how rich the algebra $H(D)$ is.

Proposition B.2.15: *Let D have an empty annulus Λ. Let w_1, w_2 be the functions defined on D by $w_1(x) = 1, w_2(x) = 0$ if $x \in \mathcal{I}(\Lambda)$, and $w_1(x) = 0, w_2(x) = 1$ if $x \in \mathcal{E}(\Lambda)$. Then w_1 and w_2 belong to $H(D)$.*

Proof. Let $\Lambda = \Gamma(a, r_1, r_2)$, with $a \in D$. With no loss of generality we may obviously assume $a = 0$. Let $\alpha \in \Lambda$ be such that $r_1 < |\alpha| < r_2$ and for each $n \in \mathbb{N}^*$, let $u_n = \dfrac{1}{1 - (\frac{x}{\alpha})^n}$. Then $\left\|1 - (\frac{x}{\alpha})^n - 1\right\|_{\mathcal{I}(\Lambda)} \leq (\frac{r_1}{|\alpha|})^n$ whereas $|1 - (\frac{x}{\alpha})^n| \geq (\frac{r_2}{|\alpha|})^n$ for every $x \in \mathcal{E}(\Lambda)$, hence finally $\|u_n - w_1\|_D \leq \max((\frac{r_1}{|\alpha|})^n, (\frac{|\alpha|}{r_2})^n)$. Thus we see that $w_1 = \lim_{n\to\infty} u_n \in H(D)$ and $w_2 = 1 - w_1 \in H(D)$. □

Theorem B.2.16: *Let E have finitely many infraconnected components $E_1, .., E_q$. For each $i = 1, \ldots, q$, the characteristic function of E_i belongs to $H(E)$.*

Proof. Let A be one of the infraconnected components of E. By Theorem A.1.21, there exist empty annuli $(\Lambda_j)_{0\leq j\leq n}$ such that A is either of the form

(α) $\mathcal{I}_E(\Lambda_0) \bigcap \Big(\bigcap_{j=1}^{n} \mathcal{E}_E(\Lambda_j)\Big)$ or of the form (β) $\bigcap_{j=1}^{n} \mathcal{E}_E(\Lambda_j)$.

But by Proposition B.2.15, the characteristic function u_j of $\mathcal{E}_E(\Lambda_j)$ belongs to $H_b(E)$, $(1 \leq j \leq n)$ and so does the characteristic function u_0 of $\mathcal{I}_E(\Lambda_0)$. Since all the u_j belong to $H_b(E)$, we see that the products $u = \prod_{j=0}^{n} u_j$ and $w = \prod_{j=1}^{n} u_j$ belong to $H(E)$. Then when A is of the form (α) (resp. (β)), its characteristic function is equal to u (resp. w) and, therefore, belongs to $H(E)$. □

B.3. Composition of analytic elements

Given A and $B \subset \mathbb{K}$, $f \in H(A)$ such that $f(A) \subset B$ and $g \in H(B)$, a basic question is whether $g \circ f \in H(A)$. There is an immediate application to the study of homomorphisms from an algebra $H(D)$ to another $H(D')$.

Lemma B.3.1: *Let A and B be subsets of $\mathbb{K}$ and let $f \in H(A)$ be such that $f(A) \subset B$. For every $\lambda \in \mathbb{K} \setminus \overline{B}$, $f - \lambda$ is invertible in $H(A)$. Moreover, if for every $h \in R(B)$, $h \circ f$ belongs to $H(A)$, then for every $g \in H(B)$, $g \circ f$ belongs to $H(A)$ and for every $\lambda \in \overline{B} \setminus B$, $f - \lambda$ is invertible in $H(A)$.*

Proof. Let $r = \delta(\lambda, B)$. If $\lambda \notin \overline{B}$ we have $r > 0$ and then $|f(x) - \lambda| \geq r$ whenever $x \in A$. Hence, by Lemma B.2.3, $f - \lambda$ is invertible in $H(A)$.

Now we assume that for every $h \in R(B)$, $h \circ f$ belongs to $H(A)$. Let $g \in H(B)$, let ϵ be > 0, and let $h \in R(B)$ satisfy $\|g-h\|_B \leq \epsilon$. It is seen that $\|g\circ f - h\circ f\|_A \leq \epsilon$. Since $h \circ f$ belongs to $H(A)$, then so does $g \circ f$.

Finally, let $\lambda \in \overline{B} \setminus B$ and let $h(u) = \dfrac{1}{u - \lambda}$. Since $h \circ f$ belongs to $H(A)$, $f - \lambda$ is invertible in $H(A)$. □

Theorem B.3.2: *Let A, B be subsets of $\mathbb{K}$ and let $f \in H(A)$ satisfy $f(A) \subset B$.*
(i) *If $f \in R(A)$, then $g \circ f \in H(A)$ whenever $g \in H(B)$.*
(ii) *If $A \in \mathrm{Alg}$, then $g \circ f \in H(A)$ for all $g \in H(B)$ if and only if $f - \lambda$ is invertible in $H(A)$ for all $\lambda \in \overline{B} \setminus B$.*

Proof. By Lemma B.3.1 we just have to show that for every $g \in R(B)$, $g \circ f$ belongs to $H(A)$ in each one of these two hypotheses:

(H_1) $f \in R(A)$.

(H_2) $f - \lambda$ is invertible in $H(A)$ for all $\lambda \in \overline{B} \setminus B$.

So we take $g(u) = \dfrac{P(u)}{Q(u)} \in R(B)$ and will show that $g\circ f \in H(A)$ in each hypothesis. Let $\lambda_1, \ldots, \lambda_q$ be the poles of g in $\mathbb{K} \setminus B$.

(H_1) For every $j = 1, \ldots, q$, $f - \lambda_j$ is invertible in $R(A)$ because $f - \lambda_j$ has no zero in A. Hence, $Q \circ f$ is invertible in $R(A)$ and then $g \circ f$ belongs to $R(A)$.

(H_2) For each $j = 1, \ldots, q$, either λ_j belongs to $\overline{B} \setminus B$, or it belongs to $\mathbb{K} \setminus \overline{B}$. In both cases, by Lemma B.3.1 each $f - \lambda_j$ is invertible in $H(A)$. Since A belongs to Alg, $Q \circ f$ is clearly invertible in $H(A)$ and $P \circ f$ belongs to $H(A)$. Hence, so does $g \circ f$. □

Corollary B.3.3: *Let $A \in \mathrm{Alg}$ and let B be a closed subset of $\mathbb{K}$. Let $f \in H(A)$ satisfy $f(A) \subset B$ and let $g \in H(B)$. Then $g \circ f$ belongs to $H(A)$.*

Example: Let r, $s \in \mathbb{R}_+^*$, let $f \in H(d(0,r))$ be such that $f(d(0,r)) \subset d(0,s)$, and let $g \in H(d(0,s))$. Then $g \circ f$ belongs to $H(d(0,r))$.

Lemma B.3.4: *Let $h \in R(D)$ and let $D' = h(D)$. Let $f \in H(D')$. If f is invertible in $H(D')$ then $f \circ h$ is invertible in $H(D)$. If h is a Moebius function, f is invertible in $H(D')$ if and only if $f \circ h$ is invertible in $H(D)$.*

Proof. First we suppose f invertible in $H(D')$. Let $g = \dfrac{1}{f}$. Then by Theorem B.3.2 $g \circ h$ belongs to $H(D)$ and is clearly equal to $\dfrac{1}{f \circ h}$. Now we assume that h is a Moebius function and we put $\ell = h^{-1}$. If $f \circ h$ is invertible in $H(D)$, $(f \circ h) \circ \ell$ is invertible in $H(D')$ and this ends the proof of Lemma B.3.4. □

We are now going to study the $\mathbb{K}$-algebra homomorphisms from $H(D)$ into $H(D')$. First we will consider homomorphisms from $R(D)$ into $R(D')$.

Proposition B.3.5: *Let D, D' be subsets of $\mathbb{K}$ and let $\gamma \in R(D')$ satisfy $\gamma(D') \subset D$. Let ϕ_γ be the mapping from $R(D)$ into $R(D')$ defined as $\phi_\gamma(f) = f \circ \gamma$ $(f \in R(D))$. Then ϕ_γ is a homomorphism from $R(D)$ into $R(D')$ and this homomorphism is injective if and only if γ is not a constant. Every $\mathbb{K}$-algebra homomorphism is of this form and the mapping $\gamma \to \phi_\gamma$ is a bijection from the set of the $\gamma \in R(D')$ such that $\gamma(D') \subset D$ onto the set of the $\mathbb{K}$-algebra homomorphisms from $R(D)$ into $R(D')$.*

Proof. Let $\gamma \in R(D')$ satisfy $\gamma(D') \subset D$. Then it is seen that ϕ_γ takes values in $R(D')$, is a $\mathbb{K}$-algebra homomorphism and is obviously injective if and only if γ is not a constant.

Conversely, let ψ be a $\mathbb{K}$-algebra homomorphism from $R(D)$ into $R(D')$ and let $\gamma = \psi(I_{D'})$ with $I_{D'}$ the identical mapping in D'. Then we have $\psi(P) = P \circ \gamma$ for every polynomial P. On the other hand, if $\alpha \notin D$ then $(x-\alpha)$ is invertible in $R(D)$ and $\psi\Big(\dfrac{1}{x-\alpha}\Big) = (\psi(x-\alpha))^{-1} = (\gamma-\alpha)^{-1}$, $\psi(h) = h \circ \gamma$ whenever $h \in R(D)$. The mapping $\gamma \to \phi_\gamma$ is obviously injective and, hence, is a bijection. □

Proposition B.3.6: *Let D, D' be sets in $\mathbb{K}$ and let $\gamma \in H(D')$ satisfy $\gamma(D') \subset D$ and $f \circ \gamma \in H(D')$ for all $f \in H(D)$. Let ϕ_γ be the mapping from $H(D)$ into $H(D')$ defined as $\phi_\gamma(f) = f \circ \gamma$. Then ϕ_γ is a linear mapping from $H(D)$ into $H(D')$ continuous with respect to the topology of uniform convergence on D for $H(D)$ and on D' for $H(D')$. Moreover, given f, $g \in H(D)$ such that $fg \in H(D)$ we have $\phi_\gamma(fg) = \phi_\gamma(f)\phi_\gamma(g)$. The restriction of ϕ_γ to $H_b(D)$ is a Banach $\mathbb{K}$-algebra homomorphism from $H_b(D)$ into $H_b(D')$.*

If γ is a bijection from D' onto D and if $\gamma^{-1} \in H(D)$ then ϕ_γ is a $\mathbb{K}$-vector space isomorphism from $H(D)$ onto $H(D')$ bicontinuous with respect to the topology of uniform convergence on D for $H(D)$ and on $H(D')$ for $H(D')$, satisfying $\big(\phi_\gamma\big)^{-1} = \phi_{\gamma^{-1}}$ and the restriction of ϕ_γ to $H_b(D)$ is a Banach $\mathbb{K}$-algebra isomorphism from $H_b(D)$ onto $H_b(D')$. Further, if $\gamma(D') = D$, then the equality $\|\phi_\gamma(f)\|_{D'} = \|f\|_D$ is true for every $f \in H(D)$ and the restriction of ϕ_γ to $H_b(D)$ is an isometric Banach $\mathbb{K}$-algebra isomorphism from $H_b(D)$ onto $H_b(D')$.

Proof. It is easily seen that ϕ_γ is linear and satisfies $\phi_\gamma(fg) = \phi_\gamma(f)\phi_\gamma(g)$ when $fg \in H(D)$. Next, ϕ_γ is clearly continuous because $\|\phi_\gamma(f)\|_{D'} = \|f \circ \gamma\|_D = \sup\limits_{x\in D'} |f(\gamma(x))| \leq \sup\limits_{u\in D} |f(u)| = \|f\|_D$. In particular, we notice that if $\gamma(D') = D$, we have $\|\phi_\gamma(f)\|_{D'} = \|f \circ \gamma\|_D = \sup\limits_{x\in D'} |f(\gamma(x))| = \sup\limits_{u\in D} |f(u)| = \|f\|_D$.

If $f \in H_b(D)$ obviously $f \circ \gamma \in H_b(D')$. Now let γ be a bijection from D' onto D such that $\gamma^{-1} \in H(D)$. It is seen that $(\phi_{\gamma^{-1}}) \circ \phi_\gamma = I_{H(D)}$ while $\phi_\gamma \circ (\phi_{\gamma^{-1}}) = I_{H(D')}$, hence ϕ_γ is an isomorphism such that $\phi_{\gamma^{-1}} = \left(\phi_\gamma\right)^{-1}$. □

We will study the $\mathbb{K}$-algebra homomorphisms from $H(D)$ into $H(D')$.

Definitions and notations: Given subsets D and D' of $\mathbb{K}$, we will denote by $\Xi(D', D)$ *the set of the* $\gamma \in H(D')$ *such that* $\gamma(D') \subset D$ *and such that for every* $\lambda \in \overline{D} \setminus D$, $\gamma - \lambda$ *is invertible in* $H(D')$.

Given two $\mathbb{K}$-algebras A and B we will denote by $\mathcal{H}om(A, B)$ the set of $\mathbb{K}$-algebra homomorphisms from A into B.

Remark: In particular $\Xi(D', D)$ contains the set of the $h \in R(D')$ such that $h(D') \subset D$.

Theorem B.3.7: *Let* $D, D' \in \text{Alg}$ *and let* $\gamma \in \Xi(D', D)$. *The mapping* ϕ_γ *defined in* $H(D)$ *by* $\phi_\gamma(f) = f \circ \gamma$ *has values in* $H(D')$ *and is a* $\mathbb{K}$*-algebra homomorphism from* $H(D)$ *into* $H(D')$. *Conversely, every* $\mathbb{K}$*-algebra homomorphism from* $H(D)$ *into* $H(D')$ *is continuous and of this form. Further, the mapping* $\gamma \to \phi_\gamma$ *from* $\Xi(D', D)$ *onto* $\mathcal{H}om(H(D), H(D'))$ *is a bijection.*

Let $D'' \in \text{Alg}$ *and let* $\tau \in \Xi(D'', D')$. *Then* $\gamma \circ \tau \in \Xi(D'', D)$ *and* $\phi_{\gamma \circ \tau} = \phi_\gamma \circ \phi_\tau$.

Further, a homomorphism ϕ_γ *from* $H(D)$ *into* $H(D')$ *is an isomorphism if and only if* γ *is a bijection from* D' *onto* D *such that* $\gamma^{-1} \in H(D)$ *and then, when it is satisfied, we have* $\left(\phi_\gamma\right)^{-1} = \phi_{\gamma^{-1}}$.

Proof. By Corollary B.3.3, $f \circ \gamma$ belongs to $H(D')$ whenever $f \in H(D)$ and then by Proposition B.3.6 ϕ_γ is a $\mathbb{K}$-algebra homomorphism from $H(D)$ into $H(D')$. Let $\psi \in \mathcal{H}om(H(D), H(D'))$ and first let us show that ψ satisfies $\|\psi(f)\|_{D'} \leq \|f\|_D$ whenever $f \in H(D)$. Indeed suppose that for some $f \in H(D)$ we have $\|\psi(f)\|_{D'} > \|f\|_{D'}$. Let $g = \psi(f)$. There exists $\alpha \in D'$ such that $|g(\alpha)| > \|f\|_{D'}$. Let $\lambda = g(\alpha)$. The series $\frac{f}{\lambda} \sum_{n=0}^{\infty} \left(\frac{f}{\lambda}\right)^n$ does converge in $H(D)$ to $(\lambda - f)^{-1}$. Thus $\lambda - f$ is invertible in $H(D)$ and then $\lambda - g = \psi(\lambda - f)$ is invertible in $H(D')$. But by hypothesis α is a zero for $\lambda - g$, hence $\lambda - g$ is not invertible in $H(D')$ and this shows that $\|\psi(f)\|_{D'} \leq \|f\|_D$. Now let $\gamma = \psi(I_D) \in H(D')$ and let us show that $\gamma \in \Xi(D', D)$. Let $\alpha \in \mathbb{K} \setminus D$. Since $\psi \in \mathcal{H}om(H(D), H(D'))$, ψ must satisfy $\psi\left(\frac{1}{x - \alpha}\right) = \frac{1}{\psi(x) - \alpha} = \frac{1}{\gamma - \alpha}$, hence $\gamma - \alpha$ has to be invertible in $H(D')$ for every $\alpha \in \mathbb{K} \setminus D$. But this just means that $\gamma \in \Xi(D', D)$.

In the same way, we see that for every $h \in R(D)$, we have $\psi(h) = h(\psi(x)) = h \circ \gamma$. Finally since ψ is continuous, the equality $\psi(f) = f \circ \gamma$ holds in all $H(D)$. Obviously, given $\gamma, \tau \in \Xi(D', D)$, if $\phi_\gamma = \phi_\tau$ then $\phi_\gamma(I_D) = \phi_\tau(I_D)$, hence $\gamma = \tau$. The mapping $\gamma \to \phi_\gamma$ is then a bijection from $\Xi(D', D)$ onto $\mathcal{H}om(H(D), H(D'))$.

Now let $D'' \in \text{Alg}$ and $\tau \in \Xi(D'', D')$. It is seen that $\gamma \circ \tau \in \Xi(D'', D)$ and $\phi_{\gamma\circ\tau}(I_D) = \gamma \circ \tau = \phi_\tau(\gamma) = \phi_\tau(\phi_\gamma(I_D)) = \phi_\tau \circ \phi_\gamma(I_D)$, hence $\phi_{\gamma\circ\tau} = \phi_\tau \circ \phi_\gamma$.

By Proposition B.3.6, if γ is a bijection from D' onto D and such that $\gamma^{-1} \in H(D)$ then ϕ_γ is an isomorphism of $\mathbb{K}$-vector space, hence it is an isomorphism of $\mathbb{K}$-algebra and then, by Proposition B.3.6, we have $(\phi_\gamma)^{-1} = \phi_{\gamma^{-1}}$. Conversely, if ϕ_γ is an isomorphism, then $(\phi_\gamma)^{-1}$ is in the form ϕ_τ, with $\tau \in \Xi(D, D')$, $\phi_\tau \circ \phi_\gamma(I_D) = \gamma \circ \tau(I_D) = I_D$, and $\phi_\gamma \circ \phi_\tau(I_{D'}) = \tau \circ \gamma(I_{D'}) = I_{D'}$. Hence, γ is a bijection from D' onto D such that $\gamma^{-1} = \tau \in H(D)$. That finishes proving Theorem B.3.7. □

Definitions and notations: Let A, B be subsets of $\mathbb{K}$. If there exists a bijection $f \in H(A)$ from A onto B such that f^{-1} belongs to $H(B)$, then f will be named *a bianalytic element from A onto B.*

Propositions B.3.8 and B.3.9 will be often useful to transform unbounded domains into bounded domains.

Proposition B.3.8: *Let $D \in \text{Alg}$ and let $h \in R(D)$ be a Moebius function. Let $D' = h(D)$. Then D' belongs to Alg and $H(D')$ is isomorphic to $H(D)$ with respect to the mapping ψ defined in $H(D')$ as $\psi(f) = f \circ h$.*

Proof. By Theorem B.3.2, for every $f \in H(D')$, $f \circ h$ belongs to $H(D)$ and by Proposition B.3.6 this mapping is a $\mathbb{K}$-vector space isomorphism, which satisfies $\psi(fg) = \psi(f)\psi(g)$ whenever $f, g \in H(D')$. Hence, the space $H(D') = \psi^{-1}(H(D))$ is a $\mathbb{K}$-algebra isomorphic to $H(D)$. In particular D' belongs to Alg and ψ is a $\mathbb{K}$-algebra isomorphism. □

Proposition B.3.9: *Let D be a set with a hole $T = d(a, r^-)$, let $\gamma(x) = \dfrac{1}{x-a}$, and let $D' = \gamma(D)$. Then $\overline{D'} \in \text{Alg}$ and $H(\overline{D'})$ is isomorphic to $H_b(D)$.*

Proof. Without loss of generality we may clearly assume D to be closed because by Corollary B.2.7, $H_b(D)$ is equal to $H_b(\overline{D})$. For every $f \in H(D)$ let $\psi(f) = f \circ \gamma \in H(D')$. Then $\psi(H_b(D))$ is a $\mathbb{K}$-algebra included in $H(D')$. If D is bounded, D' is bounded and closed like D, hence by Proposition B.3.8, ψ is an isomorphism from $H(D)$ onto $H(D')$. Now we suppose D unbounded. Then D' is bounded and $\psi(H_b(D))$ is obviously included in $H_b(D')$, which, by Corollary B.2.7, is just equal to $H(\overline{D'})$. On the other hand γ clearly maps $R_b(D)$ onto $R(\overline{D'})$, hence $\psi(H_b(D)) = H(\overline{D'})$. □

Theorem B.3.10: *Let $T = d(a, r^-)$ be a hole of D and let $\gamma(x) = \dfrac{1}{x-a}$. Let $D' = \gamma(D)$. The mapping ψ from $H_b(D')$ into $H_b(D)$ defined as $\psi(f) = f \circ \gamma$ is a $\mathbb{K}$-algebra isomorphism.*

Proof. D' is bounded, hence by Corollary B.2.7, $H_b(D')$ is equal to the Banach $\mathbb{K}$-algebra $H(\overline{D'})$. Now by Theorem B.3.2, we see that $\gamma \in \Xi(D, D')$ and $\gamma^{-1} \in \Xi(D', D)$. Hence, ψ is clearly a $\mathbb{K}$-Banach space isomorphism from $H_b(D')$ onto $H_b(D)$. Now, ψ satisfies $\psi(fg) = \psi(f)\ \psi(g)$ whenever $f, g \in H(D)$ such that $fg \in H(D')$. But both $H_b(D)$ and $H_b(D')$ are Banach $\mathbb{K}$-algebras, hence ψ is a Banach $\mathbb{K}$-algebra isomorphism. □

Theorem B.3.11: *Let a be a point in D, which is not isolated. Let $\gamma(x) = \dfrac{1}{x-a}$ and let $D' = \gamma(D \setminus \{a\})$. Then given $f \in H(D')$, $f \circ \gamma$ belongs to $H(D)$ if and only if $f(x)$ has a limit when $|x|$ tends to $+\infty$.*

Proof. If $f \circ \gamma$ belongs to $H(D)$ then we just have

$$\lim_{|x| \to +\infty} f(x) = \lim_{x \to a} f \circ \gamma(x) = f \circ \gamma(a).$$

Conversely, if f has a limit l when $|x|$ tends to $+\infty$, then $f \circ \gamma$ is bounded in certain disks $d(a, r) \setminus \{a\}$. Therefore, by Corollary B.2.6, $f \circ \gamma$ belongs to $H(D)$. □

Corollary B.3.12: *Let $D \in$ Alg, let a be a point in D, which is not isolated, such that $(D \setminus \{a\})$ belongs to Alg. Let $\gamma(x) = \dfrac{1}{x-a}$ and let $D' = \gamma(D \setminus \{a\})$. Then $H(D)$ is isomorphic to the subalgebra of $H(D')$, which consists of the f such that $|f(x)|$ is bounded when $|x|$ approaches $+\infty$.*

B.4. Multiplicative spectrum of $H(D)$

In Chapter B.1 we studied and characterized the multiplicative semi-norms on a $\mathbb{K}$-algebra $R(D)$ of rational functions. We will apply these properties to the completion $H(D)$ of $R(D)$ by considering multiplicative semi-norms that are continuous with respect to the topology of $H(D)$. On $H(D)$ as on $R(D)$, the role of circular filters is obviously crucial: each continuous multiplicative semi-norm of $H(D)$ is defined by a circular filter secant with D exactly as it was explained for rational functions. However circular filters that are not secant with D play no role with regards to $H(D)$.

Definitions and notations: Throughout the chapter, D is an infraconnected subset of $\mathbb{K}$. We will denote by *Mult*$(H(D), \mathcal{U}_D)$ the set of continuous multiplicative semi-norms ψ of the $\mathbb{K}$-vector space $H(D)$ that satisfy $\psi(fg) = \psi(f)\psi(g)$ whenever $f, g \in H(D)$ such that $fg \in H(D)$.

Remark: This notation does not require $H(D)$ to be a $\mathbb{K}$- algebra, though it coincides with the notation already introduced for any topological algebra when $H(D)$ is a normed $\mathbb{K}$-algebra. Multiplicative semi-norms appeared to be the main tool for studying analytic elements [43, 49, 68]. They also are at the basis of the Berkovich theory [6].

Theorem B.4.1: (G. Garandel) [50, 58, 61] *For every $\mathcal{F} \in \Phi(D)$, the multiplicative semi-norm $\varphi_{\mathcal{F}}$ defined on $R(D)$ extends by continuity to $H(D)$ to a continuous semi-norm of $\mathbb{K}$-vector space ${}_D\varphi_{\mathcal{F}}$ of $H(D)$ that satisfies ${}_D\varphi_{\mathcal{F}}(f.g) = {}_D\varphi_{\mathcal{F}}(f)\ {}_D\varphi_{\mathcal{F}}(g)$ whenever $f, g \in H(D)$ such that $fg \in H(D)$. Moreover, the mapping: $\mathcal{F} \to {}_D\varphi_{\mathcal{F}}$, from $\Phi(D)$ into $Mult(H(D), \mathcal{U}_D)$, is a bijection.*

Proof. We may obviously extend $\varphi_{\mathcal{F}}$ by continuity to ${}_D\varphi_{\mathcal{F}}$ satisfying ${}_D\varphi_{\mathcal{F}}(fg) = {}_D\varphi_{\mathcal{F}}(f)\ {}_D\varphi_{\mathcal{F}}(g)$ whenever $f, g \in H(D)$ such that $fg \in H(D)$. We now check that the mapping $\mathcal{F} \to {}_D\varphi_{\mathcal{F}}$ from $\Phi(D)$ into $Mult(H(D), \mathcal{U}_D)$ is a bijection. It is obviously injective because if ${}_D\varphi_{\mathcal{F}_1} = {}_D\varphi_{\mathcal{F}_2}$ then $\varphi_{\mathcal{F}_1} = \varphi_{\mathcal{F}_2}$, hence by Corollary B.1.10 $\mathcal{F}_1 = \mathcal{F}_2$. Now let $\psi \in Mult(H(D), \mathcal{U}_D)$. The restriction of ψ to $R(D)$ is an element ψ_0 of $Mult(R(D), \mathcal{U}_D)$, hence by Corollary B.1.10, ψ_0 is of the form $\varphi_{\mathcal{F}}$ and then, by continuity, we have $\psi = {}_D\varphi_{\mathcal{F}}$. □

From Theorem B.1.4, Corollary B.4.2 is immediate concerning a space $H(D)$.

Corollary B.4.2: *Let $\mathcal{F}$ be a large circular filter on $\mathbb{K}$ of diameter $s > 0$. The following three assertions are equivalent:*
(i) *$\varphi_{\mathcal{F}}(h) \leq \|h\|_D$ whenever $h \in H(D)$.*
(ii) *$\varphi_{\mathcal{F}}$ is a continuous ultrametric multiplicative norm on H(D) with respect to the topology of uniform convergence.*
(iii) *$\mathcal{F}$ is secant with D.*

By Corollary B.1.10, we have Corollary B.4.3:

Corollary B.4.3: *Let D be a closed bounded subset of $\mathbb{K}$. Then $Mult(H(D), \| \cdot \|_D)$ is compact with respect to the topology of pointwise convergence.*

Remark: If $\mathcal{F}$ is a large circular filter, we know that $\varphi_{\mathcal{F}}$ is a norm on $R(D)$. But we don't know whether ${}_D\varphi_{\mathcal{F}}$ is a norm on $H(D)$. This non-trivial question is linked to the problem of T-filters, which is such a big question that it would require another book [49].

Definitions and notations: For convenience, for every $a \in D$, we put $\varphi_a(f) = |f(a)|$ whenever $f \in H(D)$ and so we define the semi-norms $\varphi_a \in Mult(H(D), \mathcal{U}_D)$. An element $\psi \in Mult(H(D), \mathcal{U}_D)$ will be said to be *punctual* if it is of the form φ_a with $a \in D$, i.e., if its circular filter is punctual.

Let $\mathcal{F}$ be a monotonous filter on D. By Proposition A.2.14, there exists a unique circular filter $\mathcal{G}$ on D less thin than $\mathcal{F}$. Then we put ${}_D\varphi_{\mathcal{F}}(f) = {}_D\varphi_{\mathcal{G}}(f)$ for all $f \in H(D)$.

For simplicity, when $\mathcal{G}$ has center 0 and diameter r, we set $|f|(r) = {}_D\varphi_{\mathcal{G}}(f)$. So, when f belongs to $R(D)$ this is the definition already given in Chapter A.3.

Now let D be infraconnected. Let $a \in \widetilde{D}$ and let r satisfy $\delta(a, D) \leq r \leq \mathrm{diam}(D)$. The circular filter $\mathcal{F}$ of center a and diameter r is then secant with D.

We put ${}_D\varphi_{a,r} = {}_D\varphi_{\mathcal{F}}$. Let A be a bounded subset of $\widetilde{D}$ and let $\widetilde{A} = d(a,r)$. If $\delta(a,D) \leq r \leq \mathrm{diam}(D)$, we put ${}_D\varphi_A = {}_D\varphi_{a,r}$. In particular this notation applies to holes of an infraconnected set D.

Let $\mathcal{F}$ be a circular filter or a monotonous filter on D. We will denote by $\mathcal{J}(\mathcal{F})$ the set of the $f \in H(D)$ such that $\lim_{\mathcal{F}} f(x) = 0$. Hence, $\mathcal{J}(\mathcal{F})$ is equal to $Ker({}_D\varphi_{\mathcal{F}})$ and, therefore, if $D \in$ Alg, $Ker({}_D\varphi_{\mathcal{F}})$, is a closed prime ideal of $H(D)$.

If $\mathcal{F}$ is a monotonous filter on D, we will denote by $\mathcal{J}_0(\mathcal{F})$ the set of the $f \in H(D)$ such that $f(x) = 0$ whenever $x \in \mathcal{B}(\mathcal{F})$.

Finally, given $a \in D$ we will denote by $\mathcal{J}(a)$ the set of the $f \in H(D)$ such that $f(a) = 0$. Then, if $D \in$ Alg, $\mathcal{J}_0(\mathcal{F})$ and $\mathcal{J}(a)$ are closed prime ideals of $H(D)$.

Among many ultrametric properties, we notice the following.

Lemma B.4.4: *Let $\mathcal{F}$ be a circular filter or a monotonous filter on D and let $f \in H(D)$. There exists $A \in \mathcal{F}$ such that $|f(x)|$ is bounded in A. Moreover, for every sequence $(a_n)_{n\in\mathbb{N}}$ thinner than $\mathcal{F}$, we have $\lim_{n\in\mathbb{N}} \varphi_{a_n} = \varphi_{\mathcal{F}}$. Given $a,\ b \in D$ and $r \in]0, \mathrm{diam}(D)[$ such that $|a-b| \leq r$, we have ${}_D\varphi_{a,r} = {}_D\varphi_{b,r}$.*

Proof. Indeed, there does exist $A \in \mathcal{F}$ such that $|f(x)| \leq {}_D\varphi_{\mathcal{F}}(f) + 1$ for all $x \in A$. The last statements come from properties seen on $R(D)$. □

Lemma B.4.5: *Let $f \in H(D)$ be invertible in $H(D)$. Then for every $\psi \in$ Mult($H(D)$,$\mathcal{U}_D$) we have $\psi(f) \neq 0$.*

Proof. Indeed we have $\psi(f)\psi(\frac{1}{f}) = 1$. □

Lemma B.4.6: *Let ${}_D\varphi_{\mathcal{F}} \in$ Mult($H(D)$,$\mathcal{U}_D$), let $f \in H(D)$ and $g \in H(D)$ be such that $\|f-g\|_D < {}_D\varphi_{\mathcal{F}}(f)$. Then ${}_D\varphi_{\mathcal{F}}(f) = {}_D\varphi_{\mathcal{F}}(g)$.*

Proof. Indeed we know that ${}_D\varphi_{\mathcal{F}}(f-g) \leq \|f-g\|_D$, hence ${}_D\varphi_{\mathcal{F}}(f-g) < {}_D\varphi_{\mathcal{F}}(f)$ and, therefore, ${}_D\varphi_{\mathcal{F}}(g) = {}_D\varphi_{\mathcal{F}}(f)$. □

Lemma B.4.7: *Let D be unbounded and let $f \in H_b(D)$. Then $|f(x)|$ has a limit ${}_D\varphi_D(f)$ when $|x|$ tends to $+\infty$ while x lies in D and ${}_D\varphi_D$ belongs to Mult($H_b(D), \|\ .\ \|_D$).*

Proof. By Corollary B.2.7, $f(x)$ admits a limit λ when $|x|$ tends to $+\infty$, $(x \in D)$. Hence, $\lim_{|x|\to+\infty,\ x\in D} |f(x)| = |\lambda|$. Thus the mapping φ_∞, defined as $\varphi_\infty(f) = \lim_{|x|\to+\infty} |f(x)|$, belongs to *Mult*($R_b(D), \|\ .\ \|_D$) and obviously has continuation by continuity to an element ${}_D\varphi_D \in Mult(H_b(D), \|\ .\ \|_D)$, which satisfies ${}_D\varphi_D(f) = \lim_{|x|\to+\infty,\ x\in D} |f(x)|$. □

Definitions and notations: When there is no risk of confusion about the set D, we will just write $\varphi_{\mathcal{F}}$, (resp. $\varphi_{a,r}$, resp. φ_D, resp. φ_A), instead of ${}_D\varphi_{\mathcal{F}}$ (resp. ${}_D\varphi_{a,r}$, resp. ${}_D\varphi_D$, resp. ${}_D\varphi_A$). Next, when D is unbounded, ${}_D\varphi_D$ will also be denoted by ${}_D\varphi_\infty$.

Theorem B.4.8: *Let $\psi \in Mult(H_b(D), \| \, . \, \|_D) \setminus Mult(H(D), \mathcal{U}_D)$. If D is bounded, ψ is of the form φ_a with $a \in \overline{D} \setminus D$. If D is not bounded, ψ is either of the form φ_a, with $a \in \overline{D} \setminus D$ or of the form ${}_D\varphi_\infty$.*

Proof. First we suppose D bounded. By Corollary B.2.7 we have $H_b(D) = H(\overline{D})$. Hence, ψ is equal to some ${}_{\overline{D}}\varphi_{\mathcal{F}}$, with $\mathcal{F}$ a circular filter on $\overline{D}$. If $\mathcal{F}$ is large, it is a large circular filter secant with D and then ψ belongs to $Mult(H(D), \mathcal{U}_D)$. If $\mathcal{F}$ is not large, it is the filter of neighborhoods of a point $a \in \overline{D}$. But if $a \in D$, obviously φ_a belongs to $Mult(H(D), \mathcal{U}_D)$. Hence, $a \in \overline{D} \setminus D$.

Now we suppose D is unbounded. If D has no hole we just have $H_b(D) = R_b(D) = \mathbb{K}$, hence $Mult(H_b(D), \| \, . \, \|_D) = Mult(H(D), \mathcal{U}_D)$. Thus we may assume D to have a hole $T = d(a, r^-)$. Without loss of generality we may assume $a = 0$. Let $\gamma(x) = \dfrac{1}{x}$ and let $D' = \gamma(D)$. Then D' is bounded and, by Proposition B.3.9, we know that the algebra $H_b(D)$ is isomorphic to $H(D')$. By Proposition B.3.8, the mapping $f \to f \circ \gamma$ defines a $\mathbb{K}$-vector space isomorphism from $H(D)$ onto $H(D')$, and a $\mathbb{K}$-algebra isomorphism from $H_b(D)$ onto $H_b(D') = H(\overline{D'})$. Hence, we may define $\psi' \in Mult(H(\overline{D'}))$ by $\psi'(f \circ \gamma) = \psi(f)$ whenever $f \in H_b(D)$. If ψ' belonged to $Mult(H(D'), \mathcal{U}_{D'})$, then we would have $\psi(f) = \psi'(f \circ \gamma)$ whenever $f \in H(D)$ and, therefore, $\psi \in Mult(H(D), \mathcal{U}_D)$. Hence, ψ' does not belong to $Mult(H(D'), \mathcal{U}_{D'})$ and then ψ' is of the form φ_b with $b \in \overline{D'} \setminus D$. If $b \neq 0$ then $\psi = \varphi_{\frac{1}{b}}$. If $b = 0$, then $\psi = {}_D\varphi_D$ and this ends the proof. □

Theorem B.4.9: *Let $(\alpha_n)_{n\in\mathbb{N}}$ be a bounded sequence in D such that no subsequence converges to any point of $\overline{D} \setminus D$. There exists a subsequence $(\alpha_{n_s})_{s\in\mathbb{N}}$ such that the sequence $(\varphi_{\alpha_{n_s}})_{s\in\mathbb{N}}$ converges in $Mult(H(D), \mathcal{U}_D)$.*

Proof. By Theorem A.2.1 we may extract either a convergent subsequence, or a monotonous distances subsequence, or an equal distances subsequence from the sequence $(\alpha_n)_{n\in\mathbb{N}}$. Let $(\alpha_{n_s})_{s\in\mathbb{N}}$ be such a subsequence. If this subsequence converges to a point $\alpha \in \mathbb{K}$, then by hypothesis α lies in D, hence $\lim\limits_{s\to+\infty} \varphi_{\alpha_{n_s}} = \varphi_\alpha$. If this subsequence is a monotonous distances subsequence, or an equal distances subsequence, then by Proposition A.2.18, on D there exists a large circular filter $\mathcal{F}$ less thin than the sequence (α_{n_s}) and then we see that for every $f \in H(D)$ we have $\lim\limits_{s\to+\infty} |f(\alpha_{n_s})| = {}_D\varphi_{\mathcal{F}}(f)$, hence $\lim\limits_{s\to+\infty} \varphi_{\alpha_{n_s}}(f) = \varphi_{\mathcal{F}}(f)$. Thus, in every case we have proven that the subsequence $\varphi_{\alpha_{n_s}}$ converges in $Mult(H(D), \mathcal{U}_D)$. □

B.5. Power and Laurent series

A power series on a p-adic field admits a disk of convergence whose radius is defined in the same way as on $\mathbb{C}$. The difference of behavior between power series in $\mathbb{C}$ and in a field such as $\mathbb{K}$ concerns what happens when $|x|$ is equal to the radius of convergence. We show that the norm of uniform convergence in a disk $d(a,s) \subset d(0,R^-)$ is multiplicative and satisfies $\| \sum_{n=0}^{+\infty} a_n x^n \|_{d(0,s)} = \sup_{n\in\mathbb{N}} |a_n| s^n$. As a consequence, the product of two power series converging in $d(0,R^-)$ is bounded if and only if both are bounded. We show that the algebra of power series with a radius of convergence equal to R is equal to the intersection of algebras of analytic elements $H(d(0,s))$ when $s < R$. We show that all analytic elements in $d(0,R^-)$ are power series converging in $d(0,R^-)$. The converse is false. However, we will see that the analytic elements in $d(0,R)$ are exactly the power series converging in this disk.

Definitions and notations: Let $f(x) = \sum_{n=0}^{\infty} a_n x^n$ be a power series with coefficients in $\mathbb{K}$.

As usual, when $\limsup_{n\to\infty} \sqrt[n]{|a_n|} \neq 0$, we call *radius of convergence of* f the number $r = \dfrac{1}{\limsup_{n\to\infty} \sqrt[n]{|a_n|}}$ (with $r = 0$ when $\limsup_{n\to\infty} \sqrt[n]{|a_n|} = +\infty$).

When $\limsup_{n\to\infty} \sqrt[n]{|a_n|} = 0$, we define the radius of convergence of f as $+\infty$.

Examples: Let $f(x) = \sum_{n=1}^{\infty} nx^n$. The radius of convergence of this series is 1. This function obviously defines the rational function $\dfrac{x}{(1-x)^2}$ in $d(1,1^-)$.

Remark: If a sequence of positive numbers $(u_n)_{n\in\mathbb{N}}$ is such that the sequence $\left(\dfrac{u_{n+1}}{u_n}\right)_{n\in\mathbb{N}}$ converges to a limit $l \geq 0$, then so does the sequence $(\sqrt[n]{u_n})_{n\in\mathbb{N}}$. On the field $\mathbb{K}$, as in Archimedean analysis, it is a way to compute easily many radii of convergence.

Lemma B.5.1 is immediate.

Lemma B.5.1: *Let $f = \sum_{n=0}^{\infty} a_n x^n$ be a power series with coefficients in $\mathbb{K}$. The series converges if and only if $\lim_{n\to\infty} |a_n x^n| = 0$. Let r be its radius of convergence. If $|x| < r$, then the series converges. If $|x| > r$, then the series diverges.*

Definitions and notations: Power series whose radius of convergence is ∞ are called *entire functions on* $\mathbb{K}$ and the set of entire functions will be denoted by $\mathcal{A}(\mathbb{K})$.

For every $a \in \mathbb{K}$, $r \in \mathbb{R}_+^*$, similarly we will denote by $\mathcal{A}(d(a, r^-))$ the set of power series in $x - a$ whose radius of convergence is superior or equal to r and by $\mathcal{A}_b(d(a, r^-))$ the set of functions $f \in \mathcal{A}(d(a, r^-))$ that are bounded in $d(a, r^-)$. The set $\mathcal{A}(d(a, r^-)) \setminus \mathcal{A}_b(d(a, r^-))$ will be denoted by $\mathcal{A}_u(d(a, r^-))$.

Similarly, we will denote by $\mathcal{A}(\mathbb{K} \setminus d(a, r))$ the set of Laurent series converging whenever $|x-a| > r$, by $\mathcal{A}_b(\mathbb{K} \setminus d(a, r))$ the set of bounded Laurent series converging whenever $|x - a| > r$ and by $\mathcal{A}_u(\mathbb{K} \setminus d(a, r))$ the set of unbounded Laurent series converging whenever $|x - a| > r$.

Finally, given r', r'' such that $0 < r' < r''$, we will denote by $\mathcal{A}(\Gamma(a, r', r''))$ the set of Laurent series converging whenever $r' < |x - a| < r''$. And we will denote by $\mathcal{A}_b(\Gamma(a, r', r''))$ the set of functions $f \in \mathcal{A}(\Gamma(a, r', r''))$ that are bounded in $\Gamma(a, r', r'')$.

From Lemma B.5.1 we can derive Corollary B.5.2:

Corollary B.5.2: *Let* $\sum_{-\infty}^{+\infty} a_n x^n$ *be a Laurent series with coefficients in* $\mathbb{K}$, *let* $r'' = \dfrac{1}{\limsup\limits_{n\to\infty} \sqrt[n]{|a_n|}}$ *with* $r'' = 0$ *whenever* $\limsup\limits_{n\to\infty} \sqrt[n]{|a_n|} = +\infty$, *and let* $r' = \dfrac{1}{\limsup_{n\to-\infty} \sqrt[-n]{|a_n|}}$ *with* $r' = 0$ *whenever* $\limsup\limits_{n\to-\infty} \sqrt[-n]{|a_n|} = +\infty$. *If* $r' < |x| < r''$, *the series converges. If* $|x| > r''$ *or if* $|x| < r'$ *the series diverges.*

Corollary B.5.3: *Let* $r', r'' \in \mathbb{R}_+$ *satisfy* $0 < r' < r''$. *Then* $\mathcal{A}(\Gamma(0, r', r''))$ *is the set of Laurent series* $\sum_{-\infty}^{+\infty} a_n x^n$ *such that* $r' \leq \dfrac{1}{\limsup_{n\to-\infty} \sqrt[n]{|a_n|}} \leq r''$.

Corollary B.5.4: *Let* r', $r'' \in \mathbb{R}_+$ *be such that* $r' < r''$ *and let* $f(x) = \sum_{-\infty}^{+\infty} a_n x^n \in \mathcal{A}(\Gamma(a, r', r''))$. *For each* $r \in]r', r''[$, *one has*

$$\lim_{n\to+\infty} |a_n| r^n = \lim_{n\to-\infty} |a_n| r^n = 0.$$

Lemma B.5.5 will be useful in certain further problems:

Lemma B.5.5: *Let* $q \in \mathbb{N}^*$ *and let* $\mathbb{L}$ *be a complete algebraically closed extension of* $\mathbb{K}$. *Let* $f \in \mathcal{A}(d(a, R^-))$ *and suppose that there exists a power series* g *with coefficients in* $\mathbb{L}$, *with radius of convergence* $\geq R$ *such that* $(g(x))^q = f(x)$ $\forall x \in d(a, R^-)$. *Then* g *has all coefficients in* $\mathbb{K}$ *and belongs to* $\mathcal{A}(d(a, R^-))$.

Proof. Without loss of generality we can obviously suppose $a = 0$. Let $f(x) = \sum_{n=0}^{+\infty} b_n x^n$ (with $b_n \in \mathbb{K}$) and let $g(x) = \sum_{n=0}^{+\infty} a_n x^n$. Then $(a_0)^q = b_0$, hence $a_0 \in \mathbb{K}$ because $\mathbb{K}$ is algebraically closed. Now suppose we have proven that $a_n \in \mathbb{K}$ $\forall n \leq t - 1$. We can see that b_t is of the form $a_t (a_0)^{q-t} + h$ where h is a polynomial in

$a_0, a_1, \dots, a_{t-1}$. Therefore, a_t also belongs to $\mathbb{K}$. Consequently, g has all coefficients in $\mathbb{K}$, which ends the proof. □

Definitions and notations: Let $r \in \mathbb{R}_+^*$ and let $\sum_{-\infty}^{+\infty} a_n x^n \in H(C(0,r))$. By hypothesis, we have $\lim_{n\to+\infty} |a_n|r^n = \lim_{n\to-\infty} |a_n|r^n = 0$. Generalizing notation already introduced for rational functions, we denote by $\nu^+(f, \log r)$ the highest of the integers $m \in \mathbb{Z}$ such that $|a_m|r^m = \sup_{n\in\mathbb{Z}} |a_n|r^n$ and by $\nu^-(f, \log r)$ the lowest of the integers $m \in \mathbb{Z}$ such that $|a_m|r^m = \sup_{n\in\mathbb{Z}} |a_n|r^n$. Next, when $\nu^+(f, \log r) = \nu^-(f, \log r)$, we just set $\nu(f, \log r)$.

Recall that we have, given a circular filter $\mathcal{F}$ of center 0 and diameter r, for every element of $H(d(0,r))$ and particularly for every analytic function $f \in \mathcal{A}(\mathbb{K})$, we put $|f|(r) = \lim_{\mathcal{F}} |f(x)|$.

Theorem B.5.6: *Let $r \in \mathbb{R}_+^*$, let $\mathcal{F}$ be the circular filter of center 0 and diameter r on $\mathbb{K}$, and let $E = d(0,r)$. Then $H(E)$ is the set of power series $f(x) = \sum_{n=0}^{\infty} a_n x^n$ such that $\lim_{n\to\infty} |a_n|r^n = 0$ and we have $\|f\|_E = |f|(r) = \max_{n\in\mathbb{N}} |a_n|r^n = {}_E\varphi_{\mathcal{F}}(f) = \|f\|_{C(0,r)}$.*

For every $\alpha \in E$, $H(E)$ is also equal to the set of series $f(x) = \sum_{n=0}^{\infty} b_n (x-\alpha)^n$ such that $\lim_{n\to\infty} |b_n|r^n = 0$.

Let $B = \mathbb{K}\setminus d(0,r^-)$. Then $H(B)$ is the set of Laurent series $f(x) = \sum_{n=0}^{\infty} \frac{a_n}{x^n}$ such that $\lim_{n\to\infty} |a_n|r^{-n} = 0$ and we have $\|f\|_B = \max_{n\in\mathbb{N}} |a_n|r^{-n} = {}_B\varphi_{\mathcal{F}}(f) = \|f\|_{C(0,r)}$.

For every $\alpha \in d(0,r^-)$, $H(B)$ is also equal to the set of series $f(x) = \sum_{n=0}^{\infty} \frac{b_n}{(x-\alpha)^n}$ such that $\lim_{n\to\infty} |b_n|r^{-n} = 0$.

Let $r' \geq r$ and let $D = \Delta(0,r,r'))$. Then $H(D)$ is the set of Laurent series $f(x) = \sum_{-\infty}^{\infty} a_n x^n$ such that $\lim_{n\to-\infty} |a_n|r^n = 0$ and $\lim_{n\to+\infty} |a_n|(r')^n = 0$ and we have $\|f\|_D = \max(\max_{n<0} |a_n|r^n, \max_{n\geq 0} |a_n|(r')^n)$. Moreover, for every $\alpha \in d(0,r^-)$, $H(D)$ is also equal to the set of power series $f(x) = \sum_{-\infty}^{\infty} b_n (x-\alpha)^n$ such that $\lim_{n\to-\infty} |b_n|r^n = 0$ and $\lim_{n\to+\infty} |b_n|(r')^n = 0$.

Proof. Let $S(r)$ be the set of power series $f(x) = \sum_{n=0}^{\infty} a_n x^n$ such that $\lim_{n\to\infty} |a_n|r^n = 0$. Such a power series obviously is a uniform limit of polynomials because $|f(x) - \sum_{n=0^q} a_n x^n| \leq \sup_{n\geq q} |a_n|r^n$ and, hence, it belongs to $H(E)$. Moreover,

E is closed and bounded, hence by Theorem B.2.2 $H(E)$ is a $\mathbb{K}$-Banach algebra with respect to the norm of uniform convergence on E. By Lemma A.3.7, on $\mathbb{K}[x]$ the norm $\| \, . \, \|_E$ is $\varphi_{\mathcal{F}}$ and by Theorem B.4.1 that equality has continuation to $H(E)$.

Now, for a polynomial $P(x) = \sum_{n=0}^{q} a_n x^n$, by Lemma A.3.7 we have $\|P\|_E = \sup_{0 \le n \le q} |a_n| r^n$, hence this equality also has continuation to f. Consequently, $\|f\|_E = \lim_{\mathcal{F}} |f(x)| = \varphi_{\mathcal{F}}(f)$. Particularly, $S(r)$ is a subset of $H(E)$.

In order to show that $S(r) = H(E)$, we will first show that $S(r)$ is closed in $H(E)$. Since $\mathcal{F}$ is secant with $C(0,r)$ we have $\|f\|_{C(0,r)} = {}_E\varphi_{\mathcal{F}}(f)$. But we know that $\varphi_{\mathcal{F}}(P_n) \le \max_{0 \le i \le n} |a_i| r^i$. Since $\varphi_{\mathcal{F}}$ extends continuously to ${}_E\varphi_{\mathcal{F}} \in Mult\Big(H(E), \| \, . \, \|_E\Big)$, for n big enough, we have ${}_E\varphi_{\mathcal{F}}(f) = {}_E \varphi_{\mathcal{F}}(P_n) = |a_j| r^j$ with $j < n$ and $|a_n| r^m < |a_j| r^j$ whenever $m < j$, hence finally ${}_E\varphi_{\mathcal{F}}(f) = |a_j| r^j = \max_{n \in \mathbb{N}} |a_n| r^n$. Consequently, we have $\|f\|_{C(0,r)} = \|f\|_E \le |a_j| r^j = \max_{0 \le n} |a_n| r^n$ and, therefore, $\|f\|_{C(0,r)} = \|f\|_E = \max_{0 \le n} |a_n| r^n$. This finishes showing that $S(r)$ is a closed subset of $H(E)$.

Now we will show that $R(E)$ is included in $S(r)$. For this, we just have to show that, given any $\beta \in \mathbb{K} \setminus E$, $\Big(\frac{1}{x - \beta}\Big)^q = \Big(-\frac{1}{\beta} \sum_{n=0}^{\infty} \Big(\frac{x}{\beta}\Big)^n\Big)^q$ belongs to F. When developing $\Big(\sum_{n=0}^{\infty} \Big(\frac{x}{\beta}\Big)^n\Big)^q$, we see that for every fixed $q \in \mathbb{N}$, the coefficient A_q of x^q is a sum of terms of the form $\frac{s}{\beta^q}$, with $s \in \mathbb{N}$, hence finally $|A_q| \le \frac{1}{|\beta|^q}$ and, therefore, $|A_q| r^q \le \Big(\frac{r}{|\beta|}\Big)^q$. Since $|\beta| > r$, this shows that $\Big(\frac{1}{x - \beta}\Big)^q \in S(r)$. So, we have proven the inclusion $R(E) \subset S(r) \subset H(E)$. Since $S(r)$ is closed, we have $S(r) = H(E)$.

Now let $\alpha \in E$. Since $d(\alpha, r) = d(0, r)$, after the change of variable $x = \alpha + u$, the same reasoning shows that a series $f(x) = \sum_{n=0}^{\infty} a_n \, x^n \in H(E)$ is also of the form $\sum_{n=0}^{\infty} b_n(\alpha)(x - \alpha)^n$ with $\lim_{n \to \infty} |b_n(\alpha)| r^n = 0$. Conversely, $H(E)$ is clearly equal to the set of series $\sum_{n=0}^{\infty} b_n(x + \alpha)^n$ such that $\lim_{n \to \infty} |b_n| r^n = 0$ because any series g of the form $\sum_{n=0}^{\infty} b_n \, u^n$, with $\lim_{n \to \infty} |b_n| r^n = 0$, can be written as $\sum_{n=0}^{\infty} a_n (u + \alpha)^n$.

The statements about $H(B)$ are an obvious consequence of those about $H(E)$ after the change of variable $y = \frac{1}{x}$ and more generally, $y = \frac{1}{x - \alpha}$. So are the statements about $H(D)$. □

We can easily check the following corollaries:

Corollary B.5.7: *Let $f \in \mathcal{A}(\mathbb{K})$. The following three statements are equivalent:*

(i) $\lim\limits_{r\to+\infty} \dfrac{|f|(r)}{r^q} = +\infty \ \forall q \in \mathbb{N}$,

(ii) *there exists no $q \in \mathbb{N}$ such that* $\lim\limits_{r\to+\infty} \dfrac{|f|(r)}{r^q} = 0$,

(iii) *f is not a polynomial.*

Corollary B.5.8: *Let $f, g \in \mathcal{A}\mathbb{K})$. Then $f.g$ is a polynomial if and only if both f, g are polynomials.*

Corollary B.5.9: *Let $r \in \mathbb{R}_+^*$ and let $D = d(0,r)$. Then $H(D)$ is the set of power series $f(x) = \sum\limits_{n=0}^{\infty} a_n x^n$ such that $\lim\limits_{|n|_\infty\to\infty} |a_n| r^n = 0$ and we have*

$$\|f\|_D = \max_{n\in\mathbb{N}} |a_n| r^n = {}_D\varphi_{\mathcal{F}}(f).$$

Moreover, the norms $\| \, . \, \|_{C(0,r)}$, $| \, . \, |(r)$ and $\| \, . \, \|_{d(0,r)}$ are equal and are multiplicative.

Corollary B.5.10: *Let $r \in \mathbb{R}_+^*$ and let $D = d(0,r)$ (resp. $D = d(0,r^-)$). Then the norm $\| \, . \, \|_D$ on $H(D)$ is multiplicative.*

Corollary B.5.11: *Let $\alpha \in D$ and $r \in \mathbb{R}_+^*$ be such that $d(\alpha,r) \subset D$. Let $f \in H(D)$. In $d(\alpha,r)$, $f(x)$ is equal to a power series of the form $\sum\limits_{n=0}^{\infty} a_n(x-\alpha)^n$ such that $\lim\limits_{n\to\infty} |a_n| r^n = 0$. If $f(\alpha) = 0$ and if $f(x)$ is not identically zero in $d(\alpha,r)$, then there exists a unique integer $q \in \mathbb{N}^*$ such that $a_n = 0$ for every $n < q$ and $a_q \neq 0$ and α is an isolated zero of f in $d(\alpha,r)$.*

Proposition B.5.12: *Let $r \in \mathbb{R}_+^*$ and let $f(x) = \sum\limits_{n=0}^{\infty} a_n x^n$. The following statements are equivalent:*

(a) $f \in \mathcal{A}(d(0,r^-))$

(b) $f \in \bigcap\limits_{s<r} H(d(0,s))$

(c) *The series f is convergent in all of $d(0,r^-)$.*

Proof. (b) and (c) are clearly equivalent to the condition $\lim\limits_{n\to\infty} |a_n| s^n = 0$ whenever $s < r$ and, in the same way as in Archimedean analysis, it is shortly checked that this is also equivalent to $\limsup\limits_{n\to\infty} \sqrt[n]{|a_n|} \leq \dfrac{1}{r}$. □

Remark: If f is convergent for some $\alpha \in C(0,r)$, then $\lim\limits_{n\to\infty} |a_n| r^n = 0$, hence f belongs to $H(d(0,r))$.

Corollary B.5.13: *Let $r \in \mathbb{R}_+^*$ and let $f(x) = \sum_{-\infty}^{0} a_n x^n$. The following statements are equivalent:*

(a) $f \in \mathcal{A}(\mathbb{K} \setminus d(0,r))$

(b) $f \in \bigcap_{s>r} H_b(\mathbb{K} \setminus d(0,s))$

(c) *The series f is convergent in all of $\mathbb{K} \setminus d(0,r)$.*

Corollary B.5.14: *Let $r_1,\ r_2 \in \mathbb{R}_+^*$ with $r_1 < r_2$ and let $f(x) = \sum_{-\infty}^{+\infty} a_n x^n$. The following statements are equivalent:*

(a) $f \in \mathcal{A}(\Gamma(0, r_1, r_2))$

(b) $f \in \bigcap_{r_1<s_1<s_2<r_2} H(\Delta(0, s_1, s_2))$

(c) *The series f is convergent in all of $\Gamma(0, r_1, r_2)$.*

Corollary B.5.15: *Let $f \in \mathcal{A}(d(0, r^-))$ be not identically zero. For every $\alpha \in d(0, r^-)$, $f(x)$ is equal to a power series $\sum_{n=0}^{\infty} b_n(\alpha)(x - \alpha)^n$. If f is not identically zero and if α is a zero of f in $d(0, r^-)$, α is an isolated zero, and f factorizes in $\mathcal{A}(d(0, r^-))$ in the form $(x - \alpha)^q g(x)$, with $g \in \mathcal{A}(d(0, r^-))$, $q \in \mathbb{N}^*$ $g(\alpha) \neq 0$.*

Definitions and notations: Let $f \in H(D)$ and let $\alpha \in \overset{\circ}{D}$, let $r > 0$ be such that $d(\alpha, r) \subset D$ and suppose $f(x) = \sum_{n=q}^{\infty} b_n (x - \alpha)^n$ whenever $x \in d(\alpha, r)$, with $b_q(\alpha) \neq 0$ and $q > 0$. Then α is called *a zero of multiplicity order* q, or more simply, *a zero of order* q. In the same way, q will be named *the multiplicity order of* α.

Remark: In particular, these definitions apply to functions $f \in \mathcal{A}(d(a, r^-))$, at any point $\alpha \in d(a, r^-)$.

Corollary B.5.16: *Let $a \in \mathbb{K}$, $R \in \mathbb{R}_+^*$ and let $f \in \mathcal{A}(d(a, R^-))$ (resp. $f \in \mathcal{A}(\mathbb{K})$). Let $a_1, \ldots, a_q$ be zeros of f of respective order s_j and let $P(x) = \prod_{j=1}^{q} (x - aj)^{s_j}$. Then f factorizes in the form $P(x)u(x)$ with $u \in \mathcal{A}(d(a, R^-))$ (resp. $u \in \mathcal{A}(\mathbb{K})$.*

Corollary B.5.17: *Let $a \in \mathbb{K}$, R, $R' \in \mathbb{R}_+^*$ with $R < R'$ and let $\Lambda = d(a, R^-)$, (resp. $\Lambda = \mathbb{K} \setminus d(a, R)$, resp. $\Lambda = \Gamma(a, R, R')$)) and let $C(b, r)$ be a circle included in Λ. Then ${}_{\Lambda}\varphi_{b,r}$ applies to $\mathcal{A}(\Lambda)$.*

Corollary B.5.18: *Let $f(x) = \sum_{n=0}^{\infty} a_n x^n \in \mathcal{A}(\mathbb{K})$. If f is not a constant, then $\lim_{r \to +\infty} |f|(r) = +\infty$.*

Definitions and notations: Let $R \in \mathbb{R}_+^*$ and let $f \in \mathcal{A}(d(a, R^-))$. Given $r \in]0, R[$, by Proposition B.5.12 f belongs to $H(d(a,r))$, hence for every circular filter $\mathcal{F}$ secant with $d(a,r)$, $\varphi_{\mathcal{F}}(f)$ is defined. Particularly, if $a = 0$, $|f|(r)$ is defined.

Theorem B.5.19: *Let $R \in \mathbb{R}_+^*$ and let $f \in \mathcal{A}(d(a, R^-))$. Then f is invertible in $\mathcal{A}(d(a, R^-))$ if and only if f has no zero in $d(a, R^-)$.*

Proof. Suppose that f has no zero in $d(a, R^-)$. For each $r \in]0, R[$, f belongs to $H(d(a,r))$ and, hence, by Lemma B.2.3, it is invertible in $H(d(a,r))$. Consequently, the function defined in $d(a,r)$ as $g(x) = \dfrac{1}{f(x)}$ belongs to $H(d(a,r))$. This is true for all $r \in]0, R[$ and shows that f^{-1} belongs to $\mathcal{A}(d(a, R^-))$. The converse is obvious. □

Theorem B.5.20: *Let $R \in \mathbb{R}_+^*$. The $\mathbb{K}$-subalgebra $\mathcal{A}_b(d(0, R^-))$ of $\mathcal{A}(d(0, R^-))$ is a Banach $\mathbb{K}$-algebra with respect to the norm $\| \,.\, \|_{d(0,R^-)}$. Further, this norm is multiplicative and satisfies $\|f\|_{d(0,R^-)} = \lim_{r \to R} |f|(r) = \sup_{n \in \mathbb{N}} |a_n| R^n$.*

Let $f(x) = \sum_{n=0}^{\infty} a_n x^n \in \mathcal{A}(d(0, R^-))$. Then f is bounded in $d(0, R^-)$ if and only if so is the sequence $(|a_n| R^n)_{n \in \mathbb{N}}$. Moreover, if f is bounded, then $\|f\|_{d(0,R^-)} = \sup_{n \in \mathbb{N}} |a_n| R^n$.

Proof. Let $f(x) = \sum_{n=0}^{\infty} a_n x^n \in \mathcal{A}_b(d(0, R^-))$. By Theorem B.5.6 we have $\|f\|_{d(0,R^-)} = \sup_{n \in \mathbb{N}} |a_n| R^n$. The norm $\| \,.\, \|_{d(0,R^-)}$ is a norm of $\mathbb{K}$-algebra, hence $\|f\ g\|_{d(0,R^-)} \le \|f\|_{d(0,R^-)}\ \|g\|_{d(0,R^-)}$. On the other hand, by Theorem B.5.6, the norm $\| \,.\, \|_{d(0,s)}$ is multiplicative on $H(d(0,s))$ for every $s < R$, hence $\|fg\|_{d(0,R^-)} \ge \|fg\|_{d(0,s)} = \|f\|_{d(0,s)}\ \|g\|_{d(0,s)}$ whenever $s < R$, and, therefore, $\| \,.\, \|_{d(0,R^-)}$ is multiplicative on $\mathcal{A}_b(d(0, R^-))$. Now let $(f_m)_{m \in \mathbb{N}}$ be a Cauchy sequence in $\mathcal{A}_b(d(0, R^-))$. We put $f_m(x) = \sum_{n=0}^{\infty} a_{n,m} x^n$. By hypothesis, for every $\epsilon > 0$ we have an integer $N(\epsilon)$ such that $|a_{n,m} - a_{n,q}| R^n \le \epsilon$ for every $n \in \mathbb{N}$, whenever $m, q \ge N(\epsilon)$. Thus it is easily seen that each sequence $(a_{n,m})_{m \in \mathbb{N}}$ converges in $\mathbb{K}$ to a limit a_n that satisfies $|a_n - a_{n,m}| R^n \le \epsilon$ whenever $m \ge N(\epsilon)$ and then the series $f(x) = \sum_{n=0}^{\infty} a_n x^n$ satisfies $\|f - f_m\|_{d(0,R^-)} \le \epsilon$. Obviously f belongs to $H(d(0,s))$ for all $s < R$ and then the sequence (f_m) is proven to converge in $\mathcal{A}_b(d(0, R^-))$.

For every $s \in]0, R[$, we have

$$|f|(s) = \|f\|_{d(0,s)} = \sup_{n \in \mathbb{N}} |a_n| s^n \le \sup_{n \in \mathbb{N}} |a_n| R^n = \|f\|_{d(0,R^-)}$$

and hence we can check that the real increasing bounded function h defined in $]0, R[$ as $h(s) = \sup_{n\in\mathbb{N}} |a_n|s^n$ is obviously continuous at R. Consequently, $\|f\|_{d(0,R^-)} = \lim_{r\to R} |f|(r) = \sup_n |a_n|R^n$. Therefore, obviously, f is bounded in $d(0, R^-)$ if and only if so is the sequence $(|a_n|R^n)_{n\in\mathbb{N}}$. □

Corollary B.5.21: *Let $R \in \mathbb{R}_+^*$ and let $f,\ g \in \mathcal{A}(d(a, R^-))$. Then fg belongs to $\mathcal{A}_b(d(a, R^-))$ if and only if so do both $f,\ g$ and $\mathcal{A}_b(d(a, R^-))$ is $\mathbb{K}$-subalgebra of $\mathcal{A}(d(a, R^-))$.*

Theorem B.5.22: *Suppose that $\mathbb{K}$ has characteristic different from 2. Let $f,\ g \in \mathcal{A}(\mathbb{K}) \setminus \mathbb{K}$ (resp. $f,\ g \in \mathcal{A}_u(d(0, r^-))$) be distinct. Then $f^2 - g^2$ belongs to $\mathcal{A}(\mathbb{K}) \setminus \mathbb{K}$ (resp. $f^2 - g^2 \in \mathcal{A}_u(d(0, r^-))$).*

Proof. Indeed, $f^2 - g^2 = (f - g)(f + g)$. Suppose that $|f + g|(r)$ is bounded when r tends to $+\infty$ (resp. to R). Since the characteristic of $\mathbb{K}$ is different from 2, $|f - g|(r)$ is obviously unbounded when r tends to $+\infty$ (resp. to R). Consequently, since the norm $|\ .\ |(r)$ is multiplicative, $|f^2 - g^2|(r)$ cannot be bounded when r tends to $+\infty$ (resp. to R). Therefore, $f^2 - g^2$ belongs to $\mathcal{A}(\mathbb{K}) \setminus \mathbb{K}$ (resp. to $\mathcal{A}_u(d(0, r^-))$). Similarly, if $|f - g|(r)$ is bounded when r tends to $+\infty$ (resp. to R) we have symmetric proof. □

Theorem B.5.23: *For every $r \in \mathbb{R}_+^*$, $H(d(0, r^-))$ is included in $\mathcal{A}_b(d(0, r^-))$.*

Proof. Since $\mathcal{A}_b(d(0, r^-))$ is complete with respect to the norm $\|\ .\ \|_{d(0,r^-)}$, we just have to show that $R(d(0, r^-)) \subset \mathcal{A}_b(d(0, r^-))$, hence finally we just have to show that given $\alpha \in \mathbb{K} \setminus d(0, r^-)$ $\dfrac{1}{x - \alpha} \in \mathcal{A}_b(d(0, r^-))$. But we have
$\dfrac{1}{x-\alpha} = -\dfrac{1}{\alpha(1 - \frac{x}{\alpha})} = -\dfrac{1}{\alpha}\sum_{n=0}^{\infty}\left(\dfrac{x}{\alpha}\right)^n$ for all $x \in d(0, r^-)$ because $\left|\dfrac{x}{\alpha}\right| < 1$, hence $\dfrac{1}{x-\alpha} \in \mathcal{A}_b(d(0, r^-))$ and that finishes proving Theorem B.5.23. □

Remarks: We will see later that $H(d(0, r^-))$ is much smaller than $\mathcal{A}_b(d(0, r^-))$. In particular, we will see that $\sqrt[q]{1 + x}$ belongs to $\mathcal{A}_b(d(0, 1^-)$, but does not belong to $H(d(0, 1^-))$.

Let $\sum_0^{+\infty} a_n x^n$ be a power series whose radius of convergence is r. Suppose first that $r \in |\mathbb{K}|$. If there is at least one point $\alpha \in C(0, r)$ such that the series converges at α, then this implies that $\lim_{n\to+\infty} |a_n|r^n = 0$ and, hence, the series converges in all $C(0, r)$ and defines an element of $H(d(0, r))$. If r does not belong to $|\mathbb{K}|$, the power series converging in $d(0, r)$ are just the power series converging in $d(0, r^-)$. This is why we don't have to consider *analytic functions* inside a disk $d(a, r)$.

Theorem B.5.24: *Let* $f(x) = \sum_{n=0}^{\infty} a_n x^n \in \mathcal{A}(d(0, R^-))$ *and suppose that* $a_n \in \mathbb{Q}_p$ $\forall n \in \mathbb{N}$. *Then for every* $x \in d(0, R^-)$, *if* x *is algebraic over* $\mathbb{Q}_p$, *so is* $f(x)$.

Proof. Suppose x is algebraic, of degree q over $\mathbb{Q}_p$ and let $\mathbb{E} = \mathbb{Q}_p[x]$. For every $m \in \mathbb{N}$, $\sum_{n=0}^{m} a_n x^n$ belongs to $\mathbb{E}$. But since $\mathbb{E}$ is a finite extension of $\mathbb{Q}_p$, it is complete, hence $f(x)$ also belongs to $\mathbb{E}$. □

B.6. Krasner-Mittag-Leffler theorem

The wonderful Mittag–Leffler theorem for analytic elements is due to Marc Krasner who showed it on quasi-connected sets [68]. The same proof holds on infraconnected sets as it was shown by Philippe Robba [83]. The theorem shows that a Banach space $H_b(D)$ is a direct topological sum of elementary subspaces and is indispensable to have a clear image of the space $H(D)$. Further, it appears necessary when studying meromorphic functions as we will see later.

Throughout this chapter, D is supposed to be infraconnected. We remember that if D is unbounded, $H_0(D)$ denotes the set of the $f \in H(D)$ such that $\lim_{\substack{|x| \to +\infty \\ x \in D}} f(x) = 0$.

Theorem B.6.1: (M. Krasner) [68, 83] *Let D be closed and bounded (resp. unbounded) and let $f \in H_b(D)$. There exists a unique sequence of holes $(T_n)_{n \in \mathbb{N}^*}$ of D and a unique sequence $(f_n)_{n \in \mathbb{N}}$ in $H(D)$ such that $f_0 \in H(\widetilde{D})$ (resp. $f_0 \in \mathbb{K}$), $f_n \in H_0(\mathbb{K} \setminus T_n)$ $(n > 0)$, $\lim_{n \to \infty} f_n = 0$ further satisfying,*

(1) $\quad f = \sum_{n=0}^{\infty} f_n$ *and* $\|f\|_D = \sup_{n \in \mathbb{N}} \|f_n\|_D$.

For every hole $T_n = d(a_n, r_n^-)$, we have

(2) $\quad \|f_n\|_D = \|f_n\|_{\mathbb{K} \setminus T_n} = {}_D\varphi_{a_n, r_n}(f_n) \leq {}_D\varphi_{a_n, r_n}(f) \leq \|f\|_D$.

If D is bounded and if $\widetilde{D} = d(a, r)$ we have

(3) $\quad \|f_0\|_D = \|f_0\|_{\widetilde{D}} = {}_D\varphi_{a,r}(f_0) \leq {}_D\varphi_{a,r}(f) \leq \|f\|_D$.

If D is not bounded then $|f_0| = \lim_{\substack{|x| \to \infty \\ x \in D}} |f(x)| \leq \|f\|_D$.

Let $D' = \widetilde{D} \setminus \left(\bigcup_{n=1}^{\infty} T_n \right)$. Then f belongs to $H(D')$ (resp. $H_b(D')$) and its decomposition in $H(D')$ is given again by (1) and f satisfies $\|f\|_{D'} = \|f\|_D$.

Proof. Since $f \in H_b(D)$, by Corollary B.2.7 we know that $f \in H(\overline{D})$. Hence, without loss of generality we may assume that D is closed. Obviously we may also assume $0 \in \widetilde{D}$.

First we suppose $f \in R(D)$. Then f has decomposition in the form $E(x) + \sum_{j=1}^{t} \frac{\lambda_j}{(x-\alpha_j)^{q_j}}$ with $E(x) \in \mathbb{K}[x]$ and $\alpha_j \in \mathbb{K} \setminus D$. Now for each j, either α_j belongs to a hole T or α_j belongs to $\mathbb{K} \setminus \widetilde{D}$. Let $T_1, \dots T_s$ be the holes that contain some α_j. Then $\sum_{j=1}^{t} \frac{\lambda_j}{(x-\alpha_j)^{q_j}}$ is of the form $\sum_{n=1}^{s} f_n + h_0$ with $f_i \in H_0(\mathbb{K} \setminus T_i)$ and $h_0 \in H_b(\widetilde{D})$. Finally we put $f_0 = E(x) + h_0$ and we have the announced decomposition: $f = \sum_{n=0}^{s} f_i$ with $f_i \in H_0(\mathbb{K} \setminus T_i)$ and $f_0 \in H_b(\widetilde{D})$. In the case when D is unbounded, f_0 is just a constant.

For each $i = 1, \dots, s$, $f - f_i$ clearly belongs to $H_b(D \cup T_i)$ and obviously f belongs to $H_b\Big(\widetilde{D} \setminus \Big(\bigcup_{i=1}^{s} T_i\Big)\Big)$.

First we will show that for any $n \in \mathbb{N}^*$, we have $\|f_n\|_D = \|f_n\|_{\mathbb{K}\setminus T_n}$. Let $\mathcal{F}_n$ be the circular filter on $\mathbb{K}$ of center α_n and diameter r_n. By Theorem B.5.6 we have

(4) $\quad \|f_n\|_{\mathbb{K}\setminus T_n} = \lim_{\mathcal{F}_n \cap (\mathbb{K}\setminus T_n)} |f_n(x)|.$

But by Proposition A.2.17, $\mathcal{F}_n$, is secant with D, hence

(5) $\quad \lim_{\mathcal{F}_n \cap (\mathbb{K}\setminus T_n)} |f_n(x)| = \lim_{\mathcal{F}_n \cap D} |f_n(x)|$

and obviously

(6) $\quad \lim_{\mathcal{F}_n \cap D} |f_n(x)| \leq \|f_n\|_D \leq \|f_n\|_{\mathbb{K}\setminus T_n}.$

Finally by (4), (5), and (6) we obtain

(7) $\quad \|f_n\|_{\mathbb{K}\setminus T_n} = \|f_n\|_D = {}_D\varphi_{\mathcal{F}_n}(f_n).$

In the same way, when D is bounded, say $\widetilde{D} = d(0, r)$, we consider the circular filter $\mathcal{F}_0$ of center 0 and diameter r, in order to prove that

(8) $\quad \|f_0\|_D = \|f_0\|_{\widetilde{D}} = {}_D\varphi_{\mathcal{F}_0}(f).$

Now let us show that $\|f\|_D \geq \|f_n\|_D$ for any $n \in \mathbb{N}^*$. Since $f \in R(D)$, there exists an annulus $\Gamma(a_n, r_n, r'_n)$ such that f has neither any zero nor any pole inside $\Gamma(a_n, r_n, r'_n)$. We put $I =]\log(r_n), \log(r'_n)[$. By hypothesis f_n has no pole in $\mathbb{K} \setminus d(0, r_n)$. Hence, since $\lim_{|x|\to\infty} f_n(x) = 0$, by Corollary 3.17. we see that $\frac{d\Psi_{a_n}}{d\mu}(f_n, \mu) < 0$ whenever $\mu \in I$. Let $g_n = f - f_n \in R(D \cup T_n)$. Since g_n has no pole inside T_n, by Corollary A.3.17 we see that $\frac{d_r\Psi_{a_n}}{d\mu}(g_n, \mu) \geq 0$ whenever $\mu < \log(r_n)$.

Therefore, the equation $\Psi_{a_n}(f_n, \mu) = \Psi_{a_n}(g_n, \mu)$ has at most one solution in I and then $\Psi_{a_n}(f, \mu)$ is equal to $\max(\Psi_{a_n}(f_n, \mu), \Psi_{a_n}(g_n, \mu))$ whenever $\mu \in I$, hence $\Psi_{a_n}(f,\mu) \geq \Psi_{a_n}(f_n,\mu)$ whenever $\mu \in I$. It follows that the multiplicative semi-norm $\varphi_{\mathcal{F}_n}$ defined on $R(D)$ satisfies $\log(\varphi_{\mathcal{F}_n}(f_n)) = \Psi_{a_n}(f_n, \log(r_n)) \leq \Psi_{a_n}(f, \log(r_n)) = \log(\varphi_{\mathcal{F}_n})(f)$, hence

(9) $\quad \varphi_{\mathcal{F}_n}(f_n) \leq \varphi_{\mathcal{F}_n}(f)$.

But $\varphi_{\mathcal{F}_n}(f_n) = \|f_n\|_D$ and $\varphi_{\mathcal{F}_n}(f) \leq \|f\|_D$, hence by (9) we have $\|f\|_D \geq \|f_n\|_D$. Finally by (7), we see that (2) is clearly proven.

When D is bounded we put $\widetilde{D} = d(0,r)$ and we prove (3) in the same way as above when proving (2) by considering an annulus $\Gamma(0,r,r')$ such that f has neither any zero nor any pole inside $\Gamma(0,r,r')$. Then the element $g_0 = f - f_0$ is of the form $\frac{P}{Q}$ with $P, Q \in \mathbb{K}[x]$, all the zeros of Q in $\widetilde{D}$ and $\deg(P) < \deg(Q)$ because $g_0 \in R_0(\mathbb{K} \setminus \widetilde{D})$. Hence, we have $\frac{d\,\Psi}{d\mu}(g_0,\mu) < 0$ whereas $\frac{d\Psi}{d\mu}(f_0,\mu) \geq 0$ whenever $\mu \in]\log r, \log r'[$, so we have ${}_D\varphi_{\mathcal{F}_0}(f_0) \leq {}_D\varphi_{\mathcal{F}_0}(f)$, and, hence, by (8) we obtain (3).

When D is not bounded the inequality $|f_0| \leq \|f\|_D$ is obvious because $\lim_{\substack{|x|\to\infty \\ x\in D}} f(x) - f_0 = 0$. This finishes proving the Mittag–Leffler theorem when $f \in R(D)$.

Now let $f \in H(D)$ and let $(h_m)_{m\in\mathbb{N}}$ be a sequence in $R(D)$ that converges to f in $H(D)$. The set of holes of D that contains at least one pole of some h_m is clearly countable. Hence, there exists a sequence of holes $(T_n)_{n\in\mathbb{N}^*}$ such that, denoting by D', the set $\widetilde{D} \setminus \Big(\bigcup_{n\in\mathbb{N}^*}^{\infty} T_n\Big)$, then h_m belongs to $H(D')$ whenever $m \in \mathbb{N}$. For each $m \in \mathbb{N}$, h_m splits in $H(D')$ in the form $h_m = \sum_{n=0}^{\infty} h_{m,n}$ with $h_{m,0} \in H(\widetilde{D}), h_{m,n} \in H_0(\mathbb{K} \setminus T_n)$. In particular, for each fixed $n \in \mathbb{N}$, we have $\|h_{m,n} - h_{q,n}\|_D \leq \|h_m - h_q\|_D$. Thus we see that the sequence $(h_{m,n})_{m\in\mathbb{N}}$ converges in $H(\mathbb{K} \setminus T_n)$ for $n > 0$ (resp. in $H(\widetilde{D})$ for $n = 0$) to a limit f_n and then we have $f = \sum_{n=0}^{\infty} f_n$ in $H(D')$. Obviously $\|f\|_D = \sup_{n\in\mathbb{N}} \|f_n\|_D$, whereas $\|f_n\|_D = \|f_n\|_{D'}$ whenever $n \in \mathbb{N}$ and so, $\|f\|_D = \|f\|_{D'}$. This ends the proof of Theorem B.6.1. □

Corollary B.6.2: *Let $(T_i)_{i\in I}$ be the family of holes of D. Let J be a subset of I and let $S = I \setminus J$. Let $E = D\bigcup(\bigcup_{i\in J} T_i)$ and let $F = D\bigcup(\bigcup_{i\in S} T_i)$. Then we have $H(D) = H_0(E) \oplus H(F)$ and for each $g \in H_0(E)$, $h \in H(F)$, we have $\|g+h\|_D = \max(\|g\|_E, \|h\|_F)$.*

The Mittag–Leffler theorem suggests some new definitions.

Definitions and notations: Let $f \in H_b(D)$. We consider the series $\sum_{n=0}^{\infty} f_n$ obtained in Theorem B.6.1, whose sum is equal to f in $H(D)$, with $f_0 \in H(\widetilde{D})$,

$f_n \in H(\mathbb{K} \setminus T_n) \setminus \{0\}$ and with the T_n holes of D. Each T_n will be called a *f-hole* and f_n will be called the *Mittag–Leffler term of f associated to T_n*, whereas f_0 will be called the *principal term* of f. For each f-hole T of D, the Mittag–Leffler term of f associated to T will be denoted by $\overline{\overline{f_T}}$, whereas the principal term of f will be denoted by $\overline{\overline{f_0}}$. The series $\sum_{n=0}^{\infty} f_n$ will be called *the Mittag–Leffler series of f on the infraconnected set D*. More generally, let E be an infraconnected set and let $f \in H(E)$. According to Theorem B.2.5, f is of the form $g+h$ with $g \in R(\mathbb{K}\setminus(\overline{E}\setminus E))$ and $h \in H_b(\overline{E})$ whereas such a decomposition is unique, up to an additive constant. For every hole T of $\overline{E}$, we will denote by $\overline{\overline{f_T}}$ the Mittag–Leffler term of h associated to T and $\overline{\overline{f_T}}$ will still be named *the Mittag–Leffler term of f associated to T*.

Corollary B.6.3: *Let $f \in H_b(D)$, let $(T_n)_{n\in\mathbb{N}^*}$ be the sequence of the f-holes, with $T_n = d(a_n, \rho_n^-)$, let $f_0 = \overline{\overline{f_0}}$ and $f_n = \overline{\overline{f_{T_n}}}$ for every $n \in \mathbb{N}^*$. Let $\widetilde{D} = d(a,s)$, (resp. $\widetilde{D} = \mathbb{K}$). There exists $q \in \mathbb{N}$ such that $\|f\|_D = \|f_q\|_D$. If $q \geq 1$ then $\|f\|_D = {}_D\varphi_{a_q,r_q}(f) = {}_D\varphi_{a_q,r_q}(f_q)$. If $q = 0$ and if D is bounded (resp. is not bounded) then $\|f\|_D = {}_D\varphi_{a,s}(f) = {}_D\varphi_{a,s}(f_0)$ (resp. $\|f\|_D = |f_0|$). Further, given a hole T of D, if f belongs to $H_b(D)$ and if g belongs to $H_0(\mathbb{K}\setminus T)$ and satisfies $f - g \in H(D \cup T)$, then $\overline{\overline{f_T}}$ is equal to g.*

Definitions and notations: Let D be bounded, of center α, and diameter r. A circular filter $\mathcal{F}$ on D will be called *the specific filter of a hole $T = d(a, r^-)$* if it is the circular filter on D of diameter r. If D is bounded and $\widetilde{D} = d(a, R)$, the circular filter of center a and diameter R will be called *the specific filter of $\widetilde{D}$*. In general, a specific filter of a hole of D or of $\widetilde{D}$ will be called *a specific filter of D*.

Corollary B.6.4: *Let $f \in H_b(D)$. There exists a large circular filter $\mathcal{F}$ with center $\alpha \in \widetilde{D}$ secant with D such that ${}_D\varphi_{\mathcal{F}}(f) = \|f\|_D$. If D is bounded, there exists a specific filter $\mathcal{F}$ of D such that ${}_D\varphi_{\mathcal{F}}(f) = \|f\|_D$.*

Corollary B.6.5: *Let $f \in H(d(0,1^-))$ and let $(d(\alpha_m, 1^-))_{m\in\mathbb{N}^*}$ be the family of the f-holes. Then f is of the form*

$$(1)\ \sum_{n=0}^{\infty} a_{n,0}x^n + \sum_{m,n\in\mathbb{N}^*} \frac{a_{n,m}}{(x-\alpha_m)^n}$$

with $\lim_{n\to\infty} a_n = 0$, $\lim_{n\to\infty} |a_{n,m}| = 0$ whenever $m \in \mathbb{N}^*$ and $\lim_{m\to\infty} \big(\sup_{n\in\mathbb{N}^*} |a_{n,m}|\big) = 0$. On the other hand, f satisfies $\|f\|_{d(0,1^-)} = \sup_{m,n\in\mathbb{N}^*} |a_{n,m}|$.

Conversely, every function of the form (1), with the α_m satisfying $|\alpha_m| = |\alpha_j - \alpha_m| = 1$ whenever $m \neq j$, belongs to $H(d(0,1^-))$. The norm $\|\ .\ \|_{d(0,1^-)}$ is multiplicative and equal to ${}_{d(0,1^-)}\varphi_{0,1}$.

Corollary B.6.6: *Let $r_1, r_2 \in \mathbb{R}_+$ satisfy $0 < r_1 < r_2$. Then $H(\Delta(0, r_1, r_2))$ is equal to the set of the Laurent series $\sum_{-\infty}^{+\infty} a_n x^n$ with $\lim_{n\to-\infty} |a_n| r_1^n = \lim_{n\to\infty} |a_n| r_2^n = 0$ and we have $\left\| \sum_{-\infty}^{+\infty} a_n x^n \right\|_{\Delta(0,r_1,r_2)} = \max\Big(\sup_{n\geq 0} |a_n| r_1^n, \ \sup_{n<0} |a_n| r_2^n\Big)$.*

Proof. In Theorem B.5.6 we saw that $H(\Delta(0, r_1, r_2))$ is equal to the set of the Laurent series $\sum_{-\infty}^{+\infty} a_n x^n$ with $\lim_{n\to-\infty} |a_n| r_1^n = \lim_{n\to\infty} |a_n| r_2^n = 0$. Then the conclusions on the norm come from Theorem B.6.1. □

Theorem B.6.7: *Let $r \in \mathbb{R}_+$. Then $H(C(0, r))$ is equal to the set of the Laurent series $\sum_{-\infty}^{+\infty} a_n x^n$ with $\lim_{|n|_\infty \to\infty} |a_n| r^n = 0$ and we have $\| \sum a_n x^n \|_{C(0,r)} = \sup_{n\in\mathbb{Z}} |a_n| r^n$. Next, the norm $\| \ . \ \|_{C(0,r)}$ is multiplicative and equal to ${}_{C(0,r)}\varphi_{0,r}$.*

Proof. We put $\Lambda = C(0, r)$. We may apply Theorem B.5.6 by taking $r_1 = r_2 = r$ and we obtain all the conclusions but the fact that $\| \ . \ \|_\Lambda$ is multiplicative. Let us show this. Let $h \in R(\Lambda)$. Hence, h is of the form $\frac{P}{Q}$, with $P, Q \in \mathbb{K}[x]$ and $Q(x)$ has no zero in Λ. Let Θ be a class of Λ. By Lemma A.3.9 we have $|Q(x)| = \varphi_{0,r}(Q)$ whenever $x \in \Theta$, and $|P(x)| \leq \varphi_{0,r}(P)$ whenever $x \in \Theta$. Hence, we see that $\left\| \frac{P}{Q} \right\|_\Lambda \leq \varphi_{0,r}\left(\frac{P}{Q}\right)$ and, therefore, $\|h\|_\Lambda = \varphi_{0,r}(h)$ whenever $h \in R(\Lambda)$. Consequently, we have $\|f\|_\Lambda = {}_\Lambda\varphi_{0,r}(f)$ whenever $f \in H(\Lambda)$. □

Proposition B.6.8: *Let $r_1, r_2 \in \mathbb{R}_+^*$, with $r_1 < r_2$.*

(i) *$\mathcal{A}(\Gamma(0, r_1, r_2)) = \mathcal{A}(d(0, r_2^-)) \oplus \mathcal{A}_0(\mathbb{K} \setminus d(0, r_1))$ and $\mathcal{A}_b(\Gamma(0, r_1, r_2)) = \mathcal{A}_b(d(0, r_2^-)) \oplus \mathcal{A}_{0,b}(\mathbb{K} \setminus d(0, r_1))$*

(ii) *Let $f(x) = \sum_{-\infty}^{+\infty} a_n x^n \in \mathcal{A}(\Gamma(0, r_1, r_2))$. Then $f \in \mathcal{A}_b(\Gamma(0, r_1, r_2))$ if and only if $\max\Big(\sup_{n\geq 0} |a_n| r_2^n, \sup_{n<0} |a_n| r_1^n\Big) < +\infty$. Moreover, if $f \in \mathcal{A}_b(\Gamma(0, r_1, r_2))$ then*

$$\|f\|_{\Gamma(0,r_1,r_2)} = \max\Big(\sup_{n\geq 0} |a_n| r_2^n, \sup_{n<0} |a_n| r_1^n\Big)$$

(iii) *$\mathcal{A}_b(\Gamma(0, r_1, r_2))$ is a Banach $\mathbb{K}$-algebra that contains $H(\Gamma(0, r_1, r_2))$.*

Proof. (i) is obvious. We will show (ii). Let $f \in \mathcal{A}_b(\Gamma(0, r_1, r_2))$ and let $f = f_1 + f_2$ with $f_2 \in \mathcal{A}_b(d(0, r_2^-))$ and $f_1 \in \mathcal{A}_{0,b}(\mathbb{K} \setminus d(0, r_1))$. We put $\Lambda = \Gamma(0, r_1, r_2)$. It is obviously seen that $\|f\|_\Lambda \leq \max\Big(\|f_1\|_\Lambda, \|f_2\|_\Lambda\Big) \leq \max\Big(\sup_{n\geq 0} |a_n| r_2^n, \sup_{n<0} |a_n| r_1^n\Big)$.

Now for every s_1, s_2 such that $r_1 < s_1 < s_2 < r_2$ we know that f belongs to $H(\Delta(0, s_1, s_2))$ because so do both f_1, f_2. Then by Theorem B.6.1 we have

$$\|f\|_{\Delta(0,s_1,s_2)} = \max\Big(\|f_2\|_{d(0,s_2)}, \|f_1\|_{\mathbb{K}\setminus d(0,s_1^-)}\Big).$$

Finally $\|f_2\|_{d(0,s_2)} = \sup\limits_{n\in\mathbb{N}} |a_n|s_2^n$ while $\|f_1\|_{\mathbb{K}\setminus d(0,s_1^-)} = \sup\limits_{n<0} |a_n|s_1^n$. Thus we see that $\|f\|_{\Gamma(0,r_1,r_2)} \geq \|f\|_{\Delta(0,s_1,s_2)} = \max\Big(\sup\limits_{n\geq 0} |a_n|s_2^n, \sup\limits_{n<0} |a_n|s_1^n\Big)$. This is true for every $s_1, s_2 \in]r_1, r_2[$, hence finally $\|f\|_\Lambda = \max\Big(\sup\limits_{n\geq 0} |a_n|r_2^n, \sup\limits_{n<0} |a_n|r_1^n\Big)$. All statements in (ii) are then proven.

We will now prove (iii). By (ii) $\mathcal{A}_b(\Lambda)$ is just the Banach $\mathbb{K}$-algebra

$$\mathcal{A}_b(d(0, r_2^-)) \oplus \mathcal{A}_{0,b}(\mathbb{K} \setminus d(0, r_1))$$

provided with the norm $\|f_1 + f_2\|_\Lambda = \max\Big(\|f_2\|_{d(0,r_2^-)}, \|f_1\|_{\mathbb{K}\setminus d(0,r_1)}\Big)$. We saw that $R(d(0, r_2^-)) \subset \mathcal{A}_b(d(0, r_2^-))$, hence $R(d(0, r_2^-)) \subset \mathcal{A}_b(\Lambda)$.

In the same way we have $R(\mathbb{K} \setminus d(0, r_1)) \subset \mathcal{A}_b(\mathbb{K} \setminus d(0, r_1))$ and then $\mathcal{A}_b(\mathbb{K} \setminus d(0, r_1))$ is obviously included in $\mathcal{A}_b(\Lambda)$. Since $R(\Lambda) = R(d(0, r_2^-)) + R(\mathbb{K} \setminus d(0, r_1))$, $R(\Lambda)$ is included in $\mathcal{A}_b(\Lambda)$, which is complete for the norm $\| \cdot \|_\Lambda$, hence $H(\Lambda) \subset \mathcal{A}_b(\Lambda)$. This finishes proving Proposition B.6.8. □

Definitions and notations: Given a subset A of $\widehat{\mathbb{K}}$, we will denote by $\widehat{H}(\widehat{A})$ the set of the analytic elements in A, taking $\widehat{\mathbb{K}}$ as a ground field.

Now we will apply the Mittag–Leffler theorem to the analytic extension of analytic elements.

Theorem B.6.9: *For all $f \in H(D)$, f has continuation to a unique element $\widehat{f} \in \widehat{H}(\widehat{D})$. Further, if $f \in H_b(D)$ the Mittag–Leffler series of $\widehat{f}$ in $\widehat{D}$ is the same as this of f in D.*

Proof. By Theorem B.2.5 we may easily assume that f belongs to $H_b(\overline{D})$. The Mittag–Leffler series of f on D obviously converges on $\widehat{D}$, to an element of $\widehat{H}(\widehat{D})$. This is unique because so is the Mittag–Leffler series of f on D. □

Theorem B.6.10: *Let E be an infraconnected set such that $D \cap E$ is infraconnected and such that every hole of $D \cap E$ is either a hole of D or a hole of E. Let $F \in H(D)$, $G \in H(E)$, satisfying $F(x) = G(x)$ whenever $x \in D \cap E$. Then there exists $h \in H(D \cup E)$ such that $h(x) = F(x)$ whenever $x \in D$, $h(x) = G(x)$ whenever $x \in E$, such that for every h-hole V of $D \cup E$, $\overline{\overline{h_V}}$ is either of the form $\overline{\overline{F_S}}$, when V is a F-hole S of D, or of the form $\overline{\overline{G_T}}$ when V is a G-hole T of E.*

Proof. By Theorem B.2.5 it is easily seen that we may assume $F \in H_b(D)$, $G \in H_b(E)$ without loss of generality. Let $A = D \cup E$, $B = D \cap E$. Let h be the restriction of F and G to B. Let $(V_n)_{n\in\mathbb{N}^*}$ be the sequence of h-holes that are holes of D and let $(W_n)_{n\in\mathbb{N}^*}$ be the sequence of h-holes that are holes of E, but not of D. For each $q \in \mathbb{N}^*$, as $h(x)$ is equal to $F(x)$ in B, $\overline{\overline{h_{W_q}}}$ is an element of $H_0(D)$ of the form $\sum_{m=1}^{\infty} \overline{\overline{F_{S_m^q}}}$ with S_m^q some F-holes of D included in W_q. We put $f_{q,m} = \overline{\overline{F_{S_m^q}}}$ for every $(q,m) \in (\mathbb{N}^{*2})$. In the same way, for each $q \in \mathbb{N}^*$, $\overline{\overline{h_{V_q}}}$ is an element of $H_0(E)$ of the form $\sum_{m=1}^{\infty} \overline{\overline{G_{T_m^q}}}$ with T_m^q the G-holes of E included in V_q. We put $g_{q,m} = \overline{\overline{G_{T_m^q}}}$ for every $(q,m) \in (\mathbb{N}^{*2})$. Without loss of generality we may obviously assume $\widetilde{D} \subset \widetilde{E}$, we put $h_0(x) = \overline{\overline{G_0(x)}}$. We notice that A is clearly included in the set $A' = \widetilde{E} \setminus \Big(\big(\bigcup_{(q,m)\in(\mathbb{N}^{*2})} S_m^q \big) \bigcup \big(\bigcup_{(q,m)\in(\mathbb{N}^{*2})} T_m^q \big) \Big)$. Then, it is easily seen that the series $h_0(x) + \sum_{(m,q)\in\mathbb{N}^{*2}} f_{q,m} + \sum_{(m,q)\in\mathbb{N}^{*2}} g_{q,m}$ converges in $H(A')$ because by Corollary B.6.2 we have $\|f_{q,m}\|_{A'} = \|\overline{\overline{F_{S_m^q}}}\|_D$ and $\|g_{q,m}\|_{A'} = \|\overline{\overline{G_{T_m^q}}}\|_E$, whereas $\lim_{q+m\to+\infty} \|\overline{\overline{F_{S_m^q}}}\|_E = \lim_{q+m\to+\infty} \|\overline{\overline{G_{T_m^q}}}\|_E = 0$. Further, by construction, $h(x)$ is equal to $F(x)$ and $G(x)$ in B and is such that for every h-hole V of $D \cup E$, $\overline{\overline{h_V}}$ is either of the form $\overline{\overline{F_S}}$, when V is a F-hole S of D, or of the form $\overline{\overline{G_T}}$ when V is a G-hole T of E. This clearly ends the proof of Theorem B.6.10. □

Corollary B.6.11: *Let E be an infraconnected set such that $D \subset E$ and such that each hole of D contains a unique hole of E. Let $f \in H(E)$ and let $f = \overline{\overline{f_0}} + \sum_{n=1}^{+\infty} \overline{\overline{f_{T_n}}}$ be the Mittag-Leffler series of f on the infraconnected E. For every $n \in \mathbb{N}^*$, let V_n be the hole of D containing T_n. Then the Mittag-Leffler series of f on the infraconnected D is of the form $\overline{\overline{f_0}} + \sum_{n=1}^{+\infty} \overline{\overline{f_{V_n}}}$ with $\overline{\overline{f_{V_n}}} = \overline{\overline{f_{T_n}}}$ $\forall n \in \mathbb{N}^*$.*

In the particular case of affinoid subsets, we can be more accurate for Theorem B.6.12.

Theorem: B.6.12 *Let D_1, D_2 be infraconnected affinoid subsets of $\mathbb{K}$ such that $D_1 \cap D_2 \neq \emptyset$ and let $f_j \in H(D_j)$, $j = 1,2$ be such that $f_1(x) = f_2(x)$ $\forall x \in D_1 \cap D_2$. Then the function f defined in $D_1 \cup D_2$ as $f(x) = f_j(x)$ $\forall x \in D_j$, $j = 1,2$, belongs to $H(D_1 \cup D_2)$.*

Proof. Let $D = D_1 \cup D_2$. Without loss of generality, we can assume that $\widetilde{D}_1$ contains $\widetilde{D}_2$ and, hence, $\widetilde{D} = \widetilde{D}_1$. We can also assume that $0 \in D_1 \cap D_2$. Set $A = D_1 \setminus D_2$. Then A is included in a finite union of holes of D_1. Consider such a hole $T = d(a, r^-)$ of D_1 (with $r \in |\mathbb{K}|$) such that $T \cap D_2 \neq \emptyset$. Since $D_1 \cap D_2 \neq \emptyset$, both D_1, D_2 have points on $C(a,r)$. Moreover, since both are affinoid, D_1 contains all classes of $C(a,r)$ except maybe finitely many because T is a hole of D_1. On the other hand, D_2 also contains all classes of $C(a,r)$ except maybe finitely many because it has points on $C(a,r)$ and inside $d(a,r^-)$. Consequently, $f_1(x)$ and $f_2(x)$ coincide in all classes of $C(a,r)$ except maybe in finitely many: $\Lambda_1,\dots,\Lambda_q$.

Let g be the Mittag–Leffler term of f_1 relative to T. Let S_k, $1 \le k \le t$ be the holes of D_2 included in T and for each $k = 1,\dots,t$, let h_k be the Mittag–Leffler term of f_2 relative to the hole S_k. Consider now the restrictions $\underline{f_1}$ of f_1 and $\underline{f_2}$ of f_2 on the set $D' = D_1 \cap D_2$.

The two functions are equal in D' and of course have the same Mittag–Leffler term relative to the hole $d(a,r^-)$. Concerning h_2, this term is $\sum_{k=1}^t h_k$. Consequently, $g = \sum_{k=1}^t h_k$. Since g and the h_k are Laurent series converging in $\mathbb{K}\setminus d(a,r^-)$, g and $\sum_{k=1}^t h_k$ coincide in all this set. Consequently, in the Mittag–Leffler series of f_1, we can replace g by $\sum_{k=1}^t h_k$. Thus f_1 becomes an element of $D_1 \cup \big((d(a,r^-)\cap D_2\big)$. We can do the same with each hole of D_1 containing points of D_2 and, hence, after finitely many similar changes, we obtain an element f of $H(D)$ such that $f(x) = f_j(x)\ \forall x \in D_j,\ j = 1,2$. □

Definitions and notations: Let E be a $\mathbb{K}$-Banach space. We will denote by E^* the $\mathbb{K}$-Banach space of continuous linear forms of E provided with its usual norm. The dual of a Banach space $H(D)$ was thoroughly studied by Yvette Amice [2].

Theorem B.6.13: (Y. Amice) *Let $r \in \mathbb{R}_+$. Given $h(t) = \sum_{n=0}^{\infty} \frac{b_n}{t^n} \in \mathcal{A}_b(\mathbb{K}\setminus d(0,r))$ there exists a unique $\phi_h \in H(d(0,r))^*$ satisfying $\phi_h(x^q) = b_q$, $(q \in \mathbb{N})$. Moreover, on the space $\mathcal{A}_b(\mathbb{K}\setminus d(0,r))$ provided with the norm $\|\ .\ \|_{\mathbb{K}\setminus d(0,r)}$, the mapping $h \to \phi_h$ is an isometric isomorphism from $\mathcal{A}_b(\mathbb{K}\setminus d(0,r))$ onto $H(d(0,r))^*$.*

Proof. Let $F = \mathbb{K}\setminus d(0,r)$. First let $h(t) = \sum_{n=0}^{\infty} \frac{b_n}{t^n} \in \mathcal{A}_b(F)$ and let $f(x) = \sum_{n=0}^{\infty} a_n x^n \in H(d(0,r))$. Since the sequence $\frac{|b_n|}{r^n}$ is bounded and $\lim_{n\to\infty} |a_n| r^n = 0$, it is seen that $\lim_{n\to\infty} a_n b_n = 0$ and then the series $\sum_{n=0}^{\infty} a_n b_n$ is convergent. Hence, we may put $\phi_h(f) = \sum_{n=0}^{\infty} a_n b_n$. Thus, we define a linear form ϕ_h of $H(d(0,r))$ that satisfies

$$|\phi_h(f)| \le \sup_{n\in\mathbb{N}} |a_n b_n| \le \Big(\sup_{n\in\mathbb{N}} |a_n| r^n\Big)\Big(\sup_{n\in\mathbb{N}} \frac{|b_n|}{r^n}\Big) = \|f\|_{d(0,r)} \|h\|_F.$$

Therefore, with respect to the norm $\|\ .\ \|$ of $H((d(0,r))^*$, we have $\|\phi_h\| \le \|h\|_F$. Now we check that the equality is satisfied. Indeed let $q \in \mathbb{N}$. We have

$$\frac{|\phi_h(x^q)|}{\|x^q\|_{d(0,r)}} = \frac{|\phi_h(x^q)|}{r^q} = \frac{|b_q|}{r^q} \le \sup_{f\ne 0} \frac{|\phi_h(f)|}{\|f\|_{d(0,r)}}$$

for all $q \geq 0$. Hence, we have $\|h\|_F = \sup_{q\in\mathbb{N}} \frac{|b_q|}{r^q} \leq \|\phi_h\|$. So we obtain the announced equality. Thus we have defined an isometric homomorphism from $\mathcal{A}_b(F)$ into $H(d(0,r))^{\circledast}$.

Now we check that this mapping is surjective. Indeed, let $\psi \in H(d(0,r))^{\circledast}$ and for each $n \in \mathbb{N}$, let $b_n = \psi(x^n)$. Obviously we have $\|\psi\| \geq \frac{|b_n|}{r^n}$ for every $n \in \mathbb{N}$, hence the sequence $(|b_n|r^{-n})_{n\in\mathbb{N}}$ is bounded and, therefore, defines a function $f(t) = \sum_{n=0}^{\infty} \frac{b_n}{t^n} \in \mathcal{A}_b(F)$. Thus ψ is equal to ϕ_h and, therefore, the mapping $h \to \phi_h$ is surjective. This ends the proof of Theorem B.6.13. □

Remark: There obviously exists an isometric homomorphism from $H(d(0,1))$ into $H(d(0,1))^{\circledast}$ defined as follows: let $f = \sum_{n=0}^{+\infty} a_n x^n \in H(d(0,1^-))$ and let $\underline{f}(x) = f(\frac{1}{x}) \in \mathcal{A}_b(\mathbb{K} \setminus d(0,1))$. Then we have an element $f^* \in H(d(0,1))^*$ equal to $\phi_{\underline{f}}$. The question whether this homomorphism is surjective depends on the ground field $\mathbb{K}$. If $\mathbb{K}$ is spherically complete, this homomorphism is not surjective. If $\mathbb{K}$ is not spherically complete, this homomorphism is surjective [89].

Corollary B.6.14: *Let $r \in \mathbb{R}_+$. For each $h(t) = \sum_{n=0}^{\infty} b_n t^n \in \mathcal{A}_b(d(0,r^-))$, there exists a unique $\phi_h \in H(\mathbb{K}\setminus d(0,r^-))^{\circledast}$ satisfying $\phi_h(x^{-q}) = b_q$ $(q \in \mathbb{N})$. Moreover, the space $\mathcal{A}_b(d(0,r^-))$ being provided with the norm $\| \ . \ \|_{d(0,r^-)}$, the mapping $h \to \phi_h$ is an isometric isomorphism from $\mathcal{A}_b(d(0,r^-))$ onto $H(\mathbb{K} \setminus d(0,r^-))^{\circledast}$.*

Corollary B.6.15: *Let $r \in \mathbb{R}_+$. For each $h(t) = \sum_{n=0}^{\infty} b_n t^n \in \mathcal{A}_b(d(0,r^-))$ such that $h(0) = 0$ there exists a unique $\phi_h \in H_0(\mathbb{K} \setminus d(0,r^-))^{\circledast}$ satisfying $\phi_h(x^{-q}) = b_q$ $(q \in \mathbb{N}^*)$. Moreover, this mapping $h \to \phi_h$ from the subspace of the $h \in \mathcal{A}_b(d(0,r^-))$ such that $h(0) = 0$ into $H_0(\mathbb{K} \setminus d(0,r^-))^{\circledast}$, is an isometric isomorphism.*

Now applying Theorem B.6.13 to $H(\widetilde{D})$ and Corollary B.6.15 to the spaces $H_0(\mathbb{K} \setminus T_i)$ for each hole T_i of an infraconnected set D, we obtain Corollary B.6.16.

Corollary B.6.16: **(Y. Amice)** *Let D be closed bounded infraconnected. Let $(T_i)_{i\in J}$ be the family of its holes and for every $i \in J$, let $a_i \in T_i$. Let $M \in \mathbb{R}_+^*$. Let $h_0 \in \mathcal{A}_b(\mathbb{K} \setminus \widetilde{D})$ and let $(h_i)_{i\in J}$ be a family such that for each $i \in J$, h_i belongs to $\mathcal{A}_b(T_i)$ and satisfies*

(1) $h_i(a_i) = 0$

and

(2) $\|h_i\|_{T_i} \leq M$ for all $i \in J$.

There exists a unique $\psi \in H(D)^{\circledast}$ *satisfying*

$\psi(f) = \phi_{h_0}(f)$ *for every* $f \in H(\widetilde{D})$

$\psi(f) = \phi_{h_i}(f)$ *for every* $f \in H_0(\mathbb{K} \setminus T_i)$*, whenever* $i \in J$.

Further, for every element ψ *of* $H(D)^{\circledast}$ *there exists a unique* $h_0 \in \mathcal{A}_b(\mathbb{K} \setminus d(0,r))$ *and a unique family* $(h_i)_{i\in J}$ *satisfying* (1) *and* (2) *for some* $M \in \mathbb{R}_+^*$ *such that* ψ *is defined as above.*

Now we will use the continuous linear forms to define the residue of an element on a hole.

Theorem B.6.17: *Let* $f \in H_b(\mathbb{K} \setminus d(a, r^-))$ *and for each* $\alpha \in d(a, r^-)$*, let* $f(x) = \sum_{n=0}^{\infty} \frac{b_n(\alpha)}{(x-\alpha)^n}$. *Then* $b_1(\alpha)$ *does not depend on* α *in* $d(a, r^-)$.

Proof. Let $E = \mathbb{K} \setminus d(a, r^-))$. We know that $\frac{|b_1(\alpha)|}{r} \leq \|f\|_E$ and, therefore, fixing α in $d(a, r^-)$, the linear form ψ_α on $H_b(E)$ defined as $\psi_\alpha(f) = b_1(\alpha)$ is obviously continuous. We will show that $\psi_\alpha(f) = \psi_a(f)$. First, for every $q \in \mathbb{N}$, we put $f_q(x) = \frac{1}{(x-\alpha)^q}$. We have $f_q(x) = \frac{1}{(x-a)^q}\Big(\sum_{j=0}^{\infty}(\frac{\alpha-a}{x-a})^j\Big)^q$. Therefore, for every $q \geq 2$ we have $\psi_a(f_q) = 0$ and that $\psi_a(f_1) = 1$. Hence, $\psi_\alpha(f_q) = \psi_a(f_q)$ for every $q \in \mathbb{N}$. This shows that $\psi_\alpha(f) = \psi_a(f)$ for every $f \in H_b(E)$. □

Definitions and notations: Let $f \in H_b(D)$, let T be a hole of D, and let $a \in T$. Let $f_T(x) = \sum_{n=1}^{\infty} \frac{b_n(a)}{(x-a)^n}$. By Theorem B.6.17, $b_1(a)$ actually does not depend on a in T. We set $\text{res}(f, T) = b_1(a)$ and this number $\text{res}(f, T)$ will be called *the residue of* f *on the hole* T.

By Theorem B.6.1, Theorem B.6.18 is obvious.

Theorem B.6.18: *Let* $f \in H(D)$ *and let* T *be a hole of* D *of diameter* r. *Then*

$$|\text{res}(f, T)| \leq r\|\overline{\overline{f_T}}\|_{\mathbb{K}\setminus T} \leq r\|f\|_D.$$

We can now characterize $\mathbb{K}$-algebra homomorphisms among the continuous linear forms.

Theorem B.6.19: *Let* D *be closed, bounded, and infraconnected, let* $a \in \widetilde{D}$*, let* $(T_i)_{i\in J}$ *be the family of holes of* D*, and for every* $i \in J$*, let* $a_i \in T_i$. *Let* $M \in \mathbb{R}_+^*$. *Let* $h_0 \in \mathcal{A}_b(\mathbb{K} \setminus \widetilde{D})$ *and let* $(h_i)_{i\in J}$ *be a family such that for each* $i \in J$*,* h_i *belongs to* $\mathcal{A}_b(T_i)$ *and satisfies conditions* (1) *and* (2) :

(1) $h_i(a_i) = 0$

and

(2) $\|h_i\|_{T_i} \leq M$ *for all* $i \in J$.

Let $\psi \in H(D)^*$ *satisfy:*

$\psi(f) = \phi_{h_0}(f)$ *for every* $f \in H(\widetilde{D})$
$\psi(f) = \phi_{h_i}(f)$ *for every* $f \in H_0(\mathbb{K} \setminus T_i)$*, whenever* $i \in J$.

Then ψ *is a homomorphism of* $\mathbb{K}$*-algebra from* $H(D)$ *onto* $\mathbb{K}$ *if and only if there exists* $\alpha \in D$ *such that*

(3) $h_0(t) = \dfrac{t-a}{t-\alpha}$,

and for every $i \in J$,

(4) $h_i(t) = \dfrac{t-a_i}{\alpha - t}$.

Moreover, every $\mathbb{K}$*-algebra homomorphism from* $H(D)$ *to* $\mathbb{K}$ *is continuous and is of this form.*

Proof. First we suppose that ψ is a $\mathbb{K}$-algebra homomorphism from $H(D)$ onto $\mathbb{K}$ and we put $\psi(x) = \alpha$. As h_0 is of the form $h_0(t) = \sum_{n=0}^{\infty} \dfrac{b_n}{(t-a)^n}$, here for every $n \in \mathbb{N}$ we have $b_n = \psi((x-a)^n) = (\alpha - a)^n$ and, therefore, $h_0(t) = \sum_{n=0}^{\infty} \Big(\dfrac{\alpha - a}{t-a}\Big)^n = \dfrac{t-a}{t-\alpha}$.

Next, we fix $i \in J$. Then h_i is of the form $h_i(t) = \sum_{n=1}^{\infty} b_{i,n}(t-a_i)^n$. Hence, for every $n \in \mathbb{N}^*$, we have $b_{i,n} = \psi\Big(\dfrac{1}{(x-a_i)^n}\Big) = \Big(\dfrac{1}{\alpha - a_i}\Big)^n$ and, therefore, $h_i(t) = \sum_{n=1}^{\infty} \Big(\dfrac{t-a_i}{\alpha - a_i}\Big)^n = \dfrac{t-a_i}{\alpha - t}$.

Conversely, we suppose (3) and (4) are satisfied. Then it is easily checked that $h_0(t) = \sum_{n=0}^{\infty} \big(\dfrac{\alpha - a}{t-a}\big)^n$ and, therefore, for every $n \in \mathbb{N}$, we have $\psi((x-a)^n) = (\alpha - a)^n$. Hence, for every $f \in H(\widetilde{D})$, we have $\psi(f) = f(\alpha)$.

In the same way, we check that, fixing $i \in J$, we have $\psi\Big(\dfrac{1}{(x-a_i)^n}\Big) = \Big(\dfrac{1}{\alpha - a_i}\Big)^n$, hence $\psi(f) = f(\alpha)$ for every $f \in H_0(\mathbb{K} \setminus T_i)$. This clearly finishes proving that $\psi(f) = f(\alpha)$ for every $f \in H(D)$.

Now, let ψ be a $\mathbb{K}$-algebra homomorphism from $H(D)$ to $\mathbb{K}$. By Corollary B.1.10, ψ is continuous, hence belongs to $H(D)^*$. Consequently it is of the form defined by the theorem. □

B.7. Factorization of analytic elements

In $\mathbb{C}$, it is well known that when a (not identically zero) holomorphic function admits a zero at a point α, this zero has a finite order of multiplicity. Actually this is a generalization of a property of rational functions. In the non-Archimedean context, we find that property again among analytic elements and it is essential [43]. In this chapter, D is just a subset of $\mathbb{K}$.

Lemma B.7.1: *Let $\alpha \in \overset{\circ}{D}$. Let $q \in \mathbb{N}^*$ and let (g_n) be a sequence of $H(D)$ such that the sequence $(x-\alpha)^q g_n$ converges in $H(D)$. Then the sequence $(g_n)_{n\in\mathbb{N}}$ also converges in $H(D)$.*

Proof. Without loss of generality, we can assume $\alpha = 0$. Set $f_n = x^q g_n$, $n \in \mathbb{N}$. Since 0 lies in $\overset{\circ}{D}$, there exists a disk $d(0,r) \subset D$. Let $E = D \setminus d(0,r)$. Clearly, we have $\|f_s - f_n\|_{d(0,r)} = \|x^q\|_{d(0,r)}\|g_s - g_n\|_{d(0,r)} = r^q\|g_s - g_n\|_{d(0,r)}$, and, hence, $\|g_s - g_n\|_E \le \dfrac{\|f_s - f_n\|}{r^q}$. Consequently, $\|g_s - g_n\|_D \le \frac{\|f_s - f_n\|}{r^q}$ and, therefore, the sequence (g_n) is a Cauchy sequence, which ends the proof. □

Theorem B.7.2: *Let α belong to $D \cap \overset{\circ}{\overline{D}}$ and let $f \in H(D)$ be such that $f(\alpha) = 0$. Then f has factorization in $H(D)$ in the form $(x-\alpha)g$ with $g \in H(D)$. If there is no neighborhood V of α such that $f(x) = 0$ whenever $x \in V$, then there exists a unique integer $q \in \mathbb{N}$ and $h \in H(D)$ such that $f(x) = (x-\alpha)^q\, h(x)$ and $h(\alpha) \neq 0$, and then α is a zero of order q of f.*

Proof. First we will prove the main factorization in the form $(x-\alpha)g$. We may obviously assume $\alpha = 0$. By hypothesis, there exists a disk $d(0,s)$ included in $\overline{D}$. And then, by Theorem B.2.5, there exists a disk $d(0,r)$ included in $d(0,s)$ such that f has no pole in $d(0,r)$. Consequently, f belongs to $H(D \cup d(0,r))$. So we can assume that 0 is interior to D and that $d(0,r) \subset D$ without loss of generality. By Corollary B.5.11, the restriction of f to $d(0,r)$ is equal to a power series $\displaystyle\sum_{n=1}^{\infty} a_n x^n$ for all $x \in d(0,r)$.

Now let t_n be a sequence in $R(D)$ such that $\lim\limits_{n\to\infty} t_n = f$. Clearly $|t_n(0)| \le \|t_n - f\|_D$ because $f(0) = 0$, so we have $\|t_n - t_n(0) - f_n\|_D \le \|t_n - f\|_D$. We put $h_n = t_n - t_n(0)$ $(n \in \mathbb{N})$. The sequence (h_n) of $R(D)$ approaches f in $H(D)$ and satisfies $h_n(0) = 0$ whenever $n \in \mathbb{N}$, hence h_n has factorization in $R(D)$ in the form xg_n. By Lemma B.7.1 the sequence $(g_n)_{n\in\mathbb{N}}$ converges in $H(D)$. Let g be its limit. We will show that $\lim\limits_{q\to\infty} \|xg_q - xg\|_D = 0$. Let $\varepsilon > 0$ and let $N \in \mathbb{N}$ be such that $\|h_n - f\|_D \le \varepsilon$ whenever $n \ge N$. We fix $q \ge N$. Then $\|h_n - h_q\|_D \le \varepsilon$, hence $|xg_n(x) - xg_q(x)| \le \varepsilon$ whenever $x \in D$. So, when n tends to $+\infty$, we see that $|xg(x) - xg_q(x)| \le \varepsilon$ whenever x in D. Thus we have $\|xg - xg_q\|_D \le \varepsilon$, and, therefore, $\lim\limits_{q\to\infty} \|xg_q - xg\|_D = 0$. But by hypothesis we have $\lim\limits_{q\to\infty} \|f - xg_q\|_D = 0$ and then $f = xg$.

Now we suppose that f is not identically zero in $d(0,r)$. Then at least one of the coefficients a_n of its power series is not zero. By Corollary B.5.11, f admits 0 as a zero of order q and then q is the smallest integer such that $a_q \neq 0$. In $d(0,r)$ we have $f(x) = \sum_{n=q}^{\infty} a_n x^n = xg(x)$, hence $g(x) = \sum_{n=q}^{\infty} a_n x^{n-q}$ whenever $x \in d(0,r)$. Suppose that f has been proven to be factorized in the form $x^s g_s$ with $s < q$ and $g_s \in H(D)$. Clearly $g_s(x) = \sum_{n=q}^{\infty} a_n x^{n-s}$ whenever $x \in d(0,r)$, hence $g_s(0) = 0$ and, therefore, g_s has factorization in the form xg_{s+1} with $g_{s+1} \in H(D)$. Thus by induction we obtain $f = x^q g_q(x)$ with $g_q(x) = \sum_{n=q}^{\infty} a_n x^{n-q}$ and then $g_q(0) = a_q \neq 0$. That finishes proving Theorem B.7.2. □

Definitions and notations: Let $a \in \overset{\circ}{D}$ and let $f \in H(D)$ be such that $f(a) = 0$, $f(x) \neq 0$ in a disk $d(a,r)$. The order of the zero a of f will be denoted by $\omega_a(f)$.

Corollary B.7.3: *Let $\overline{D}$ be open, let $f \in H(D)$, and let α be a zero of f in D. Either there exists a disk $d(\alpha,r)$ such that $f(x) \neq 0$ whenever $x \in d(\alpha,r) \setminus \{\alpha\}$, or there exists a disk $d(\alpha,r)$ such that $f(x) = 0$ whenever $x \in d(\alpha,r)$.*

Corollary B.7.4: *Let $f \in H(D)$ have a zero of order q at a point $\alpha \in \overset{\circ}{D}$. Then for every $s = 1, \dots, q$, f factorizes in the form $(x-\alpha)^s g_s$, with $g_s \in H(D)$ having a zero of order $q - s$ at α.*

Corollary B.7.5: *Let $\alpha \in \overset{\circ}{D}$ and let $f \in H(D)$. Let $\sum_{n=0}^{\infty} a_n(x-\alpha)^n$ be its power series in a disk $d(\alpha,r) \subset D$. Let $P(x) = \sum_{n=0}^{q} a_n(x-\alpha)^n$ and let $g(x) = f(x) - P(x)$. Then g factorizes in the form $(x-\alpha)^q h(x)$, with $h \in H(D)$.*

Definitions and notations: Let $A \subset D$ be an open subset of $\mathbb{K}$, let $f \in H(D)$ have finitely many zeros $a_1, \dots, a_n$ in A of multiplicity order of $q_1, \dots, q_n$ respectively. The polynomial $\prod_{i=1}^{n}(x - a_i)^{q_i}$ will be named *the polynomial of zeros of f in A.*

We are now able to give the Corollary B.7.6.

Corollary B.7.6: *Let A be a subset of D open in $\mathbb{K}$, let $f \in H(D)$ have finitely many zeros in A, and let P be the polynomial of its zeros in A. Then f have a factorization in the form $f = Pg$, with $g \in H(D)$ and $g(x) \neq 0$ whenever $x \in A$.*

Definitions and notations: An element $f \in H(D)$ will be said to be *semi-invertible* (resp. *quasi-invertible*) if it factorizes in the form $P(x)\ g(x)$, with g invertible in $H(D)$ and with P a polynomial whose zeros belong to D (resp. to $D \cap \overset{\circ}{\overline{D}}$).

An element $f \in H(D)$ will be said to be *quasi-minorated* if for every bounded sequence $(a_n)_{n\in\mathbb{N}}$ of D such that $\lim_{n\to\infty} f(a_n) = 0$ we can extract a subsequence that converges in $\mathbb{K}$.

Remarks: (1) If a semi-invertible element of $H(D)$ has no zero in D, it is invertible in $H(D)$. (2) Let D belong to Alg. If f_1, f_2 are semi-invertible (resp. quasi-invertible) elements of $H(D)$, then f_1f_2 is also semi-invertible, (resp. quasi-invertible). However when D does not belong to Alg, counter-examples show that the product of two semi-invertible (resp. quasi-invertible) elements is not always semi-invertible (resp. quasi-invertible). Such counterexamples will be given in a further remark.

Lemma B.7.7: *Let $D \in$ Alg, let $f \in H(D)$ be quasi-invertible (resp. quasi-minorated) and $h \in R(D)$ be a Moebius function. Let $D' = h(D)$ and let $g = f \circ h^{-1}$. Then g is a quasi-invertible (resp. quasi-minorated) element of $H(D')$.*

Proof. Suppose first f to be quasi-invertible in $H(D)$. Let $u = h(x)$. So, f is of the form $P(x)\phi(x)$ with P a polynomial whose zeros are interior to D and ϕ is an invertible element of $H(D)$. Then $\phi \circ h^{-1}$ is invertible in $H(D')$ and $P \circ h^{-1}$ belongs to $R(D')$ and is of the form $\dfrac{Q(u)}{(u-b)^s}$, where b is the unique pole of h^{-1}. Consequently g is of the form $Q(u)\dfrac{\phi(u)}{(u-b)^s}$. Thus $\dfrac{\phi(u)}{(u-b)^s}$ is invertible in $H(D')$ and, hence, g is quasi-invertible.

Now suppose f is quasi-minorated. Let $(a_n)_{n\in\mathbb{N}}$ be a sequence in D' such that $\lim_{n\to\infty} g(a_n) = 0$ and let $b_n = h^{-1}(a_n)$, $(n \in \mathbb{N})$. Then $\lim_{n\to\infty} f(b_n) = 0$. Since f is quasi-minorated, one can extract a subsequence $(b_{q(m)})_{m\in\mathbb{N}}$ that either converges or satisfies $\lim_{m\to\infty} |b_{q(m)}| = \infty$. But then, the sequence $(a_{q(m)})$ either converges or satisfies $\lim_{m\to\infty} |a_{q(m)}| = \infty$. Hence, f is quasi-minorated. □

Theorem B.7.8: *Let D be bounded, open, closed and let $f \in H(D)$. If f is quasi-minorated then it is quasi-invertible.*

Proof. We suppose f is not quasi-invertible and we will prove that f is not quasi-minorated either.

First we suppose f to have finitely many zeros. Since D is open, by Corollary B.7.6, f has factorization in the form $P(x)\ g(x)$, with P a polynomial whose zeros are interior to D and g an element of $H(D)$, which has no zero in D, but is not invertible in $H(D)$ since f is not quasi-invertible. Hence, there exists a bounded sequence $(\alpha_n)_{n\in\mathbb{N}}$ in D such that $\lim_{n\to\infty} g(\alpha_n) = 0$. If f were quasi-minorated we could extract a convergent subsequence from the sequence $(\alpha_n)_{n\in\mathbb{N}}$ whose limit would belong to D and would be a zero of g. Hence, f is not quasi-minorated when it has finitely many zeros in D.

Now we suppose that f has a sequence of (distinct) zeros (α_n) in D and that f is quasi-minorated. Hence, we may extract a convergent subsequence of limit α. Obviously α is another zero of f, hence by Corollary B.7.3, $f(x)$ is equal to zero inside a disk $d(\alpha, r)$ and then f is not quasi-minorated, a contradiction. That ends the proof of Theorem B.7.8. □

Theorem B.7.9: *Let D be closed and bounded. Let $f \in H(D)$ be quasi-minorated and have no zero in D. Then f is invertible in $H(D)$.*

Proof. Assume that $\inf_{x\in D} |f(x)| = 0$ and let $(a_n)_{n\in\mathbb{N}}$ be a sequence in D such that $\lim_{n\to\infty} f(a_n) = 0$. Since D is bounded and since f is quasi-minorated, we can extract a subsequence (a_n), which converges in $\mathbb{K}$ to a point $a \in \overline{D}$. Since D is closed, a belongs to D and satisfies $f(a) = 0$, which contradicts the hypothesis. Thus there exists $\lambda > 0$ such that $|f(x)| \geq \lambda$ whenever $x \in D$ and then by Lemma B.2.3, f is invertible in $H(D)$. □

B.8. Algebras $H(D)$

We have seen that $H(D)$ is a Banach $\mathbb{K}$-algebra if and only if D is closed and bounded. But studying analytic elements, analytic functions require to know algebras of analytic elements, which are not necessarily bounded. Thus we have to examine the class Alg of subsets D of $\mathbb{K}$ such that $H(D)$ is a $\mathbb{K}$-algebra with respect to usual laws [43].

Definitions and notations: Throughout the chapter, D denotes a subset of $\mathbb{K}$. Let $f \in H(D)$. According to Theorem B.2.5, f is of the form $f^* + \overline{f}$, with $f^* \in R_0(\mathbb{K} \setminus (\widetilde{D} \setminus D))$ and $\overline{f} \in H(\overline{D})$. We will keep that notation throughout the chapter.

Proposition B.8.1: *Let α belong to $\overset{\circ}{\overline{D}}$ and let $f \in H(D\setminus\{\alpha\})$. For every $q \in \mathbb{N}$, $\dfrac{f}{(x-\alpha)^q}$ belongs to $H(D \setminus \{\alpha\})$.*

Proof. Since α belongs to $\overset{\circ}{\overline{D}}$, there exists a disk $d(\alpha, s)$ included in $\overline{D}$. On the other hand, there exists $r \in]0, s[$ such that f^* has no pole in $d(\alpha, r) \setminus \{\alpha\}$. Hence, by Theorem B.2.10 f is of the form $\dfrac{g}{(x-\alpha)^t}$ with $g \in H(D \cup d(\alpha, r))$ and $t \in \mathbb{N}$. Then $\dfrac{f}{(x-\alpha)^q} = \dfrac{g}{(x-\alpha)^{q+t}}$. Thus without loss of generality we may assume that α belongs to D and that f belongs to $H(D \cup d(\alpha, r))$.

By Corollary B.5.11, in $d(\alpha, r)$, $f(x)$ is equal to a power series $\sum_{n=0}^{\infty} a_n(x-\alpha)^n$.

Let $P(x) = \sum_{n=0}^{q} a_n(x-\alpha)^n$. By Theorem B.7.2, $f(x) - P(x)$ factorizes in the form

$(x-\alpha)^q h$ with $h \in H(D)$. Hence, we see that $\dfrac{f(x)}{(x-\alpha)^q} = \dfrac{P(x)}{(x-\alpha)^q} + h$. Since $\dfrac{P}{(x-\alpha)^q} \in R(D)$ it is clear that $\dfrac{f}{(x-\alpha)^q}$ belongs to $H(D)$. □

Corollary B.8.2: *Let $f \in H(D)$ and let P be a polynomial whose zeros are interior to D. Let $a_1, \dots, a_n$ be the zeros of P. Then $\dfrac{f}{P}$ belongs to $H(D \setminus \{a_1, \dots, a_n\})$.*

Proposition B.8.3: *If D is bounded and satisfies $\overline{D} \setminus D \subset \left(\overset{\circ}{\overline{D}}\right)$ then $D \in$ Alg.*

Proof. Let $f, g \in H(D)$ and let us show that $fg \in H(D)$. By Theorem B.2.5, we have $f = f^* + \overline{f}\, g = g^* + \overline{g}$ with $f^*, g^* \in R_0(\mathbb{K} \setminus (\overline{D} \setminus D))$ and $\overline{f}, \overline{g} \in H(\overline{D})$. Since D is bounded, by Corollary B.2.7, we have $H(\overline{D}) = H_b(D)$ and then $\overline{f}\overline{g}$ obviously belongs to $H(D)$ while $f^* g^* \in R_0(\mathbb{K} \setminus (\overline{D} \setminus D))$. Finally by Corollary B.8.2, both $f^*\overline{g}$ and $g^*\overline{f}$ belong to $H(D)$ and, therefore, so does fg. □

Definitions and notations: Let $\mathcal{F}$ be a filter in D. An element $f \in H(D)$ will be said to be *vanishing along* $\mathcal{F}$ if $\lim_{\mathcal{F}} f(x) = 0$. Further f will be said to be *properly vanishing along* $\mathcal{F}$ if $\lim_{\mathcal{F}} f(x) = 0$ and if $\|f\|_A \neq 0$ whenever $A \in \mathcal{F}$.

Proposition B.8.4 is a polyvalent result that not only helps us characterize the sets $D \in$ Alg, but also find conditions for $H(D)$ not to be a Noetherian algebra.

Proposition B.8.4: *Let $\mathcal{F}$ be a pierced filter on D, let $(T_n)_{n \in \mathbb{N}}$ be a sequence of holes of D that runs $\mathcal{F}$, and let $E = \mathbb{K} \setminus (\bigcup_{n=0}^{\infty} T_n)$. Let $g_1, \dots, g_q \in H_b(E)$ be vanishing along $\mathcal{F}$, with g_1 properly vanishing. For every $x \in E$ let $S(x) = \sup_{1 \le i \le q} |g_i(x)|$, let $\mathcal{J}$ be the ideal generated by $g_1, \dots, g_q$ in $H_b(E)$, and let $\overline{\mathcal{J}}$ be its closure in $H_b(D)$.*

There exists a sequence $(z_n)_{n \in \mathbb{N}}$ in D, thinner than $\mathcal{F}$, such that $g_1(z_n) \neq 0$ and an element $G \in \overline{\mathcal{J}}$ such that $\lim_{n \to \infty} \dfrac{|G(z_n)|}{S(z_n)} = +\infty$.

Proof. Without loss of generality we may assume $\mathcal{F}$ to be a decreasing filter or a Cauchy filter. Indeed if $\mathcal{F}$ is an increasing filter of center α and diameter R, consider a hole of D $T(b, \rho)$ included in $d(\alpha, R^-)$, take $\gamma(x) = \dfrac{1}{x-b}$, and set $D' = \gamma(D)$. Then by Theorem A.2.11 D' admits a decreasing pierced filter $\mathcal{F}'$, image of $\mathcal{F}$ by γ. Next, D' is clearly bounded. By Theorem B.3.7, the mapping ϕ from D onto D' defined by $\phi(f) = f \circ \gamma^{-1}$ is an isomorphism from $H(\overline{D'})$ onto $H_b(D)$. Hence, $\mathcal{J}$ is isomorphic to the ideal generated by $\{g_j \circ \gamma^{-1} | 1 \le j \le q\}$ in $H(\overline{D'})$. Hence, we will assume $\mathcal{F}$ to be a decreasing pierced filter or a Cauchy pierced filter.

Without loss of generality we may now clearly assume $D = E$. Since the g_j are bounded, we may obviously assume $\|g_j\|_D \le 1$ whenever $j = 1, \dots, q$. Let $R = \text{diam}(\mathcal{F})$ and let $(x_m)_{m\in\mathbb{N}}$ be a sequence in D thinner than $\mathcal{F}$, such that $g_1(x_m) \neq 0$ whenever $m \in \mathbb{N}$, with $|x_{m+2} - x_{m+1}| < |x_{m+1} - x_m|$. Since $\mathcal{F}$ is pierced, there exists a subsequence $(x_{m_q})_{q\in\mathbb{N}}$ of the sequence (x_m) together with a sequence of holes $(T_q)_{q\in\mathbb{N}}$ of D such that

$$T_q \subset d(x_{m_{q+1}}, d_{m_q}) \setminus d(x_{m_{q+2}}, d_{m_{q+1}}).$$

Hence, without loss of generality we may assume that we have a sequence of holes $(T_m)_{m\in\mathbb{N}}$ of D such that $T_m \subset d(x_{m+1}, d_m) \setminus d(x_{m+2}, d_{m+1})$.

We put $D_m = d(x_{m+1}, d_m) \cap D$ and $A_n = D_{2n+1} \setminus D_{2n+3}$. For each n, let $u_n \in A_n$ be such that $|g_1(u_n)| \ge \|g_1\|_{A_n}\Big(\dfrac{n}{n+1}\Big)$. For each $j = 1, \dots, t$, let $M_n^j = \|g_j\|_{A_n}$, and let $M_n = \max\limits_{1\le j\le t} M_n^j$. Since $g_1(x_m) \neq 0$ we have $M_n > 0$ whenever $n \in \mathbb{N}$ and since $\|g_j\|_D \le 1$ for all j, we have $M_n \le 1$ whenever $n \in \mathbb{N}$.

We will construct a sequence (U_n) in $H_b(D)$ satisfying:

(1) $\quad |U_n(x)| \le \dfrac{1}{n+1}$ whenever $x \in D \setminus A_n$.

(2) $\quad \sqrt{M_n}\Big(\dfrac{n+1}{n}\Big) > \|g_1 U_n\|_{A_n} > \sqrt{M_n}$.

For every $n \in \mathbb{N}$, set $T_n = d(\beta_n, \rho_n^-)$, $u_n = x_{2n+2}$, $a_n = \beta_{n+1}$, $b_n = \beta_{n+2}$, $c_n = \beta_{2n+3}$, and set $\epsilon_n \in d(0, \frac{1}{n})$. Let us fix $n \in \mathbb{N}$. It is seen that $|u_n - a_n| > |u_n - b_n|$, hence there exists $q_n \in \mathbb{N}$ such that

(3) $\quad |\epsilon_n|\Big|\dfrac{u_n - a_n}{u_n - b_n}\Big|^{q_n} g(u_n) > \sqrt{M_n}$

and of course there exists q'_n such that

(4) $\quad \Big(\dfrac{d_{2n+1}}{d_{2n+2}}\Big)^{q_n} \Big(\dfrac{d_{2n+3}}{d_{2n+2}}\Big)^{q'_n} < 1.$

We put $h_n(x) = \epsilon_n\Big(\dfrac{x - a_n}{x - b_n}\Big)^{q_n} \Big(\dfrac{x - c_n}{x - b_n}\Big)^{q'_n}$.

Then by (4) we see that:

when $|x - c_n| > d_{2n+1}$ we have $|h_n(x)| = |\epsilon_n| < \frac{1}{n}$,

when $|x - c_n| \le d_{2n+3}$ we have $|x - a_n| = |a_n - c_n| = d_{2n+1}$ and $|x - b_n| = |b_n - c_n| = d_{2n+2}$, hence $|h_n(x)| \le |\epsilon_n|\Big(\dfrac{d_{2n+1}}{d_{2n+2}}\Big)^{q_n} \Big(\dfrac{d_{2n+3}}{d_{2n+2}}\Big)^{q'_n} < \dfrac{1}{n}$.

But now we notice that x belongs to $D \setminus A_n$ if and only if it satisfies either $|x - c_n| > d_{2n+1}$or $|x - c_n| \le d_{2n+3}$, hence we have proven that $|h_n(x)| < \frac{1}{n}$ whenever $x \in D \setminus A_n$. This shows h_n satisfies (1).

When $x \in A_n$, i.e., when $d_{2n+3} < |x - c_n| < d_{2n+1}$, we see that $\|g_1 h_n\|_{A_n} \ge |g_1(u_n) h_n(u_n)|$, hence by (3) we have $\|g_1 h_n\|_{A_n} \ge \sqrt{M_n}$. Hence, there trivially exists $\lambda_n \in d(0, 1)$ such that $\Big(\dfrac{n+1}{n}\Big)\sqrt{M_n} > |\lambda_n g_1 h_n|_{A_n} > \sqrt{M_n}$.

Now we put $U_n = \lambda_n h_n$ and we see that U_n satisfies (1) and (2). In particular we have $\|g_1 U_n\|_D \leq \max\Big(\sqrt{M_n}\Big(\frac{n+1}{n}\Big), \frac{\|g_1\|_D}{n+1}\Big)$, hence $\lim_{n\to\infty} \|g_1 U_n\|_D = 0$. Let $T = \sum_{n=0}^{\infty} g_1 U_n$. By definition T belongs to $\overline{F}$ because for every $t \in \mathbb{N}$, $g\sum_{n=0}^{t} U_n$ belongs to F.

By (2) there exists a sequence $(z_n)_{n\in\mathbb{N}}$ in D satisfying $z_n \in A_n$ and

$$(5) \qquad \sqrt{M_n} < |g_1(z_n)U(z_n)| < M_n\Big(\frac{n+1}{n}\Big),$$

hence we have

$$(6) \qquad |U_n(z_n)| > \frac{\sqrt{M_n}}{|g_1(z_n)|} \geq \frac{1}{\sqrt{M_n}} \text{ because } |g_1(z_n)| \leq M_n.$$

When $j \neq n$, z_n belongs to $D \setminus A_j$, hence by (1) and (6) we have $|U_j(z_n)| < \frac{1}{j+1} < \frac{1}{\sqrt{M_n}} < |U_n(z_n)|$ whenever $j \neq n$. Hence, we see that $|T(z_n)| = |g_1(z_n)U_n(z_n)|$ whenever $n \in \mathbb{N}$. But then, by (5) we see that $\frac{|T(z_n)|}{S(z_n)} = \frac{|T(z_n)|}{M_n} > \frac{1}{\sqrt{M_n}}$. Consequently, $\lim_{n\to\infty} \frac{|T(z_n)|}{S(z_n)} = +\infty$ and this finishes the proof of Proposition B.8.4. □

Corollary B.8.5: *Let $a \in D\setminus \overset{\circ}{D}$ and let $\mathcal{F}$ be the pierced filter of the neighborhoods of a. There exists a sequence $(z_n)_{n\in\mathbb{N}}$ in D of limit a and an element $G \in H_b(D)$ vanishing along $\mathcal{F}$ such that* $\limsup_{n\to\infty} \frac{|G(z_n)|}{|z_n - a|} = +\infty$.

Lemma B.8.6: *Let D have a hole $T = d(a, r^-)$. Let $\gamma(x) = b + \frac{\lambda}{x-a}$ with $\lambda \in \mathbb{K}$ and let $D' = \gamma(D)$. For every $\alpha \in \overline{D}, \alpha$ belongs to $\Big(\overset{\circ}{\overline{D}}\Big)$ if and only if $\gamma(\alpha)$ belongs to $\Big(\overset{\circ}{\overline{D'}}\Big)$. Moreover if D is not bounded then $\mathbb{K} \setminus \overline{D}$ is bounded if and only if b belongs to $\Big(\overset{\circ}{\overline{D'}}\Big)$.*

Proof. γ is obviously a bicontinuous bijection from $\mathbb{K} \setminus \{a\}$ onto $\mathbb{K} \setminus \{b\}$. Let $\alpha \in \Big(\overset{\circ}{\overline{D}}\Big)$. There exists a disk $d(\alpha, r)$ included in $\overline{D}$. Since $a \notin d(\alpha, r)$ γ is bounded in $d(\alpha, r)$. Since γ is bicontinuous, $\gamma(d(\alpha, r))$ is open in $\mathbb{K} \setminus \{a\}$, hence it is clearly open in $\mathbb{K}$. So $\gamma(\alpha)$ belongs to $\gamma\Big(\overset{\circ}{\overline{D}}\Big)$. But $\gamma(\overline{D}) \subset \overline{\gamma(D)}$, hence $\gamma(\alpha) \in \Big(\overset{\circ}{\overline{D'}}\Big)$. Let $\xi = \gamma^{-1}$. Then $\xi(u) = a + \frac{\lambda}{u-b}$ and then what is true for γ is also true for ξ. Hence, conversely, if $\gamma(\alpha) \in \Big(\overset{\circ}{\overline{D'}}\Big)$ we see that $\alpha = \xi(\gamma(\alpha)) \in \overset{\circ}{\overline{D'}}$ because $D = \xi(D')$.

We now suppose D is unbounded. If $\mathbb{K} \setminus \overline{D}$ is bounded then $\overline{D}$ contains a set E of the form $\{|x| \; |x-a| \geq s\}$ with $s > |a-b|$ whose image E' is $d(a, \frac{|\lambda|}{s}) \setminus \{a\}$, hence $\overline{D'}$ contains $d(a, \frac{|\lambda|}{s})$ and so does $\overset{\circ}{\overline{D'}}$.

Finally we suppose that $\mathbb{K} \setminus \overline{D}$ is not bounded. Then $\gamma((\mathbb{K} \setminus \{a\}) \setminus \overline{D})$ is an open set E in $\mathbb{K}$ whose closure contains b. Since $b \in \overline{D'} \setminus D'$, it is easily seen that there is a sequence of holes of D', $(T_n)_{n \in \mathbb{N}}$, which approaches b (each one is obviously included in E), hence $b \notin \left(\overset{\circ}{\overline{D'}}\right)$. This ends the proof of Lemma B.8.6. □

Corollary B.8.7: *Let D have a hole $T = d(a, r^-)$, satisfy $\overline{D} \setminus D \subset \left(\overset{\circ}{\overline{D}}\right)$ and be such that $\mathbb{K} \setminus \overline{D}$ is bounded. Let $\gamma = \dfrac{1}{x-a}$ and let $D' = \gamma(D)$. Then $\overline{D'} \setminus D' \subset \left(\overset{\circ}{\overline{D'}}\right)$.*

Lemma B.8.8: *The following two conditions are equivalent:*

(A) $\widetilde{D} \setminus \overline{D}$ *is bounded*
(A') *either D is bounded or $\mathbb{K} \setminus \overline{D}$ is bounded.*

Proof. If D is bounded, (A) and (A') are clearly satisfied, hence we have nothing to show. Now we suppose D to be unbounded. Hence, $\widetilde{D} = \mathbb{K}$ and then $\widetilde{D} \setminus \overline{D} = \mathbb{K} \setminus \overline{D}$, so $\widetilde{D} \setminus \overline{D}$ is bounded if and only if $\mathbb{K} \setminus \overline{D}$ is bounded. Finally (A) and (A') are equivalent. □

Theorem B.8.9: *D belongs to* Alg *if and only if it satisfies the following two conditions:*

(A) $\widetilde{D} \setminus \overline{D}$ *is bounded*
(B) $\overline{D} \setminus D \subset \left(\overset{\circ}{\overline{D}}\right)$.

Proof. Suppose first that D satisfies (A) and (B). By Lemma B.8.8 it satisfies (A') and (B). Suppose first that D is bounded. Since D satisfies (B), by Proposition B.8.3, $D \in$ Alg. Suppose now that $\mathbb{K} \setminus \overline{D}$ is bounded. We may obviously assume D to have a hole $T = d(a, r^-)$ because if D has no hole then by Corollary B.2.8, we have $H(D) = R(D)$. Let $\gamma = \dfrac{1}{x-a}$ and let $D' = \gamma(D)$. The set D' is then bounded and by Corollary B.8.7, D' satisfies $\overline{D'} \setminus D' \subset \left(\overset{\circ}{\overline{D'}}\right)$, and, hence, $D' \in$ Alg, therefore by Proposition B.8.3, D belongs to Alg.

We now suppose that (B) is not satisfied and will prove that $D \notin$ Alg. Indeed let $\alpha \in (\overline{D} \setminus D) \setminus \left(\overset{\circ}{\overline{D}}\right)$. By definitions D has a Cauchy pierced filter $\mathcal{F}$ that converges in $\mathbb{K}$ to α. Let $T = d(a, r^-)$ be a hole of D and let $f = \dfrac{x-\alpha}{x-a}$. Then $f \in R_b(D)$.

By Proposition B.8.4 there exists $S \in H_b(D)$ such that $S(\alpha) = 0$ together with a sequence $(z_n)_{n\in\mathbb{N}}$ in D such that $\lim_{n\to\infty} z_n = \alpha$, while $\lim_{n\to\infty} \left|\frac{S(z_n)}{f(z_n)}\right| = +\infty$.

Let us assume $D \in$ Alg. Then $\frac{S(x)}{f(x)} \in H(D)$ because $\frac{x-a}{x-\alpha} \in R(D)$. Since $\left|\frac{S(x)}{f(x)}\right|$ is not bounded in any neighborhood of α, by Theorem B.2.10 and Corollary B.2.9 there exists an integer $n \geq 1$ such that $(x-\alpha)^n \frac{S(x)}{f(x)}$ has a non-zero limit ℓ at α. But when $x \in D \cap d(\alpha, |a|)$, we have $\left|(x-\alpha)^n\left(\frac{S(x)}{f(x)}\right)\right| = \left|(x-\alpha)^{n-1}(x-a)S(x)\right| = |a|\ |x-\alpha|^{n-1}\ |S(x)|$, hence $\ell = 0$, a contradiction. Consequently $D \notin$ Alg.

Suppose now that D satisfies (B) but does not satisfy (A). Both D and $\mathbb{K} \setminus \overline{D}$ are unbounded. Since $\mathbb{K} \setminus \overline{D}$ is unbounded, D has a hole $T = d(\alpha, r^-)$, and then by Lemma B.8.6, the inversion $\gamma(x) = \frac{1}{x-\alpha}$ maps D onto a bounded set D' such that $\alpha \in (\overline{D'} \setminus D') \subset \left(\overset{\circ}{\overline{D'}}\right)$. Hence, D' does not belong to Alg and then neither does D. This ends the proof of Theorem B.8.9. □

Corollary B.8.10: *If D is bounded and if $\overline{D}$ is open, then D belongs to* Alg.

Definitions and notations: Throughout the book, Conditions (A) and (B) will be those given in Theorem B.8.9.

Theorem B.8.11: *Let D belong to* Alg *and have a hole $T = d(a, r^-)$. Let $h(x) = \frac{1}{x-a}$ and let $D' = h(D)$. Then D' is bounded and belongs to* Alg.

Proof. Since $D \in$ Alg, by Theorem B.8.10, D satisfies Conditions (A) and (B) and, hence, we can we can check that D', being obviously bounded, satisfies (A) and (B) too. If D is bounded, this is immediate. If D is not bounded, then 0 is the unique point that might not belong to $\overline{D}' \setminus D' \subset \left(\overset{\circ}{\overline{D}'}\right)$. But if 0 is the limit of a sequence of holes of D', then $\widetilde{D} \setminus \overline{D}$ is not bounded, which contradicts Condition (A). □

Lemma B.8.12: *Let D be open. Then D satisfies Condition (B) if and only if $\overline{D}$ is open.*

Proof. Since D is open, $\overline{D}$ is open if and only if for every $\alpha \in \overline{D} \setminus D$, α is interior to D. This is just equivalent to Condition (B). □

Definitions and notations: Let $D \in$ Alg and let $a \in D$. We will denote by $\mathcal{J}(a)$ the ideal of the $f \in H(D)$ such that $f(a) = 0$.

Theorem B.8.13: *Let $D \in \text{Alg}$ and let $a \in D$. If a belongs to $\overset{\circ}{\overline{D}}$, then $\mathcal{I}(a) = (x-a)H(D)$. Else, $\mathcal{J}(a)$ is not of finite type.*

Proof. Suppose first $a \in \overset{\circ}{\overline{D}}$. By Theorem B.7.2 it is clearly seen that $\mathcal{J}(a) = (x-a)H(D)$. Now let $a \notin \overset{\circ}{\overline{D}}$. Then the filter of the neighborhoods of a is a Cauchy pierced filter. We will denote it by $\mathcal{F}$. Suppose that $\mathcal{J}(a)$ is of finite type and let $\{g_1, \dots, g_q\}$ be a system of generators. For each $j = 1, \dots, q$ let Q_j be the polynomial of poles of g_j in $\overline{D} \setminus D$. Now, let $d(b, r^-)$ be a hole of D. By Corollary B.2.7, for each $j = 1, \dots, q$ there exists a rational function of the form $\frac{1}{(x-b)^{t_j}}$ such that the function $h_j = \frac{g_j}{Q_j(x-b)^{t_j}}$ belongs to $H_b(\overline{D})$.

Of course, at least one of the g_j is properly vanishing along $\mathcal{F}$, otherwise all the elements of $\mathcal{J}(a)$ would be equal to 0 inside a neighborhood of a and then $\mathcal{J}(a)$ couldn't contain $x-a$. Consequently, we can assume that g_1 is properly vanishing and so is h_1. For every $x \in D$, we put $S(x) = \max_{1 \le j \le q} |h_j(x)|$. We notice that $\mathcal{J}(a)$ is obviously closed in $H(D)$, hence by Proposition B.8.4, there exists $f \in \mathcal{J}(a)$ together with a sequence $(z_n)_{n \in \mathbb{N}}$ in D, of limit a, such that $S(z_n) \neq 0$ whenever $n \in \mathbb{N}$ and that $\lim_{n \to \infty} \frac{|f(z_n)|}{S(z_n)} = +\infty$. This obviously contradicts the fact that f should be of the form $\sum_{j=1}^{q} f_j g_j$ with the f_j in $H(D)$. Thus we have shown that $\mathcal{J}(a)$ is not of finite type and this ends the proof of Theorem B.8.13. □

Corollary B.8.14: *Let $D \in \text{Alg}$. If $D \setminus \overset{\circ}{\overline{D}} \neq \emptyset$ then $H(D)$ is not Noetherian.*

B.9. Derivative of analytic elements

Given an infraconnected set, the main question we consider here is whether an element f of $H(D)$ has a derivative that belongs to $H(D)$ and when it does, whether its Mittag–Leffler series is obtained by deriving that of f [58]. Another question is whether an analytic element on D whose derivative is identically zero is a constant. Both questions are answered on an infraconnected clopen set.

Throughout this chapter D is a subset of $\mathbb{K}$ and is supposed to be open and infraconnected, and we fix $R > 0$.

Theorem B.9.1: *Let $f(x) = \sum_{n=0}^{+\infty} a_n x^n \in H(d(0,R))$. Then f has a derivative $f'(\alpha)$ at each point $\alpha \in d(0,R)$ and $f'(0) = a_1$.Moreover, the function f' also* *belongs to* $H(d(0,R))$ *and is equal to* $\sum_{n=0}^{+\infty} na_n x^{n-1}$, *and satisfies*

$$|f'|(r) \leq \frac{|f|(r)}{r} \ \forall r \in]0,R].$$

Further, f *is indefinitely derivable in* $d(0,R)$ *and*

$$f^{k)}(x) = \sum_{n=k}^{\infty} n(n-1)\cdots(n-k+1)a_n x^{n-k}.$$

Proof. Without loss of generality, we can suppose that $R \leq 1$. Obviously, $\lim_{x\to 0} \frac{f(x)-a_0}{x} = a_1$, hence $f'(0)$ exists and is equal to a_1. More generally, take $\alpha \in d(0,R) \setminus \{0\}$ and consider

$$\frac{f(x)-f(\alpha)}{x-\alpha} = \frac{\sum_{n=1}^{\infty} a_n(x^n - \alpha^n)}{x-\alpha} = \sum_{n=1}^{\infty} a_n(x^{n-1} + \alpha x^{n-2} + \cdots + \alpha^{n-1}).$$

Then $|(x^{n-1} + \alpha x^{n-2} + \cdots + \alpha^{n-1}) - n\alpha^{n-1}| \leq |x-\alpha|\big(\max(|\alpha|,|x|)\big)^{n-1}$. Particularly, when x is close enough to α, since $\alpha \neq 0$, we have $|x| = |\alpha|$, hence

$$|a_n||(x^{n-1} + \alpha x^{n-2} + \cdots + \alpha^{n-1}) - n\alpha^{n-1}| \leq |a_n||\alpha|^{n-1}|x-\alpha|.$$

That proves that $\Big(a_n(x^{n-1} + \alpha x^{n-2} + \cdots + \alpha^{n-1})\Big) - na_n\alpha^{n-1}$ converges to 0 uniformly with respect to n and uniformly with respect to x inside a disk of center α. Consequently, $f'(\alpha) = \sum_{n=1}^{\infty} na_n(x-\alpha)^n \ \forall \alpha \in d(0,R)$. Then

$$|f'|(r) = \sup_{n\in\mathbb{N}}(|na_n|r^{n-1}) \leq \sup_{n\in\mathbb{N}}(|a_n|r^{n-1}) = \frac{1}{r}|f|(r).$$

The last statement concerning $f^{(k)}$ is then immediate. □

More generally, we can derive the following:

Theorem B.9.2: *Let* $f \in H(\Delta(0,R,R'))$. *Then* $f^{(k)}$ *also belongs to* $H(\Delta(0,R,R'))$ *for every* $k \in \mathbb{N}^*$ *and satisfies* $|f'|(r) \leq \frac{|f|(r)}{r} \ \forall r \in]R,R'[$. *Moreover, if the residue characteristic does not divide* $\nu^+(f,\log r)$ *or* $\nu^-(f,\log r)$, *then* $|f'|(r) = \frac{|f|(r)}{r}$.

Proof. $f(x)$ is equal to a Laurent series $\sum_{-\infty}^{+\infty} a_n x^n$ with $\lim_{n\to+\infty} |a_n|R'^n = 0$ and $\lim_{n\to-\infty} |a_n|R^n = 0$, hence obviously $\lim_{n\to-\infty} |na_n|R^{n-1} = 0$ and $\lim_{n\to+\infty} |na_n|R'^{n-1} = 0$.

Consequently, $f'(x)$ belongs to $H(\Delta(0,R,R'))$. Then

$$|f'|(r) = \sup_{n\in\mathbb{Z}}(|na_n|r^{n-1}) \leq \sup_{n\in\mathbb{Z}}(|a_n|r^{n-1}) = \frac{1}{r}|f|(r).$$

Suppose now that the residue characteristic p does not divide $\nu^+(f,\log r)$ or $\nu^-(f,\log r)$. If $\nu^+(f,\log r) = \nu^-(f,\log r)$ is an integer q, it is obvious that $|f|(r) = |a_q|r^q$ and $|qa_q| = |a_q|$, hence $|f'|(r) = \dfrac{|f|(r)}{r}$, provided $q \neq 0$. Next, the property has continuation by continuity to the points μ such that $\nu^+(f,\mu) \neq \nu^-(f,\mu)$. □

Corollary B.9.3: *Let $R,\ R' \in]0,+\infty[(R < R')$ and let $f \in \mathcal{A}(d(0,R^-))$ (resp. $f \in \mathcal{A}(\Gamma((0,R,R')$. Then $|f'|(r) \leq \dfrac{|f|(r)}{r}$ $\forall r \in]0,R[$ (resp. $\forall r \in]R,R'[$).*

Corollary B.9.4: *Suppose $\mathbb{K}$ has characteristic zero and f belongs to $H(d(0,R))$. Then $a_n = \dfrac{f^{(n)}(0)}{n!}$ for every $n \in \mathbb{N}$ and if α is a zero of multiplicity order q of f, then we have $f^{(j)}(\alpha) = 0$ for every $j < q$ and $f^{(q)}(\alpha) \neq 0$.*

Corollary B.9.5: *Let $h \in \mathbb{K}(x)$. Then for all $r > 0$, we have $|h'|(r) \leq \dfrac{|h|(r)}{r}$.*

Theorem B.9.6: *Let $f(x) = \sum_{n=0}^{\infty} a_n x^n \in \mathcal{A}(d(0,r^-))$. The power series $\sum_{n=1}^{\infty} na_n x^{n-1}$ also belongs to $\mathcal{A}(d(0,r^-))$ and is equal to the derivative of f in $d(0,r^-)$. The radius of convergence of f' is superior or equal to the one of f. Further, if $\mathbb{K}$ has characteristic 0, the radius of convergence of f' is the same as that of f.*

Proof. By Theorem B.9.1, the first statement is clear. Now we suppose that $\mathbb{K}$ has characteristic zero. If $\mathbb{K}$ has residue characteristic zero, then $|n| = 1$ for all $n \in \mathbb{N}^*$ and, therefore, the last statement is clear. Now, we assume that $\mathbb{K}$ has residue characteristic $p \neq 0$. By Lemma A.5.17, we have $\dfrac{1}{n} \leq |n| \leq 1$ for every $n \in \mathbb{N}^*$ and, therefore, $\lim_{n\to+\infty} \sqrt[n]{|n|} = 1$. Consequently, $\limsup_{n\to+\infty} \sqrt[n]{|a_n|} = \limsup_{n\to+\infty} \sqrt[n-1]{|na_n|}$ and finally f' has the same radius of convergence as f. □

Corollary B.9.7: *Suppose $\mathbb{K}$ has characteristic 0. Let $f(x) = \sum_{n=0}^{\infty} a_n x^n \in \mathcal{A}(d(0,r^-))$. The power series $\sum_{n=0}^{\infty} \dfrac{a_n}{n+1} x^{n+1}$ also belongs to $\mathcal{A}(d(0,r^-))$ and has the same radius of convergence as that of f and is a primitive of f in $d(a,r^-)$.*

Remark: Unlike in Archimedean analysis, when the characteristic p of $\mathbb{K}$ is not zero, there do exist power series f whose derivatives have a radius of convergence bigger than that of f. For example, let $f(x) = \sum_{n=0}^{\infty} x^{p^n}$: the radius of convergence of f is 1 while this of f' is $+\infty$.

Theorem B.9.8: *Let $a \in \mathbb{K}$ and let $R \in \mathbb{R}_+$. Let $(f_m)_{m\in\mathbb{N}^*}$ be a sequence of $H(d(a,R))$ converging uniformly to a function f. Then the sequence $(f'_m)_{m\in\mathbb{N}^*}$ converges uniformly to f' in $H(d(a,R))$ and we have $\|f'_m - f'\|_{d(a,R)} \leq \frac{\|f_m-f\|_{d(a,R)}}{R}$ $\forall m \in \mathbb{N}$.*

Proof. For each $m \in \mathbb{N}$, set $f_m(x) = \sum_{n=0}^{\infty} a_{n,m}x^n$ and let $f(x) = \sum_{n=0}^{\infty} b_n x^n$. Then for each $m \in \mathbb{N}$, we have $\lim_{n\to+\infty} |a_{n,m}|R^n = 0$, $\lim_{n\to+\infty} |b_n|R^n = 0$. Now, the Banach norm of $f_m - f$ tends to zero when m goes to ∞, hence $\lim_{m\to+\infty} \left(\sup_{n\in\mathbb{N}}(|a_{n,m} - b_n|R^n)\right) = 0$. Consequently, considering the respective derivatives, we have

$$\lim_{m\to+\infty} \left(\sup_{n\in\mathbb{N}}(|na_{n,m} - nb_n|R^n)\right) = 0$$

and, therefore, by Theorem B.9.1, we have $\|f'_m - f'\|_{d(a,R)} \leq \frac{\|f_m - f\|_{d(a,R)}}{R}$. We are done. □

Corollary B.9.9: *Let $a \in \mathbb{K}$ and let $R \in \mathbb{R}_+$. Let $(f_m)_{m\in\mathbb{N}^*}$ be a sequence of $H(\mathbb{K}\setminus d(a,R^-))$ converging uniformly to a function f. Then the sequence $(f'_m)_{m\in\mathbb{N}^*}$ converges uniformly to f' in $H(\mathbb{K}\setminus d(a,R^-))$ and we have $\|f'_m - f'\|_{\mathbb{K}\setminus d(a,R^-)} \leq R\|f_m - f\|_{\mathbb{K}\setminus d(a,R^-)}$ $\forall m \in \mathbb{N}$.*

Theorem B.9.10: *Suppose $\mathbb{K}$ has characteristic 0. Let $a \in \mathbb{K}$ and let $R \in \mathbb{R}_+$. Let $(f_m)_{m\in\mathbb{N}^*}$ be a sequence of $\mathcal{A}(d(a,R^-))$ such that the sequence $(f'_m)_{m\in\mathbb{N}^*}$ converges uniformly to a function h in $H(d(a,r))$ $\forall r \in]0,R[$ and such that the sequence $f_m(a)$ converges in $\mathbb{K}$. Then the sequence $(f_m)_{m\in\mathbb{N}^*}$ converges to a function $f \in \mathcal{A}(d(a,R^-))$ such that $f' = h$ and the convergence is uniform in $d(a,r)$ for every $r \in]0,R[$.*

Proof. Without loss of generality, we can assume $a = 0$. Let us fix $r \in]0,R[$ and let us show that the sequence $(f_m)_{m\in\mathbb{N}^*}$ converges uniformly to a function $f \in H(d(a,r))$ such that $f' = h$. For every $m \in \mathbb{N}$, let $f'_m = \sum_{n=0}^{+\infty} b_{n,m}x^n$ and let $h(x) = \sum_{n=0}^{+\infty} b_n x^n$. Take $s \in]r,R[$. By hypothesis, we have

(1) $\lim_{m\to+\infty} \sup_{n\in\mathbb{N}} |b_{n,m} - b_n|s^n = 0.$

But since $|n| \geq \frac{1}{n}$, we have $\lim_{n\to+\infty} \frac{\rho^{n+1}}{|n+1|} = 0\ \forall \rho < 1$, therefore, by (1) we have $\lim_{m\to+\infty} \left(\sup_{n\in\mathbb{N}} |\frac{b_{n,m} - b_n}{n+1}|r^{n+1} \right) = 0$. Consequently, the sequence $(f_m - f_m(0))_{m\in\mathbb{N}}$ converges uniformly to the function $g(x) = \sum_{n=0}^{+\infty} \frac{b_n}{n+1} x^{n+1}$. Set $\lim_{m\to+\infty} f_m(0) = b_0$. Then the sequence $(f_m)_{m\in\mathbb{N}}$ converges to $g(x) + b_0$ uniformly in $d(0,r)$ for every $r < R$. □

Corollary B.9.11: *Suppose $\mathbb{K}$ has characteristic 0. Let $\alpha \in \mathbb{K}$, let $R \in \mathbb{R}_+$, and let $a \in \mathbb{K} \setminus d(\alpha, R)$. Let $(f_m)_{m\in\mathbb{N}^*}$ be a sequence of $\mathcal{A}(\mathbb{K} \setminus d(\alpha, R))$ be such that the sequence $(f'_m)_{m\in\mathbb{N}^*}$ converges uniformly to a function h in $H(\mathbb{K} \setminus d(\alpha, r^-))\ \forall r > R$ and that the sequence $(f_m(a))_{m\in\mathbb{N}}$ converges in $\mathbb{K}$. Then the sequence $(f_m)_{m\in\mathbb{N}^*}$ converges to a function $f \in \mathcal{A}(\mathbb{K} \setminus d(\alpha, R))$ such that $f' = h$ and the convergence is uniform in $\mathbb{K} \setminus d(\alpha, r^-)$ for every $r > R$.*

Theorem B.9.12: *Let $r > 0$. If $\mathbb{K}$ has characteristic 0, then an element $f \in H(d(0, r))$ has a derivative identically equal to 0 if and only if it is equal to a constant.*

If $\mathbb{K}$ has a characteristic $p \neq 0$, then an element $f \in H(d(0, r))$ has a derivative identically equal to 0 if and only if there exists $g \in H(d(0, r))$ such that $f(x) = (g(x))^p$.

Proof. By Theorem B.5.6, each element of $H(d(0, r))$ is a convergent power series, hence the statement about the case when $\mathbb{K}$ has characteristic zero is obvious. Now, suppose that $\mathbb{K}$ has a characteristic $p \neq 0$. If there exists $g \in H(d(0, r))$ such that $f(x) = (g(x))^p$, obviously we have $f'(x)$ identically equal to 0.

Now, we suppose $f'(x)$ identically equal to 0. Hence, $f(x)$ is of the form $\sum_{j=0}^{\infty} b_j x^{jp}$, with $\lim_{j\to\infty} |b_j| r^{jp} = 0$. For each $j \in \mathbb{N}$, we can take $c_j \in \mathbb{K}$ such that $(c_j)^p = b_j$. Then, it is seen that $\lim_{j\to\infty} |c_j| r^j = 0$. Now, we can put $g(x) = \sum_{n=0}^{\infty} c_n x^n$ and, therefore, g belongs to $H(d(0, r))$. Since $\mathbb{K}$ has characteristic p, we have $(g(x))^p = f(x)$. This ends the proof. □

Corollary B.9.13: *Let $r > 0$. If $\mathbb{K}$ has characteristic 0, then an element $f \in H(\mathbb{K} \setminus d(0, r^-))$ (resp. $f \in H_0(\mathbb{K} \setminus d(0, r^-))$) has a derivative identically equal to 0 if and only if it is equal to a constant (resp. to 0).*

If $\mathbb{K}$ has a characteristic $p \neq 0$, then an element $f \in H(\mathbb{K} \setminus d(0, r^-))$ (resp. $f \in H_0(\mathbb{K} \setminus d(0, r^-))$) has a derivative identically equal to 0 if and only if there exists $g \in H(\mathbb{K} \setminus d(0, r^-))$ (resp. $g \in H_0(\mathbb{K} \setminus d(0, r^-))$) such that $f(x) = (g(x))^p$.

Corollary B.9.14: *Let $r > 0$. If $\mathbb{K}$ has characteristic 0, then a power series $f(x) \in \mathcal{A}(d(0, r^-))$ has a derivative identically equal to 0 if and only if it is equal to a constant.*

If $\mathbb{K}$ has a characteristic $p \neq 0$, then a power series $f(x) \in \mathcal{A}(d(0, r^-))$ has a derivative identically equal to 0 if and only if there exists $g \in \mathcal{A}(d(0, r^-))$ such that $f(x) = (g(x))^p$.

Theorem B.9.15 improves Theorems B.9.1 and B.9.2 concerning derivatives of order $k > 1$.

Theorem B.9.15: *Let $a \in \mathbb{K}$, let $R, R', R'' \in \mathbb{R}_+^*$ with $R' < R''$, and let $f \in H(d(0, r))$ (resp. let $f \in H(\mathbb{K} \setminus d(0, R^-))$, resp. let $f \in H(\Gamma(0, R', R'')))$. Then, for every $k \in \mathbb{N}^*$, for every $r < R$ (resp. $r > R$, resp. $r \in]R',\ R''[$), we have*

$$|f^{(k)}|(r) \leq |k!| \frac{|f|(r)}{r^k}.$$

Proof. Let $f(x) = \sum_{n=0}^{\infty} a_n x^n \in H(d(0, R^-))$. By Theorem B.5.6 we have $|f|(r) = \sup_{n \in \mathbb{N}} |a_n| r^n$ and $|f^{(k)}|(r) = \sup_{n \geq k} \left| \frac{(n!)}{((n-k)!)} a_n \right| r^{n-1}$. But $\frac{(n!)}{((n-k)!)}$ is an integer multiple of $k!$ because the combination $\binom{n}{k}$ belongs to $\mathbb{N}$. Consequently, $\left| \frac{(n!)}{((n-k)!)} \right| \leq |k!|$ and, therefore, we obtain $|f^{(k)}|(r) \leq \sup_{n \geq k} |k!||a_n| r^{n-k}$. The proof is similar when f belongs to $H(\mathbb{K} \setminus d(0, r))$ or to $H(\Gamma(0, r', r''))$. □

Corollary B.9.16: *Let $a \in \mathbb{K}$, let $r \in \mathbb{R}_+^*$, and let $f \in H(d(a, r))$. Then*

$$\|f^k\|_{d(a,r)} \leq |k!| \frac{\|f\|_{d(a,r)}}{r^k}.$$

Theorem B.9.17: *Let $f \in H_b(D)$ and let $\rho = \delta(D, (\mathbb{K} \setminus \overline{D}))$. If $\rho > 0$ then f' belongs to $H_b(D)$ and satisfies $\|f'\|_D \leq \frac{1}{\rho} \|f\|_D$.*

Proof. Let $(T_n)_{n \in \mathcal{S}}$ be the sequence of the f-holes and let $D' = \widetilde{D} \setminus \left(\bigcup_{n \in \mathcal{S}} T_n \right)$. By Theorem B.6.1 we know that $f \in H_b(D')$ and that

(1) $\|f\|_{D'} = \|f\|_D$.

By Theorem B.9.1, f has a derivative f' in D' and we will first check that the function f' satisfies $\|f'\|_{D'} \leq \frac{1}{\rho} \|f\|$. Let $a \in D'$. The disk $d(a, \rho^-)$ is obviously included in D because if a point $b \in d(a, \rho^-)$ belonged to $\mathbb{K} \setminus D'$, since D is closed there would be a disk $d(b, r^-) \subset \mathbb{K} \setminus D'$ with $r < \rho$. Thus, when $x \in d(a, \rho^-)$, $f(x)$ is of the form $\sum_{n=0}^{\infty} a_n (x-a)^n$ and, hence, the conclusion comes from Theorem B.9.1. □

Corollary B.9.18: *Let $f \in H(D)$ be such that the set of diameters of the f-holes has a strictly positive lower bound. Then $f' \in H(D)$.*

Proof. By Theorem B.2.5 we know that f is in the form $g + h$ with $h \in R(D)$ and $g \in H_b(\overline{D})$. Obviously h' belongs to $R(D)$ and by Theorem B.9.17, we have $g' \in H_b(\overline{D})$. □

Theorem B.9.19: *Let $D \in \text{Alg}$ be closed and open. Let $f \in H_b(D)$, let $(V_n)_{n \in \mathbb{N}^*}$ be the set of f-holes, and let $\sum_{n=0}^{\infty} f_n$ be its Mittag–Leffler series on D defined as $f_0 = \overline{\overline{f_0}}$ and for every $n \in \mathbb{N}^*$, $f_n = \overline{\overline{f_{V_n}}}$. The following three conditions are equivalent:*

(a) *f' belongs to $H(D)$*

(b) *the series $\sum_{n=0}^{\infty} f'_n$ converges in $H(D)$*

(c) *the series $\sum_{n=0}^{\infty} f'_n$ converges to f' in $H_b(D)$.*

Proof. We will first prove the equivalence between (b) and (c). For each $q \in \mathbb{N}$, the sum $\sum_{n=0}^{q} f'_n$ clearly belongs to $H_b(D)$. Thus we assume that this series $\sum_{n=0}^{\infty} f'_n$ converges to an element $h \in H(D)$ and we will prove that $h = f' \in H_b(D)$. Let $\alpha \in D$. There exists a disk $d(\alpha, r) \subset D$. We will show that $h(\alpha) = f'(\alpha)$. For every $\psi \in H(D)$, $\widetilde{\psi}$ will denote the restriction of ψ to $d(\alpha, r)$. Since the sequence $\widetilde{\left(\sum_{j=0}^{n} f_j\right)}$ converges to $\widetilde{f}$, by Theorem B.9.17 the sequence of the derivatives $\widetilde{\left(\sum_{j=0}^{n} f'_j\right)}$ does converge to $\widetilde{f'}$, hence it is clearly seen that $f'(\alpha) = h(\alpha)$ and, therefore, $f' = h$. We will check that f' is bounded in D. The sequence $\|f'_n\|_D$ has limit 0, hence is obviously bounded and, therefore, its sum f' is bounded in D. Thus (b) and (c) are equivalent.

Since (c) trivially implies (a), we just have to prove that (a) implies (b). Thus we suppose (a) to be true and will prove (b).

First, we suppose D bounded. For each hole T of D that is either a f-hole or a f'-hole, we denote by $\overline{\overline{f_T}}$ (resp. $\overline{\overline{g_T}}$) the Mittag–Leffler term of f (resp. f'). Let $\mathcal{S}$ be the set of holes T such that $\overline{\overline{(f_T)'}} \neq \overline{\overline{g_T}}$ and let $\mathcal{T}$ be the set of f-holes such that $\overline{\overline{(f_T)'}} = \overline{\overline{g_T}}$. If we can show that $\mathcal{S} = \emptyset$, then (b) is clearly proven.

Hence, we suppose $\mathcal{S} \neq \emptyset$. All the $\overline{\overline{g_T}}$ are equal to zero except maybe a countable family of them. The series $\sum_{T \in \mathcal{T}} \overline{\overline{g_T}}$ and $\sum_{T \in \mathcal{S}} \overline{\overline{g_T}}$ obviously converge in $H(D)$, and then we have $f' = \sum_{T \in \mathcal{T}} \overline{\overline{(f_T)'}} + \sum_{T \in \mathcal{S}} \overline{\overline{g_T}}$. Since (b) implies (c), the series $\sum_{T \in \mathcal{T}} \overline{\overline{(f_T)'}}$ is clearly equal to the derivative of $\sum_{T \in \mathcal{T}} \overline{\overline{f_T}}$. Let $h = \sum_{T \in \mathcal{S}} \overline{\overline{f_T}} = f - \sum_{T \in \mathcal{T}} \overline{\overline{f_T}}$. Then $h' = f' - \sum_{T \in \mathcal{T}} \overline{\overline{(f_T)'}} = \sum_{T \in \mathcal{S}} \overline{\overline{g_T}}$. Let $\mathcal{D}$ be the family of diameters of the holes T that belong to $\mathcal{S}$ and let λ be its lower bound. Suppose $\lambda > 0$. By Theorem B.9.17, the series $\sum_{T \in \mathcal{S}} \overline{\overline{(f_T)'}}$ converges to h', hence $\sum_{T \in \mathcal{S}} \overline{\overline{(f_T)'}}$ is the Mittag–Leffler series of h' on D, hence $\overline{\overline{(f_T)'}} = \overline{\overline{g_T}}$ for all $T \in \mathcal{S}$ and that contradicts the definition of $\mathcal{S}$. Hence, $\lambda = 0$.

Now, we will prove that there exists a hole $V = d(a, r^-) \in \mathcal{S}$ with an annulus $\Gamma(a, r, s)$ such that the set $\mathcal{U}$ of the diameters ρ of the f-holes included in $\Gamma(a, r, s)$ has a strictly positive lower bound. Indeed, suppose such a hole V does not exist. Then we can easily construct a sequence of f-holes $(T_n)_{n \in \mathbb{N}^*}$ of the form $T_n = d(a_n, r_n^-)$ with (1) $r_n \leq \frac{1}{n}$ and

(2) $|a_{n+1} - a_n| \leq \frac{2}{n}$.

For example, assume the sequence has just been constructed up to the rank q, satisfying (1) and (2) for $n \leq q$. Since V does not exist, then in $\Gamma(a_q, r_q, \frac{2}{q})$ we can find a f-hole $T_{q+1} = d(a_{q+1}, r_{q+1}^-)$ with $r_{q+1} < \frac{2}{q+1}$ and then the sequence is clearly constructed by induction by taking first any f-hole $T_1 = d(a_1, r_1^-)$. The sequence $(T_n)_{n \in \mathbb{N}^*}$ clearly converges to a point $w \in D$ and that contradicts the hypothesis "D is closed and open." Hence, we have now proven the existence of the f-hole V with an annulus $\Gamma(a, r, s)$ and a number $\xi > 0$ such that every f-hole $T \subset \Gamma(a, r, s,)$ satisfies

(3) diam $(T) \geq \xi$.

Let $\mathcal{L}$ be this family of f-holes included in $\Gamma(a, r, s)$. Let $l = \sum_{T \in \mathcal{L}} \overline{\overline{f_T}}$. By Theorem B.9.17 the series $\sum_{T \in \mathcal{L}} \overline{\overline{(f_T)'}}$ converges to l' in $H(D)$. Now let $\psi = h - l - \overline{\overline{f_V}}$. Clearly ψ belongs to $H(D)$ and no hole T (of D) included in $d(a, s)$ is a ψ-hole. Hence, ψ extends to an element of $H(D \cup d(a, s))$. In $d(a, s)$, $\psi(x)$ is equal to a power series $\phi(x) \in H(d(a, s))$, hence $\phi' \in H(d(a, s))$. Thus in $D \cap d(a, s)$, $\psi'(x)$ is equal to the series $\phi'(x)$ and then for every hole T of D included in $d(a, s)$ the Mittag–Leffler term of ψ associated to T (with respect to D) is zero.

On the other hand, we have $\psi' = h' - l' - (\overline{\overline{f_V}})' = \sum_{T\in\mathcal{S}} \overline{\overline{g_T}} - \sum_{T\in\mathcal{L}} (\overline{\overline{f_T}})' - (\overline{\overline{f_V}})'$ and then the Mittag–Leffler term of ψ' associated to V (with respect to D) is $\overline{\overline{g_V}} - (\overline{\overline{f_V}})' \neq 0$. Hence, we have a contradiction with $\psi \in H(d(a,s))$. This finishes proving that (b) is true when D is bounded.

Now, we suppose D unbounded. Since $D \in \text{Alg}$, there exists a disk $d(0,S)$ such that all holes of D are included in this disk. Then for every element $h \in H(D)$, its Mittag–Leffler series in $H(D)$ is the same as in $H(D')$. This is true for both f, f', and, therefore, (b), which is true in $H(D')$, is obviously true in $H(D)$. That ends the proof of Theorem B.9.19. □

When $\mathbb{K}$ has characteristic zero, in most of the cases, we are now able to answer the question "does $f' = 0$ implies $f = ct$." When D is not infraconnected, it admits an empty annulus $\Lambda = \Gamma(a, r', r")$ and, hence, by Proposition B.2.15 we know that there exists $w \in H(D)$ such that $w(x) = 1$ whenever $x \in \mathcal{I}(\Lambda)$, while $w(x) = 0$ whenever $x \in \mathcal{E}(\Lambda)$. Thus the condition "$D$ is infraconnected" is certainly necessary to be able to answer "yes" the question above.

The two Theorems that follow show this condition to be sufficient too, provided D satisfies a little extra condition like to be closed or to belong to Alg.

By Theorems B.6.1, B.9.19, B.9.8 and Corollary B.9.9, we can derive the following:

Corollary B.9.20: *Let D be an open closed bounded infraconnected subset of $\mathbb{K}$ and let $(f_n)_{n\in\mathbb{N}}$ be a sequence of $H(D)$ converging uniformly to a function f. Then the sequence $(f'_n)_{n\in\mathbb{N}}$ converges uniformly to f'.*

Remark: Let D be an open closed infraconnected bounded subset of $\mathbb{K}$, let $\alpha \in D$, and let $(f_n)_{n\in\mathbb{N}}$ be a sequence of $H(D)$ such that the sequence $f_n(\alpha)$ converges in $\mathbb{K}$ and that the sequence $(f'_n)_{n\in\mathbb{N}}$ converges in $H(D)$ to a function $h \in H(D)$. Comparatively to the Archimedean context, we could expect that the sequence $(f_n)_{n\in\mathbb{N}}$ converges in $H(D)$ to a function f such that $f' = h$. Actually, that's wrong. For example, define the sequence $(f_n)_{n\in\mathbb{N}}$ as $f_n(x) = \sum_{k=0}^{n} \frac{x^{p^{2k}}}{p^k}$. Then $f'_n(x) = 1 + \sum_{k=1}^{n} p^k x^{p^{2k-1}}$ and, hence, the sequence $(f'_n)_{n\in\mathbb{N}}$ converges in $H(d(0,1))$ to the function $1 + \sum_{k=1}^{+\infty} p^k x^{p^{2k-1}}$, whereas $f_n(0) = 1 \ \forall n \in \mathbb{N}$. However, the function $\sum_{k=0}^{+\infty} \frac{x^{p^{2k}}}{p^k}$ is unbounded in $d(0,1)$ and, hence, does not belong to $H(d(0,1))$. That remark does not contradict Theorem B.9.10, which only concerned analytic functions in an "open" disk $d(0,r^-)$.

Theorem B.9.21: *$\mathbb{K}$ is supposed to have characteristic zero. Let E be an open subset of $\mathbb{K}$ such that $\overline{E}$ also is open. Then E is infraconnected if and only if for every $f \in H(E)$ such that $f'(x) = 0$ whenever $x \in E$, we have $f = ct$.*

Proof. If E is not infraconnected, it admits at least an empty annulus Λ and then by Proposition B.2.15, the characteristic function u of $\mathcal{I}(\Lambda)$ belongs to $H(E)$. Hence, there do exist nonconstant elements $f \in H(E)$ whose derivative is identically 0. Now let E be infraconnected and let $f \in H(E)$ satisfy $f'(x) = 0$ whenever $x \in E$. We just have to prove that f is a constant.

Suppose first that E is bounded. By Theorem B.2.5, f is in the form $f^* + \overline{f}$ with $f^* \in R_0(\mathbb{K} \setminus (\overline{E} \setminus E))$ and $\overline{f} \in H(\overline{E})$, so $f^{*\prime}(x) + \overline{f}'(x) = 0$ whenever $x \in E$ and, therefore, $f^{*\prime} = -\overline{f}'$. Hence, $\overline{f}' \in R_0(\mathbb{K} \setminus (\overline{E} \setminus E)) \cap H(\overline{E})$. Thus f^* is a rational function that has no pole in $\mathbb{K}$ and then it is a polynomial. Now, as an element of $R_0(\mathbb{K} \setminus (\overline{E} \setminus E))$ it tends to 0 when $|x|$ goes to $+\infty$, hence $f^* = 0$ and, therefore, $f^{*\prime}$ is identically 0. Since f^* belongs to $R_0(\mathbb{K} \setminus (\overline{E} \setminus E))$ clearly, $f^* = 0$, and, therefore, f belongs to $H(\overline{E})$.

Let $a \in E$ and let $\sum_{n \in \mathcal{S}} h_n$ be the Mittag–Leffler series of f in $\overline{E}$, with $h_0 = \overline{\overline{f_0}}$ and for each $n \in \mathcal{S}$, set $h_n = \overline{\overline{f_{T_n}}}$, for any f-hole T_n. Since $\overline{E}$ is open, we can apply Theorem B.9.19 to E and then we have $(h_n)' = 0$ for every $n \in \mathbb{N}$. Since $\overline{E}$ is bounded of diameter r, then by Theorem B.7.8 we know that h_0 is a constant. In the same way, by Corollary B.9.13, for each $q \in \mathbb{N}$, we know that $h_q = 0$. Hence, f is a constant.

Finally let E be unbounded. Then for all $r > 0$ the set $E_r = E \cap d(0, r)$ is such that $\overline{E_r}$ is open, hence f is constant in E_r and, therefore, in all of E. □

Remark: In particular, Theorem B.9.21 applies to open closed sets.

Corollary B.9.22: *$\mathbb{K}$ is supposed to have characteristic zero. Let E be open and belong to* Alg. *Then E is infraconnected if and only if for every $f \in H(E)$ such that $f'(x) = 0$ whenever $x \in E$, f is a constant in E.*

Proof. Indeed, since E belongs to Alg, it satisfies Condition (B) in Theorem B.8.9: $\overline{D} \setminus D \subset \left(\overset{\circ}{\overline{D}}\right)$. But then, as it is also open; we check that $\overline{E}$ is open. □

We will now study thoroughly the question whether all analytic elements in a set D have derivative in $H(D)$.

Definitions and notations: We call *piercing of D* the number $\delta(\overline{D}, \mathbb{K} \setminus D) > 0$ and D will be said to be *well pierced* if $\delta(\overline{D}, \mathbb{K} \setminus D) > 0$.

Theorem B.9.23: *Let $\overline{D}$ be open. Then D is well pierced if and only if for every $f \in H(D)$, f' also belongs to $H(D)$.*

Proof. If D is well pierced, by Corollary B.9.18 we know that for every $f \in H(D)$, f' belongs to $H(D)$. Now let us suppose D has piercing zero and let $(T_n)_{n\in\mathbb{N}}$ be a sequence of holes $T_n = d(\alpha_n, \rho_n^-)$ such that $\lim_{n\to\infty} \rho_n = 0$. Let λ_n be a sequence in $\mathbb{K}$ such that $\lim_{n\to\infty} \frac{|\lambda_n|}{\rho_n} = 0$ while $\lim_{n\to\infty} \frac{|\lambda_n|}{\rho_n^2} = +\infty$. It is seen that the series $\sum_{n=1}^{\infty} \frac{\lambda_n}{x-\alpha_n}$ converges in $H(D)$ to an element f, while the series $\sum_{n=1}^{\infty} \frac{\lambda_n}{(x-\alpha_n)^2}$ does not. On the other hand, $\sum_{n=1}^{\infty} \frac{\lambda_n}{x-\alpha_n}$ obviously is the Mittag–Leffler series of f. If f' belongs to $H(D)$, by Theorem B.9.19 its Mittag–Leffler series must be $\sum_{n=1}^{\infty} \frac{\lambda_n}{(x-\alpha_n)^2}$. Since this series does not converge this is just impossible. $\square$

Before closing this chapter we will notice the following result that may be sometimes helpful in differential equations.

Theorem B.9.24: *Let $\overline{D}$ be open. We suppose that both f and f' belong to $H(D)$. For every $\epsilon > 0$ there exists $h \in R(D)$ such that $\|f - h\|_D \leq \epsilon$ together with $\|f' - h'\|_D \leq \epsilon$.*

Proof. First we suppose that f belongs to $H_b(\overline{D})$. We have to introduce a notation. Let $g \in H_b(\widetilde{D})$. If D is bounded, $\widetilde{D}$ is a disk $d(a,r)$ and g is of the form $\sum_{m=0}^{\infty} \lambda_m (x-a)^m$. Then for every $q \in \mathbb{N}$ we put $(g)_q = \sum_{m=0}^{q} \lambda_m (x-a)^m$. If D is unbounded, then g is a constant $\lambda_\circ$ and we put $(g)_q = g$ whenever $q \in \mathbb{N}$. Let $T = d(b, r^-)$ be a hole of D and let $l(x) = \sum_{m=1}^{\infty} \frac{\mu_m}{(x-b)^m}$. For each $q \in \mathbb{N}^*$, we put $(l)_q = \sum_{m=1}^{q} \frac{\mu_m}{(x-b)^m}$. Now let $\sum_{n=0}^{\infty} f_n$ be the Mittag–Leffler series of f, with $f_0 = \overline{\overline{f_0}}$ and for each $n \in \mathbb{N}^*$, $f_n = \overline{\overline{f_{T_n}}}$, for any f-hole T_n. By Theorem B.9.19, the Mittag–Leffler series of f' is $\sum_{n=0}^{\infty} f'_n$ and, therefore, there exists an integer $N(\epsilon)$ such that

$$(1) \quad \|\sum_{n=0}^{N(\epsilon)} f_n - f\|_D \leq \epsilon$$

and

$$(2) \quad \|\sum_{n=0}^{N(\epsilon)} f'_n - f'\|_D \leq \epsilon.$$

Obviously we have an integer $Q(\epsilon)$ such that $\|f_n - (f_n)_{Q(\epsilon)}\|_D \leq \epsilon$ whenever $n = 0, \dots, N(\epsilon)$ and then by (1) and (2) it is easily seen that $\|\sum_{n=0}^{N(\epsilon)} (f_n)_{Q(\epsilon)} - f\|_D \leq \epsilon$ and $\|\sum_{n=0}^{\infty}(f_n')_{Q(\epsilon)} - f'\|_D \leq \epsilon$. By putting $h = \sum_{n=0}^{N(\epsilon)} (f_n)_{Q(\epsilon)}$ we obtain the $h \in R(D)$ we want.

Now we can consider the general case. By Corollary B.2.7, f is of the form $l + \psi$, with $l \in H_b(\overline{D})$ and $\psi \in R(D)$. Hence, $f' = l' + \psi'$. Since f' belongs to $H(D)$, so does ψ'. Hence, by Theorem B.9.21, ψ' belongs to $H_b(\overline{D})$. We have just proven that there exists $t \in H_b(\overline{D})$ such that $\|l - t\|_D \leq \epsilon$ and $\|l' - t'\|_D \leq \epsilon$. Hence, we just have to consider $h = t + \psi$ and this ends the proof. □

Theorem B.9.25: *Let $\overline{D}$ be open. Suppose $0 \in D$. Let $f \in H(D)$ and let $r \in [\delta(0, D), \mathrm{diam}(D)]$. Then $|f'|(r) \leq \dfrac{|f|(r)}{r}$.*

Proof. Suppose first that $|f|(r) \neq 0$. Let $\epsilon > 0$ and let $\eta > 0$ be such that $\dfrac{|f| + \eta}{r} + \eta < \dfrac{|f|}{r} + \epsilon$. Now, by Theorem B.9.24 we can find $h \in R(D)$ such that $|h'|(r) \leq |f'| + \eta$ and $|h|(r) \leq |f| + \eta$. By Corollary B.9.3 we have $|h'|(r) \leq \frac{|h|(r)}{r}$, hence

$$|f'|(r) \leq |h'|(r) + \eta \leq \frac{|h|(r)}{r} + \eta \leq \frac{|f|(r) + \eta}{r} + \eta \leq \frac{|f|(r)}{r} + \epsilon.$$

Now suppose $|f|(r) = 0$. Then by Theorem B.9.24 we can find $h \in R(D)$ such that $\max(|h'|(r), |h|(r)) \leq \epsilon$, hence we have again $|f'| \leq \dfrac{|f|(r)}{r}$. This is true for all $\epsilon > 0$ and, hence the claim is proven. □

In the case when $\mathbb{K}$ has a characteristic $p \neq 0$, we have Theorem B.9.26.

Theorem B.9.26: *Let $\mathbb{K}$ have characteristic $p \neq 0$, let D be closed, and let $f \in H_b(D)$. Then $f'(x)$ is identically 0 if and only if there exists $g \in H_b(D)$ such that $f = g^p$.*

Proof. Indeed, if there exists $g \in H_b(D)$ such that $f = g^p$, of course we have $f' = 0$. Now, suppose that $f'(x)$ is identically 0. Let $a \in D$ and let $\sum_{n \in \mathcal{S}} h_n$ be the Mittag–Leffler series of f in D, with $h_0 = \overline{\overline{f_0}}$ and for each $n \in \mathcal{S}$, $h_n = \overline{\overline{f_{T_n}}}$, for any f-hole T_n. By Theorem B.9.19 we have $(h_n)' = 0$ for every $n \in \mathbb{N}$.

If D is unbounded, h_0 is a constant and then we can find $g_0 \in \mathbb{K}$ such that $(g_0)^p = h_0$. If D is bounded of diameter r, then by Corollary B.9.13 we can find $g_0 \in H(d(a, r))$ such that $(g_0)^p = h_0$. In the same way, for each $q \in \mathbb{N}$, by Corollary B.9.13 we can find $g_q \in H_0(\mathbb{K} \setminus T_q)$ such that $(g_q)^p = h_q$ and then, it is seen that

$\lim_{n\to\infty} \|g_n\|_D = 0$ because for each $n \in \mathbb{N}^*$, we have $(\|g_n\|_D)^p = \|h_n\|_D$. So, the series $(\sum_{n=0}^{\infty} g_n)$ converges in $H_b(D)$ to an element g, which clearly satisfies $g^p = f$. This ends the proof of Theorem B.9.26. □

B.10. Properties of the function Ψ for analytic elements

Throughout this chapter D is infraconnected.

The function $\Psi(f,\mu)$ was defined for rational functions in Chapter A.3. Here we will generalize that function to analytic elements. Its interest is to transform the multiplicative property of the norm $|\,.\,|$ into an additive property. Overall, Ψ is piecewise affine. Long ago, such a function was first defined in classical works such as the valuation function of an analytic element [2, 58, 72] denoted by $v(f,\mu)$. However the function $v(f,\mu)$ has the inconvenience of being contravariant: $\mu = -\log(|x|)$ and $v(f, -\log(|x|)) = -\log(|f|(r))$. Here we will change both senses of variation: $\Psi(f,\mu) = -v(f,-\mu)$.

Among applications, we can show that a set E is infraconnected if and only if for all $f \in H(E)$, $f(E)$ is infraconnected and that an analytic element converges along a monotonous filter $\mathcal{F}$ if and only if f' is vanishing along $\mathcal{F}$.

Definitions and notations: For every $a \in \widetilde{D}$, we put $\lambda(a) = \log(\delta(a,D))$ if $\delta(a,D) > 0$ and $\lambda(a) = -\infty$ if $\delta(a,D) = 0$. We denote by S the diameter of D, with $S = +\infty$ if D is not bounded.

Let $a \in \widetilde{D}$ and let $\mathcal{F}$ be a circular filter of center a and diameter $r \in [\delta(a,D), S] \cap \mathbb{R}$. By Proposition A.2.17, $\mathcal{F}$ is secant with D and then defines an element ${}_D\varphi_{\mathcal{F}}$ of *Mult*$(H(D), \mathcal{U}_D)$.

For every $f \in H(D)$ such that ${}_D\varphi_{\mathcal{F}}(f) \neq 0$ we put $\Psi_a(f, \log r) = \log\big({}_D\varphi_{\mathcal{F}}(f)\big)$. Next, for $f \in H(D)$ such that ${}_D\varphi_{\mathcal{F}}(f) = 0$ we put $\Psi_a(f,\log r) = -\infty$.

When $a = 0$ for simplicity we just put $\Psi(f,\mu) = \Psi_0(f,\mu)$. Then by definition, we have $\Psi(f,\log r) = \log(|f|(r))$.

In the same way, consider an annulus $\Gamma(0,r,t)$ and $f \in \mathcal{A}(\Gamma(0,r,t))$. Then for any $s \in]r,t[$, f belongs to $H(C(0,s)$, so we can consider $\Psi(f,\log(s)) = \log(|f|(s))$. If we consider $f \in \mathcal{A}(a,r^-)$, so much the more, we can consider $\Psi_a(f,\ell)$ for each $\ell < \log(r)[$.

Remark: Let $f(x) = \sum_{-\infty}^{+\infty} a_n x^n \in H(C(0,r))$ for some $r > 0$. By Theorem B.6.7 we have $\Psi(f,\log r) = \log\big({}_{C(0,r)}\varphi_{0,r}(f)\big) = \log\|f\|_{C(0,r)} = \sup_{n\in\mathbb{Z}} \Psi(a_n) + n\log r$.

Proposition B.10.1: *Let $a \in \widetilde{D}$, let $\mu \in [\lambda(a), \log(S)] \cap \mathbb{R}$, and let $f, g \in H(D)$. Then $\Psi_a(f+g,\mu) \leq \max(\Psi_a(f,\mu), \Psi_a(g,\mu))$ and when $\Psi_a(f,\mu) > \Psi_a(g,\mu)$, then $\Psi_a(f+g,\mu) = \Psi_a(f,\mu)$. Moreover, $\Psi_a(fg,\mu) = \Psi_a(f,\mu) + \Psi_a(g,\mu)$.*

Let $r,\ t\ \in]0,\ +\infty[$ be such that $r \leq t$. Let $f \in H(D)$ be such that $\Psi_a(f,\mu)$ is bounded in $[\log(r),\log(t)]$. Then $\Psi_a(f,\mu)$ is continuous and piecewise affine in $[\log(r),\log(t)]$. Further, there exists $h \in R(D)$ such that $\Psi_a(f,\mu) = \Psi_a(h,\mu)\ \forall \mu \in [\log(r),\log(t)]$.

Inside $D \cap \Gamma(a,r,t)$, the relation $\Psi(f(x)) = \Psi_a(f,\Psi(x-a))$ holds in all classes of all circles $C(a,s)$, except maybe in finitely many classes of finitely many circles $C(a,s)$.

Moreover, if $\Gamma(a,r,t) \subset D$, the function $\Psi_a(f,\mu)$ is convex in $[\log r, \log t]$.

Proof. Without loss of generality, we can assume $a = 0$. The first statements concerning operations and inequalities come directly from those of multiplicative seminorms ${}_D\varphi$. Now, suppose that $\Psi(f,\mu)$ is bounded in $[\log(r),\log(t)]$, hence there exists $\epsilon > 0$ such that $\Psi(f,\mu) > \log\epsilon\ \forall\mu \in [\log r, \log t]$.

Let $h \in R(D)$ satisfy $\|f-h\|_D < \epsilon$. Particularly, for every circular filter $\mathcal{F}$ secant with D, we have ${}_D\varphi_{\mathcal{F}}(f-h) < \epsilon$ and particularly ${}_D\varphi_{a,\rho}(f-h) < \epsilon\ \forall\rho \in [r,t]$, i.e., $\Psi(f-h,\mu) < \log(\epsilon) < \Psi(f,\mu)\ \forall\mu \in [\log r, \log t]$. Consequently, $\Psi(f,\mu) = \Psi(h,\mu)\ \forall\mu \in [\log r,\log t]$. Now, by Corollary A.3.18, the function $\Psi(h,\mu)$ is continuous, piecewise in $[\log r, \log t]$ and so is $\Psi(f,\mu)$. Moreover, if $\Gamma(a,r,t) \subset D$, the function $\Psi(h,\mu)$ is convex in $[\log r,\log t]$, hence so is $\Psi(f,\mu)$.

By Lemma A.3.13 the relation $\Psi(h(x)) = \Psi_a(h,\Psi(x-a))$ holds in all classes of all circles $C(a,s)$, except maybe in finitely many classes of finitely many circles $C(a,s)$. Therefore the same relation holds for f. □

Proposition B.10.2: *Let $a \in D$ and let $f \in H(D)$ satisfy $f(a) \neq 0$. There exists $\mu_\circ \in \mathbb{R}$ such that $\Psi_a(f,\mu) = \Psi(f(a))$ whenever $\mu \leq \mu_\circ$. Let $r \in \mathbb{R}_+^*$, let $\Lambda = C(0,r)$, and let f and $g \in H(\Lambda)$ satisfy $\|f-g\|_\Lambda < \|f\|_\Lambda$. Then we have $\nu^+(f,\log r) = \nu^+(g,\log r)$, $\nu^-(f,\log r) = \nu^-(g,\log r)$.*

Proof. Indeed let us take $r > 0$ such that $|f(x) - f(a)| < |f(a)|$ whenever $x \in d(a,r)\cap D$, hence $|f(x)| = |f(a)|$ whenever $x \in d(a,r)\cap D$ and, therefore, $\Psi_a(f,\mu) = \Psi(f(a))$ whenever $\mu \leq \log(r)$.

Let $f(x) = \sum_{-\infty}^{+\infty} a_n x^n$ and let $g(x) = \sum_{-\infty}^{+\infty} b_n x^n$. From the hypothesis we see that $\|f\|_\Lambda = \|g\|_\Lambda$. By Corollary B.5.9 we have

(1) $\sup_{n\in\mathbb{Z}} |a_n|r^n = \|f\|_\Lambda = \sup_{n\in\mathbb{Z}} |b_n|r^n$

and $\|f-g\|_\Lambda = \sup_{n\in\mathbb{Z}} |a_n - b_n|r^n$.
Let $s = \nu^-(f,\log r)$ and let $t = \nu^+(f,\log r)$. We see that

$|a_s - b_s|r^s \leq \|f-g\|_\Lambda < \|f\|_\Lambda = |a_s|r^s$, hence

(2) $|b_s| = |a_s|$.

In the same way we have

(3) $|a_t| = |b_t|$.

Now for every $n < s$ and for every $n > t$ we have $|a_n|r^n < |a_s|r^s = \|f\|_\Lambda$, hence $|b_n|r^n < \|f\|_\Lambda$. Finally by (1), (2), (3) we see that $\nu^-(g, \log r) = s$, $\nu^+(g, \log r) = t$. □

By Propositions B.10.1, B.10.2, and Corollary A.3.19 we can derive Corollary B.10.3.

Corollary B.10.3: *Let $f(x) \in H(\Gamma(0, r_1, r_2))$ (resp. $f(x) \in H(\Delta(0, r_1, r_2))$) (with $0 < r_1 < r_2$) and let $\sum_{-\infty}^{+\infty} a_n x^n$ be its Laurent series. The function $\mu \to \Psi(f, \mu)$ is bounded in $]\log r_1, \log r_2[$ (resp. in $[\log(r_1), \log(r_2)]$) and equal to $\sup_{n\in\mathbb{Z}}(\Psi(a_n) + n\mu)$. Next, we have $\Psi(f(x)) \leq \Psi(f, \Psi(x))$ whenever $x \in \Gamma(0, r_1, r_2)$ (resp. whenever $x \in \Delta(0, r_1, r_2)$) and the equality holds in all of $\Gamma(0, r_1, r_2)$ (resp. un all of $\Delta(0, r_1, r_2)$) except in finitely many classes of finitely many circles $C(0, r)$ ($r_1 < r < r_2$) (resp. $r_1 \leq r \leq r_2$). The right-side derivative (resp. the left-side derivative) of the function $\Psi(f, .)$ at μ is equal to $\nu^+(f, \mu)$ (resp. to $\nu^-(f, \mu)$). Moreover, if the function in μ $\Psi(f, \mu)$ is not derivable at μ, then μ lies in $\Psi(\mathbb{K})$.*

Further, the function $\Psi(f, .)$ is convex in $]\log r_1, \log r_2[$ (resp. in $[\log r_1, \log r_2]$). Next, given another $g \in H(\Gamma(0, r_1, r_2))$, (resp. $g \in H(\Delta(0, r_1, r_2))$) the functions ν^+ and ν^- satisfy $\nu^+(fg, \mu) = \nu^+(f, \mu) + \nu^+(g, \mu)$, $\nu^-(fg, \mu) = \nu^-(f, \mu) + \nu^-(g, \mu)$. Further, the function $\nu^+(f, .)$ is continuous on the right and the function $\nu^-(f, .)$ is continuous on the left at each point μ. They are continuous at μ if and only if they are equal.

Proposition B.10.4: *Let $a \in \widetilde{D}$ and let $f \in H(D)$. If $f(a) \neq 0$, there exists $s > 0$ such that $\Psi(f, \mu) = \Psi(f(a))$ $\forall \mu \leq s$. Let $b \in D$ be such that $|a - b| = r$ and $d(b, r^-) \subset D$. Then we have $\Psi_b(f, \mu) = \Psi_a(f, \mu)$ $\forall \mu \leq \Psi(b - a)$.*

Proof. Since $f(a) \neq 0$, the first statement is immediate since $|f(x)|$ is a constant inside a disk of center a. Next, by Lemma A.3.14 the relation $\Psi_a(f, \mu) = \Psi_b(f, \mu)$ when $\Psi(a - b) \leq \mu$ is true for every $f \in R(D)$, hence by (2), is obviously generalized to every $f \in H(D)$. □

Proposition B.10.5: *Let $\mu \in \mathbb{R}$ and let $f(x) = \sum_{-\infty}^{+\infty} a_n x^n \in H(C(0, \theta^\mu))$. Then $\Psi(f, \mu)$ is equal to $\sup_{n\in\mathbb{Z}} \Psi(a_n) + n\mu$ and we have $\Psi(f(x)) \leq \Psi(f, \mu)$ for all $x \in C(0, \theta^\mu)$. Moreover, the equality holds in every class except in finitely many classes where f admits zeros. Further, if $\nu^+(f, \mu) = \nu^-(f, \mu)$, then $\Psi(f(x)) = \Psi(f, \mu)$ whenever $x \in C(0, \theta^\mu)$.*

If $h \in H(C(0, \theta^\mu))$ satisfies $\Psi(f - h, \mu) < \Psi(f, \mu)$, then $\nu^+(f, \mu) = \nu^+(h, \mu)$ and $\nu^-(f, \mu) = \nu^-(h, \mu)$.

Proof. Let $\Lambda = C(0, \theta^\mu)$, let $s = \nu^-(f,\mu)$ and let $t = \nu^+(f,\mu)$. By the Remark above $\Psi(f,\mu)$ is obviously equal to $\sup_{n\in\mathbb{Z}}(\Psi(a_n) + n\mu)$. Let $x \in C(0,\theta^\mu)$. The inequality $\Psi(f(x)) \leq \Psi(f,\mu)$ is true because $\Psi(f,\mu) = \log \|f\|_\Lambda \geq \Psi(f(x))$. Finally by Proposition B.10.1 the equality holds in all the classes except in finitely many. If $\nu^+(f,\mu) = \nu^-(f,\mu)$ then $\Psi(a_s x^s) = \Psi(a_s) + s\mu > \Psi(a_n x^n)$ whenever $n \neq s$, hence $\Psi(f(x)) = \Psi(f,\mu)$.

Now, let $h \in H(\Lambda)$ satisfy $\Psi(f - h, \mu) < \Psi(f,\mu)$ and let $h(x) = \sum_{-\infty}^{+\infty} b_n x^n$. We have $\Psi(a_n - b_n) + n\mu < \Psi(a_s) + s\mu$ whenever $n \in \mathbb{Z}$, hence $\Psi(b_s) = \Psi(a_s)$, $\Psi(b_t) = \Psi(a_t)$, $\Psi(b_n) + n\mu < \Psi(a_s) + s\mu$ whenever $n < s$, and $n > t$ and $\Psi(b_n) + n\mu \leq \Psi(a_s) + s\mu$ whenever $n \in [s,t]$, hence finally $\nu^+(h,\mu) = \nu^+(f,\mu)$ and $\nu^-(h,\mu) = \nu^-(f,\mu)$. □

Corollary B.10.6: *Let* $f(x) = \sum_{-\infty}^{+\infty} a_n x^n \in \mathcal{A}(\Gamma(0,r_1,r_2))$ *(with* $0 < r_1 < r_2$*). The function* $\mu \to \Psi(f,\mu)$ *defined in* $]\log r_1, \log r_2[$ *is equal to* $\sup_{n\in\mathbb{Z}}(\Psi(a_n) + n\mu)$. *Next, we have* $\Psi(f(x)) \leq \Psi(f, \Psi(x))$ *whenever* $x \in \Gamma(0,r_1,r_2)$ *and the equality holds in all of* $\Gamma(0,r_1,r_2)$ *except in finitely many classes of each circle* $C(0,r)$ $(r_1 < r < r_2)$. *The right-side derivative (resp. the left-side derivative) of the function* $\Psi(f,.)$ *at* μ *is equal to* $\nu^+(f,\mu)$ *(resp. to* $\nu^-(f,\mu)$*). Moreover, if the function in* μ $\Psi(f,\mu)$ *is not derivable at* μ, *then* μ *lies in* $\Psi(\mathbb{K})$.

Further, the function $\Psi(f,.)$ *is convex in* $]\log r_1, \log r_2[$. *Next, given another* $g \in \mathcal{A}(\Gamma(0,r_1,r_2))$ *the functions* ν^+ *and* ν^- *satisfy*

$$\nu^+(fg,\mu) = \nu^+(f,\mu) + \nu^+(g,\mu), \quad \nu^-(fg,\mu) = \nu^-(f,\mu) + \nu^-(g,\mu).$$

Moreover, the function $\nu^+(f,.)$ *is continuous on the right and the function* $\nu^-(f,.)$ *is continuous on the left at each point* μ. *They are continuous at* μ *if and only if they are equal.*

Proof. All statements hold in all annuli $\Gamma(0,r',r'')$ with $r_1 < r' < r'' < r_2$ because the restriction of f to $\Gamma(0,r',r'')$ belongs to $H(\Gamma(0,r',r''))$. □

Proposition B.10.7: *Let* $\mu \in \mathbb{R}$ *and let* $f, g \in H(C(0,\theta^\mu))$. *Then* $\nu^+(fg,\mu) = \nu^+(f,\mu) + \nu^+(g,\mu)$ *and* $\nu^-(fg,\mu) = \nu^-(f,\mu) + \nu^-(g,\mu)$.

Proof. By Proposition B.10.2 the relations are obvious when f and $g \in R(C(0,\theta^\mu))$ because there is an annulus $\Gamma(0,r_1,r_2) \supset C(0,\theta^\mu)$ such that $f, g \in R(\Gamma(0,r_1,r_2))$. Now by Corollary B.10.6, we may extend them to $H(C(0,\theta^\mu))$ by taking h and $\ell \in R(C(0,\theta^\mu))$ such that $\Psi(f-h,\mu) < \Psi(f,\mu)$ and $\Psi(g-\ell,\mu) < \Psi(g,\mu)$. □

Proposition B.10.8: *Let* r_1, $r_2 \in \mathbb{R}$ *and let* f, $g \in \mathcal{A}(\Gamma(0,r_1,r_2))$ *having no zero in* $\Gamma(0,r_1,r_2)$ *and satisfying* $\nu(f,\mu) \neq \nu(g,\mu)$, $\forall \mu \in]\log r_1, \log r_2[$. *Then both* $\nu^+(f+g,\mu)$ *and* $\nu^-(f+g,\mu)$ *are equal either to* $\nu(f,\mu)$ *or to* $\nu(g,\mu)$.

Proof. Let $\mu_j = \log(r_j)$, $j = 1, 2$. Since both f, g have no zero in $\Gamma(0, r_1, r_2)$, $\nu(f, \mu)$ is a constant integer s and $\nu(g, \mu)$ is a constant integer $t \neq s$. Consequently, $\Psi(f, \mu)$ is of the form $a + s\mu$, $\Psi(g, \mu)$ is of the form $b + t\mu$, therefore the two functions in μ can coincide at most at one point in $[\mu_1, \mu_2]$. So, by Proposition B.10.1, we have $\Psi(f + g, \mu) = \max(\Psi(f, \mu), \Psi(g, \mu))$ for all $\mu \in [\log(r_1), \log(r_2)]$ except maybe at all point. But then, by continuity, the equality holds in all $[\log(r_1), \log(r_2)]$.

Let us fix $\mu_0 \in]\mu_1, \mu_2[$. Suppose $\Psi(f+g, \mu) = \Psi(f, \mu)$ in a neighborhood $]\mu_1, \mu_2[$ of μ_0. Then of course, $\nu(f + g, \mu_0) = \nu(f, \mu_0)$. Suppose now that $\Psi(f + g, \mu) = \Psi(f, \mu)$ in a left neighborhood $]\mu_1, \mu_0]$ of μ_0 and $\Psi(f + g, \mu) = \Psi(g, \mu)$ in a right neighborhood $[\mu_0, \mu_2[$ of μ_0, which implies $\Psi(f, \mu) > \Psi(g, \mu)$ $\forall \mu \in]\mu_1, \mu_0[$ and $\Psi(f, \mu) < \Psi(g, \mu)$ $\forall \mu \in]\mu_0, \mu_2[$. Then we have $\nu(f + g, \mu) = \nu(f, \mu) \forall \mu \in]\mu_1, \mu_0[$ and $\nu(f + g, \mu) = \nu(g, \mu) \forall \mu \in]\mu_0, \mu_2[$. Consequently, since ν^+ is continuous on the left and ν^- is continuous on the right, we can check that both $\nu^+(f + g, \mu_0)$ and $\nu^-(f + g, \mu_0)$ are equal either to $\nu(f \mu_0)$ or to $\nu(g, \mu_0)$. □

Theorem B.10.9: *Let $f \in \mathcal{A}(\mathbb{K} \setminus d(0, R))$. There exists $q \in \mathbb{N}$ such that*

$$\lim_{r \to +\infty} |f|(r) r^q = +\infty.$$

Proof. Let $s \in]R, +\infty[$ be such that $\nu^+(f, \log s) = \nu^-(f, \log s)$ and let $\tau = \nu^+(f, \log s)$. Thus $\Psi(f, \mu)$ has a derivative at $\log s$ equal to τ. Consequently, since by Proposition B.10.1 $\Psi(f, \mu)$ is convex, we have $\Psi(f, \mu) - \Psi(f, \log s) \geq \tau(\mu - \log s)$. Therefore,

$$\lim_{\mu \to +\infty} [\Psi(f, \mu) + (1 - \tau)\mu] = +\infty,$$

i.e., $\lim_{r \to +\infty} |f|(r) r^{(1-\tau)} = +\infty$. Finally we can take $q = \max(0, 1 - \tau)$. □

B.11. Vanishing along a monotonous filter

Throughout this chapter, the set D is supposed to be infraconnected.

By Chapter A.7 we know that there exists a spherically complete algebraically closed extension $\widehat{\mathbb{K}}$ of $\mathbb{K}$ whose residue class field is not countable and whose valuation group is equal to $\mathbb{R}$. Given a subset D of $\mathbb{K}$, we will denote by $\widehat{D}$ the subset

$$D \cup \left\{ \bigcup_{a \in \overset{\circ}{D}} \widehat{d}(a, \delta(a, (\mathbb{K} \setminus D))) \right\}.$$

The question whether an analytic element can tend to zero along a monotonous filter is known to be one of the main problems, which happens with p-adic analytic functions [49, 50, 58]. Here we will not describe T-filters [49]. However, we will describe sufficient conditions to prevent analytic elements to vanish along a monotonous filter, which is sufficient to study analytic and meromorphic functions inside disks or annuli.

We will apply results to characteristic functions and to the image of an infraconnected set.

Definitions and notations: Let $f \in H(D)$ and let $\mathcal{F}$ be a monotonous filter on D. When $\mathcal{F}$ is decreasing (resp. increasing) of center a and diameter S, f will be said to be *strictly vanishing* along $\mathcal{F}$ if $\lim_{\mathcal{F}} f(x) = 0$ and if there exists $S' > S$ (resp. $S' < S$) such that for every $r \in]S, S']$ (resp. $r \in [S', S[$) we have ${}_D\varphi_{a,r}(f) \neq 0$.

When $\mathcal{F}$ is decreasing with no center in $\mathbb{K}$, it admits a canonical basis $(D_n)_{n\in\mathbb{N}}$ with $D_n = d(a_n, r_n) \cap D$ and then f will be said to be *strictly vanishing* along $\mathcal{F}$ if $\lim_{\mathcal{F}} f(x) = 0$ and if there exists $S' > S$ such that ${}_D\varphi_{a_{n+1},r}(f) \neq 0$ whenever $r \in [r_n, S']$ and whenever $n \in \mathbb{N}$. Actually $\mathcal{F}$ admits a center α in $\widehat{\mathbb{K}}$ and then the definition given for decreasing filters with a center in $\mathbb{K}$ also applies and is obviously equivalent.

Lemma B.11.1 just translates these definitions by using the function Ψ.

Lemma B.11.1: *Let $f \in H(D)$ and let $\mathcal{F}$ be a decreasing (resp. an increasing) filter of center a and diameter S on D. Then f is strictly vanishing along $\mathcal{F}$ if and only if there exists $S' > S$ (resp. $S' \in]0, S[$) such that $\Psi_a(f, \log S) = -\infty, \Psi_a(f, \mu) > -\infty$ whenever $\mu \in]\log S, \log S']$ (resp. $[\log S', \log S[$).*

Let $\mathcal{G}$ be a decreasing filter with no center of diameter S and canonical basis $(D_n)_{n\in\mathbb{N}}$, with $D_n = d(a_n, r_n) \cap D$. Then f is strictly vanishing along $\mathcal{G}$ if and only if there exists $S' > S$ such that $\lim_{n\to\infty} \Psi_{a_n}(f, \log r_n) = -\infty$ and $\Psi_{a_n}(f, \log r) > -\infty$ for $r_n \leq r < S'$, $\forall n \in \mathbb{N}$.

Lemma B.11.2: *Let $f \in H(D)$ and let $\mathcal{F}$ be a monotonous filter such that f is strictly vanishing along $\mathcal{F}$. Then f is properly vanishing along $\mathcal{F}$.*

Proof. Let $S = \text{diam}(\mathcal{F})$. Let (D_n) be a canonical basis of $\mathcal{F}$ and suppose that f is not properly vanishing along $\mathcal{F}$. Since f is vanishing along $\mathcal{F}$, there exists $q \in \mathbb{N}$ such that $f(x) = 0$ whenever $x \in D_q$. But then, we can check that f does not satisfy the definition of an analytic element strictly vanishing along $\mathcal{F}$, because for every multiplicative semi-norm ${}_D\varphi_{a,r}$ whose circular filter is secant to D_q, we have ${}_D\varphi_{a,r}(f) = 0$. In particular, this applies to ${}_D\varphi_{a,r}$, for $r_q < r < S$, (resp. $S < r < r_q$) when $\mathcal{F}$ is increasing, (resp. decreasing) of center a, and $D_q = D \cap \Gamma(a, r_q, S)$, (resp. $D_q = D \cap \Gamma(a, S, r_q)$) and this applies to φ_{a_{q+1},r_q}, when $\mathcal{F}$ has no center, whereas $D_q = D \cap d(a_{q+1}, r_q)$. $\square$

Proposition B.11.3: *Let $a \in \widetilde{D}$ and $b \in D$, let $f \in H(D)$ satisfy $f(b) \neq 0$, and ${}_D\varphi_{a,r}(f) = 0$ for some $r \in]0, |a-b|]$. If ${}_D\varphi_{a,|a-b|}(f) = 0$, then f is strictly vanishing along an increasing filter of center b and diameter $S \leq |a - b|$. If ${}_D\varphi_{a,|a-b|}(f) \neq 0$ then f is strictly vanishing along a decreasing filter of center a and diameter $S \in [r, |a - b|[$.*

Proof. First suppose ${}_D\varphi_{a,|a-b|}(f) = 0$, hence we have

$$\Psi_b(f, \log|a-b|) = \Psi_a(f, \log|a-b|) = -\infty.$$

Since $f(b) \neq 0$, by Proposition B.10.4 we know that $\lim_{\mu \to -\infty} \Psi_b(f, \mu) = \Psi(f(b))$, hence there exists a unique $\gamma \leq \log|a-b|$ such that $\Psi_b(f, \gamma) = -\infty$ and $\Psi_b(f, \mu) > -\infty$ whenever $\mu \in]\gamma, \log|a-b|[$. Therefore, f is strictly vanishing along the increasing filter of center b and diameter $S = \theta^\gamma$.

Now suppose ${}_D\varphi_{a,|a-b|}(f) \neq 0$. Since ${}_D\varphi_{a,r}(f) = 0$ we have $\Psi_a(f, \log r) = -\infty$ and $\Psi_a(f, \log|a-b|) > -\infty$, hence there exists a unique $\gamma \in [\log r, \log|a-b|[$ such that $\Psi_a(f, \gamma) = -\infty$ and $\Psi_a(f, \mu) > -\infty$ whenever $\mu \in]\gamma, \log|a-b|]$, so f is strictly vanishing along the decreasing filter of center a and diameter $S = \theta^\gamma$. □

Proposition B.11.4: *Let $a \in \widetilde{D}$ and $b \in D$, let $f \in H(D)$ satisfy $f(b) \neq 0$, and ${}_D\varphi_{a,r}(f) = 0$ for some $r \in \mathbb{R}_+$. Then f is strictly vanishing along a monotonous filter with a center.*

Proof. If $r \leq |a-b|$ the statement comes directly from Proposition B.11.3. If $r > |a-b|$, then we have $\Psi_b(f, \log r) = \Psi_a(f, \log r) = -\infty$ whereas $\lim_{\mu \to -\infty} \Psi_b(f, \mu) = \Psi(f(b))$, hence there exists a unique $\gamma \leq \log r$ such that $\Psi_b(f, \gamma) = -\infty$ and $\Psi_b(f, \mu) > -\infty$ whenever $\mu < \gamma$. Thus f is strictly vanishing along the increasing filter of center b and diameter $S = \theta^\gamma$. □

Proposition B.11.5: *Let $\mathcal{F}$ be a monotonous filter on D and let $f \in H(D)$ be strictly vanishing along $\mathcal{F}$. Then $\mathcal{F}$ is pierced.*

Proof. Suppose that $\mathcal{F}$ is increasing (resp. decreasing) of center a and diameter S and is not pierced. There exists an annulus $\Gamma(a, S, S')$ (resp. $\Gamma(a, S', S)$) included in D such that $\Psi_a(f, \log S) = -\infty$ and $\Psi_a(f, \mu) > -\infty$ whenever $\mu \in [\log S', \log S[$ (resp. $\mu \in]\log S, \log S']$). Hence, by Corollary B.10.3 we know that $\Psi_a(f, \mu)$ is bounded in $]\log S', \log S[$ (resp. $]\log S, \log S'[$), a contradiction to the hypothesis.

When $\mathcal{F}$ has no center, we consider a center a of $\mathcal{F}$ in $\widehat{\mathbb{K}}$ and we consider f as an element of $\widehat{H}(\widehat{D})$. Then the disks $\widehat{d}(a_n, r_n)$ contain no hole of $\widehat{D}$ when n is big enough. Therefore, the restriction of $\widehat{f}$ to $d(a_n, r_n)$ is a power series and, therefore, we have the same conclusion. □

Corollary B.11.6: *Let $a, b \in D$ and let $f \in H(D)$ satisfy $f(b) \neq 0$ and ${}_D\varphi_{a,r}(f) = 0$ for some $r \in \mathbb{R}_+$. Then f is strictly vanishing along a pierced monotonous filter with a center.*

Proposition B.11.7: *Let $a \in \widetilde{D}$. Let $r, r' \in]\log(\delta(a, D)), \log(\operatorname{diam}(D))[$, with $r < r'$. Let $f \in H(D)$ be such that the function $\Psi_a(f, \mu)$ is neither bounded nor identically equal to $-\infty$ in $[\log r, \log r']$. Then there exists a monotonous filter $\mathcal{F}$ of center a and diameter $s \in [r, r']$ such that f is strictly vanishing along $\mathcal{F}$.*

Proof. For convenience we assume $a = 0$. By compacity of $[\log r, \log r']$ there exists $\mu \in [\log r, \log r']$ such that $\Psi(f, \mu) = -\infty$. Since the function $\Psi(f, \mu)$ is not identically $-\infty$ in $[\log r, \log r']$, by continuity, either there exists $\xi, \zeta \in [\log r, \log r']$ with $\xi < \zeta$ such that $\Psi(f, \xi) = -\infty$, $\Psi(f, \mu) > -\infty$ whenever $\mu \in]\xi, \zeta]$ and then f is strictly vanishing along a decreasing filter of center 0 and diameter θ^{ξ}, or there exist $\xi, \zeta \in [\log r, \log r']$ with $\xi < \zeta$, such that $\Psi(f, \zeta) = -\infty$, $\Psi(f, \mu) > -\infty$ whenever $\mu \in [\xi, \zeta[$ and then f is strictly vanishing along an increasing filter of center 0 and diameter θ^{ζ}. This ends the proof. □

Proposition B.11.8: *Let $f \in H(D)$ be vanishing along an increasing (resp.a decreasing) filter $\mathcal{F}$ of diameter s. Let a be a center of $\mathcal{F}$ in $\widehat{\mathbb{K}}$ and let $E = \widehat{D} \cup (\widehat{\mathbb{K}} \setminus \widehat{d}(a, s^-))$ (resp. $E = \widehat{D} \cup \widehat{d}(a, s)$). Then f has continuation to an element F of $\widehat{H}(E)$ such that $F(x) = 0$ whenever $x \in \widehat{\mathbb{K}} \setminus \widehat{d}(a, s^-)$, (resp. $x \in \widehat{d}(a, s)$).*

Proof. We suppose $\mathcal{F}$ increases. By Theorem B.6.9 f has an extension $\widehat{f}$ to $\widehat{D}$. For every $r > 0$, the set of classes of $C(a, r)$ that contain $\widehat{f}$-holes is countable. Since the residue class field of $\widehat{\mathbb{K}}$ is not countable, in $\widehat{C}(a, r)$ there exist classes $\Lambda = d(b, r^-)$ that contain no $\widehat{f}$-holes. Thereby $\widehat{f}$ has continuation to an infraconnected set D', which contains $\widehat{D}$ and satisfies $\widehat{D'} = \widehat{\mathbb{K}}$, such that every hole is of the form $\widehat{d}(\alpha, \rho^-)$, with $d(\alpha, \rho^-)$ a hole of D. In this set, we have

(1) $_{D'}\varphi_{a,s}(\widehat{f}) = {}_{D}\varphi_{\mathcal{F}}(f) = 0.$

First, we suppose that $\mathcal{F}$ is increasing. Let $V = d(a, s^-)$ and let $D'' = D' \setminus V$. Clearly V is a hole of D''. Then, as an element of $\widehat{H}(D'')$, by Theorem B.6.1,

(1) implies:

(2) $\overline{\overline{(\widehat{f})}}_V = 0.$

Now, by Theorem B.6.1, $\widehat{f}$ has a decomposition of the form $g + h$ with $g \in H_0(D' \cup (\widehat{\mathbb{K}} \setminus \widehat{d}(a, s^-)))$ and $h \in H(D' \cup \widehat{d}(a, s^-))$. By (2), it is seen that $_{D'}\varphi_{a,s}(h) = 0$, hence $h = 0$ because h belongs to $\widehat{H}(\widehat{d}(a, s))$. As a consequence, $\widehat{f}$ belongs to $H_0(D' \cup (\widehat{\mathbb{K}} \setminus \widehat{d}(a, s^-)))$ and, therefore, in $\mathbb{K}$, f belongs to $H(D \cup (\mathbb{K} \setminus d(a, s^-)))$. Moreover, by (2) we have $\widehat{f}(x) = 0$ whenever $x \in D''$, hence $\widehat{f}(x) = 0$ whenever $x \in \mathbb{K} \setminus d(a, s^-)$.

If $\mathcal{F}$ is decreasing, we can easily perform a symmetric proof. □

Theorem B.11.9: *Let $f \in H(D)$ be vanishing along an increasing (resp. a decreasing) filter $\mathcal{F}$ of center a and diameter s. Let $E = D \cup (\mathbb{K} \setminus d(a, s^-))$ (resp. $E = D \cup d(a, s)$). Then f has continuation to an element of $H(E)$ such that $f(x) = 0$ whenever $x \in \mathbb{K} \setminus d(a, s^-)$, (resp. whenever $x \in d(a, s)$).*

Proof. By Proposition B.11.8, f has continuation to an element $\widehat{f} \in \widehat{H}(\widehat{D} \cup (\widehat{\mathbb{K}} \setminus \widehat{d}(a, s^-)))$, (resp. to $H(\widehat{D} \cup \widehat{d}(a, s)))$. Therefore in $\mathbb{K}$, f belongs to $H(D \cup (\mathbb{K} \setminus d(a, s^-)))$. Moreover, we have $\widehat{f}(x) = 0$ whenever $x \in \widehat{\mathbb{K}} \setminus \widehat{d}(a, S^-)$, (resp. $x \in \widehat{d}(a,s)$), hence $f(x) = 0$ whenever $x \in \mathbb{K} \setminus d(a, s^-)$, (resp. $f(x) = 0$ whenever $x \in d(a,s)$). □

Definitions and notations: Let $\mathcal{F}$ be a monotonous filter on D and let $f \in H(D)$. Then, f is said to be *collapsing along* $\mathcal{F}$ if there exists $b \in \mathbb{K}$ such that $f - b$ is vanishing along $\mathcal{F}$.

Theorem B.11.10: *Let D be open and closed, and let $\mathcal{F}$ be a monotonous filter on D. Let $f \in H(D)$ be such that $f' \in H(D)$. Then f is collapsing along $\mathcal{F}$ if and only if f' is vanishing along $\mathcal{F}$.*

Proof. Without loss of generality, we can suppose that D is bounded because $\mathcal{F}$ is obviously secant with a bounded subset of $\mathbb{K}$.

Suppose first that $\mathcal{F}$, of center a and diameter R, is decreasing. Without loss of generality we can obviously assume $a = 0$. Suppose f is collapsing along $\mathcal{F}$ of limit ℓ. Since $\lim_{r \to R} \frac{|f - \ell|(r)}{r} = 0$, by Theorem B.9.25 we have $\lim_{r \to R} |f'|(r) = 0$, hence f' is vanishing along $\mathcal{F}$.

Conversely, suppose that f' is vanishing along $\mathcal{F}$. Let $(T_i)_{i \in I}$ be the family of holes of D included in $\mathbb{K} \setminus d(0,R)$ and let $(T_i)_{i \in J}$ be the family of holes of D included in $d(0,R)$. Then the pair (I, J) makes a partition of the set of holes of D. Let $D_1 = D \cup d(0,R)$ and let $D_2 = D \cup (\mathbb{K} \setminus d(0,R))$. By Corollary B.6.2, we have $H(D) = H(D_1) \oplus H(D_2)$. For every $g \in H(D)$ we set $g = g_1 + g_2$ with $g_k \in H(D_k)$, $k = 1, 2$. Then $\|g\|_D = \max(\|g_1\|_{D_1}, \|g_2\|_{D_2})$. By Theorem B.9.19, we can check that the decomposition of f' in the form $f' = (f')_1 + (f')_2$ is such that $(f')_k = (f_k)'$, $k = 1, 2$. When we take numbers $S' > R$ and $S'' < R$, we have $\lim_{S' \to R^+} \|f'\|_{D \cap d(0,S')} = 0 = \lim_{S'' \to R^-} \|f'\|_{D \setminus d(0,S'')}$, hence $\lim_{S' \to R^-} \|f_1'\|_{D \cap d(0,S')} = 0 = \lim_{S'' \to R^+} \|f_1'\|_{D \setminus d(0,S'')}$ and $\lim_{S' \to R^-} \|f_2'\|_{D \cap d(0,S')} = 0 = \lim_{S'' \to R^+} \|f_2'\|_{D \setminus d(0,S'')}$. Consequently, passing to the limit, we get $\varphi_{0,R}(f_1') = \varphi_{0,R}(f_2') = 0$. Therefore, by Corollary B.11.6, f_2', which belongs to $H(\mathbb{K} \setminus d(0,R))$ and satisfies $\varphi_{0,R}(f_2') = 0$ is identically zero in $\mathbb{K} \setminus d(0,R)$. Consequently, $f'(x) = f_1'(x)\ \forall x \in D \cap (\mathbb{K} \setminus d(0,R))$. On the other hand, since f_1' belongs to $H(d(0,R))$ and satisfies $\varphi_{0,R}(f_1') = 0$, f_1' is identically zero in $d(0,R)$ and, hence, $f_1(x)$ is a constant C in $d(0,R)$. Therefore, $\varphi(f_1 - C) = 0$, hence $f_1 - C$ is vanishing along $\mathcal{F}$. But since $f_2(x) = 0\ \forall x \in D \setminus d(0,R)$, actually, $f - C$ is vanishing along $\mathcal{F}$. This finishes showing that f is collapsing along $\mathcal{F}$ when $\mathcal{F}$ is a decreasing filter with a center.

Now suppose that $\mathcal{F}$ has no center. We can place ourselves in an algebraically closed spherically complete extension of $\mathbb{K}$ and prove the same property for the expansion of f, hence it holds for f. Finally, if $\mathcal{F}$ is increasing, we can make an inversion and prove the same. □

Thanks to monotonous filters we are now able to complete the study of the characteristic functions.

Theorem B.11.11: *Let E be a subset of $\mathbb{K}$ whose interior is not empty. Then E is not infraconnected if and only if there exists a proper subset B of E whose characteristic function belongs to $H(E)$.*

Proof. If E is not infraconnected, it admits an empty annulus $\Gamma(a,r',r'')$ and then by Proposition B.2.15 the characteristic functions of $\mathcal{I}_E(\Gamma(a,r',r''))$ and $\mathcal{E}_E(\Gamma(a,r',r''))$ belong to $H(E)$.

Now we suppose E to be infraconnected and assume that there is a subset B of E, $B \neq E$ and $B \neq \emptyset$, whose characteristic function u belongs to $H(E)$. Since u, by definition, belongs to $H_b(E)$, it belongs to $H_b(\overline{E})$. And of course, $\overline{E}$ has an interior that is not empty. Hence, without loss of generality, we can assume that E is closed. Let $A = E \setminus B$. Suppose A and B are different from $\emptyset$. Since u is locally constant in all E, at least one of the two subsets A and B has an interior that is not empty. Without loss of generality we can assume that the interior of A is not empty and, hence, there exists $a \in A$ and $r > 0$ such that $d(a,r) \subset A$ and $u(x) = 0$ whenever $x \in d(a,r)$ and then we have ${}_D\varphi_{a,r}(u) = 0$. By Proposition B.11.3 there exists a monotonous filter $\mathcal{F}$ with center $\alpha \in E$ such that u is strictly vanishing along $\mathcal{F}$, hence by Lemma B.11.2 f is properly vanishing along $\mathcal{F}$. But this contradicts the hypothesis "$f(x) = 0$ or 1 for all $x \in E$." This finishes proving Theorem B.11.11. □

Corollary B.11.12: *Let $E \in$ Alg. The algebra $H(E)$ has non-trivial idempotents if and only if E is not infraconnected.*

Theorem B.11.13: *Let $f \in H(D)$. Then $f(D)$ is infraconnected.*

Proof. Let $D' = f(D)$ and let us suppose that D' admits an empty annulus $\Gamma(a,r',r'')$. Let $D'' = \overline{D'}$. It is seen that $\Gamma(a,r',r'')$ also is an empty annulus of D''.

Let $A'' = \mathcal{I}_{D''}(\Gamma(a,r',r''))$ and let $B'' = \mathcal{E}_{D''}(\Gamma(a,r',r''))$. Let u be the characteristic function of A''. By Proposition B.2.15 we know that u belongs to $H(D'')$. Since D'' is closed and contains $f(D)$, by Corollary B.3.3, $u \circ f$ belongs to $H(D)$. Let $A = f^{-1}(A'')$ and $B = f^{-1}(B'')$. Obviously we have $A \cap B = \emptyset$ and $A \cup B = D$. Now, $u \circ f(x) = 1\ \forall x \in A$, $u \circ f(x) = 0\ \forall x \in B$. But since D is infraconnected, by Theorem B.11.11, $H(D)$ contains no characteristic function of any proper subset. This ends the proof of Theorem B.11.13. □

B.12. Quasi-minorated elements

Throughout this chapter, the set D is supposed to be infraconnected.

The main results given here were published in [43, 49, 58]. According to the definition of quasi-minorated elements, Theorem B.12.1 is easy.

Theorem B.12.1: *Let $f \in H(D)$. Then, f is not quasi-minorated if and only if there exists a large circular filter $\mathcal{F}$ secant with D, such that ${}_D\varphi_{\mathcal{F}}(f) = 0$.*

Proof. By Lemmas B.8.12 and B.7.7, without loss of generality we can assume that D is bounded.

Suppose first that there exists a large circular filter $\mathcal{F}$ secant with D, such that ${}_D\varphi_{\mathcal{F}}(f) = 0$. Let (a_n) be a monotonous distances sequence thinner than $\mathcal{F}$ such that $\lim_{n\to+\infty} f(a_n) = 0$. Then f is not quasi-minorated.

Conversely, suppose that f is not quasi-minorated. Then there exists a bounded sequence (a_n) of D such that $\lim_{n\to+\infty} f(a_n) = 0$ and such that one can't extract a sequence converging in $\mathbb{K}$. By Theorem A.2.1 we can extract from the sequence (a_n) either a monotonous distances sequence or a constant distances sequence. In both cases, there exists a circular filter $\mathcal{F}$ less thin than this subsequence and, hence, we have $\lim_{\mathcal{F}} f(x) = 0$. □

Theorem B.12.2: *Let f be a non-identically zero element of $H(D)$. Then f is not quasi-minorated if and only if there exists a pierced monotonous filter $\mathcal{F}$ on D such that f is strictly vanishing along $\mathcal{F}$.*

Proof. By Theorem B.12.1, if f is vanishing along a monotonous filter, it is not quasi-minorated. Now suppose that f is not quasi-minorated. Since f is not identically zero, we can find a bounded sequence $(a_n)_{n\in\mathbb{N}}$ in D such that $\lim_{n\to\infty} f(a_n) = 0$ such that no subsequence converges in $\mathbb{K}$.

Suppose first we can extract from the sequence $(a_n)_{n\in\mathbb{N}}$ a constant distances sequence $(a_{q(m)})m \in \mathbb{N}$, let $c = a_{q(1)}$, let $r = |a_{q(1)} - a_{q(2)}|$, and let $\phi = \varphi_{b,r}$. Since $\lim_{m\to+\infty} f(a_{q(m)}) = 0$, we have $\varphi_{c,r}(f) = 0$. Since f is not identically zero in D, there exists $b \in D$ such that $f(b) \neq 0$. If $b \in d(0, r)$ then by Proposition B.11.3, f is strictly vanishing along an increasing filter of center c and diameter $\rho \in]0, r]$. If $b \notin d(c, r)$, by Proposition B.11.3, if $\varphi_{c,|b-c|}(f) \neq 0$, then f is strictly vanishing along a decreasing filter of center c and diameter $\rho \in d(c, |b-c|)$. Finally, if $\varphi_{c,|b-c|}(f) = 0$, then f is strictly vanishing along an increasing filter of center b and diameter $\rho \in d(0, |b-c|)$, and by Proposition B.11.5, that filter is pierced.

Suppose now we can't extract from the sequence $(a_n)_{n\in\mathbb{N}}$ a constant distances sequence. Then we can extract from the sequence $(a_n)_{n\in\mathbb{N}}$ a monotonous distances sequence $(a_{q(m)})_{m\in\mathbb{N}}$. There exists a unique monotonous filter $\mathcal{F}$ less thin than the subsequence $(a_{q(m)})_{m\in\mathbb{N}}$. Suppose first that $\mathcal{F}$ has a center c and let $r = \operatorname{diam}(\mathcal{F})$. Then we have $\varphi_{c,r}(f) = 0$ and, hence, the same reasoning shows that f is strictly vanishing along a monotonous filter, which by Proposition B.11.5, is pierced. Finally, suppose $\mathcal{F}$ is decreasing with no center. We can find a center in a spherically closed extension, and make the same reasoning again. □

Corollary B.12.3: *If D has no monotonous pierced filter, every element of $H(D)$ different from zero is quasi-minorated and takes every value finitely many times.*

Definitions and notations: A subset D of $\mathbb{K}$ is said to be *analytic* if for every disk $d(a,r)$ included in D and for every $f \in H(D)$, the property $f(x) = 0$ $\forall x \in d(a,r)$ implies that f is identically zero in D.

Corollary B.12.4: *If D has no monotonous pierced filter, D is analytic. Particularly, if D has finitely many holes, D is analytic. Particularly, if D is infra-connected affinoid, it is analytic.*

Theorem B.12.5: *Let f_1, f_2 be quasi-minorated elements of $H(D)$. If $f_1 f_2$ belongs to $H(D)$, then it is also quasi-minorated.*

Proof. Indeed suppose that $f_1 f_2$ is not quasi-minorated. By Theorem B.12.2 there exists a pierced monotonous filter $\mathcal{F}$ on D such that ${}_D\varphi_{\mathcal{F}}(f_1 f_2) = 0$. Hence, either ${}_D\varphi_{\mathcal{F}}(f_1) = 0$ or ${}_D\varphi_{\mathcal{F}}(f_2) = 0$. But by Theorem B.12.1 both options are impossible because both f_1, f_2 are quasi-minorated. Hence, so is $f_1 f_2$. □

Theorem B.12.6: *Let $f \in H(D)$. If f is semi-invertible then it is quasi-minorated.*

Proof. Let f be semi-invertible and of the form $P(x)g(x)$, with g invertible in $H(D)$ and P a polynomial whose zeros $a_1, \dots, a_q$ belong to D. We then suppose f is not quasi-minorated. By Theorem B.12.5, g is not quasi-minorated either. Hence, there exists a pierced monotonous filter $\mathcal{F}$ on D such that ${}_D\varphi_{\mathcal{F}}(g) = 0$. But by Lemma B.2.3 that contradicts the hypothesis "g invertible in $H(D)$." Hence, f is quasi-minorated. □

Theorem B.12.7: *Suppose that $\overline{D}$ is open. Then an element of $H(D)$ is quasi-minorated if and only if it is quasi-invertible.*

Proof. If f is quasi-invertible, it is semi-invertible and then by Theorem B.12.6, it is quasi-minorated. Now, assume f to be quasi-minorated. We will prove it to be quasi-invertible. As in Theorem B.12.1, by Lemmas B.8.12 and B.7.7, without loss of generality we can assume that D is bounded. Let $S(D)$ be the set of polynomials whose zeros belong to $\overline{D} \setminus D$. If $D \in$ Alg then $H(D) = S(D)^{-1}H(\overline{D})$. By Corollary B.2.13, there exists $Q \in S(D)$ and $h \in H(\overline{D})$ such that $f = \dfrac{h}{Q}$. If h is not quasi-minorated, there exists a monotonous filter $\mathcal{F}$ on D such that ${}_D\varphi_{\mathcal{F}}(h) = 0$ and then we have ${}_D\varphi_{\mathcal{F}}(f) = 0$; this contradicts the hypothesis "f quasi-minorated" and therefore h is quasi-minorated. Then by Theorem B.7.8, h is quasi-invertible, in the form Pg with P a polynomial whose zeros are interior to $\overline{D}$ and g an element invertible in $H(\overline{D})$. Since $\overline{D}$ is bounded and open, by Corollary B.8.10, D belongs to

Alg, hence $\frac{g}{Q}$ is invertible in $H(D)$. Let $P = P_1 P_2$ with P_1 (resp. P_2) the polynomial of the zeros of P in $\overset{\circ}{\overline{D}} \cap D$ (resp. in $\overline{D} \setminus D$). Then $\frac{P_2 g}{Q}$ is invertible in $H(D)$ and then f is quasi-invertible. □

Theorem B.12.8: *If D belongs to* Alg *and has no pierced filter, every element of $H(D)$ different from zero is quasi-invertible.*

Proof. As in Theorem B.12.1, by Lemmas B.8.12 and B.7.7, without loss of generality we can assume that D is bounded. Now, since D has no pierced filter, by Corollary B.12.3, every element of $H(D)$ is quasi-minorated. But since D has no pierced filter, $\overline{D}$ is open, hence by Theorem B.12.7 every element of $H(D)$ is quasi-invertible. □

Corollary B.12.9: *Let D be closed and let $\mathcal{T}$ be the set of holes of D. If $\{\widetilde{T} | T \in \mathcal{T}\}$ is finite, then every element of $H(D)$ different from zero is quasi-invertible.*

Corollary B.12.10: *If D is closed and has finitely many holes then every element of $H(D)$ different from zero is quasi-invertible.*

Corollary B.12.11: *If D is a disk $d(a, r)$ or $d(a, r^-)$, or if D is an annulus $\Gamma(a, r_1, r_2)$ (with $0 < r_1 < r_2$) or $\Delta(a, r_1, r_2)$ (with $0 < r_1 < r_2$) or a circle $C(a, r)$, then every element of $H(D)$ different from zero is quasi-invertible.*

Proof. (Corollaries B.12.9, B.12.10, B.12.11) Indeed D has no pierced filter, hence the elements different from zero are quasi-minorated and are quasi-invertible because D is open. □

We will see that when D belongs to Alg, a quasi-minorated element that has no zero in D, actually is invertible in $H(D)$.

Lemma B.12.12: *Let $f \in H(\overline{D})$ be quasi-minorated in $H(D)$. Then f is quasi-minorated in $H(\overline{D})$.*

Proof. Indeed let $(a_n)_{n\in\mathbb{N}}$ be a bounded sequence in $\overline{D}$ such that $\lim_{n\to\infty} f(a_n) = 0$. There obviously exists a sequence $(b_n)_{n\in\mathbb{N}}$ in D such that $\lim_{n\to\infty} a_n - b_n = 0$ and $\lim_{n\to\infty} f(b_n) = 0$. So, the sequence (b_n) is bounded as the sequence (a_n). Since f is quasi-minorated in $H(D)$, then we can extract a Cauchy sequence $(b_{q(m)})_{m\in\mathbb{N}}$ from the sequence $(b_n)_{n\in\mathbb{N}}$ and then the sequence $(a_{q(m)})_{m\in\mathbb{N}}$ is a Cauchy subsequence of the sequence $(a_n)_{n\in\mathbb{N}}$. Thus we have proven that f is quasi-minorated in $H(\overline{D})$. □

Theorem B.12.13: *Let $D \in$ Alg. Let $f \in H(D)$ be quasi-minorated and have no zero in D. Then f is invertible in $H(D)$.*

Proof. If D is closed, the statement is given by Theorem B.7.9. Consider now the general case. As in Theorem B.12.1, by Lemmas B.8.12 and B.7.7, without loss of generality we can assume that D is bounded. Let Q be the polynomial of the poles of f in $\overline{D} \setminus D$ and let $h(x) = Q(x)f(x)$. By Theorem B.12.5 h is quasi-minorated in $H(D)$. But by Corollary B.12.11 h belongs to $H(\overline{D})$ and, therefore, by Lemma B.12.12, it is quasi-minorated in $H(\overline{D})$. We will prove that h has finitely many zeros in $\overline{D}$. Indeed we assume that h admits infinitely many zeros in $\overline{D}$. So we can find a sequence $(a_n)_{n\in\mathbb{N}}$ in $\overline{D} \setminus D$ such that $h(a_n) = 0$ whenever $n \neq m$, $n, m \in \mathbb{N}$. Since h is quasi-minorated in $H(\overline{D})$ and since D is bounded, we can extract a cauchy subsequence from the sequence $(a_n)_{n\in\mathbb{N}}$. This Cauchy subsequence obviously converges to a point $a \in \overline{D}$ and, therefore, we have $h(a) = 0$. But, as h has no zero in D, a belongs to $\overline{D} \setminus D$. And now, since D belongs to Alg, then a belongs to $\overset{\circ}{\overline{D}}$. But by Corollary B.7.3, a zero of h, which is interior to $\overline{D}$, is isolated in $\overline{D}$ and this contradicts the definition of a. Thus we have proven that h has finitely many zeros $(b_j)_{1\leq j\leq q}$ in $\overline{D}$, all of them in the set $\overline{D} \setminus D$, which is included in $\overset{\circ}{\overline{D}}$. Then each zero b_j has a multiplicity order n_j, $(1 \leq j \leq q)$. Let $P(x) = \prod_{j=1}^{q}(x - b_j)^{n_j}$ be the polynomial of the zeros of h in $\overset{\circ}{\overline{D}}$. By Corollary B.7.6 the function $g(x) = \dfrac{h(x)}{P(x)}$ belongs to $H(\overline{D})$ and obviously has no zero in $\overline{D}$. As we have already seen when D is closed, g is invertible in $H(\overline{D})$. Now, since both P, Q have all their zeros in $\overline{D} \setminus D$, they are invertible in $R(D)$ and so is $\dfrac{P}{Q}$. But then $f = \dfrac{P}{Q}g$ and, hence, f is invertible in $H(D)$. □

Theorem B.12.14: *Let D be closed, bounded, having finitely many holes and let $f \in H(D)$. Then $f(D)$ is an infraconnected closed and bounded subset of $\mathbb{K}$.*

Proof. Since D has finitely many holes, by Corollary B.12.10, f is quasi-minorated. By Theorem B.11.13 $f(D)$ is infraconnected. Since f is bounded on a closed bounded subset, $f(D)$ is bounded. Let b belong to the closure of $f(D)$ and let $(a_n)_{n\in\mathbb{N}}$ be a sequence of D such that $\lim_{n\to+\infty} f(a_n) = b$. Since the sequence (a_n) is bounded, there exists a subsequence thinner than a circular filter $\mathcal{F}$ secant with D. If $\mathcal{F}$ is large, then we have ${}_D\varphi_{\mathcal{F}}(f) = 0$ and, hence, by Theorem B.12.1 f is not quasi-minorated, a contradiction. Consequently, $\mathcal{F}$ converges to a point a. Since D is closed, a belongs to D. Consequently, $f(a) = b$ and, hence, b belongs to $f(D)$. Therefore $f(D)$ is closed. □

Theorem B.12.15 shows an example of very simple increasing T-filter, without describing the general theory of T-filters [49, 50].

Theorem B.12.15: *Let $(r_n)_{n\in\mathbb{N}}$ be a sequence in $|\mathbb{K}|$ such that $0 < r_n < r_{n+1}$ and $\lim_{n\to+\infty} r_n = R$, let $(q_n)_{n\in\mathbb{N}}$ be a sequence of $\mathbb{N}$ prime to the characteristic of $\mathbb{K}$, such that $q_n \leq q_{n+1}$, and $\lim_{n\to+\infty} (\frac{r_n}{r_{n+1}})^{q_n} = 0$. Let $l \in]0, R[$ and for each $n \in \mathbb{N}$, let $b_n \in C(0, (r_n)^{q_n})$, let $a_{n,1}, \dots, a_{n,q_n}$ be the q_n-th roots of b_n, and let $E = d(0, R^-) \setminus \Big(\bigcup_{n\in\mathbb{N}} (\cup_{j=1}^{q_n} d(a_{n,j},\, l^-))\Big)$. Set $f_n(x) = \prod_{k=1}^{n}\prod_{j=1}^{q_k}\Big(\frac{1}{1-(\frac{x}{a_{k,j}})}\Big)$. Then each f_n belongs to $R(E)$ and the sequence $(f_n)_{n\in\mathbb{N}}$ converges in $H(E)$ to an element strictly vanishing along the pierced increasing filter of center 0 and diameter R.*

Proof. Let us first show that the sequence $(f_n)_{n\in\mathbb{N}}$ converges in $H(E)$. We notice that each pole $a_{k,j}$ of f_n lies in a hole of diameter l and is unique in that hole and in the class. Moreover, since each q_n is prime to the residue characteristic of $\mathbb{K}$, each pole $a_{k,j}$ of f_n is unique in the class $d(a_{k,j}, r_k^-)$. Consequently, we have $|f_n(x)| \leq \frac{|x||f_n|(|x|)}{l}$ $\forall x \in E$ and, hence,

$$|f_n(x)| \leq \frac{R|f_n|(|x|)}{l} \ \forall x \in E. \tag{1}$$

Let us now show that the sequence $(f_n)_{n\in\mathbb{N}}$ converges in $H(E)$. We first notice that each factor $Q_k = \prod_{j=1}^{q_k}\Big(\frac{1}{1-(\frac{x}{a_{k,j}})}\Big)$ satisfies $|Q_k|(r) \leq 1$ $\forall r < R$ because $|Q_k|(r) = 1$ $\forall r \leq r_k$ and $|Q_k|(r) < r$ $\forall r \in]r_k, R[$. Consequently, we have

$$|f_n|(r) \leq \prod_{k=0}^{n}\prod_{j=1}^{q_k}\Big(\frac{r_k}{r}\Big)^{q_k} \leq \Big(\frac{r_{n-1}}{r_n}\Big)^{q_{n-1}} \ \forall r \geq r_n. \tag{2}$$

On the other hand, when $r < r_n$, we have

$$|1 - Q_{n+1}|(r) = \left|1 - \frac{1}{1-(\frac{x}{b_{n+1}})^{q_{n+1}}}\right|(r) \leq \Big(\frac{r_n}{r_{n+1}}\Big)^{q_{n+1}},$$

hence

$$|f_{n+1} - f_n|(r) = |f_n|(r)|1 - Q_{n+1}|(r) \leq |f_n|(r)\Big(\frac{r_n}{r_{n+1}}\Big)^{q_{n+1}} \ \forall r < r_n. \tag{3}$$

By (2) and (3), we can see that $\lim_{n\to+\infty}\Big(\sup_{r<R}|f_{n+1}(r) - f_n(r)|\Big) = 0$ and, hence, by (1) we have $\lim_{n\to+\infty}\|f_{n+1} - f_n\|_E = 0$, hence the sequence f_n converges to an element $f \in H(E)$. Moreover, by (1) and (2) we can see that $\lim_{|x|\to R^-,\ x\in E} f(x) = 0$, so f is vanishing along the increasing pierced filter $\mathcal{F}$ of center 0 and diameter R. Further, we may notice that $|Q_n|(r) = 1$ $\forall r \leq r_n$, hence, when $r \leq r_n$, we have $|f|(r) = |f_s|(r)$ $\forall s \geq n$. Consequently, $|f|(r) \neq 0$ $\forall r < R$ and, hence, f is strictly vanishing along $\mathcal{F}$. □

B.13. Zeros of power series

Most of classical results on zeros of polynomials will now be extended to power series. In particular, power series converging inside a disk satisfies a Schwarz Lemma that is even simpler than in $\mathbb{C}$.

Throughout this chapter, r is a strictly positive real number and r', r'' are strictly positive real numbers satisfying $r' < r''$.

Theorem B.13.1: *Let $f \in H(C(0,r))$. The number of zeros of f in $C(0,r)$ is equal to $\nu^+(f,\log r) - \nu^-(f,\log r)$, (taking multiplicity into account).*

Proof. This equality was given for a polynomial in Theorem A.3.16. First we prove it when f is an element of $H(C(0,r))$ invertible in $H(C(0,r))$. By Proposition B.10.7, we have $\nu^+\left(\frac{1}{f},\log r\right) = -\nu^+(f,\log r)$ and $\nu^-\left(\frac{1}{f},\log r\right) = -\nu^-(f,\log r)$. Since any $h \in H(C(0,r))$ satisfies $\nu^+(h,\log r) \geq \nu^-(h,\log r)$ we see that $\nu^+(f,\log r) = \nu^-(f,\log r)$.

We now consider the general case. By Corollary B.12.11, f has a factorization of the form Pg with P a polynomial whose zeros belong to $C(0,r)$ and g invertible in $H(C(0,r))$. Then $\nu^+(f,\log r) - \nu^-(f,\log r) = \nu^+(P,\log r) - \nu^-(P,\log r) = \deg(P)$ and this just ends the proof of Theorem B.13.1. □

Corollary B.13.2: *Let $a \in \mathbb{K}$ and $r > 0$. Let $f(x) \in H(C(a,r^-))$. Let $\widehat{\mathbb{K}}$ be a complete algebraically closed extension of $\mathbb{K}$ and let $\widehat{C}(a,r) = \{x \in \widehat{\mathbb{K}} \mid |x-a| = r\}$. Then the zeros of f in $\widehat{C}(a,r)$ are exactly those of f in $C(a,r)$ (taking multiplicity into account). Similarly, the zeros of f in $\widehat{d}(a,r) = \{x \in \widehat{\mathbb{K}} \mid |x-a| \leq r\}$ (resp. in $\widehat{d}(a,r^-) = \{x \in \widehat{\mathbb{K}} \mid |x-a| < r\}$) are exactly those of f in $d(a,r)$ (resp. in $d(a,r^-)$) (taking multiplicity into account).*

Corollary B.13.3: *Let $f \in H(C(0,r))$ have t zeros in $C(0,r)$. Let $q = \nu^+(f,\log r) - \nu^-(f,\log r)$. Then $r = \sqrt[q]{\left|\frac{a_t}{a_{q+t}}\right|}$.*

Theorem B.13.4: *Let $\Lambda = C(0,r)$ and let $f(x) = \sum_{-\infty}^{+\infty} a_n x^n$ be a convergent Laurent series in Λ, having no zero in Λ. Let $\mu = \log r$. Then $\nu^+(f,\mu) = \nu^-(f,\mu) = q \in \mathbb{Z}$ and $|f(x)| = |a_q x^q|$ whenever $x \in \Lambda$. Moreover, if $q \neq 0$, then $f(\Lambda) = C(0,|a_q|r^q)$.*

Proof. By Proposition B.10.7, we see that $\nu^+(f,\mu) = \nu^-(f,\mu)$. Thus we have $\Psi(f,\mu) = \Psi(a_q) + q\mu$ whenever $\mu \in I$. Consequently, $f(C(0,r)) \subset C(0,|a_q|r^q)$. Now, suppose $q \neq 0$ and let $s = |a_q|r^q$. Let $b \in C(0,s)$ and let $g = f - b$. So, by definition, $g(x) = \sum_{-\infty}^{-1} a_n x^n + (a_0 - b) + \sum_{1}^{+\infty} a_n x^n$ and by hypothesis $|a_0| < |a_q|r^q$, hence

$|a_0 - b| = |a_q|r^q$. Consequently, $\nu^-(g,\mu) < \nu^+(g,\mu)$, therefore by Theorem B.13.1, g admits at least one zero in $C(0,r)$ and, hence, b lies in $f(C(0,r))$. This proves that $f(C(0,r)) = C(0,|a_q|r^q)$. □

Corollary B.13.5: *Let* $f(x) = \sum_{-\infty}^{+\infty} a_n x^n$ *be a convergent Laurent series in* $\Lambda = \Gamma(a,R',R'')$ *(resp. in* $\Lambda = \Delta(a,R',R'')$*), having no zero in* $\Gamma(a,R',R'')$*. Then we have* $\nu^+(f,\mu) = \nu^-(f,\mu) = q \in \mathbb{Z}$ *whenever* $\mu \in]\log r', \log r''[$ *(resp.* $\mu \in [\log r', \log r'']$*) and* $|f(x)| = |a_q x^q|$ *whenever* $x \in \Lambda$*. Moreover, if* $q > 0$*, putting* $s' = |a_q|r'^q$, $s'' = |a_q|r''^q$ *we have* $f(\Lambda) = \Gamma(0,s',s'')$ *(resp.*$f(\Lambda) = \Delta(0,s',s'')$*) and if* $q < 0$*, putting* $s' = |a_q|r''^q$, $s'' = |a_q|r'^q$ *we have* $f(\Lambda) = \Gamma(0,s',s'')$ *(resp.*$f(\Lambda) = \Delta(0,s',s'')$*).*

Corollary B.13.6: *Let* $\Lambda = \Gamma(0,r',r'')$ *and let* $f(x)$, $g(x)$ *be convergent Laurent series in* Λ*, having no zero in* Λ *such that* $|f|(r) = |g|(r)$ $\forall r \in]r',r''[$*. Then* $\nu(f,\log r) = \nu(g,\log r)$ $\forall r \in]r',r''[$*.*

Proof. By hypothesis, we have $\Psi(f,\mu) = \Psi(g,\mu)$ $\forall \mu \in]\log r', \log r''[$. But $\nu(f,\log r) = \dfrac{d\Psi(f)}{d\mu}(\log r)$ and $\nu(g,\log r) = \dfrac{d\Psi(g)}{d\mu}(\log r)$, hence $\nu(f,\log r) = \nu(g,\log r)$. □

Theorem B.13.7: *Let* $f \in H(d(0,r))$*. The number of zeros of* f *in* $d(0,r)$ *is equal to* $\nu^+(f,\log r)$*, (taking multiplicity into account).*

Proof. This equality was given for a polynomial in Corollary A.3.17. First we prove it when f is an invertible element of $H(d(0,r))$. By Theorem B.13.1 we have $\nu^+(f,\mu) = \nu^-(f,\mu) = 0$ $\forall \mu \le \log(r)$. Consequently, $\nu^+(f,\log r) = 0$. Consider now the general case. Since by Corollary B.12.11, f is quasi-invertible, it has a factorization of the form $P(x)h(x)$ with P a polynomial whose zeros lie in $d(0,r)$ and h is an invertible element of $H(d(0,r))$. Then by Corollary A.3.17 $\nu^+(P,\log(r))$ is the number of zeros of P (hence of f) and by Proposition B.10.7 we have $\nu^+(f,\log(r)) = \nu^+(P,\log(r)) + \nu^+(h,\log(r)) = \nu^+(P,\mu)$, which ends the proof. □

Corollary B.13.8: *Let* $\Lambda = \Gamma(0,r',r'')$ *and let* $f(x)$, $g(x)$ *be convergent Laurent series in* Λ*, such that* $|f|(r) = |g|(r)$ $\forall r \in]r',r''[$*. Then for each* $r \in]r',r''[$*,* f *and* g *have the same number of zeros in* $C(0,r)$ *(taking multiplicity into account).*

Theorem B.13.9: *Let* $f(x) = \sum_{n=0}^{\infty} a_n(x-a)^n \in H(d(a,r))$ *and let* $s = \sup_{n\ge 1} |a_n|r^n$*. Then* $f(d(a,r)) = d(a_0,s)$ *and* $\Psi_a(f - a_0, \log r) = \log s$*.*

Proof. Without loss of generality we can suppose that $a = 0$. Let $b \in d(a_0,s)$ and consider $f(x) - b = a_0 - b + \sum_{n=1}^{+\infty} a_n x^n$. By hypothesis, $|a_0 - b| \le s$, hence

$\nu^+(f-b,\log(r)=\nu^+(f-a_0,\log(r))\geq 1$ and, hence, $f-b$ has at least one zero in $d(0,r)$. Consequently, $d(0,s)\subset f(d(0,r))$. Conversely, when $x\in d(0,r)$, we have $|f(x)-a_0|\leq s$, hence $d(0,s)=f(d(0,r))$. As a consequence, $\Psi_a(f-a_0,\log r)=\log s$. □

Corollary B.13.10: *Let* $f(x)=\sum_{n=0}^{\infty}a_n(x-a)^n\in\mathcal{A}_b(d(a,r^-))$ *(not identically zero) and let* $s=\sup_{n\geq 1}|a_n|r^n$. *Then* $f(d(a,r^-))=d(a_0,s^-)$.

Proof. On the one hand $f(d(a,r^-))$ is obviously included in $d(a_0,s^-)$. On the other hand, given $b\in d(a_0,s^-)$ and $\rho\in]0,r[$ such that $\sup_{n\geq 1}|a_n|\rho^n\geq|b-a_0|$, by Theorem B.13.9, b belongs to $f(d(a,\rho))$ because $f\in H(d(a,\rho))$. □

Corollary B.13.11: *Let* $f(x)=\sum_{n=0}^{\infty}a_n(x-a)^n\in H(d(a,r^-))$ *(not identically zero) and let* $s=\sup_{n\geq 1}|a_n|r^n$. *Then* $f(d(a,r^-))=d(a_0,s^-)$.

Lemma B.13.12: *Let* $f\in H(d(0,r))$ *satisfy* $\nu^+(f,\log r)\geq 1$, *let* $b\in f(d(0,r))$, *and let* $g(x)=f(x)-b$. *Then we have* $\nu^+(g,\log r)=\nu^+(f,\log r)$.

Proof. Let $f(x)=\sum_{n=0}^{\infty}a_nx^n$ and let $t=\nu^+(f,\log r)$. By hypothesis we have $|a_t|r^t\geq|a_n|r^n$ whenever $n<t$ and $|a_t|r^t>|a_n|r^n$ whenever $n>t$. Now let $g(x)=\sum_{n=0}^{\infty}b_nx^n$. Hence, $b_0=a_0-b$, $b_n=a_n$ whenever $n\geq 1$. By hypothesis we have $|b_0|\leq\sup_{n\geq 1}|a_n|r^n$, hence $|b_0|\leq|b_t|r^t$ and finally $\nu^+(g,\log r)=t$. □

Lemma B.13.13: *Let* $f\in H(d(a,r))$ *have* t *zeros in* $d(a,r)$ *with* $t\geq 1$ *(taking multiplicity into account) and let* $b\in f(d(a,r))$. *Then* $f-b$ *also admits* t *zeros in* $d(a,r)$ *(taking multiplicity into account).*

Proof. We assume $a=0$. By Lemma B.13.12 we know that $\nu^+(f,\log r)=t$. Hence, we have $\nu^+(f,\log r)=t$ and $\Psi(f,\log r)=\Psi(a_t)+t\log r$. Next, $\Psi(f(x))\leq\Psi(f,\log r)$ for all $x\in d(0,r)$, hence $\Psi(b)\leq\Psi(f,\log r)$ and, therefore, $\Psi(b)\leq\Psi(a_t)+t\log r$. Hence, $\nu^+(f-b,\log r)=t$. That ends the proof. □

Theorem B.13.14: *If an entire function* $f\in\mathcal{A}(\mathbb{K})$ *is bounded or has no zero, it is a constant.*

Proof. Let $f(x)=\sum_{j=0}^{\infty}a_jx^j$. By Lemma B.13.12, the number of zeros of f in any disk $d(0,r)$ is equal to the biggest of the integers l such that $|a_l|r^l=\sup_{j\in\mathbb{N}}|a_j|r^j$ and this is also equal to $\lim_{|x|\to r, |x|\neq r} |f(x)|$. Hence, if f is bounded, or has no zero in $\mathbb{K}$, obviously we have $a_n = 0\ \forall n > 0$. □

Theorem B.13.15: *Let $f(x) \in \mathcal{A}(\mathbb{K}) \setminus \mathbb{K}$. Then f admits at least one zero in $\mathbb{K}$. Moreover, if f is not a polynomial, then f has infinitely many zeros in $\mathbb{K}$ and the zeros make a sequence $(\alpha_n)_{n\in\mathbb{N}}$ such that $\lim_{n\to+\infty} |\alpha_n| = +\infty$.*

Proof. Suppose first that f has finitely many zeros in $\mathbb{K}$. Then by Theorem B.5.16, there exists a polynomial P such that f factorizes in the form Pg with $g \in \mathcal{A}(\mathbb{K})$ and $g(x) \neq 0 \forall x \in \mathbb{K}$. Hence, by Theorem B.13.14, g is a constant, hence f is a polynomial. Next, by Theorem B.13.7, f has finitely many zeros in each disk, hence the sequence $(\alpha_n)_{n\in\mathbb{N}}$ tends to $+\infty$. □

Theorem B.13.16: *Let Λ be a disk of the form $d(a, r)$ (resp. $d(a, r^-)$) and let $f \in H(\Lambda)$ have no zero in Λ. Then $|f(x)|$ is equal to a constant in Λ and f is invertible in $H(\Lambda)$.*

Proof. By Corollary B.12.11 and Theorem B.12.13 f is invertible in $H(\Lambda)$. We may obviously assume $a = 0$. By Lemma B.13.12, we have $\nu^+(f, \mu) = \nu^-(f, \mu) = 0$ whenever $\mu \leq \log r$ (resp. $\mu < \log r$), hence by Proposition B.10.7, $\Psi(f, \mu)$ has a derivative equal to 0 and, therefore, is equal to a constant in $]-\infty, \log r[$. Then by Corollary B.10.6, we have $\Psi(f(x)) = \Psi(f, \Psi(x))$, hence $\Psi(f(x))$ is equal to a constant in Λ. □

Theorem B.13.17: *Let Λ be a set in one of the following forms:*

(i) $\Lambda = d(0, r)$
(ii) $\Lambda = d(0, r^-)$
(iii) $\Lambda = C(0, r)$.

Let $f \in H(\Lambda)$ (f not identically 0) and let $h \in H(\Lambda)$ satisfy $\|f - h\|_\Lambda < \|f\|_\Lambda$. Then f and h have the same number of zeros in Λ (taking multiplicity into account).

Proof. Regardless of the case (i), (ii), and (iii), we know that $\|f\|_\Lambda = {}_\Lambda\varphi_{0,r}(f)$ and then we have $\log \|f\|_\Lambda = \Psi(f, \log r)$. Since $\|f - h\|_\Lambda < \|f\|_\Lambda$ then $\Psi((f - h, \log r) < \Psi(f, \log r)$. Hence, by Proposition B.10.5, we know that $\nu^+(f, \log r) = \nu^+(h, \log s)$ and $\nu^-(f, \log r) = \nu^-(h, \log r)$. Consequently, f has as many zeros as h in Λ, by Theorem B.13.7 if $\Lambda = d(0, r)$ and, by Theorem B.13.1 if $\Lambda = C(0, r)$.

We now suppose $\Lambda = d(0, r^-)$. By Corollary B.12.11 both f, h are quasi-invertible in $H(\Lambda)$. Let $\rho \in]0, r]$ be such that $d(0, \rho)$ contains all zeros of h in Λ. According to statements (i) and (ii) already proven, we see that f has as many zeros as h in $d(0, \rho)$ (taking multiplicity into account) and has no zero in $C(0, s)$ whenever $s \in]\rho, r[$. This ends the proof. □

Theorem B.13.18: *If $f \in H(C(a,r))$, it satisfies $|f(x)| \leq {}_D\varphi_{a,r}(f)$ $\forall x \in C(a,r)$ and the equality $|f(x)| = {}_D\varphi_{a,r}(f)$ holds in all classes except finitely many that are the classes where f has at least one zero.*

Proof. Let $f \in H(C(a,r))$ and $\Lambda = C(a,r)$. We can find $h \in R(\Lambda)$ such that $\|f - h\|_\lambda < \|f\|_\Lambda$, hence, by Theorem B.13.17, f has the same number of zeros as h in each class of Λ (taking multiplicity into account). So, in each class of $\Xi = d(b, r^-)$ where f has no zero, $|f(x)|$ is equal to ${}_D\varphi_{b,r}(f)$. But by Lemma B.4.4 we have ${}_D\varphi_{b,r}(f) = {}_D\varphi_{a,r}(f)$, which ends the proof. □

Corollary B.13.19: *Let $f(x) \in \mathcal{A}(d(0,r^-))$ have infinitely many zeros in $d(0,r^-)$. Then the set of zeros of f in $d(0,r^-)$ is a sequence $(\alpha_n)_{n\in\mathbb{N}}$, such that $\lim_{n\to+\infty} |\alpha_n| = r$.*

Proof. Indeed, by Proposition B.5.12, for each $\rho \in]0,r[$, f belongs to $H(d(0,\rho))$ and, hence, has finitely many zeros in $d(0,\rho)$. □

Theorem B.13.20: *Let $f(x) = \sum_{n=0}^{\infty} a_n x^n \in \mathcal{A}(d(0,r^-))$. Then f has finitely many zeros in $d(0,r^-)$ if and only if there exists $q \in \mathbb{N}$ such that $|a_q|r^q \geq \sup_{n\in\mathbb{N}} |a_n|r^n$. Moreover, if t is the smallest of the integers q such that $|a_q|r^q \geq \sup_{n\in\mathbb{N}} |a_n|r^n$, then f has exactly t zeros in $d(0,r^-)$. Further, the three following statements are equivalent:*

(i) *f has no zero in $d(0,r^-)$,*
(ii) *f is invertible in $\mathcal{A}(d(0,r^-))$,*
(iii) *$|f(x)|$ is a non-zero constant.*

Proof. First we suppose that there exists $q \in \mathbb{N}$ such that $|a_q|r^q \geq \sup_{n\in\mathbb{N}} |a_n|r^n$. Let $t \in \mathbb{N}$ be the smallest of the integers q such that $|a_q|r^q \geq \sup_{n\in\mathbb{N}} |a_n|r^n$. There exists a unique $s \in]0,r[$ such that $|a_t|s^t \geq |a_n|s^n$ for every $n < t$. Then for all $\rho \in]s,r[$, in $H(d(0,\rho))$ we have $\nu^+(f,\log\rho) = \nu^-(f,\log\rho) = t$, hence f admits exactly t zeros in $d(0,\rho)$, whenever $\rho \in]s,r[$.

Conversely, suppose that f admits exactly t zeros in $d(0,r^-)$. There then exists $s \in]0,r[$ such that f admits exactly t zeros in $d(0,s)$ and of course in each disk $d(0,\rho)$ for every $\rho \in]s,r[$. Hence, we have $\nu^+(f,\log\rho) = \nu^-(f,\log\rho) = t$ for every $\rho \in]s,r[$. Therefore, we have $|a_t|\rho^t > |a_n|\rho^n$ for every $n \neq t$ and for every $\rho \in]s,r[$. Finally we see that $|a_t|r^t \geq \sup |a_n|r^n$.

Further, the equivalence between (i) and (ii) is shown at Theorem B.5.19. The equivalence of (i) with (iii) comes from the fact that f has no zero if and only if $|a_0| > |a_n|s^n$ $\forall n \in \mathbb{N}$, $\forall s \in]0,r[$. □

Theorem B.13.21: *Let $f(x) = \sum_{-\infty}^{+\infty} a_n x^n \in \mathcal{A}(\mathbb{K} \setminus d(0,R))$ be not identically zero, let $R' > R$ and let $q = \nu^+(f, \log(R'))$. The family of zeros of f in $\mathbb{K} \setminus d(0, R'^-)$ either is finite or is a sequence $(\alpha_n)_{n\in\mathbb{N}}$ such that $\lim_{n\to+\infty} |\alpha_n| = +\infty$.*

Proof. By definition, f is of the form $g + h$ with $g(x) = \sum_{-\infty}^{-1} a_n x^n \in \mathcal{A}(\mathbb{K} \setminus d(0,R))$ and $h(x) = \sum_{0}^{+\infty} a_n x^n \in \mathcal{A}(\mathbb{K})$. Particularly, by Theorem B.5.6 for every R', $R'' \in]R, +\infty[$ (with $R' < R''$) both g, h belong to $H(\Delta(0, R', R''))$. Then, so does f. Therefore, by Corollary B.13.3, f has finitely many zeros in $\Delta(0, R', R'')$, which are on the circles $C(0,r)$ such that $\nu^+(f, \log r) > \nu^-(f, \log r)$. Thus, the family of zeros of f in $\mathbb{K} \setminus d(0, R'^-)$ either is finite or is a sequence $(\alpha_n)_{n\in\mathbb{N}}$ such that $\lim_{n\to+\infty} |\alpha_n| = +\infty$. □

Theorem B.13.22: *Let R, $S \in]0, +\infty[$, $R < S$ and let*

$$f(x) = \sum_{-\infty}^{+\infty} a_n x^n \in \mathcal{A}(\mathbb{K} \setminus d(0,R))$$

have infinitely many zeros in $\mathbb{K} \setminus d(0,S))$. Then for every fixed $t \in \mathbb{N}$, we have $\lim_{r\to+\infty} \frac{|f|(r)}{r^t} = +\infty$.

Proof. By Theorem B.13.21, the sequence of zeros $(\alpha_n)_{n\in\mathbb{N}}$ is such that $\lim_{n\to+\infty} |\alpha_n| = +\infty$. Set $r_n = |\alpha_n|$. The sequence $\nu + (f, \log(r_n)$ is strictly increasing and, hence, there exists $q \in \mathbb{N}$ such that $\nu^+(f, \log(r_q) > t$, therefore, $\nu^+(f, \mu) \geq 1\ \forall \mu > \log(r_q)$. Set $g(x) = \frac{f(x)}{x^t}$. Then clearly $|g|(r) = \frac{|f|(r)}{r^t}$ and, hence, $\nu^+(g, \mu) \geq 1\ \forall \mu \geq \log(r_q)$. Thus, the function (in μ) $\Psi(f, \mu)$ is a convex function, piecewise affine, whose derivative, when it is derivable, is greater than 1 whenever $\mu \geq \log(r_q)$. Consequently $\lim_{r\to+\infty} |g|(r) = +\infty$ and, therefore, $\lim_{r\to+\infty} \frac{|f|(r)}{r^t} = +\infty$. □

Corollary B.13.23: *Let $f(x) \in \mathcal{A}(\mathbb{K}) \setminus \mathbb{K}[x]$. Then for every fixed $t \in \mathbb{N}$, we have $\lim_{r\to+\infty} \frac{|f|(r)}{r^t} = +\infty$.*

Theorem B.13.24: *Let $r \in \mathbb{R}_+$, let $r_1, r_2 \in]0, R[)$ satisfy $r_1 < r_2$, and let $f \in \mathcal{A}(d(0, R^-))$. If f admits exactly q zeros in $d(0, r_1)$ (taking multiplicity into account) and has no zeros in $\Gamma(0, r_1, r_2)$, then f satisfies*

$$\Psi(f, \log r_2) - \Psi(f, \log r_1) = q(\log r_2 - \log r_1).$$

Proof. By Lemma B.13.13, we have $\nu^+(f, \log r_1) = q$ and by Theorem B.13.4, we have $\Psi(f, \mu) - \Psi(f, \log r_1) = q(\mu - \log r_1)$ for every $\mu \in [\log r_1, \log r_2[$, hence by continuity we have $\Psi(f, \log r_2) - \Psi(f, \log r_1) = q(\log r_2 - \log r_1)$. □

Theorem B.13.25: *Let $f(x) \in \mathcal{A}(d(a, r^-))$. If f is not bounded, then f has infinitely many zeros in $d(a, r^-)$.*

Proof. Without loss of generality, we can suppose $a = 0$. Suppose f has t zeros in $d(0, r^-)$. Let $d(0, s)$ be a disk containing all zeros of f, with $s < r$. Then by Theorem B.13.22, for all $\rho \in [s, r[$, we have $\Psi(f, \log(\rho)) \leq \Psi(f, \log(s)) + t(\log(\rho) - \log(s)) \leq \Psi(f, \log(s)) + t(\log(r) - \log(s))$. So, $\psi(f, \mu)$ is bounded, hence, by Theorem B.5.20, f is bounded in $d(0, r^-)$, a contradiction. □

Theorem B.13.26: *Let $f \in \mathcal{A}(d(0, R^-))$ and let $r_1, r_2 \in]0, R[$ satisfy $r_1 < r_2$. If f admits exactly q zeros in $d(0, r_1)$ (taking multiplicity into account) and t different zeros α_j, of respective multiplicity order m_j $(1 \leq j \leq t)$ in $\Gamma(0, r_1, r_2)$, then f satisfies*

$$\Psi(f, \log r_2) - \Psi(f, \log r_1) = \sum_{j=1}^{t} m_j(\log(r_2) - \Psi(\alpha_j)) + q(\log r_2 - \log r_1).$$

Proof. Let $C(0, \rho_h)$ $1 \leq h \leq s$ be the circles containing at least one zero of f. For each $h = 1, \ldots, s$, let $\alpha_{n(h)}, \ldots, \alpha_{n(h+1)-1}$ be the zeros of f in $C(0, \rho_h)$. Let $u = \sum_{j=1}^{t} m_j$.

First by Theorem B.13.24, we notice that

(1) $\Psi(f, \log(\rho_1)) - \Psi(f, \log(r_1)) = q(\log(\rho_1) - \log(r_1))$

and similarly

(2) $\Psi(f, \log(r_2)) - \Psi(f, \log(\rho_s)) = (q + u)(\log(r_2) - \log(\rho_s))$.

Next, for each $h = 1, \ldots, s - 1$, set $l_h = q + \sum_{j=1}^{n(h+1)-1} m_j$. Then, f has no zero in $\Gamma(0, \rho_h, \rho_{h+1})$ and has l_h zeros in $d(0, \rho_h)$, hence by Theorem B.13.24 we have

(3) $\Psi(f, \log(\rho_{h+1})) - \Psi(f, \log(\rho_h)) = l_h(\log(\rho_{h+1}) - \log(\rho_h))$.

Then by (1), (2), and (3) we can check the conclusion. □

Corollary B.13.27: *Let $f(x) \in \mathcal{A}(d(0, R^-))$ be such that $f(0) \neq 0$, let $r \in]0, R[$, and let a_j, $1 \leq j \leq q$ be the zeros of f in $d(0, r)$, of respective multiplicity m_j. Then*

$$\Psi(f, \log r) = \Psi(f(0)) + \sum_{j=1}^{q} m_j(\log r - \Psi(a_j)).$$

Corollary B.13.28: *Let $f(x)=\sum_{n=0}^{\infty} a_n x^n \in \mathcal{A}(d(0,r^-))$ have a set of zeros in $d(0,r^-)$ that consists of a sequence $(\alpha_n)_{n\in\mathbb{N}}$, such that $\alpha_n \neq 0\ \forall n\in\mathbb{N}$ and where each α_n is of order u_n. Then f is unbounded if and only if the sequence $(\alpha_n)_{n\in\mathbb{N}}$ satisfies $\prod_{n=0}^{\infty}\Big(\frac{|\alpha_n|}{r}\Big)^{u_n}=0$.*

Corollary B.13.29: *Let $f(x)=\sum_{n=0}^{\infty} a_n x^n \in \mathcal{A}(d(0,r^-))$ have a set of zeros in $d(0,r^-)$ that consists of a sequence $(\alpha_n)_{n\in\mathbb{N}}$ such that $\alpha_n \neq 0\ \forall n\in\mathbb{N}$ and where each α_n is of order u_n. Then $\|f\|_D=|f(0)|\prod_{n=0}^{\infty}\Big(\frac{r}{|\alpha_n|}\Big)^{u_n}$.*

Corollary B.13.30: (Schwarz Lemma) *Let $D=d(a,s)$, let $f\in H(D)$ have at least (resp. at most) q zeros in $d(a,r)$ with $q>0$ and $0<r<s$. Then we have $\frac{\varphi_{a,s}(f)}{\varphi_{a,r}(f)}\geq\Big(\frac{s}{r}\Big)^q$,* (resp. $\frac{\varphi_{a,s}(f)}{\varphi_{a,r}(f)}\leq\Big(\frac{s}{r}\Big)^q$.

Corollary B.13.31: *Let $f\in\mathcal{A}(\mathbb{K})$. The following two statements are equivalent: f is a polynomial of degree q, there exists $q\in\mathbb{N}$ such that $\frac{|f|(r)}{r^q}$ has a finite limit when r tends to $+\infty$.*

Corollary B.13.32: *Let r, s, $R\in]0,+\infty[$ satisfy $0<r<s<R$ and let $f\in H((0,R))$. Then*

$$\Psi(f,\log(s))-\Psi(f,\log(r))\leq\Big(\Psi(f,\log(R))-\Psi(f,\log(s))\Big)\Big(\frac{\log(s)-\log(r)}{\log(R)-\log(s)}\Big).$$

Proof. Let q be the total number of zeros of f in $d(0,s)$, each counted with its multiplicity. Then by Theorem B.13.30, we have $\Psi(f,\log(s))-\Psi(f,\log(r))\leq q(\log(s)-\log(r))$. On the other hand, $\Psi(f,\log(R))-\Psi(f,\log(s))\geq q(\log(R)-\log(s))$. Consequently,

$$q\leq\frac{\Psi(f,\log(R))-\Psi(f,\log(s))}{\log(R)-\log(s)}$$

and, hence, the proof is over. □

Theorem B.13.33: *Let $f(x)=\sum_{n=0}^{\infty} a_n x^n\in\mathcal{A}(\mathbb{K})$ (resp. $f(x)=\sum_{n=0}^{\infty} a_n x^n\in\mathcal{A}(d(0,r^-))$). All zeros of f are of order one and the set of zeros of f is a sequence $(\alpha_n)_{n\in\mathbb{N}}$ such that $|\alpha_n|<|\alpha_{n+1}|$ if and only if the sequence $\Big|\frac{a_n}{a_{n+1}}\Big|$ is strictly increasing. Moreover, if these properties are satisfied, then the sequence of zeros* *of f in $\mathbb{K}$ (resp. in $d(0,r^-)$) is a sequence $(\alpha_n)_{n\in\mathbb{N}^*}$ such that $\lim_{n\to+\infty}|\alpha_n| = +\infty$ (resp. $\lim_{n\to+\infty}|\alpha_n| = r$) and $|\alpha_n| = \left|\dfrac{a_n}{a_{n+1}}\right|$.*

Proof. Suppose first that $f \in \mathcal{A}(d(0,r^-))$. First we suppose that the set of zeros of f in $d(0,r^-)$ is an increasing distances sequence $(\alpha_n)_{n\in\mathbb{N}^*}$. Then by Corollary B.13.19, we know that $\lim_{n\to+\infty}|\alpha_n| = r$. By Corollary B.13.3, for each $q \in \mathbb{N}^*$, we have $\nu^+(f,\Psi(\alpha_q)) - \nu^-(f,\Psi(\alpha_q)) = 1$ and $|\alpha_0| = |\dfrac{a_0}{a_1}|$. Then by an immediate induction we deduce that $|\alpha_n| = \left|\dfrac{a_n}{a_{n+1}}\right|$ for every $n \in \mathbb{N}^*$ and, therefore, the sequence $\left|\dfrac{a_n}{a_{n+1}}\right|$ is strictly increasing.

Conversely, we suppose that $\left|\dfrac{a_n}{a_{n+1}}\right|$ is a strictly increasing sequence. Hence, we have $\left|\dfrac{a_n}{a_{n+1}}\right| < \left|\dfrac{a_{n+1}}{a_{n+2}}\right|$ for every $n \in \mathbb{N}$. For each $m \in \mathbb{N}^*$, we put $s_m = \Psi(a_m) - \Psi(a_{m-1})$ and $r_m = \theta^{s_m}$. Clearly, we have $\nu^+(f,s_m) - \nu^-(f,s_m) = 1$ for every $m \in \mathbb{N}^*$ and $\nu^+(f,\mu) = \nu^-(f,\mu)$, for every $\mu \in \big(]-\infty,\log r[\setminus\{s_m|\ m\in\mathbb{N}^*\}\big)$. Hence by Theorem B.13.1, f admits exactly one zero in each circle $C(0,r_m)$ and no other zero in $d(0,r^-)$.

Suppose now that $f \in \mathcal{A}(\mathbb{K})$. The same proof applies with $\lim_{n\to+\infty}|\alpha_n| = +\infty$. □

Corollary B.13.34: *Let $f(x) = \sum_{n=0}^{\infty} a_n x^n$ and suppose that the sequence $\left(\dfrac{|a_n|}{|a_{n+1}|}\right)_{n\in\mathbb{N}}$ is strictly increasing, of limit $+\infty$ (resp. of limit R). Then f belongs to $\mathcal{A}(\mathbb{K})$ (resp. $\mathcal{A}(d(0,R^-))$). Moreover, putting $r_n = \dfrac{|a_n|}{|a_{n+1}|}$, $n \in \mathbb{N}$, f admits a unique zero on each circle $C(0,r_n)$ and has no other zero in $\mathbb{K}$ (resp. in $d(0,R)$).*

Proof. Indeed, thanks to the remark at the beginning of Chapter B.5, the radius of convergence of f is $+\infty$ (resp. R). Then conclusion comes from Theorem B.13.33. □

Remark: We can easily construct a sequence $(r_n)_{n\in\mathbb{N}}$ satisfying the hypothesis of Corollary B.13.34 and thereby the function h.

It is often uneasy to determine whether a function defined in an infraconnected set is an analytic element. The following example may be useful.

Theorem B.13.35: *Let $R > 0$ and let $(a_n)_{n\in\mathbb{N}}$ be a sequence of $d(a,R^-)$ such that $|a_n| < |a_{n+1}$, $\lim_{n\to+\infty}|a_n| = R$, $\prod_{k=0}^{+\infty}\dfrac{|a_n|}{R} = 0$ and set $r_n = |a_n|$, $n \in \mathbb{N}$. Let*

$f \in \mathcal{A}(d(a, R^-))$ admit each a_n as a zero of order 1 and no other zero in $d(a, r_n)$. Let $E = d(a, R^-) \setminus \Big(\bigcup_{n\in\mathbb{N}} C(a, r_n)\Big)$. Then the function $\frac{1}{f}$ defined in E belongs to $H(E)$.

Proof. Without loss of generality we can obviously suppose $a = 0$ and $f(0) = 1$. By Corollary B.13.28 f belongs to $\mathcal{A}_u(d(0, R^-))$. For each $n \in \mathbb{N}$, set $r_n = |a_n|$. Let $r \in]0, R[$. If $r \neq r_n\ \forall n \in \mathbb{N}$, then $|f(x)| = |f|(r)\ \forall x \in C(0, r)$. Now let $x \in C(0, r_n)$. If $|x - a_n| = r_n$, we have $|f(x)| = r_n$. And if x belongs to E so that $|x - a_n| < r_n$, then $|f(x)| \geq |f|(r_n)\frac{r_n}{s}$. Consequently,

$$(1) \qquad |f(x)| = |f|(|x|)\ \forall x \in E.$$

Let $f(x) = \sum_{n=0}^{+\infty} c_n x^n$ and for each $k \in \mathbb{N}$, let $P_k(x) = \sum_{n=0}^{k} c_n x^n$. Of course, the sequence $(P_k)_{k\in\mathbb{N}}$ converges to f uniformly in every disk $d(0, r)$ with $r \in]0, R[$. Moreover, by Theorem B.13.34, each zero a_k of f satisfies $|a_k| = \frac{|c_{k-1}|}{|c_k|}$. Consequently, for $n > k$, P_n also admits a unique zero $a_{n,k}$ in $C(0, r_n)$ and has no other zero in E. Therefore, we have $|P_n|(r) = |f|(r)\ \forall r \leq r_n$. We will show that the sequence $\left(\frac{1}{P_n}\right)_{n\in\mathbb{N}}$ converges to $\frac{1}{f}$ in $H(E)$. Indeed, let us fix $\epsilon > 0$ and let us choose $q \in \mathbb{N}$ such that $|f|(r_q) > \frac{1}{\epsilon}$. Consider now, integers $n > q$ such that $|f(x) - P_n(x)| < \epsilon\ \forall x \in d(0, r_q)$. Then obviously,

$$\frac{1}{f(x)} - \frac{1}{P_n(x)}| < \epsilon\ \forall x \in E \cap d(0, r_q).$$

On the other hand, given $x \in E \setminus d(0, r_q)$, we have $|P_n|(r) \geq |P_n|(r_q) > \frac{1}{\epsilon}$, hence $\left|\frac{1}{P_n(x)} - \frac{1}{f(x)}\right| < \epsilon$. Therefore, $\left\|\frac{1}{P_n(x)} - \frac{1}{f(x)}\right\|_E \leq \epsilon$. This finishes proving that $\frac{1}{f(x)}$ belongs to $H(E)$. □

Definitions and notations: Let $f \in \mathcal{A}(\mathbb{K})$ and let $n \in \mathbb{N}^*$. We will denote by $f^{<n>}$ the function $f \circ f \circ f \circ \cdots \circ f$, n times.

Theorem B.13.36: *Suppose the residue characteristic p is different from 0. There exists functions $\mathcal{A}(\mathbb{K})$ and points $a \in \mathbb{K}$ such that $\lim_{n\to+\infty} |f^{<n>}(a)| = +\infty$.*

Proof. Let $t(n) = n^2$, $n \in \mathbb{N}$. Let $f(x) = \sum_{n=0}^{+\infty} p^{2t(n)} x^{2n+1}$. For each $n \in \mathbb{N}$, put $r_n = \frac{p^{t(n-1)}}{p^{t(n)}}$. Then, we can check that f admits exactly two zeros (taking multiplicity into account) in the circle $C(0, r_n)$ and has no zeros outside $\bigcup_{n=1}^{\infty} C(0, r_n)$, except $\{0\}$.

Consider now a number $s \in]r_n, r_{n+1}[$ of the form $q\sqrt{p}$, with $q \in \mathbb{Q}$. Inside the disk $d(0,s)$, f admits $2n+1$ zeros (taking multiplicity into account). Consequently, $|f|(s) = |p^{2t(n)}|s^{2n+1} = \dfrac{p^{-2t(n)}|q|^{2n+1}p^{-2n}}{\sqrt{p}} = \dfrac{p^{-2t(n)-2n}|q|^{2n+1}}{\sqrt{p}}$. Thus we can see that $|f|(s)$ is of the form $q'\sqrt{p}$ with $q' \in \mathbb{Q}$. Moreover, since q belongs to $\mathbb{Q}$, f has no zero in $C(0,s)$, hence $|f(x)| = |f|(s) = \dfrac{|q'|}{\sqrt{p}}$ $\forall x \in C(0,s)$. Now consider $\dfrac{|f|(s)}{s} = \dfrac{|p^{2t(n)}|s^{2n+1}}{s} = p^{-2t(n)}s^{2n}$.

On the other hand, by (1) $r_n = p^{t(n)-t(n-1)}$. Since $r_n < s$, we can derive

$$p^{-2t(n)}s^{2n} > p^{-2t(n)}p^{2n(t(n)-t(n-1)}.$$

Now $-2t(n) + 2n(t(n) - t(n-1) = 2\big(-t(n) + n(t(n) - t(n-1))\big) = 2(-n^2 + n(2n-1)) = 4n^2 - 2n$. Consequently, when $s > r_n$, we have $\dfrac{|f|(s)}{s} > p^{4n^2-2n}$. Now, let us take $s_2 > r_2$ and for each $n \in \mathbb{N}$, $n \geq 2$, let us define by induction $s_{n+1} = |f|(s_n)$. So, we have $\dfrac{s_{n+1}}{s_n} \geq p^8$, hence $\lim\limits_{n\to+\infty} s_n = +\infty$. But by construction, for all $x \in C(0, s_n)$, we have $|f(x)| = |f|(s_n)$. Consequently, taking $a \in C(0, s_2)$, we have $|f^{<k>}(a)| = s_k$ and, hence, $\lim\limits_{n\to+\infty} |f^{<n>}(a)| = +\infty$. □

B.14. Image of a disk

In this chapter D is just an open subset of $\mathbb{K}$.

Theorem B.14.1: *Let $f(x) = \sum_{n=0}^{\infty} a_n(x-a)^n \in H(d(a,r))$. Then the following statements are equivalent:*

(a) $|a_0| > |a_n|r^n$ *for all* $n > 1$,
(b) $\|f - f(a)\|_{d(a,r)} < |f(a)|$,
(c) f *has no zero in* $d(a,r)$,
(d) $|f(x)|$ *is constant and different from* 0 *in* $d(a,r)$,
(e) f *is invertible in* $H(d(a,r))$.

Proof. First, (a) and (b) are equivalent by Theorem B.5.6. Second, (a), (c), and (d) are equivalent by Theorems B.13.7 and B.13.16. Third, by Lemma B.2.3, (d) implies (e) and finally (e) obviously implies (c). □

Corollary B.14.2: *Let $f(x) = \sum_{n=0}^{\infty} a_n(x-a)^n \in \mathcal{A}(d(a,r^-))$ (resp. let $f(x) = \sum_{n=0}^{\infty} a_n(x-a)^n \in H((d(a,r^-)))$. Then the following statements are equivalent:*

(a) $|a_0| \geq |a_n| r^n$ *for all* $n > 1$,
(b) $|f - f(a)| < |f(a)|\ \forall x \in d(a, r^-)$,
(c) f *has no zero in* $d(a, r^-)$,
(d) $|f(x)|$ *is constant and different from* 0 *in* $d(a, r^-)$,
(e) f *is invertible in* $\mathcal{A}(d(a, r^-))$ (*resp.* f *is invertible in* $H(d(a, r^-))$).

Proof. Concerning $\mathcal{A}(d(a, r^-))$, we just have to apply Theorem B.14.1 to f in $H(d(a, \rho))$ for every $\rho \in]0, r[$. Concerning $H(d(a, r^-))$, we can use Lemma B.2.3 to check that an element of $H(d(0, r^-))$ having no zero is invertible. □

Theorem B.14.3: *Let* a, $b \in \mathbb{K}$ *and* $r, s \in \mathbb{R}_+^*$, *and let* $f \in \mathcal{A}(d(a, r^-))$, $g \in \mathcal{A}(d(b, s^-))$ *be such that* $f(d(a, r^-)) \subset d(b, s^-)$. *Then* $g \circ f$ *belongs to* $\mathcal{A}(d(a, r^-))$.

Proof. Without loss of generality we can clearly assume $a = b = 0$. First, suppose that f has no zero in $d(0, r^-)$. Then $|f(x)|$ is equal to a constant c in $d(0, r^-)$, with $c < s$. Hence, of course, $f(d(0, r^-))$ is included in $d(0, c)$. Now, let $\rho \in]0, r[$. The restriction of f to $d(0, \rho)$ belongs to $H(d(0, \rho^-))$ and the restriction of g to $d(0, c)$ belongs to $H(d(0, c))$. Hence, the restriction of $g \circ f$ to $d(0, \rho)$ belongs to $H(d(0, \rho))$. Consequently, by Corollary B.3.3 $g \circ f$ belongs to $H(d(0, \rho))$. This is true for every $\rho \in]0, r[$ and, therefore, this shows that $g \circ f$ belongs to $\mathcal{A}(d(0, r^-))$.

Now, we suppose that f admits at least one zero in $d(0, r^-)$. Hence, there exists $r' \in]0, r[$ such that f has at least one zero in $d(0, r')$. Therefore, by Corollary B.13.30, $\|f\|_{d(0,\rho)}$ is strictly increasing in ρ in the interval $[r', r[$. Now, let $\rho \in]0, r[$ and let $\sigma = \|f\|_{d(0,\rho)}$. The restriction of f to $d(0, \rho)$ belongs to $H(d(0, \rho))$ and further, $f(d(0, \rho))$ is included in $d(0, \sigma)$. Since g belongs to $H(d(0, \sigma))$, $g \circ f$ belongs to $H(d(0, \rho))$. As previously, this is true for every $\rho \in]0, r[$, hence $g \circ f$ belongs to $\mathcal{A}(d(0, r^-))$. □

Theorem B.14.4: (Dieudonn´e–Dwork) *Let* $f \in \mathcal{A}(d(0, R^-))$ *satisfy* $f(0) = 1$ *and have no zero in* $d(0, R^-)$. *There exists a sequence* $(u_k)_{k \in \mathbb{N}^*}$ *in* $d(0, R)$ *such that* $f(x) = \prod_{k=1}^{\infty}(1 - u_k x^k)$ *whenever* $x \in d(0, R^-)$.

Proof. Since $f(0) = 1$, we can write $f(x)$ in the form $1 + \sum_{n=1}^{\infty} b_n x^n$. Since f has no zero in $d(0, R^-)$, by Corollary B.14.2 we have $|b_n| R^n \leq 1$ for every $n \in \mathbb{N}^*$. Now, suppose that we have already found $u_1, \dots, u_k$ such that $f(x)$ factorizes in the form

$(\mathcal{R}_k)$ $\prod_{j=1}^{k}(1 - u_j x^j)\Big(1 + x^{k+1} \sum_{n=0}^{\infty} \beta_{n,k+1} x^n\Big)$ with $|\beta_{n,k+1}| \leq \frac{1}{R}$, for every $n \in \mathbb{N}$.

Actually, we have $(1+\beta_{0,k+1}x^{k+1})\Big(1+\sum_{n=1}^{\infty}(-\beta_{0,k+1}x^{k+1})^n\Big) = 1$ and, therefore, we can factorize $\Big(1+x^{k+1}\sum_{n=0}^{\infty}\beta_{n,k+1}x^n\Big)$ in the form

$$(1+\beta_{0,k+1}x^{k+1})\Big(1+\sum_{n=1}^{\infty}(-\beta_{0,k+1}x^{k+1})^n\Big)\Big(1+x^{k+1}\sum_{n=0}^{\infty}\beta_{n,k+1}x^n\Big).$$

Now, consider the function g_{k+1} defined as

$$g_{k+1}(x) = \Big(1+\sum_{n=1}^{\infty}(-\beta_{0,k+1}x^{k+1})^n\Big)\Big(1+x^{k+1}\sum_{n=0}^{\infty}\beta_{n,k+1}x^n\Big).$$

In g_{k+1}, it is seen that the term in x^{k+1} is equal to 0, so g_{k+1} is of the form $\Big(1+x^{k+2}\sum_{n=0}^{\infty}\beta_{n,k+2}x^n\Big)$, with $|\beta_{n,k+2}| \leq \frac{1}{R}$, for every $n \in \mathbb{N}$.

Now we just put $u_{k+1} = \beta_{0,k+1}$ and then we have proven $(\mathcal{R}_{k+1})$. Since $(\mathcal{R}_0)$ is trivially satisfied, by induction we can construct a sequence $(u_k)_{k\in\mathbb{N}^*}$ in $d(0,R^-)$ and a sequence $(g_k)_{k\in\mathbb{N}^*}$ in $\mathcal{A}(d(0,R^-))$, such that for each $k \in \mathbb{N}$, g_k is of the form $1+x^{k+1}\sum_{n=0}^{\infty}\beta_{n,k+1}x^n$ and satisfies $f(x) = \prod_{j=1}^{k}(1-u_jx^j)g_k(x)$. For each $k \in \mathbb{N}^*$, let $f_k(x) = \prod_{j=1}^{k}(1-u_jx^j)$.

It is seen that for each $r \in]0,R[$, we have $\|g_k - 1\|_{d(0,r)} \leq r^{k+1}$. As a consequence, for every $r \in]0,R[$, the sequence $(f_k)_{k\in\mathbb{N}^*}$ converges to f in $H(d(0,r))$ and, therefore, we have $f(x) = \prod_{k=1}^{\infty}(1-u_kx^k)$ for all $x \in d(0,R^-)$. That ends the proof. □

Proposition B.14.5: *Let D be a set of the form $\bigcup_{i\in I} d(\alpha_i, r^-)$ with $|\alpha_i - \alpha_j| = r$ whenever $i \neq j$. Let ℓ be fixed in I, let $f \in H(D)$ be such that $f(x) = \sum_{n=0}^{\infty} a_n (x-\alpha_\ell)^n \in H(D)$ whenever $x \in d(\alpha_\ell, r^-)$, and let $s = \sup_{n\geq 1}|a_n|r^n$. For every $i \in I$, let $\beta_i = f(\alpha_i)$. Then we have $f(d(\alpha_i, r^-)) = d(\beta_i, s^-)$ for every $i \in I$ and $|\beta_i - \beta_j| \leq s$ whenever $i, j \in I$. Moreover, if I is not finite, the equality $|\beta_i - \beta_\ell| = s$ holds for every $i \in I$ but finitely many.*

Proof. We set $a = \alpha_\ell$ and $g = f - \beta_\ell$. By Theorem B.13.9, for every $\rho \in]0,r[$ we have $\Psi_a(g, \log\rho) = \sup_{n\geq 1}\Psi(a_n) + n\log\rho$, hence by continuity:

$$\Psi_a(g, \log r) = \sup_{n\geq 1}\Psi(a_n) + n\log r = \log s.$$

Now let $j \in I$, $j \neq \ell$ and set $b = \alpha_j$. Since $|b-a| = r$, by Proposition B.10.4, we have $\Psi_a(g, \log r) = \Psi_b(g, \log r)$, hence $\Psi(g(x)) \leq \Psi_a(g, \log r) = \log s$ for all $x \in d(b, r^-)$. Thus we see that $|f(x) - \beta_\ell| \leq s$ for all $x \in d(b, r^-)$. Hence, by Corollary B.13.11 $f(d(b, r^-))$ is a disk $d(\beta_j, t^-)$ with $t \leq s$. Since α_ℓ and α_j play the same role, in the same way we show that $s \leq t$ and, therefore, $s = t$. Thus we have proven that $f(d(\alpha_i, r^-)) = d(\beta_i, s^-)$ for every $i \in I$ and $|\beta_i - \beta_j| \leq s$ whenever $i, j \in I$.

Now we suppose that I is infinite. By Proposition B.10.1, the equality $\Psi(g(x)) = \Psi_a(g, \log r) = \log s$ holds in all the classes of $C(a, r)$ but finitely many ones, hence we have $|\beta_j - \beta_\ell| = s$ for every $j \in I$ but finitely many and this ends the proof of Proposition B.14.5. □

Theorem B.14.6: *Let D be an open analytic subset of $\mathbb{K}$, let $f \in H(D)$, and let $D' = f(D)$. Then D' is open and satisfies* $\text{codiam}(D') \geq \text{codiam}(D) \inf_{x \in D} |f'(x)|$.

Proof. Let $b \in f(D)$ and let $a \in D$ be such that $f(a) = b$. Since D is open, there exists a disk $d(a, r)$ included in D. Let $f(x) = \sum_{j=0}^{\infty} a_n (x-a)^n \ \forall x \in d(a, r)$. Since D is an analytic set, f is not identically zero in any disk included in D, hence, by Theorem B.13.9, $\|f - b\|_{d(a,r)}$ is a number $s = \sup_{n \geq 1} |a_n| r^n$. Then, by Theorem B.13.9, $f(d(a, r))$ is the disk $d(b, s)$, hence $d(b, s) \subset D$, which proves that $f(D)$ is open. Particularly, $|f'(a)| = |a_1|$, hence $\|f - b\|_{d(a,r)} \geq r|f'(a)|$, and, therefore, $\delta(D', (\mathbb{K} \setminus D')) \geq \delta(D, (\mathbb{K} \setminus D)) \inf_{x \in D} |f'(x)|$. □

Theorem B.14.7: *Let $f \in H(d(0, r))$, let $t = \nu^+(f, \log r)$, and assume $t \geq 1$. Suppose that f' is not identically zero and let $\alpha_1, \dots, \alpha_q$ be the zeros of f' in $d(0, r)$. For every $b \in f(d(0, r)) \setminus f(\{\alpha_1, \dots, \alpha_q\})$, $f - b$ admits exactly t zeros of order 1 in $d(0, r)$.*

Proof. Let $b \in f(d(0, r)) \setminus f(\{\alpha_1, \dots \alpha_q\})$. By Theorem B.13.7, $f - b$ admits t zeros in $d(0, r)$ (taking multiplicity into account). But for each zero α of $f - b$, (as $\alpha \neq \alpha_j$ whenever $j = 1, \dots, q$) we have $f'(\alpha) \neq 0$, hence the t zeros of $f - b$ are of order one. □

Definitions: Let $f \in H(D)$. Then f will be said to be *strictly injective* in D if f is injective and if $f'(x) \neq 0$ whenever $x \in D$.

In the same way, given $a \in \mathbb{K}$ and $r > 0$, an analytic function $f(x) \in \mathcal{A}(d(a, r^-))$ will be said to be *strictly injective* in $d(a, r^-)$ if f is injective and if $f'(x) \neq 0$ whenever $x \in D$.

Theorem B.14.8: *Let $\mathbb{K}$ have characteristic zero and let $f \in H(D)$ be injective in D. Then f is strictly injective.*

Proof. Suppose that f is not strictly injective and let $\alpha \in D$ be a zero of f'. Let $d(\alpha, r)$ be a disk included in D. Without loss of generality we may assume $\alpha = 0$. Hence, in $d(0, r)$, $f(x)$ is equal to a series of the form $a_0 + \sum_{n=q}^{\infty} a_n x^n$ with $q \geq 2$ and $a_q \neq 0$. Therefore, we have $\nu^+(f, \log r) \geq q \geq 2$. Let $t = \nu^+(f, \log r)$. Since $\mathbb{K}$ has characteristic zero, f' is not identically zero and, therefore, admits finitely many zeros $\alpha_1 = 0, \alpha_2, \ldots, \alpha_s$ in $d(0, r)$. Let $b \in f(d(0, r)) \setminus f(\{\alpha_1, \ldots \alpha_s\})$. By Theorem B.14.7, $f - b$ admits t simple zeros in $d(0, r)$ and this contradicts the hypothesis "f injective in $d(0, r)$." □

Theorem B.14.9: *Let $a \in \mathbb{K}$, $r \in \mathbb{R}_+$, let $f(x) = \sum_{n=0}^{\infty} a_n (x - a)^n \in H(d(a, r))$, and let $s = \sup_{n \geq 1} |a_n| r^n$ be > 0. Then the following statements are equivalent:*

(α) $|a_1| > |a_n| r^{n-1}$ *whenever* $n > 1$
(β) $|f(x) - f(y)| = |x - y||a_1|$ *whenever* $x, y \in d(a, r)$
(γ) f *is strictly injective in* $d(a, r)$.

Moreover when Conditions $(\alpha), (\beta)$, and (γ) are satisfied, then we have $s = |a_1| r$ and $|f'(x)| = |a_1|$ whenever $x \in d(a, r)$.

Proof. Without loss of generality we may obviously assume $a = 0$.
First, we suppose (α) is satisfied and consider

$$f(x) - f(y) = (x - y)\left(a_1 + \sum_{n=2}^{\infty} a_n \left(\sum_{j=0}^{n-1} x^j y^{n-1-j}\right)\right).$$

For every $n \geq 2$ it is seen that $|x^j y^{n-1-j}| \leq r^{n-1}$. Hence, we have $|a_1| > \left|\sum_{n=2}^{\infty} a_n \left(\sum_{j=0}^{n-1} x^j y^{n-1-j}\right)\right|$ and thereby $|f(x) - f(y)| = |a_1|\ |x - y|$. At the same time we notice that (α) implies $|f'(x)| = |a_1|$ whenever $x \in d(a, r)$ while, by Theorem B.13.9, we have $s = |a_1| r$. So (α) implies (β).

Second, we suppose (β) is satisfied. Since $s > 0$, by Corollary B.13.10, f is not a constant, hence $a_1 \neq 0$. Then by (β) we have $f(x) \neq f(y)$ whenever $x \neq y$. Moreover, $|f'(x)| = |a_1|\ \neq 0$, hence (γ) is satisfied.

Third, we suppose (γ) is satisfied. Let $b \in f(d(a, r))$, let $g = f - b$, and let $t = \nu^+(g, \log r)$. If $t \geq 2$ either g admits several different zeros or g admits a zero α of order t. In both cases we see that g is not strictly injective, hence neither is f. Finally we have $t = 1$ and, hence, (α) is satisfied. This ends the proof of Theorem B.14.9. □

Theorem B.14.9 is easily applied to analytic functions inside a disk.

Corollary B.14.10: *Let $f(x) = \sum_{n=0}^{\infty} a_n(x-a)^n \in \mathcal{A}_b(d(a,r^-))$ and suppose that the number $s = \sup_{n\geq 1} |a_n|r^n$ is strictly positive. Then Conditions $(\alpha), (\beta), (\gamma)$, and (δ) are equivalent.*

(α) $|a_1| \geq |a_n|r^{n-1}$ *whenever* $n > 1$
(β) $|f(x) - f(y)| = |x-y|\ |a_1|$ *whenever* $x, y \in d(a, r^-)$
(γ) f *is strictly injective in* $d(a, r^-)$
(δ) $s = |a_1|r$.

Moreover, when Conditions $(\alpha), (\beta), (\gamma)$, and (δ) are satisfied, we have $|f'(x)| = |a_1|$ whenever $x \in d(a, r^-)$.

Proof. For every $\rho \in]0, r[$ we apply Theorem B.14.9 to $f \in H(d(a, \rho))$. □

Lemma B.14.11: *Let $f \in \mathcal{A}(d(\alpha, r^-))$ be injective such that f' is not identically zero. Then f is strictly injective.*

Proof. We may obviously assume that $f'(\alpha) \neq 0$ and $\alpha = 0$. Hence, f is of the form $\sum_{n=0}^{\infty} a_n x^n$ with $a_1 \neq 0$. If f' has a zero β, there exists an integer $q > 1$ such that

(1) $|qa_q|\ |\beta|^{q-1} = |a_1|$.

Let $g(x) = f(x) - f(0)$. Then g is also injective and has a simple zero at 0. But by (1) we have $|a_q|\ |\beta|^q \geq |a_1|\ |\beta|$, hence we have $\nu^+(g, \mu) \geq q$ when μ is close enough to $\log(r)$. Then, by Theorem B.13.7 we know that g has at least q zeros in $d(0, r^-)$ and then admits another zero $\gamma \neq 0$, which contradicts the fact that it is injective. Consequently, f' has no zero in $d(\alpha, r^-)$, i.e., f is strictly injective. □

We are now able to study the inverse functions of an analytic element inside a disk $d(a, r)$.

Theorem B.14.12: *Let $a \in \mathbb{K}$, $r \in \mathbb{R}_+$, let $f(x) \in H(d(a,r))$ be strictly injective in $d(a,r)$, let $s = |f'(a)|r$, and let $b = f(a)$. The homomorphism Θ from $H(d(b,s))$ into $H(d(a,r))$ defined as $\Theta(h) = h \circ f$ is an isometric isomorphism from $H(d(b,s))$ onto $H(d(a,r))$.*

Proof. Without loss of generality we may obviously suppose $a = b = f(a) = 0$ and $f'(0) = 1$. We put $E = d(0,r)$. Hence, by Theorem B.14.9, in $d(0,r)$ $f(x)$ is equal to a series of the form $x + \sum_{j=2}^{\infty} a_j x^j$ with $\sup_{j\geq 2} |a_j|r^j < r$. Next, by Theorem B.13.9, r is equal to s. It is obviously seen that $\|\Theta(h)\|_E = \|h\|_E$ for every $h \in H(E)$. So, we only have to prove that Θ is surjective. Let $\lambda = \inf_{P\in\mathbb{K}[x]} \|\Theta(P) - x\|_E$ and suppose $\lambda > 0$.

Set $\lambda = \frac{\inf_{P\in\mathbb{K}[x]}(\|P(f) - x)\|)}{r}$ and suppose that $\lambda > 0$. By definition, $\lambda < 1$ because $\|f - x\| < r$. Let $P \in \mathbb{K}[x]$ be such that $\|P(f) - x\| < r\lambda^{\frac{2}{3}}$. Then we can write $x = P(f) + h(x)$ with $h \in H(d(0,r))$ and $h(x) = \sum_{k=0}^{\infty} c_k x^k$ with $|c_n| < r\lambda^{\frac{2}{3}}\ \forall n \in \mathbb{N}$ and $\lim_{n\to+\infty} c_n = 0$.

Let $q \in \mathbb{N}$ be such that $|c_n < r\lambda^2\ \forall n > q$ and let $\omega(x) = \sum_{n=0}^{q} c_n x^n$. We first notice that for all $n \in \mathbb{N}$, we have

(1) $\|(P(f) + h(x))^n - (P(f) + \omega(x))^n\| \leq \|h - \omega\| \leq r\lambda^{\frac{2}{3}}$.

Now, for each $n \in \mathbb{N}$, set $(P(f) + h(x))^n = (P(f) + \omega(x))^n + \omega_n(x)$. Then by (1), we have

(2) $\|\omega_n\| \leq r\lambda^2\ \forall n \in \mathbb{N}$.

On the other hand, $\|(P(f) + \omega(x))^n - (P(f))^n| \leq \|\omega\| \leq r\lambda^{\frac{2}{3}}\ \forall n \in \mathbb{N}$. Consequently, we can write

(3) $\|(P(f) + \omega(x))^n = (P(f))^n + \ell_n(x)$ with $\|\ell_n\| \leq r\lambda^{\frac{2}{3}}\ \forall n \in \mathbb{N}$.

Now, we have

$$x - P(f) = \sum_{n=0}^{q} c_n((P(f) + \omega(x))^n + \omega_n(x)) + \sum_{n=q+1}^{\infty} c_n((P(f) + \omega(x))^n + \omega_n(x))$$

therefore, by (2) and (3) we can write

$$x - P(f) = \sum_{n=0}^{q} c_n(P(f))^n + \sum_{n=0}^{q} c_n(\ell_n(x) + \omega_n(x)) + \sum_{n=q+1}^{\infty} c_n((P(f) + \omega(x))^n + \omega_n(x)) \tag{4}$$

with $\|c_n\big((P(f) + \omega(x))^n + \omega_n(x)\big)\| < r\lambda^2\ \forall n \geq q$ and $\|\ell_n + \omega_n\| < r\lambda^{\frac{2}{3}}\ \forall n = 0,\dots,q$, hence $\|\sum_{n=0}^{q} c_n\Big(\ell_n(x) + \omega_n(x)\Big)\| \leq r\lambda^{\frac{4}{3}}$. Consequently, $\Big\|x - P(f) - \sum_{n=0}^{q} c_n(P(f))^n\Big\| \leq r\lambda^{\frac{4}{3}}$. Set $Q(x) = P(x) - \sum_{n=0}^{q} c_n(P(x))^n$. Then the polynomial Q satisfies $\|x - Q(f)\| \leq r\lambda^{\frac{4}{3}}$, a contradiction to the hypothesis $\lambda > 0$. Consequently, $\lambda = 0$ and, therefore, x does belong to the closure of $\Theta(H(E))$.

But since Θ is isometric, $\Theta(H(E))$ is obviously closed in $H(E)$ and, therefore, x belongs to $\Theta(H(E))$. As a consequence $\Theta(H(E)) = H(E)$. □

Corollary B.14.13: *Let $f \in H(d(a,r))$ be strictly injective and let $d(b,s) = f(d(a,r))$. Then $\overset{-1}{f}$ belongs to $H(d(b,s))$.*

Corollary B.14.14: *Let $\mathbb{K}$ have characteristic 0, let $f \in H(a,r))$ be injective, and let $d(b,s) = f(d(a,r))$. Then $\overset{-1}{f}$ belongs to $H(d(b,s))$.*

Corollary B.14.15: *Let $f \in \mathcal{A}(d(0,r^-))$ be strictly injective in $d(0,r^-)$ and let $s = r|f'(0)|$. Then $\overset{-1}{f}$ belongs to $\mathcal{A}(d(0,s^-))$.*

Proof. Indeed, by Corollary B.14.13 $\overset{-1}{f}$ belongs to $H(d(0,u))$ for every $u \in [0,s[$. □

Corollary B.14.16: *Let $\mathbb{K}$ have characteristic 0, let $f \in \mathcal{A}(d(0,r^-))$ be injective in $d(0,r^-)$, and let $s = r|f'(0)|$. Then $\overset{-1}{f}$ belongs to $\mathcal{A}(d(0,s^-))$.*

Definitions and notations: An injective analytic function $f \in \mathcal{A}_b(d(a,r^-))$ will be said to be *bianalytic* if f^{-1} belongs to $\mathcal{A}(f(d(a,r^-)))$.

We will use the Lemma B.14.17 in topology.

Lemma B.14.17: *Let E be a topological space and let F and G be subsets dense in E. If F is open then $F \cap G$ is dense in E.*

Proof. Indeed let $a \in E$, let V be an open neighborhood of a and let $u \in V \cap F$. Since F is open, $V \cap F$ is a neighborhood of u. Hence, as G is dense in E, there exists $x \in (V \cap F) \cap G$. Therefore, $V \cap (F \cap G) \neq \emptyset$. □

Lemma B.14.18: *Let $\overline{D}$ be open. Let a be a point of $\overline{D} \setminus D$ and let $f \in H(D \cup \{a\})$ be strictly injective in D. There exists an open set E satisfying:*

$\overline{E}$ is open,
$D \cup \{a\} \subset E \subset \overline{D}$,
f belongs to $H(E)$ and is strictly injective in E.

Proof. By Theorem B.2.5 we know that f is of the form $g + h$ with $g \in H(\overline{D})$ and $h \in R(\mathbb{K}\setminus(\overline{D}\setminus(D\cup\{a\})))$. Since $\overline{D}$ is open, there exists $\sigma > 0$ such that $d(a,\sigma) \subset \overline{D}$. Now, as $h \in R(D \cup \{a\})$ there exists $\tau > 0$ such that $h \in R(D \cup d(a,\tau))$. Let $\rho = \min(\sigma,\tau)$ and let $E = D \cup d(a,\rho)$. It is seen that $\overline{E} = \overline{D} \cup d(a,\rho)$ and, therefore, $\overline{E}$ is open. Next, both g, h belong to $H(E)$, hence so does f.

We suppose that f is not injective in E. Let b and $c \in E$ be such that $f(b) = f(c)$ and let $\omega = f(b)$. Let $r \in \mathbb{R}_+^*$ be such that $d(b,r) \cup d(c,r) \subset E$ whereas $d(b,r) \cap d(c,r) = \emptyset$. Then $f(d(b,r))$ is a disk $d(\omega,s)$ whereas $f(d(c,r))$ is a disk $d(\omega,t)$.

We may obviously assume $s \leq t$. Let $\Sigma = d(b,r) \cap D$ and let $\Lambda = d(c,r) \cap D$. Obviously, Σ is dense in $d(b,r)$ whereas Λ is dense in $d(c,r)$, hence $f(\Sigma)$ is dense in $d(\omega, s)$ whereas $f(\Lambda)$ is dense in $d(\omega, t)$. Hence, $f(\Lambda) \cap d(\omega, s)$ is dense in $d(\omega, s)$. Since Σ and Λ are open sets in $\mathbb{K}$, by Corollary B.13.10, both $f(\Sigma)$ and $f(\Lambda)$ are open sets in $\mathbb{K}$ because f is not a constant in $d(b,r)$ or in $d(c,r)$. Hence, both $f(\Lambda) \cap d(\omega, s)$, $f(\Sigma)$ are dense open subsets of $d(\omega, s)$. Therefore, by Lemma B.14.17 we see that $f(\Sigma) \cap f(\Lambda)$ is dense in $d(\omega, s)$ and certainly is not empty. Let $y \in f(\Sigma) \cap f(\Lambda)$ and $\alpha \in \Sigma$, $\beta \in \Lambda$ satisfy $f(\alpha) = f(\beta) = y$. By definition of Σ and Λ we see that α and $\beta \in D$ whereas $\alpha \neq \beta$. This contradicts the hypothesis "f is injective in D" and finally shows f to be injective in E.

Now since f is strictly injective in $d(a, r^-) \cap D$ we have $f'(x) \neq 0$ whenever $x \in d(a, r^-) \cap D$, hence by Lemma B.14.11 f is strictly injective in $d(a, r^-)$. Finally, we have $f'(x) \neq 0$ whenever $x \in E$ and this ends the proof of Lemma B.14.18. □

We remember that Condition (B) was defined in Chapter B.8.

Theorem B.14.19: *Let $\overline{D}$ be open, let D' satisfy $D \subset D' \subset \overline{D}$, and let $f \in H(D')$ be strictly injective in D. There exists an open set D'' satisfying $D' \subset D'' \subset \overline{D}$ such that f belongs to $H(D'')$ and is strictly injective in D''.*

Proof. By Lemma B.14.18, for every $a \in D'$ there exists an open set D_a such that $\overline{D_a}$ is open, satisfying Condition (B), such that $D \cup \{a\} \subset D_a \subset \overline{D}$ and such that f belongs to $H(D_a)$ and is strictly injective in D_a. Let $D'' = \bigcup_{a \in D'} D_a$.
Then D'' is open and such that $D' \subset D'' \subset \overline{D}$. By Theorem B.2.5 f has a unique decomposition in the form $g + h$ with $g \in H(\overline{D})$ and $h \in R(\mathbb{K} \setminus (\overline{D} \setminus D'))$. Since $f \in H(D_a)$ obviously $h \in H(D_a)$. Hence, h has no pole in D_a whenever $a \in D'$, therefore, $h \in R(D'')$. Hence, f belongs to $H(D'')$. Moreover, by Lemma B.14.18 we have $f'(x) \neq 0$ whenever $x \in D_a$, hence whenever $x \in D''$.

Now we just have to check that f is injective in D''. Let $a, b \in D'$ satisfy $f(a) = f(b)$. Since D_a satisfies Condition (B), we may apply Lemma B.14.18 to D_a and b and then we have an open set E such that $D_a \cup \{b\} \subset E \subset \overline{D}_a = \overline{D}$ and such that f belongs to $H(E)$ and is strictly injective in E. Hence, the hypothesis $f(a) = f(b)$ is impossible and, therefore, f is injective in D'. But the hypothesis made on D' actually is satisfied on D''. Hence, f is injective in D'' and this finishes proving Theorem B.14.19. □

Proposition B.14.20: *Let $(d(\alpha_i, r_i^-))_{i \in I}$ be a partition of D. Let $h \in H(D)$ be injective in D and let $f \in H(D)$ satisfy*

(i) $|f'(\alpha_i) - h'(\alpha_i)| < |h'(\alpha_i)|$ *whenever* $i \in I$
(ii) $\|f - h\|_{d(\alpha_i, r_i^-)} < |h'(\alpha_i)| r_i$ *whenever* $i \in I$.

Then f is strictly injective. Furthermore, for all $i \in I$ we have $f(d(\alpha_i, r_i^-)) = h(d(\alpha_i, r_i^-))$.

Proof. By (i) h' is not identically zero in $d(\alpha_i, r_i^-)$, hence by Lemma B.14.11 h is strictly injective in $d(\alpha_i, r_i^-)$. For every $i \in I$ we put $s_i = |h'(\alpha_i)|r_i$. By Corollary B.13.10, we have $h(d(\alpha_i, r_i^-)) = d(h(\alpha_i), s_i^-)$, hence by (ii) it is seen that $f(d(\alpha_i, r_i^-)) \subset d(h(\alpha_i), s_i^-)$. Let $f(d(\alpha_i, r_i^-)) = d(f(\alpha_i), t_i^-)$. Therefore, we have

(1) $t_i \leq s_i$

while obviously

(2) $t_i \geq |f'(\alpha_i)|r_i$.

But by (i) we have $|f'(\alpha_i)| = |h'(\alpha_i)|$, hence by (1) and (2) we can derive $t_i = s_i$, hence $t_i = |f'(\alpha_i)|r_i$ and, therefore, by Theorem B.14.9 f is strictly injective in $d(\alpha_i, r_i^-)$. Suppose that f is not injective in all of D. Then there exists a and $b \in D$ such that $f(a) = f(b)$. Since f is injective in each disk $d(\alpha_i, r_i^-)$ we see that there exist j and $m \in I$ with $j \neq m$ such that $a \in d(\alpha_j, r_j^-)$ and $b \in d(\alpha_m, r_m^-)$. Since we have just proven that $f(d(\alpha_i, r_i^-)) = h(d(\alpha_i, r_i^-))$ for all $i \in I$, we see that there exist $a' \in d(\alpha_j, r_j^-)$ and $b' \in d(\alpha_m, r_m^-)$ such that $h(a') = f(a), h(b') = f(b)$. This clearly contradicts the hypothesis "h is injective in D." Consequently, f is injective in all of D. Therefore by Lemma B.14.11 f is strictly injective. □

Proposition B.14.21 shows the set of the strictly injective elements to be open in $H(D)$.

Proposition B.14.21: *Let D be such that $\delta(D, \mathbb{K} \setminus D) = \rho > 0$. Let $\lambda \in]0, +\infty[$. Let $h \in H(D)$ be injective and satisfy $|h'(x)| \geq \lambda$ whenever $x \in D$. For every $f \in H(D)$ such that $\|f - h\|_D < \lambda\rho$, f is strictly injective in D and satisfies $f(d(\alpha, \rho^-)) = h(d(\alpha, \rho^-))$ for all $\alpha \in D$.*

Proof. Let $h \in H(D)$ satisfy $\|f - h\|_D < \lambda\rho$. Since the distance from D to $\mathbb{K} \setminus D$ is $\rho > 0$, there exists a partition of D in the form $(d(\alpha_i, \rho^-))_{i \in I}$ and then we have $\|f - h\|_{d(\alpha_i, \rho^-)} < \lambda\rho \leq |h'(\alpha_i)|\rho$. Moreover, given any $g \in H(d(\alpha, \rho^-))$, by Theorem B.9.1, we have $\|g'\|_{d(\alpha, \rho^-)} \leq \frac{1}{\rho}\|g\|_{d(\alpha, \rho^-)}$, hence

$$|f'(\alpha_i) - h'(\alpha_i)| \leq \frac{1}{\rho}\|f - h\|_{d(\alpha_i, \rho^-)} < \lambda \leq |h'(\alpha_i)|.$$

So Conditions (i) and (ii) of Proposition B.14.20 are clearly satisfied. □

Remark: As we know, when D has holes, ρ is just the lower bound of the diameters of the holes.

Theorem B.14.22: *Let $d(\alpha_i, r_i^-)_{i \in I}$ be a partition of D, let $h \in H(D)$, and let $u \in]0, 1[$. Let $\phi \in H(D)$ satisfy $\|\phi\|_D < 1$ and $\|h\phi\|_{d(\alpha_i, r_i^-)} \leq ur_i\|h'\|_{d(\alpha_i, r_i^-)}$ for all $i \in I$. Then for every $t \in]\max(u, \|\phi\|_D), 1[$ there exists a family $(\beta_i)_{i \in I}$ with $\beta_i \in d(\alpha_i, r_i^-)$ such that*

(i) $\quad |h(\beta_i)\phi'(\beta_i)| \leq t|h'(\beta_i)|$ *whenever* $i \in I$

(ii) $\quad \|h\phi\|_{d(\alpha_i,r_i^-)} \leq t|h'(\beta_i)|r_i$ *whenever* $i \in I$.

Let h *be strictly injective and let* $f = h(1+\phi)$. *Then* f *is strictly injective and satisfies* $f(d(\alpha_i, r_i^-)) = h(d(\alpha_i, r_i^-))$ *for all* $i \in I$.

Proof. Let us fix $i \in I$ and put $\alpha = \alpha_i, r = r_i$. For every $g \in H(d(\alpha, r^-))$ we have $\|g\|_{d(\alpha,r^-)} = \lim_{|x-\alpha| \to r^-} |g(x)|$, hence there clearly exists $\beta \in d(\alpha, r^-)$ such that

(1) $\quad |h'(\beta)| \geq \frac{u}{t} \|h'\|_{d(\alpha,r^-)}$, hence the hypothesis implies

(2) $\quad \|h\phi\|_{d(\beta,r^-)} \leq t\frac{u}{t} |h'(\beta)|r$.

Moreover, we know that $\|\phi'\|_{d(\alpha,r^-)} \leq \frac{1}{r} \|\phi\|_{d(\alpha,r^-)}$, hence we have

$$|h(\beta)\phi'(\beta)| \leq |h(\beta)|\frac{1}{r}\|\phi\|_{d(\alpha,r^-)} \leq \frac{1}{r}\|h\|_{d(\alpha,r^-)}\|\phi\|_{d(\alpha,r^-)}.$$

As the norm $\| \,.\, \|_{d(a,r^-)}$ is multiplicative, we have $|h(\beta)\phi'(\beta)| \leq \frac{1}{r}\|h\phi\|_{d(\alpha,r^-)}$, hence by the above hypothesis $|h(\beta)\phi'(\beta)| \leq u\|h'\|_{d(\alpha,r^-)}$ and finally by (1) we obtain

(3) $\quad |h(\beta)\phi'(\beta)| \leq t|h'(\beta)|$.

Hence, we just have to put $\beta_i = \beta$ and do this for every $i \in I$ in order to obtain (i) and (ii) from (2) and (3). Since $\|\phi\|_D \leq 1$ we may take $t \geq \|\phi\|_D$. We see that $f' - h' = h'\phi + h\phi'$, hence by Condition (i) we obtain $|f'(\beta_i) - h'(\beta_i)| \leq t|h'(\beta_i)|$. Then, as $t < 1$, and as $h'(x) \neq 0$ whenever $x \in D$, we see that $|f'(\beta_i) - h'(\beta_i)| < |h'(\beta_i)|$ whenever $i \in I$. This is just Condition (i) in Proposition B.14.20. Moreover, Condition (ii) implies Condition (ii) in Proposition B.14.20. Hence, by Proposition B.14.20, f is strictly injective and satisfies $f(d(\alpha_i, r_i^-) = h(d(\alpha_i, r_i^-))$ for all $i \in I$. □

Theorem B.14.23: *Let D be analytic and let F be the set of the injective elements of H(D). The closure of F in H(D) is equal to* $F \cup \mathbb{K}$.

Proof. Let $\overline{F}$ be the closure of F and let $f \in \overline{F} \setminus \mathbb{K}$. Suppose that f is not injective and let $a, b \in D$ be such that $f(a) = f(b)$. Without loss of generality we may obviously assume $f(a) = 0$. Now let $r \in]0, |a-b|[$ be such that $d(a,r) \cup d(b,r)$ is included in D. Suppose that f is not identically zero in D. Since D is an analytic set, the restriction of f to $d(a,r)$ (resp. $d(b,r)$), is not identically zero. Let $h \in F$ satisfy $\|f - h\|_D < \min(\|f\|_{d(a,r)}, \|f\|_{d(b,r)})$. By Theorem B.13.16, h admits a zero in $d(a,r)$ and another in $d(b,r)$. But since $r < |a-b|$, these two zeros are different and, therefore, this contradicts the hypothesis "$h \in F$." That ends the proof. □

B.15. Quasi-invertible analytic elements

Throughout this chapter D is supposed to be infraconnected.

Some of the results given here were obtained in [43]. We will show that when an ideal of an algebra $H(D)$ contains a quasi-invertible element, this ideal is principal and generated by a polynomial.

Lemma B.15.1: *Let T be a hole of D and let $f \in H(D \cup T)$ be invertible in $H(D)$. If f has no zero in T then f is invertible in $H(D \cup T)$.*

Proof. Let $(f_n)_{n\in\mathbb{N}}$ be a sequence in $R(D \cup T)$, which converges to f in $H(D \cup T)$. Let $T = d(a, r^-)$. Since f has no zero in T, by Theorem B.13.15 we have $|f(x)| = |f(a)|$ for all $x \in T$ and, therefore, $|f(x)| = {}_D\varphi_{a,r}(f)$ for all $x \in T$. Now since D is infraconnected, by Corollary B.4.2, we have ${}_D\varphi_{a,r}(g) \leq \|g\|_D$ whenever $g \in H(D)$, hence for n big enough:

$$\left|\frac{1}{f(x)} - \frac{1}{f_n(x)}\right| = \frac{|f_n(x) - f(x)|}{|f(x)|^2} \leq \frac{\|f_n - f\|_D}{({}_D\varphi_{a,r}(f))^2}$$

whenever $x \in T$. Hence, we see that the sequence $\dfrac{1}{f_n}$ converges to $\dfrac{1}{f}$ in $H(D \cup T)$. □

Theorem B.15.2: *Let $D \in$ Alg. If an ideal contains a quasi-invertible element, then it is generated by a polynomial whose zeros belong to $\overset{\circ}{\overline{D}} \cap D$.*

Proof. Let $\mathcal{J}$ be an ideal of $H(D)$ that contains a quasi-invertible element f that is of the form Pg with g invertible in $H(D)$ and $P(x) \in \mathbb{K}[x]$, all the zeros of P lying inside $D \cap \overset{\circ}{\overline{D}}$. Then P belongs to $\mathcal{J}$. Consequently, the set $\mathcal{J}_0$ of polynomials that belong to $\mathcal{J}$ is not empty and, hence, is an ideal of $\mathbb{K}[x]$; hence $\mathcal{J}_0$ is of the form $Q(x)\mathbb{K}[x]$ with $Q \in \mathbb{K}[x]$. On the other hand, by hypothesis f factorizes in $H(D)$ in the form Pg with g invertible in $H(D)$ and $P(x) \in \mathbb{K}[x]$, all the zeros of P lying inside $D \cap \overset{\circ}{\overline{D}}$. Since fg^{-1} belongs to $\mathcal{J}$, obviously P belongs to $\mathcal{J}_0$. Hence, Q divides P and then all zeros of Q lie in $D \cap \overset{\circ}{\overline{D}}$. We will show that $\mathcal{J} = QH(D)$.

First we suppose that D is bounded. Let $\alpha_1, \dots, \alpha_q$ be the zeros of Q and suppose that there exists some $h \in \mathcal{J} \setminus QH(D)$. Since D is bounded, by Theorem B.2.12, h is of the form $\dfrac{\ell}{S}$ with $\ell \in H(\overline{D})$ and S a polynomial whose zeros belong to $\overline{D} \setminus D$. Hence, ℓ belongs to $\mathcal{J}$. Now we can find $r > 0$ such that $d(\alpha_i, r) \subset \overline{D}$, whenever $i = 1, \dots, q$. Let $\Lambda = \displaystyle\bigcup_{i=1}^{q} d(\alpha_i, r)$ and let $D' = D \cup \Lambda$. Since the zeros of Q lie in Λ, there exists $\lambda > 0$ such that $|Q(x)| \geq \lambda$ whenever $x \in D \setminus \Lambda$. Now since $\overline{D}$ is closed and bounded, there exists $b \in \mathbb{K}$ such that $\|b\ell\|_D < \lambda$. We put $\phi = Q + b\ell$. Clearly outside Λ, we have $|\phi(x)| \geq \lambda$. Next, by Theorem B.13.18, in each disk

$d(\alpha_i, r), \phi$ has finitely many zeros, hence in D', ϕ has finitely many zeros, all of them in Λ. Therefore, by Theorem B.7.2, it factorizes in the form $V(x)W(x)$ with $W \in H(D'), W(x) \neq 0$ whenever $x \in D'$ and V a polynomial whose zeros belong to Λ. By Theorem B.13.16, $|W(x)|$ has a strictly positive lower bound in Λ and another non-zero lower bound in $D' \setminus \Lambda$ because $|V(x)|$ is obviously bounded in D'. Finally, W has a non-zero lower bound in D', therefore it is invertible in $H(D')$. Hence, V belongs to $\mathcal{J}$. But then Q divides V in $\mathbb{K}[x]$. But since $Q + b\ell$ is equal to VW, then Q divides $Q + b\ell$ and, hence, it divides ℓ and h too. This contradicts the hypothesis $h \in \mathcal{J} \setminus QH(D)$ and finishes proving that Q generates $\mathcal{J}$ when D is bounded.

Now we suppose D unbounded. We may obviously assume D to have at least one hole $d(a, r^-)$ and without loss of generality, we may assume $a = 0$. Let $\gamma(x) = \dfrac{1}{x}$ and let $D'' = \gamma(D)$. Then D'' is a bounded set that belongs to Alg, such that $0 \notin D'$. We also have $\gamma = \gamma^{-1}$ and $\gamma(D'') = D$. Let ψ be the mapping from $H(D)$ onto $H(D'')$ defined as $\psi(f) = f \circ \gamma$. Then ψ is a $\mathbb{K}$-algebra isomorphism from $H(D)$ onto $H(D'')$. Moreover, $\psi(\mathcal{J})$ is an ideal $\mathcal{J}''$ of $H(D'')$. Let $u = \frac{1}{x}$ and $T(u) = Q(x)$. Then we can check that $\mathcal{J}'' = TH(D'')$, which ends the proof. $\square$

Definitions and notations: For any integer $n \in \mathbb{N}$ we will denote by $\mathcal{Q}_n(D)$ the set of the quasi-invertible elements $f \in H(D)$ that have exactly n zeros, taking multiplicity into account and by $\mathcal{Q}(D)$ the set $\displaystyle\bigcup_{n=0}^{\infty} \mathcal{Q}_n(D)$.

Theorem B.15.3 shows that if two analytic elements f, g are close enough, then the zeros of f and g also are respectively close, once correctly ordered. It is known as the convergence of zeros theorem.

Theorem B.15.3: *Let D be closed and bounded. Let $n \in \mathbb{N}$, let $f \in \mathcal{Q}_n(D)$, and let $\alpha_1, \dots, \alpha_n$ be the zeros of f (taking multiplicity into account). For every $\epsilon > 0$ there exists $\eta > 0$ such that for every $h \in H(D)$ satisfying $\|f - h\|_D \leq \eta, h$ belongs to $\mathcal{Q}_n(D)$ and the zeros $\beta_1, \dots, \beta_n$ of h, once correctly ordered, satisfy $|\alpha_i - \beta_i| \leq \epsilon$ $(1 \leq i \leq n)$.*

Proof. Let $f = Pg \in \mathcal{Q}_n(D)$ with g invertible in $H(D)$ and P a n-degree monic polynomial whose zeros are interior to D. Let $\gamma_1, \dots, \gamma_q$ be the different zeros of P, each γ_j of order s_j (with obviously $\displaystyle\sum_{j=1}^{q} s_j = n$). Let $\xi = \inf_{j \neq \ell} |\gamma_j - \gamma_\ell|$, let $\epsilon \in]0, \xi[$, let $\Lambda_j(\epsilon) = d(\gamma_j, \epsilon)$, and let $\Lambda(\epsilon) = \displaystyle\bigcup_{j=1}^{q} d(\gamma_j, \xi)$. It is easily seen that $|P(x)|$ has a non-zero lower bound in $D \setminus \Lambda(\epsilon)$. Since D is closed and bounded, $|g(x)|$ has a non-zero lower bound in D. Hence, $|f(x)|$ has a lower bound $\lambda > 0$ in $D \setminus \Lambda(\epsilon)$. Let $\eta = \min(\lambda, \min_{1 \leq j \leq q} \|f\|_{\Lambda_j(\epsilon)})$ and let $h \in H(D)$ satisfy

(1) $\|f-h\|_D<\eta$.

Obviously we have $|f(x)|=|h(x)|\geq\lambda$ whenever $x\in D\backslash\Lambda(\epsilon)$. But then by (1) and by Theorem B.13.17, we see that h has exactly s_i zeros like f in $\Lambda_i(\epsilon)$, $(1\leq i\leq q)$ taking multiplicity into account. Thus we have already proven the statement when all zeros of f have order 1.

Now extending this to the general case is just a question of writing. We may assume the α_i to be ordered in such a way that

$$\alpha_1=\cdots=\alpha_{s_1}=\gamma_1,\quad \alpha_{s_1+1}=\cdots=\alpha_{s_1+s_2}=\gamma_2,$$
$$\alpha_{s_1+\cdots s_{q-1}+1}=\cdots=\alpha_{s_1+\cdots+s_q}=\gamma_q.$$

Thus, for every $j=1,\cdots,q$ in $\Lambda_j(\epsilon)$, we can check that γ_j is equal to $\alpha_{s_1+\cdots+s_{j-1}+k}$ whenever $k=1,\cdots s_j$. Since f admits s_j zeros in $\Lambda_j(\epsilon)$, as does h, we may denote them by $\beta_{s_1+\cdots s_{j-1}+1},\ldots,\beta_{s_1+\cdots+s_j}$ (some of them being eventually equal). So we obtain $|\alpha_i-\beta_i|\leq\epsilon$ whenever $i=1,\ldots,n$. □

Corollary B.15.4: *Let D be closed and bounded. For every $n\in\mathbb{N}$, $\mathcal{Q}_n(D)$ is open in $H(D)$ and so is $\mathcal{Q}(D)$.*

Lemma B.15.5: *Let $a\in\overline{D}$ satisfy $a\notin\overset{\circ}{\overline{D}}$. There exists a quasi-minorated element $f\in H_b(D)$, which is not semi-invertible, satisfying $\lim\limits_{\substack{x\to a\\x\in D}}f(x)=0$ and $\limsup\limits_{\substack{x\to 0\\x\in D}}|\dfrac{f(x)}{x}|=+\infty$.*

Proof. Without loss of generality we may obviously assume $a=0$. Then the Cauchy filter $\mathcal{F}$ of base $\{d(0,r)\cap D\mid r>0\}$ is pierced. Let $(T_m)_{m\in\mathbb{N}}$ be a sequence of holes of D that runs $\mathcal{F}$ and let $D'=\mathbb{K}\setminus\Big(\bigcup\limits_{m=0}^{\infty}T_m\Big)$. By Corollary B.8.5, there exists $f\in H_b(D')$ such that $\lim\limits_{\substack{x\to 0\\x\in D'}}f(x)=0$ and such that $\limsup\limits_{\substack{x\to 0\\x\in D}}|\dfrac{f(x)}{x}|=+\infty$. We check that D' has no monotonous pierced filter because its only holes are the T_m. Hence, by Corollary B.12.3, f is quasi-minorated. If 0 belongs to D it is seen that $f(0)=0$ while $\limsup\limits_{\substack{x\to 0\\x\in D}}|\dfrac{f(x)}{x}|=+\infty$, hence f can't factorize in the form $xg(x)$ with $g\in H(D)$ and, therefore, f is not semi-invertible.

Now, suppose $0\notin D$ and that f semi-invertible. Then it factorizes in the form $P(x)g(x)$ with P the polynomial of the zeros of f in $\overset{\circ}{\overline{D}}$ and g an invertible element in $H(D)$. Since $0\notin D$, we have $P(0)\neq 0$, hence $\lim\limits_{\substack{x\to 0\\x\in D}}g(x)=0$. But since $\lim\limits_{x\to 0}|g(x)|=+\infty$, by Corollary B.2.9, $\dfrac{1}{g(x)}$ admits a pole at 0. Let n be its order. Then by Corollary B.2.9, $\dfrac{x^n}{g(x)}$ has a finite limit different from zero at 0. But since

$\frac{f(x)}{x}$ is unbounded in any set $d(0, r) \cap (D' \setminus \{0\})$, so is $\frac{g(x)}{x}$ and, therefore, we have $\liminf_{x \to 0} |\frac{x^n}{g(x)}| = 0$. Hence, g can't be invertible. This finally shows that f is not semi-invertible and finishes the proof of Lemma B.15.5. □

Lemma B.15.6: *Let D be such that $\widetilde{D} \setminus \overline{D}$ is not bounded. Then there exists a quasi-minorated element $f \in H_b(D)$ satisfying*

$$(1) \quad \lim_{\substack{|x| \to \infty \\ x \in D}} f(x) = 0$$

and

$$(2) \quad \limsup_{\substack{|x| \to \infty \\ x \in D}} |xf(x)| = +\infty.$$

Moreover, xf does not belong to $H(D)$.

Proof. Since D has holes, we may obviously assume that 0 belongs to a hole. Let $\gamma(x) = \frac{1}{x}$ and let $D' = \gamma(D)$. Then D' is bounded and 0 belongs to $\overline{D'} \setminus \overset{\circ}{\overline{D'}}$. By Lemma B.15.5, there exists a quasi-minorated element $h \in H_b(D')$ satisfying

$$(3) \quad \lim_{\substack{x \to 0 \\ x \in D}} f(x) = 0$$

and

$$(4) \quad \limsup_{\substack{x \to 0 \\ x \in D}} |\frac{f(x)}{x}| = +\infty.$$

Then we set $f = h \circ \gamma$. By (3), f satisfies (1), by (4), f satisfies (2). Now, by Lemma B.7.7, f is quasi-minorated. Finally we check that xf does not belong to $H(D)$. Indeed suppose $xf \in H(D)$. By Theorem B.2.5, xf is of the form $g(x) + P(x)$, with $g \in H_b(D)$ and $P \in \mathbb{K}[x]$. Let $q = \deg(P)$. Since xf is not bounded, we have $q > 0$. Then $x^{1-q} f$ has a limit different from 0 when $|x|$ tends to $+\infty$ and this contradicts (1). This ends the proof of Lemma B.15.6. □

Theorem B.15.7: *If D does not belong to* Alg, *there exist invertible elements $f, g \in H(D)$ such that fg belongs to $H(D)$ but is not semi-invertible.*

Proof. First we suppose that there exists $a \in (\widetilde{D} \setminus D) \setminus \overset{\circ}{\widetilde{D}}$. Without loss of generality, we assume $a = 0$. By Lemma B.15.5, there exists $f \in H_b(D)$ such that $\lim_{\substack{x \to 0 \\ x \in D}} f(x) = 0$ while $\frac{f(x)}{x}$ is not bounded in any set $D \cap d(0, r)$ $(r > 0)$. Since $f \in H_b(D)$, we can find $A \in \mathbb{K}$ such that $|A| > \|f\|_D$. Let $g = A + f$. Then g is invertible in $H(D)$.

Let $T = d(b, \rho^-)$ be a hole of D and let $F = \dfrac{x}{(x-b)g}$. Then both $\dfrac{x}{x-b}$, g^{-1} belong to $H_b(D)$, hence so does F. But by definition, F is the product of two invertible elements of $H(D)$. We also notice that F has no zero in D. Next, we notice that $\dfrac{f(x)(x-b)}{x}$ is not bounded in any set $D \cap d(0, r)$ $(r > 0)$, although $\lim\limits_{\substack{x \to 0 \\ x \in D}} f(x)(x-b) = 0$. Thus, $\dfrac{f(x)(x-b)}{x}$ cannot admit a pole at 0 and, therefore, by Corollary B.2.9, $\dfrac{f(x)(x-b)}{x}$ does not belong to $H(D)$. But then, since $\dfrac{A(x-b)}{x}$ does belong to $H(D)$, we see that F^{-1} does not belong to $H(D)$. Since F has no zero in D, it is not semi-invertible although both $\dfrac{x}{x-b}$, g^{-1} are invertible in $H(D)$.

Now we suppose that $\widetilde{D} \setminus \overline{D}$ is not bounded. Since D has holes, we may obviously assume that 0 belongs to a hole. Let $\gamma(x) = \dfrac{1}{x}$ and let $D' = \gamma(D)$. Then D' is bounded, and 0 belongs to $\in \overline{D'} \setminus \overset{\circ}{\overline{D'}}$. Hence, as we just saw, there exist invertible elements h, $g \in H_b(D')$ such that hg belongs to $H_b(D')$ and has no zero in D' but is not invertible in $H(D')$. Then we put $\tau = h \circ \gamma$, $\psi = g \circ \gamma$, $\phi = (hg) \circ \gamma$. By Theorem B.3.7, both τ, ψ are invertible in $H(D)$, ϕ belongs to $H(D)$ and has no zero in D. Since hg is not invertible in $H(D')$, by Theorem B.3.7 again, ϕ is not invertible in $H(D)$. Since it has no zero in D, it is not semi-invertible in $H(D)$ and that finishes the proof of Theorem B.15.7. □

Theorem B.15.8: *The following three statements are equivalent:*

(i) *D belongs to* Alg *and $\overline{D}$ is open,*

(ii) *$\widetilde{D} \setminus \overline{D}$ is bounded and $\overline{D}$ is open,*

(iii) *The set of the quasi-minorated elements of $H(D)$ is equal to the set of the quasi-invertible elements.*

Proof. By Theorem B.8.9 we know that (i) implies (ii). Conversely, suppose (ii) is satisfied. Particularly $\overline{D}$ is open. Suppose (i) is not satisfied. Then D does not belong to Alg. Since $\widetilde{D} \setminus \overline{D}$ is bounded, there must exist $a \in \overline{D} \setminus D$ that does not belong to $\overset{\circ}{\overline{D}}$ and, therefore, this contradicts (ii). Hence, (i) and (ii) are equivalent. Now, since (ii) implies (i), we can apply Theorem B.12.7, hence (ii) implies (iii). Finally it just remains to show that if $\widetilde{D} \setminus \overline{D}$ is not bounded or if $\overline{D}$ is not open then there exist quasi-minorated elements that are not quasi-invertible.

On one hand, if $\overline{D}$ is not open, by Lemma B.15.5 such an element does exist. On the other hand, if $\widetilde{D} \setminus \overline{D}$ is not bounded then D does not belong to Alg and, therefore, by Theorem B.15.7, there exist invertible elements f, g in $H(D)$ such that fg is not semi-invertible, hence is not quasi-invertible. But, by Theorem B.12.6, both f, g are quasi-minorated and then by Theorem B.12.5, fg also is quasi-minorated. That ends the proof of Theorem B.15.8. □

Theorem B.15.9: *Let D be closed, let T be a hole of D, and let $f \in H(D \cup T)$, have no zero in T. There exists a bounded closed infraconnected set E, such that $T \subset E \subset D \cup T$, $T \neq E$ and such that the restriction of f to E is invertible in $H(E)$.*

Proof. Let $T = d(a, r^-)$ and let $\lambda =_D \varphi_{a,r}(f)$. By Theorem B.13.16, we know that $|f(x)| = \lambda$ for all $x \in T$. Moreover, since the restriction of f to T is not identically zero, we have $\lambda > 0$. Let $\ell = \dfrac{\lambda}{2}$.

First, we suppose $C(a,r) \cap D \neq \emptyset$. Let $b \in D \cap C(a,r)$. Then we have $\lim\limits_{\substack{|x-b|\to r,\ |x-b|<r, \\ x\in D}} |f(x)| =_D \varphi_{b,r}(f) =_D \varphi_{a,r}(f) = \lambda$. Hence, there exists $s \in]0, r[$ such that $|f(x)| \geq \ell$ for every $x \in d(b, r^-) \cap D$. Then, the set $E = T \cup \big(d(b, r^-) \cap D\big)$ is clearly infraconnected, closed and bounded and we have $|f(x)| \geq \ell$ for all $x \in E$. Hence, the restriction of f to E is invertible in $H(E)$.

Now, we suppose $C(a,r) \cap D = \emptyset$. There exists $s > r$ such that $|f(x)| \geq \ell$ for every $x \in \Gamma(a, r, s) \cap D$. So, we consider the set $E = T \cup \big(\Gamma(a,r,s) \cap D\big)$. It is infraconnected, closed and bounded and we have $|f(x)| \geq \ell$ for all $x \in E$. Hence, the restriction of f to E is invertible in $H(E)$.

In both cases, we can see that T is strictly included in E. That ends the proof of Theorem B.15.9. □

We can now briefly examine the ideals of an algebra $H(D)$ when all elements are quasi-invertible.

Theorem B.15.10: *Let $D \in \mathcal{A}$. If an ideal contains a quasi-invertible element, then it is generated by a polynomial whose zeros belong to $\overset{\circ}{\overline{D}} \cap D$.*

Proof. Let $\mathcal{H}$ be an ideal of $H(D)$ that contains a quasi-invertible element f, and let $\mathcal{H}_0$ be the set of polynomials that belong to $\mathcal{H}$. By hypothesis f factorizes in $H(D)$ in the form Pg with g invertible in $H(D)$ and $P(x) \in K[x]$, all the zeros of P lying inside $D \cap \overset{\circ}{\overline{D}}$. Since fg^{-1} belongs to $\mathcal{H}$, obviously P belongs to $\mathcal{H}_0$. Hence, T divides P, and then all the zeros of T lie in $D \cap \overset{\circ}{\overline{D}}$. We will show that $\mathcal{H} = TH(D)$.

First we suppose that D is bounded. It is clearly seen that $\mathcal{H}_0$ is an ideal of $K[x]$, hence there exists $T(x) \in K[x]$ such that $\mathcal{H}_0 = T(x)K[x]$.

Let $\alpha_1, \ldots, \alpha_q$ be the zeros of T. Now we suppose that there exists some $h \in \mathcal{H} \setminus TH(D)$. Since D is bounded, by Theorem B.2.12, h is of the form $\dfrac{\ell}{S}$ with $\ell \in H(\overline{D})$ and S a polynomial whose zeros belong to $\overline{D} \setminus D$. Hence, ℓ belongs to $\mathcal{H}$. Now we can find $r > 0$ such that $d(\alpha_i, r) \subset \overline{D}$, whenever $i = 1, \ldots, q$. Let $\Lambda = \bigcup\limits_{i=1}^{q} d(\alpha_i, r)$ and let $D' = D \cup \Lambda$. Since the zeros of T lie in Λ there exists $\lambda > 0$ such that $|T(x)| \geq \lambda$ whenever $x \in D \setminus \Lambda$. Now since $\overline{D}$ is closed and bounded, there exists $b \in K$ such that $\|b\ell\|_D < \lambda$. We put $\phi = T + b\ell$. Clearly outside Λ, we have $|\phi(x)| \geq \lambda$. Besides by Theorem B.13.18, in each disk $d(\alpha_i, r), \phi$ has finitely many zeros, hence in D', ϕ has finitely many zeros, all of them in Λ. Hence, it factorizes in the form $Q(x)W(x)$ with $W \in H(D'), W(x) \neq 0$ whenever $x \in D'$ and Q a polynomial whose zeros belong to Λ. By Theorem B.13.16, $|W(x)|$ has a strictly positive lower bound in Λ and another non-zero lower bound in $D' \setminus \Lambda$ because $|Q(x)|$ is obviously bounded in D'. Finally, W has a non-zero lower bound in D', therefore it is invertible in $H(D')$. Hence, Q belongs to $\mathcal{H}$. But then T divides Q in $K[x]$. Since $T + b\ell$ is equal to WQ, then T divides $T + b\ell$, and ℓ, and h too. This contradicts the hypothesis $h \in \mathcal{H} \setminus TH(D)$, and finishes proving that T generates $\mathcal{H}$ when D is bounded.

Now we suppose D unbounded. We may obviously assume D to have at least one hole $d(a, r^-)$, and without loss of generality, we may assume $a = 0$. Let $\gamma(x) = \dfrac{1}{x}$ and let $D' = \gamma(D)$. Then D' is a bounded set that belongs to $\mathcal{A}$, such that $0 \notin D'$. We also have $\gamma = \gamma^{-1}$, and $\gamma(D') = D$. Let ψ be the mapping from $H(D)$ into $H(D')$ defined as $\psi(f) = f \circ \gamma$. Then ψ is a K-algebra isomorphism from $H(D)$ onto $H(D')$. Besides, $\psi(\mathcal{H})$ is an ideal $\mathcal{H}'$ of $H(D')$. Let $P(x) = \prod_{j=1}^{q}(x - a_i)$, for every $j = 1, \dots, q$, let $a'_i = \dfrac{1}{a_i}$, and let $B(u) = \prod_{j=1}^{q}(u - a'_i)$. Clearly, $\psi(P) = \dfrac{B(u)}{u^q}$. As $0 \notin D'$, u is invertible in $H(D')$, and B belongs to $\mathcal{H}'$. Hence, $\mathcal{J}'$ is generated by a polynomial whose zeros lie inside $D' \cap \overset{\circ}{\overline{D}'}$. Now, let $W(u) = \prod_{j=1}^{t}(u - c_j)$. Since the c_j lie in D', they are different from 0. For every $j = 1, \dots, t$, let $e_j = \dfrac{1}{c_j}$, and let $S(x) = \prod_{j=1}^{t}(x - e_j)$. It is seen that for each $j = 1, \dots, t$, e_j does belong to $D \cap \overset{\circ}{\overline{D}}$. Now, let $h \in \mathcal{H}$. Then $\psi(h)$ belongs to $\mathcal{H}'$ and is of the form $W(u)G(u)$, with $G \in H(D')$. Putting $F = \psi^{-1}(G)$, in $H(D)$ we have $h = (-1)^t S(x) \dfrac{F(x)}{x^t \prod_{i=1}^{t} e_j}$. As $\dfrac{F(x)}{x^t}$ is an invertible element of $H(D)$, this finishes showing that S generates $\mathcal{H}$. □

Corollary B.15.11: *Suppose $D \in$ Alg and all elements are quasi-invertible, except 0. Then $H(D)$ is a principal ring and each ideal is generated by a polynomial whose zeros lie in the opening of D and every maximal ideal of $H(D)$ is of the form $(x - a)H(D)$ with $a \in D$.*

By Corollaries B.12.9, B.12.10, and B.15.11 we can derive again Corollaries B.15.12 and B.15.13:

Corollary B.15.12: *Suppose $D \in \mathrm{Alg}$ be closed and let $\mathcal{T}$ be the set of holes of D. If $\{\widetilde{T} | T \in \mathcal{T}\}$ is finite, $H(D)$ is principal, each ideal is generated by a polynomial whose zeros lie in D and every maximal ideal of $H(D)$ is of the form $(x-a)H(D)$, with $a \in D$.*

Corollary B.15.13: *If D is a disk $d(a,r)$ or $d(a,r^-)$, or if D is an annulus $\Gamma(a, r_1, r_2)$ (with $0 < r_1 < r_2$) or $\Delta(a, r_1, r_2)$ (with $0 < r_1 < r_2$) or a circle $C(a,r)$, then $H(D)$ is principal, each ideal is generated by a polynomial whose zeros lie in D and every maximal ideal of $H(D)$ is of the form $(x-b)H(D)$ with $b \in D$.*

B.16. Logarithm and exponential in a p-adic field

In this chapter the field $\mathbb{K}$ is supposed to have characteristic zero.

We will define the p-adic logarithm and the p-adic exponential and will shortly study them, in connection with the study of the roots of 1 made in Chapter B.15. Both functions are also defined in [2]. Here, as in [58], we compute the radius of convergence of the p-adic exponential by using results on injectivity seen in Chapter B.14.

Lemma B.16.1: *$\mathbb{K}$ is supposed to have residue characteristic $p \neq 0$. Let $r \in]0,1[$ and for each $n \in \mathbb{N}$, let $h_n(x) = (1+x)^{p^n}$. The sequence h_n converges to 1 with respect to the uniform convergence on $d(0,r)$.*

Proof. Without loss of generality, we may assume $|p| = \frac{1}{p}$. Let $E = d(0,r)$ and for each $n \in \mathbb{N}$, let $u_n = p^n$, and let q_n be the integral part of $\frac{n}{2}$. Now we put $t_n = p^{q_n}$ and we denote by h the identical function on E. Then $h_n - 1 = \sum_{j=1}^{u_n} \binom{u_n}{j} h^j$. By Lemma A.6.1 we have $\dfrac{\left|\binom{u_n}{j}\right| \leq p^{-n}}{|j|}$, hence

$$(1) \qquad \left|\binom{u_n}{j}\right| \leq j\, p^{-n} \leq t_n\, p^{-n} \leq p^{-\frac{n}{2}} \text{ whenever } j = 1, \dots, t_n.$$

Next, we have

$h_n - 1 = \sum_{j=1}^{t_n} \binom{u_n}{j} h^j + \sum_{j=t_n+1}^{u_n} \binom{u_n}{j} h^j$. By (1) it is seen that

$$(2) \qquad \left\| \sum_{j=1}^{t_n} \binom{u_n}{j} h^j \right\|_E \leq p^{-\frac{n}{2}}$$

while

$$(3) \qquad \Big\| \sum_{j=t_n+1}^{u_n} \binom{u_n}{j} h^j \Big\|_E \leq \|h\|_E^{t_n+1}.$$

Now by (2) and (3) we see that $\lim_{n\to\infty} \|h_n - 1\|_E = 0$. □

Definitions and notations: As previously defined, for each $q \in \mathbb{N}^*$ we denote by R_q the positive number such that $\log_p(R_q) = -\frac{1}{p^{q-1}(p-1)}$. We denote by $f(x)$ the series $\sum_{n=1}^{\infty} (-1)^{n-1} \frac{x^n}{n}$.

Theorem B.16.2: *f has a radius of convergence equal to 1. If the residue characteristic of $\mathbb{K}$ is $p \neq 0$, then f is unbounded in $d(0,1^-)$. If the residue characteristic is zero, then $|f(x)|$ is bounded by 1 in $d(0,1^-)$. The function defined in $d(1,1^-)$ as $Log(x) = f(x-1)$ has a derivative equal to $\frac{1}{x}$ and satisfies $Log(ab) = Log(a) + Log(b)$ whenever $a, b \in d(1,1^-)$.*

Proof. It is clearly seen that the radius of f is 1, because $|n| \geq \frac{1}{n}$. As in the Archimedean case the property $Log(ab) = Log(a) + Log(b)$ comes from the fact that both Log and the function h_a defined as $h_a(x) = Log(ax)$ have the same derivative. The other statements are immediate. □

In Chapter A.6. when $\mathbb{K}$ has residue characteristic $p \neq 0$, we have introduced the group W of the p^s-th roots of 1, i.e., the set of the $u \in \mathbb{K}$ satisfying $u^{p^s} = 1$ for some $s \in \mathbb{N}$.

Theorem B.16.3: *$\mathbb{K}$ is supposed to have residue characteristic $p \neq 0$ (resp. 0). All zeros of Log are of order 1. The set of zeros of the function Log is equal to W, (resp. 1 is the only zero of Log). The restriction of Log to the disk $d(1,(R_1)^-)$ (resp. $d(1,1^-)$) is injective and is a bijection from $d(1,(R_1)^-)$ onto $d(0,(R_1)^-)$ (resp. from $d(1,1^-)$ onto $d(0,1^-)$).*

Proof. It is obvious that the zeros of Log are of order 1 because the derivative of Log has no zero. First, we suppose $\mathbb{K}$ to have residue characteristic $p \neq 0$. Each root of 1 in $d(1,1^-)$ is a zero of Log. Moreover, by Theorem A.6.8, we know that the only roots of 1 in $d(1,1^-)$ are the p^n-th roots. Now we can check that Log admits no zero other than the roots of 1. Indeed, suppose that a is a zero of Log but is not a root of 1, and for each $n \in \mathbb{N}$, let $b_n = a^{p^n}$. Since b_n belongs to $d(1,1^-)$, by Lemma B.16.1 we have $\lim_{n\to\infty} b_n = 1$. But obviously $Log(b_n) = 0$ for every $n \in \mathbb{N}$, hence this contradicts the fact that 1 is an isolated zero of Log.

Thus, Log has no zero in the disk $d(1,(R_1)^-)$, except 1 and, therefore, by Theorem B.13.7 the series $f(x) = \sum_{n=1}^{\infty}(-1)^{n-1}\frac{x^n}{n}$ satisfies $\nu^+(f,\log_p r) = 1$ for every $r \in]0, R_1[$, hence $r > \frac{r^n}{|n|}$ for all $r \in]0, R_1[$, for every $n \in \mathbb{N}^*$. Therefore, by Corollary B.14.10 it is injective in $d(0, R_1^-)$. Then, by Corollary B.13.10, we see that $Log(d(1, R_1^-)) = d(0, R_1^-)$.

Now we suppose that $\mathbb{K}$ has residue characteristic zero. Then, the function $f(x) = \sum_{n=1}^{\infty}(-1)^{n-1}\frac{x^n}{n}$ satisfies $\nu^+(f,\log_p r) = 1$ for every $r \in]0,1[$, hence $r > \frac{r^n}{n}$ for all $r \in]0,1[$, for every $n \in \mathbb{N}^*$. Therefore, f has no zero different from 1 in $d(0,1^-)$ and, by Corollary B.14.10, is injective in $d(0,1^-)$. Then by Corollary B.13.10 we see that $Log(d(1,1^-)) = d(0,1^-)$. This ends the proof. □

Corollary B.16.4: $\mathbb{K}$ *is supposed to have residue characteristic* 0. *There is no root of* 1 *in* $d(1,1^-)$, *except* 1.

Proof. Indeed any root of 1 should be a zero of Log in $d(1,1^-)$. □

Definitions and notations: If $\mathbb{K}$ has residue characteristic $p \neq 0$, we denote by exp the inverse (or reciprocal) function of the restriction of Log to $d(1, R_1^-)$, which obviously is a function defined in $d(0, R_1^-)$, with values in $d(1, R_1^-)$. If $\mathbb{K}$ has residue characteristic 0 we denote by exp the inverse function of Log, which is obviously defined in $d(0,1^-)$ and takes values in $d(1,1^-)$.

Theorem B.16.5: $\mathbb{K}$ *is supposed to have residue characteristic* $p \neq 0$ *(resp.* $p = 0$*). The function* exp *belongs to* $\mathcal{A}_b(d(0,R_1^-))$ *(resp.* $\mathcal{A}_b(d(0,1^-))$*), is a bijection from* $d(0,R_1^-)$ *onto* $d(1,R_1^-)$ *(resp. from* $d(0,1^-)$ *onto* $d(1,1^-)$*), and satisfies* $exp(x) = exp'(x) = \sum_{n=0}^{\infty}\frac{x^n}{n!}$ *whenever* $x \in d(0,R_1^-)$ *(resp.* $x \in d(0,1^-)$*). Moreover, the disk of convergence of its series is equal to* $d(0,R_1^-)$ *(resp.* $d(0,1^-)$*). Further, if* $p \neq 0$, *then* exp *does not belong to* $H(d(0,R_1^-))$.

Proof. By Corollary B.14.15 we know that the function exp belongs to $\mathcal{A}_b(d(0,R_1^-))$ (resp. $\mathcal{A}_b(d(0,1^-))$) and is obviously a bijection from $d(0,R_1^-)$ onto $d(1,R_1^-)$ (resp. from $d(0,1^-)$ onto $d(1,1^-)$). As it is the reciprocal of Log, it must satisfy $exp(x) = exp'(x)$ for all $x \in d(0,R_1^-)$ (resp. $x \in d(0,1^-)$) and, therefore, $exp(x) = \sum_{n=0}^{\infty}\frac{x^n}{n!}$ whenever $x \in d(0,R_1^-)$ (resp. $x \in d(0,1^-)$). Thus the radius of convergence r is at least R_1 (resp. 1). If the residue characteristic is 0, it is obviously seen that the series cannot converge for $|x| = 1$, hence the disk of convergence is $d(0,1^-)$.

Now we suppose that the residue characteristic is $p \neq 0$. Suppose that the power series of *exp* converges in $d(0, R_1)$. Then *exp* has continuation to an element of $H(d(0, R_1))$. On the other hand, since $\nu(f, \log_p r) = 1$ for all $r \in]0, R_1[$, we have $\nu^-(f, \log_p R_1) = 1$ and then by Theorem B.13.9 $Log(d(1, R_1))$ is equal to $d(0, R_1)$. Hence, we can consider $exp(Log(x))$ in all the disk $d(0, R_1)$. By Corollary B.3.3 this is an element of $H(d(1, R_1))$. But this element is equal to the identity in all of $d(1, R_1^-)$ and, therefore, in all of $d(1, R_1)$. Of course this contradicts the fact that *Log* is not injective in the circle $C(1, R_1)$. This finishes proving that the disk of convergence of *exp* is just $d(0, R_1^-)$.

Let us show that *exp* does not belong to $H(d(0, R_1^-))$. Indeed, suppose *exp* belongs to $H(d(0, R_1^-))$. Consider the Mittag–Leffler decomposition of *exp* on the infraconnected set $d(0, R_1^-)$. It is of the form $\sum_{n=0}^{\infty} g_n$ with $g_0 \in H(d(0, R_1)$ and $g_n \in H_0(\mathbb{K} \setminus d(a_n, R_n^-))$ with $a_n \in C(0, R_1)$. Set $T_n = d(a_n, R_1^-)$, $n \in \mathbb{N}^*$ and $S = \bigcup_{n=1}^{\infty} T_n$.

Let $\mathcal{K}$ be the residue class field of $\mathbb{K}$. By Theorem A.7.4, we can consider a complete algebraically closed extension $\widehat{\mathbb{K}}$ of $\mathbb{K}$ whose residue class field $\widehat{\mathcal{K}}$ is not countable. Thus we can find a class G of $\widehat{C}(0, R_1)$ that has an empty intersection with S and then, by Theorem B.6.1, *exp* has continuation to an element of $H(d(0, R_1^-) \cup G$. Let $c \in G$. Since $|c| = R_1$, by Theorem B.16.3, the function $h(x) = Log(1 + x) - c = -\Big(c + \sum_{n=1}^{\infty} \frac{(-x)^n}{n}\Big)$ satisfies $\nu^-(h, \log_p(R_1) = 0$, $\nu^+(h, \log_p(R_1) > 1$, hence h admits a zero $a \in \widehat{C}(0, R_1)$. Then a does not belong to $\mathbb{K}$ because if $a \in \mathbb{K}$, then $Log(1 + a) \in \mathbb{K}$, a contradiction. Now, let ζ be a p-th root of 1 different from 1, let $t = \zeta(1 + a)$. Since $|\zeta - 1| = R_1$, t is of the form $1 + b$ with $b \in \widehat{C}(0, R_1)$. We then have $Log(1 + a) = Log(1 + b)$. Set $E = \widehat{d}(a, R_1^-)$, $F = \widehat{d}(b, R_1^-)$, $D' = d(0, R_1^-) \cup E \cup F$, and $D'' = d(0, R_1^-) \cup G$. Since the image of $d(0, R_1^-)$ by $Log(1 + x)$ is $d(0, R_1^-)$ and since the derivative of $Log(1+x)$ has no zero, by Corollary B.13.10 we can check that for each $u \in \widehat{C}(0, R_1)$, the image of $d(u, R_1^-)$ by $Log(1 + x)$ is $d(Log(1 + u), R_1^-)$. Consequently, both images of E and F by $Log(1 + x)$ are equal to G. Now, $Log(1 + x)$ belongs to $\widehat{H}(D')$, the image of D' by the function $f(x) = Log(1 + x)$ is D'' and *exp* belongs to $\widehat{H}(D'')$. Consequently, by Corollary B.3.3, $exp \circ Log(1 + x)$ belongs to $\widehat{H}(D')$ and we have $exp \circ Log(1 + x) = 1 + x \ \forall x \in d(0, R_1^-)$. Finally, since D' has no pierced filter, by Proposition B.11.5 it is an analytic set. Consequently, the equality $exp \circ Log(1 + x) = 1 + x \ \forall x \in d(0, R_1^-)$ holds in all D', a contradiction since $Log(1 + x)$ is not injective in D'. That finishes showing that *exp* does not belong to $H(d(0, R_1^-))$. □

Remark: The exponential function admits an extension to a continuous group isomorphism defined in $\mathbb{C}_p$ onto a subgroup E of $d(1, 1^-)$ [87].

Definitions and notations: Henceforth, we put $e^x = exp(x)$.

Theorem B.16.6: *Suppose that $p \neq 0$. Let $x \in d(0, R_1^-)$. Then e^x is algebraic over $\mathbb{Q}_p$ if and only if so is x. Let $u \in d(0, 1^-)$. Then $\log(1+u)$ is algebraic over $\mathbb{Q}_p$ if and only if so is u.*

Proof. By Theorem B.5.24, if x is algebraic over $\mathbb{Q}_p$, so is e^x. Similarly, if u is algebraic over $\mathbb{Q}_p$, so is $\log(1+u)$. Consequently, suppose that e^x is algebraic over $\mathbb{Q}_p$. Then e^x is of the form $1+t$ with $|t| < 1$, hence $\log(1+t)$ is algebraic over $\mathbb{Q}_p$. But then, $\log(1+t) = \log(e^x) = x$, hence x is algebraic over $\mathbb{Q}_p$. Now, more generally, suppose $\log(1+u)$ is algebraic over $\mathbb{Q}_p$, with $|u| < 1$. Take $q \in \mathbb{N}$ such that $|p^q \log(1+u)| < R_1$. We have $p^q \log(1+u) = \log((1+u)^{p^q})$. Since $|p^q \log(1+u)| < R_1$, we have $|\log((1+u)^{p^q})| < R_1$, hence $exp(\log((1+u)^{p^q})) = (1+u)^{p^q}$. Consequently, $(1+u)^{p^q}$ is algebraic over $\mathbb{Q}_p$ and hence so is u. □

We can show a similar result when $p = 0$.

Theorem B.16.7: *Suppose that $p = 0$. Let $x \in d(0, 1^-)$. Then e^x is algebraic over $\mathbb{Q}_p$ if and only if so is x. Let $u \in d(0, 1^-)$. Then $\log(1+u)$ is algebraic over $\mathbb{Q}_p$ if and only if so is u.*

B.17. Problems on p-adic exponentials

Most of the results presented in this chapter come from [91]. The author is grateful to Michel Waldschmidt for his advices. On the other hand, most of the results first proven in the field $\mathbb{K}$ also hold (with slide changes) in an ultrametric field $\mathcal{K}$ of residue characteristic 0, as, e.g., the Levi-Civita field [88].

We will use the following classical notation:

Definitions and notations: Throughout Chapter B.17, we will denote by $\mathcal{K}$ an algebraically closed complete ultrametric field of residue characteristic 0.

Given three functions ϕ, ψ, ζ defined in an interval $J =]a, +\infty[$ (resp. $J =]a, R[$), with values in $[0, +\infty[$, we shall write $\phi(r) \leq \psi(r) + O(\zeta(r))$ if there exists a constant $b \in \mathbb{R}$ such that $\phi(r) \leq \psi(r) + b\zeta(r)$. We shall write $\phi(r) = \psi(r) + O(\zeta(r))$ if $|\psi(r) - \phi(r)|$ is bounded by a function of the form $b\zeta(r)$.

Hermite-Lindemann's theorem is well known in complex analysis. The same holds in p-adic analysis. We will need Siegel's Lemma in all the following theorems of this chapter. We will choose a particular form of this lemma [91].

Lemma B.17.1 (Siegel): *Let E be a finite extension of $\mathbb{Q}$ of degree q and let $\lambda_{i,j}$ $1 \leq i \leq m$, $1 \leq j \leq n$ be elements of E integral over $\mathbb{Z}$. Let $M = \max(\overline{|\lambda_{i,j}|}\ 1 \leq i \leq m,\ |\ 1 \leq j \leq n)$ and let $(\mathcal{S})$ be the linear system $\left\{\sum_{j=1}^{n} \lambda_{i,j} x_j = 0,\ 1 \leq i \leq m\right\}$.*

There exists solutions $(x_1, \dots, x_n)$ *of* $(\mathcal{S})$ *such that* $x_j \in \mathbb{Z}\ \forall j = 1, \dots, n$ *and*

$$\log(|x_j|_\infty) \le \log(M)\frac{qm}{n - qm} + \frac{\log(2)}{2}\ \forall j = 1, \dots, n.$$

The p-adic version of Hermite-Lindemann's theorem was proved by K. Mahler [73]. Here we give another proof, using specific ultrametric tools.

Definitions and notations: We denote by D_0 the disk $d(0, 1^-)$ and if the residue characteristic of $\mathbb{K}$ is $p > 0$ we put $R_1 = p^{\frac{-1}{p-1}}$ and denote by D_1 the disk $d(0, R_1^-)$.

Given a positive real number a, we denote by $[a]$ the biggest integer n such that $n \le a$.

Remark: In particular Levi-Civita's fields have residue characteristic 0 [3].

Theorem B.17.2: *Suppose that* $\mathbb{K}$ *has residue characteristic* $p > 0$. *Let* $\alpha \in D_1$ *be algebraic. Then* e^α *is transcendental.*

Proof. We suppose that α and e^α are algebraic. Let $h = |\alpha|$. Let E be the field $\mathbb{Q}[\alpha, e^\alpha]$, let $q = [E : \mathbb{Q}]$, and let w be a common denominator of α and e^α. We will construct a sequence of polynomials $(P_N(X, Y))_{N \in \mathbb{N}}$ in two variables such that $\deg_X(P_N) = \left[\frac{N}{\log(N)}\right]$, $\deg_Y(P_N) = [(\log N)^3]$ and such that the function $F_N(x) = P_N(x, e^x)$ satisfy further, for every $s = 0, \dots, N-1$ and for every $j = 0, \dots, [\log(N)]$.

$$\frac{d^s}{dx^s}F_N(j\alpha) = 0.$$

According to computations in the proof of Hermite-Lindemann's theorem in the complex context (Theorem 3.1.1 in [91]), we have

(1)

$$\frac{d^M F_N(\gamma_N)}{dx^M} = \sum_{l=0}^{u_1(N)} \sum_{m=0}^{u_2(N)} b_{l,m,N} \sum_{\sigma=0}^{u_1(N)} \Big(\frac{u_1(N)!}{\sigma!(u_1(N) - \sigma)!}\Big)\Big(\frac{l!}{(u_1(N) - \sigma)!}\Big) m^{u_1(N)-\sigma}$$

$$j^{u_1(N)-\sigma}.(\alpha)^{u_1(N)-\sigma}.(e^\alpha)^{ju_2(N)}.$$

We put $u_1(N) = \deg_X(P_N)$, $u_2(N) = \deg_Y(P_N)$. We will solve the system

$$w^{u_1(N)+u_2(N)}\frac{d^s}{dx^s}F_N(j\alpha) = 0, \quad 0 \le s \le N-1,\ j = 0, \dots, [\log(N)],$$

where the undeterminates are the coefficients $b_{l,m,N}$ of P_N. We then write the system under the form

$$\sum_{l=0}^{u_1(N)} \sum_{m=0}^{u_2(N)} b_{l,m,N} \sum_{\sigma=0}^{\min(s,l)} \Big(\frac{s!}{\sigma!(s-\sigma)!}\Big)\Big(\frac{l!}{(l-\sigma)!}\Big) m^{s-\sigma}.j^{l-\sigma}.$$

$$(w\alpha)^{l-\sigma}(we^\alpha)^{jm}.w^{u_1(N)-(l-\sigma)+u_2(N)-jm} = 0. \qquad (2)$$

That represents a system of $N[\log(N)]$ equations of at least $N([\log(N)])^2$ undeterminates, with coefficients in E, integral over $\mathbb{Z}$.

According to computations of Hermite-Lindemann's theorem in the complex context (Theorem 3.1.1 in [91]), it appears that in the system (2), each factor $\left(\frac{s!}{\sigma!(s-\sigma)!}\right)$, $\left(\frac{l!}{(l-\sigma)!}\right)$, $m^{s-\sigma}$, $j^{l-\sigma}$, $(w\alpha)^{l-\sigma}$, $(we^{\alpha})^{jm}$, $w^{u_1(N)-(l-\sigma)+u_2(N)-jm}$ admits a bounding of the form $SN(\log(\log(N))$ when N goes to $+\infty$. On one hand, $w^{u_1(N)+u_2(N)}$ is a common denominator and we have

$$\log(w^{u_1(N)+u_2(N)}) \leq \log(\omega)\left(\frac{N}{\log(N)}\right) + (\log(N)^3)$$

and hence we have a constant $T > 0$ such that

$$\log(w^{u_1(N)+u_2(N)}) \leq \frac{TM}{\log M}. \tag{3}$$

Next we notice that

$$\log\left(\frac{u_1(N)!}{\sigma!(u_1(N)-\sigma)!}\right) \leq u_1(N)\log(u_1(N)) \leq \frac{N}{\log(N)}\log\left(\frac{N}{\log(N)}\right) \leq N \tag{4}$$

and similarly,

$$\log\left(\frac{l!}{(u_1(N)-\sigma)!}\right) \leq u_1(N)\log(u_1(N)) \leq N \tag{5}$$

and

$$\log(m^{u_1(N)-\sigma}) \leq \frac{3N}{\log(N)}\log(\log(N)). \tag{6}$$

Now, we check that

$$\log\left(j^{u_1(N)-\sigma}.(|\overline{\alpha}|)^{u_1(N)-\sigma}.(|\overline{e^{\alpha}}|)^{ju_2(N)}\right)$$
$$\leq N + \frac{N}{\log(N)}\log(|\overline{\alpha}|) + \log(N)(\log(N))^3\log(|\overline{e^{\alpha}}|)$$

and hence there exists a constant $L > 0$ such that

$$\log\left(j^{u_1(N)-\sigma}.(|\overline{\alpha}|)^{u_1(N)-\sigma}.(|\overline{e^{\alpha}}|)^{ju_2(N)}\right) \leq LN. \tag{7}$$

Therefore, by (2), (3), (4), (5), (6), and (7), we have a constant $C > 0$ such that each coefficient a of the system satisfies

$$s(a) \leq CN(\log(\log(N)). \tag{8}$$

By Siegel's Lemma B.17.1 and by (8) there exist integers $b_{l,m,N}$, $0 \leq l \leq u_1(N)$, $0 \leq m \leq u_2(N)$ in $\mathbb{Z}$ such that

(9)
$$0 < \max_{l \leq u_1(N),\ m \leq u_2(N)} \log(|b_{l,m,N}|_\infty) \leq \frac{qN\log(N)}{N(\log(N))^2 - qN\log(N)}(CN\log(\log(N))$$

and such that the function

$$F_N(x) = \sum_{l=0}^{u_1(N)} \sum_{m=0}^{u_2(N)} b_{l,m;N} x^l e^{mx} \tag{10}$$

satisfies

$$\frac{d^s}{dx^s} F_N(j\alpha) = 0, \quad 0 \leq s \leq N-1,\ j = 0, 1, \ldots, [\log(N)].$$

Now, by (9), we can check that there exists a constant $G > 0$ such that

$$\max_{l \leq u_1(N),\ m \leq u_2(N)} (\log(|b_{l,m,N}|_\infty) \leq \frac{GN\log(\log(N))}{\log(N)}. \tag{11}$$

The function F_N defined in (10) belongs to $\mathcal{A}(D_1)$ and is not identically zero, hence at least one of the numbers $\frac{d^s}{dx^s}F_N(0)$ is not null. Let M be the biggest of the integers such that $\frac{d^s}{dx^s}F_N(j\alpha) = 0\ \forall s = 0, \ldots, M-1,\ j = 0, 1, 2, \ldots, [\log(N)]$. Thus, we have $M \geq N$ and there exists $j_0 \in \{0, 1, \ldots, [\log(N)]\}$ such that $\frac{d^M}{dx^M}F_N(j_0\alpha) \neq 0$. We put $\gamma_N = \frac{d^M}{dx^M}F_N(j_0\alpha)$.

Let us now give an upper bound of $s(\gamma_N)$. On the one hand, $w^{u_1(N)+u_2(N)}$ is a common denominator and by (2) we have a constant $T > 0$ such that

$$\log(w^{u_1(N)+u_2(N)}) \leq \frac{TM}{\log M}.$$

On the other hand, by (1) we have

$$\frac{d^M F_N(\gamma_N)}{dx^M} = \sum_{l=0}^{u_1(N)} \sum_{m=0}^{u_2(N)} b_{l,m,N} \sum_{\sigma=0}^{u_1(N)} \Big(\frac{u_1(N)!}{\sigma!(u_1(N)-\sigma)!}\Big)\Big(\frac{l!}{(u_1(N)-\sigma)!}\Big) m^{u_1(N)-\sigma}.$$

$$j^{u_1(N)-\sigma}.(\alpha)^{u_1(N)-\sigma}.(e^\alpha)^{ju_2(N)}.$$

Now by (2), (3), (6), (7), (8), and (10) and taking into account that the number of terms is bounded by $N(\log N)^2$, we can check whether there exists a constant B such that

$$s(\gamma_N) \leq BN. \tag{12}$$

Let us now give an upper bound of $|\gamma_N|$. For convenience, we first suppose that $j_0 = 0$, hence $\frac{d^M}{dx^M}F_N(0) \neq 0$. Set $h = |\alpha|$. Then by Theorem B.9.1, we have $|\gamma_N| \leq \frac{|F_N|(h)}{h^M}$. Moreover, we notice that F_N admits at least $M[\log(M)]$ zeros in $d(0,h)$ and therefore by Corollary B.13.30 we have $|F_N|(h) \leq \left(\frac{h}{R_1}\right)^{M[\log(M)]}$, because $|F_N|(r) \leq 1\ \forall r < R_1$. Consequently, $|\gamma_N| \leq \frac{h^{M(\log(M-1)}}{(R_1)^{M\log M}}$ and hence

$$\log(|\gamma_N|) \leq M(\log(M) - 1)(\log(h)) - M\log(M)(\log(R_1))).$$

Let $\lambda = \log(h) - \log(R_1)$. Then $\lambda < 0$. And we have $\log(|\gamma_N|) \leq \lambda M\log(M) - M\log(h)$, therefore there exists a constant $A > 0$ such that

$$\log(|\gamma_N|) \leq -AM\log(M). \tag{13}$$

Let us now stop assuming that $j_0 = 0$. Putting $z = x - j\alpha$ and $g(z) = f(x)$, since all points $j\alpha$ belong to $d(0,h)$, it is immediate to go back to the case $j_0 = 0$, which confirms (13) in the general case. But now, by Lemma A.8.10, Relations (12) and (13) make a contradiction to the relation $-2qs(\gamma_N) \leq \log(|\gamma_N|)$ satisfied by algebraic numbers and shows that γ_N is transcendental. But then, so is e^α. □

In $\mathcal{K}$ we have a similar version.

Theorem B.17.3: *Let $\alpha \in \mathcal{K}$ be algebraic, such that $|\alpha| < 1$. Then e^α is transcendental over $\mathbb{Q}$.*

Proof. Everything works in $\mathcal{K}$ as in a field of residue characteristic $p \neq 0$ up to Relation (8) in the proof of Theorem B.17.2. Here, we can replace R_1 by 1 and therefore the conclusion is the same as in Theorem B.17.2. □

The six-exponential problem is well known on $\mathbb{C}$ and was solved by Serge Lang [71] and K. Ramachandra. The problem is as follows: *let a_1, a_2, a_3, (resp. b_1, $b_2 \in \mathbb{C}$) be $\mathbb{Q}$-linearly independent. Then at least one of the six numbers $e^{a_ib_j}$ is transcendental.* Next, consider the same problem with only four exponentials: *let a_1, a_2 (resp. b_1, $b_2 \in \mathbb{C}$) be $\mathbb{Q}$-linearly independent.* The question is whether one of the numbers $e^{a_ib_j}$ is transcendental: this is the four-exponentials conjecture on $\mathbb{C}$ due to Serge Lang.

The problem, however, has a solution somewhat similar to that of the six-exponentials problem, in the particular case when one of the ratios $\frac{a_1}{a_2}$ and $\frac{b_1}{b_2}$ is algebraic.

The same problems make sense on a p-adic field such as $\mathbb{C}_p$ (provided the numbers a_ib_j lie in the disk of convergence of the exponential). Here, we give the solution of the six p-adic exponentials problem on the field $\mathbb{C}_p$ and this of the four

p-adic exponentials problem when one of the ratios $\frac{a_1}{a_2}$ or $\frac{b_1}{b_2}$ is algebraic. This was described by Jean-Pierre Serre [90].

Theorem B.17.4: *Let a_1, a_2, a_3 (resp. b_1, $b_2 \in \mathbb{C}_p$) be $\mathbb{Q}$-linearly independent and such that $\max_{i=1,2,3\ j=1,2} |a_i b_j| < R$. Then at least one of the numbers $e^{a_i b_j}$ is transcendental.*

Proof. Assume that all numbers $e^{a_i b_j}$ are algebraic, put $E = \mathbb{Q}[(e^{a_i b_j})_{i=1,2,3,\ j=1,2}]$ and $q = [E : \mathbb{Q}]$. Without loss of generality, we can assume that $a_1 = 1$, $|a_2|$, $|a_3| \leq 1$ and that $\max(|b_1|, |b_2|) \leq \dfrac{1}{p^2}$.

Let $t \in \mathbb{N}^*$ be such that $te^{a_i b_j}$ is integral over $\mathbb{Z}$ for every $i = 1, 2, 3$ and every $j = 1, 2$ and let $B = \log\left(t \max\{\overline{|e^{a_i b_j}|},\ i = 1, 2, 3,\ j = 1, 2\}\right)$. Let $\ell \in \mathbb{N}$ be such that $\ell > 9\sqrt{2}B(q + 1)$.

Consider now the linear system of $\ell^2 N^2$ equations with coefficients in E:

$$(\mathcal{S}_N) \qquad \sum_{1 \leq m \leq N, 1 \leq n \leq N, 1 \leq s \leq N} c_{m,n,s,N} e^{(ma_1 + na_2 + sa_3)(ib_1 + jb_2)} = 0,\ 1 \leq i \leq \ell N,$$

$1 \leq j \leq \ell N$. We notice that the coefficients $e^{(ma_1+na_2+sa_3)(ib_1+jb_2)}$ of $(\mathcal{S}_N)$ satisfy

$$\log\left(\overline{|e^{(ma_1+na_2+sa_3)(ib_1+jb_2)}|}\right) \leq 6B\ell N^2$$
$$1 \leq m \leq N, 1 \leq n \leq N,\ 1 \leq s \leq N,\ 1 \leq i \leq \ell N,$$

$1 \leq j \leq \ell N$. Now, by Siegel's Lemma B.17.1, there exists a family of solutions $(c_{m,n,s,N})_{1 \leq m \leq N, 1 \leq n \leq N, 1 \leq s \leq N}$, in $\mathbb{Z}$ such that

$$(1) \qquad \log |c_{m,n,s,N}|_\infty \leq \frac{\log 2}{2} + 6B\frac{(q\ell^2 N^2)\ell N^2}{(N^3 - q\ell^2 N^2)}.$$

Let $f_N(x) = \displaystyle\sum_{1 \leq m \leq N,\ 1 \leq n \leq N,} c_{m,n,s,N} e^{(ma_1+na_2+sa_3)x}$. Then by definition of $(\mathcal{S}_N)$, we have $f_N(ib_1 + jb_2) = 0\ \forall i = 1, \ldots, \ell N,\ j = 1, \ldots, \ell N$, hence f_N admits at least $\ell^2 N^2$ zeros in the disk $d\Big(0, \dfrac{1}{p^2}\Big)$. Let u be a point of the form $ib_1 + jb_2$ with i, $j \in \mathbb{N}$, such that $f_N(ib_1 + jb_2) \neq 0$ and such that $i + j$ is minimum and let h be this minimum: say $h = i_0 + j_0$. Thus, we can check that when i and j are two positive integers such that $i + j < h$, then $f_N(ib_1 + jb_2) = 0$. Consequently, by construction, we have $h > 2\ell N$ and the number of zeros of f_N in $d(0, \frac{1}{p^2})$ is at least $\dfrac{(h-1)^2}{2}$. We notice that $\|f_N\| \leq 1$ because $|e^x| = 1\ \forall x \in d(0, R_1^-)$ and $c_{m,n} \in \mathbb{Z}\ \forall m, n \in \mathbb{N}$. Consequently, by Corollary B.13.30 we have $\log(|f_N|(\frac{1}{p})) \leq -(h-1)N^2$ and therefore

$$(2) \qquad \log |f_N(u)| \leq -\frac{(h-1)^2}{2}.$$

Consider now some $c_{m,n,s,N}e^{(ma_1+na_2+sa_3)(ib_1+jb_2)}$ at the point u. By (1), we have $\log(|c_{m,n,s,N}|_\infty \leq \dfrac{6B(q\ell^2N^2)\ell N^2}{N^3-q\ell^2N^2}+$ and $\log\overline{|e^{(ma_1+na_2+sa_3)(i_0b_1+j_0b_2)}|} \leq 6BNh$. Consequently, by (1) we can derive

$$\log\overline{|c_{m,n,s,N}e^{(ma_1+na_2+sa_3)(i_0b_1+j_0b_2)}|} \leq 6BNh + \frac{\log 2}{2} + \frac{6Bq\ell^3N^4}{N^3-q\ell^2N^2},$$

therefore

$$\log(\overline{|f_N(u)|} \leq 6BNh + \frac{\log 2}{2} + \frac{6Bq\ell^3N^4}{N^3-q\ell^2N^2} + 3\log N. \tag{3}$$

Here, we notice that the denominator of $F_N(u)$ is bounded by t^{3Nh} because t^{3Nh} is clearly multiple of the denominator of each term $e^{(ma_1+na_2+sa_3)(ib_1+jb_2)}$ whenever $i+j \leq h$. Therefore, by Corollary A.8.12, we can derive

$$\log(|F_N(u)|) \geq -3Nh(q+1)\log(t) - q\big(6BNh + \frac{\log 2}{2} + \frac{6B\ell^3N^4}{N^3-q\ell^2N^2} + 3\log(N)\big).$$

Consequently, by (2) and (3) we obtain

$$\frac{(h-1)^2}{2} \leq (q+1)\big(9BNh + \frac{\log 2}{2} + \frac{6B\ell^3N^4}{N^3-q\ell^2N^2} + 3\log(N)\big). \tag{3}$$

Therefore, $\frac{(h-1)^2}{2} \leq (q+1)\big(9BNh + O(N)\big)$ and hence

$$h-2 \leq \frac{(h-1)^2}{h} \leq 18B(q+1)N + O(1)$$

and hence $h-1 \leq 18B(q+1)N + O(1)$. Now, since $\frac{(h-1)^2}{2} \geq \ell^2N^2$, we have $\sqrt{2}\ell N \leq 18B(q+1)N + O(1)$, hence $\ell \leq 9\sqrt{2}B(q+1)$ when N is big enough, a contradiction to the hypothesis on ℓ. □

And similarly:

Theorem B.17.5: *Let a_1, a_2, a_3 (resp. b_1, $b_2 \in \mathcal{K}$) be $\mathbb{Q}$-linearly independent and such that $\max_{i=1,2,3\ j=1,2}|a_ib_j| < 1$. Then at least one of the numbers $e^{a_ib_j}$ is transcendental over $\mathbb{Q}$.*

As explained above, when reducing to 4 exponentials $e^{a_ib_j}$, $i=1,2$, $j=1,2$, the transcendence of one of the four numbers is just a conjecture in the general case. Here, we give a proof in the particular case when one of the ratios $\frac{a_1}{a_2}$ or $\frac{b_1}{b_2}$ is algebraic.

Theorem B.17.6: *Let a_1, a_2 (resp. b_1, $b_2 \in \mathbb{C}_p$) be $\mathbb{Q}$-linearly independent and such that $\max_{i=1,2\ j=1,2}|a_ib_j| < R$ and such that $\frac{a_1}{a_2}$ is algebraic. Then at least one of the numbers $e^{a_ib_j}$ is transcendental.*

Proof. Assume that all numbers $e^{a_i b_j}$ are algebraic. Without loss of generality, we can assume that $|b_i| \leq \frac{1}{p^2}$, $i = 1,2$. Put $a = \dfrac{a_1}{a_2}$. Let $t \in \mathbb{N}^*$ be such that all the $te^{a_i b_j}$ and ta are integral over $\mathbb{Z}$ for every $i = 1,2$ and every $j = 1,2$ and let $B = \log\left(t \max(\overline{|a|}, \max\{\overline{|e^{a_i b_j}|}\ i = 1,2,\ j = 1,2\}\right)$. Without loss of generality, we can assume that $a_1 = 1$, hence $a = a_2$. Since a is algebraic, we can put $E = \mathbb{Q}[a, (e^{a_i b_j})_{i=1,2,\ j=1,2}]$ and $q = [E : \mathbb{Q}]$. We can find integers $s,\ l \in \mathbb{N}$ satisfying

$$s > (q+1)Bl\left(12 + \frac{qs}{l^2 - qs}\right). \tag{1}$$

Consider now the linear system of sN^2 equations with coefficients in E.

$$(\mathcal{S}_N) \qquad \sum_{1 \leq m \leq lN, 1 \leq n \leq lN} c_{m,n,N}(m+na)^k e^{(m+na)(ib_1+jb_2)} = 0,\ 1 \leq i \leq \mathrm{N},$$

$1 \leq j \leq N$. We notice that the coefficients $(m+na)^k e^{(m+na)(ib_1+jb_2)}$ of $(\mathcal{S}_N)$ satisfy

$$\log\left(\overline{|(m+na)^k e^{(m+na)(ib_1+jb_2)}|}\right) \leq 2BlN^2 + k\log(2N+B) \tag{2}$$

$$1 \leq m \leq lN, 1 \leq n \leq lN,\ 1 \leq i \leq N,\ 1 \leq j \leq N,\ 1 \leq k \leq s.$$

By Siegel's Lemma B.17.1 and by (2), there exists a family of solutions $(c_{m,n,N})_{1 \leq m \leq lN, 1 \leq n \leq lN}$ in $\mathbb{Z}$ such that

$$\log|c_{m,n,N}|_\infty \leq \frac{\log 2}{2} + (BlN^2 + s(\log 2N + B))\frac{qsN^2}{(l^2N^2 - qsN^2)},$$

therefore

$$\log|c_{m,n,N}|_\infty \leq \frac{qsBlN^4}{l^2N^2 - qsN^2 + O(\log(N))}. \tag{3}$$

Now, let h be the smallest integer such that $(f_N)^{(h)}(ib_1 + jb_2) \neq 0$ for some pair (i_0, j_0) such that $i_0 \leq N,\ j_0 \leq N$. By definition, $s < h$. Consequently, the number of zeros of f_N in $d(0, \frac{1}{p^2})$ is at least $(h-1)N^2$, taking multiplicity into account. Set $u = i_0 b_1 + j_0 b_2$. Consider now some $c_{m,n,N}e^{(m+na)(ib_1+jb_2)}$ at the point u. First, we have

$$\log\overline{|e^{(m+na)(i_0b_1+j_0b_2)}|} \leq 4BlN^2.$$

Consequently, by (3), when N is big enough we can derive

$$\log\overline{|c_{m,n,N}e^{(m+na)(i_0b_1+j_0b_2)}|} \leq \frac{qsBlN^4}{l^2N^2 - qsN^2} + 4BlN^2 + O(\log(N)).$$

Hence and therefore,

$$\log\overline{|f_N^{(h)}(u)|} \leq \frac{qsBlN^4}{l^2N^2 - qsN^2} + 4BlN^2 + O(\log(N)).$$

On the other hand, we can check that

$$\log(\mathrm{den}((m+na)^h e^{(m+na)(i_0b_1+j_0b_2)}) \leq 2BlN^2 + hB + O(\log(N)).$$

Hence, $\log(\mathrm{den}(f_N^{(h)}(u))) \leq 2BlN^2 + sB + O(\log(N))$. Consequently, by Corollary A.8.11, we can derive

(4)

$$\log(|f_N^{(h)}(u)|) \geq -(q+1)\left(\frac{qsBlN^4}{l^2N^2 - qsN^2} + 4BlN^2 + 8BlN^2 + 2hB + O(\log(N))\right).$$

As in Theorem B.17.4, we have $\|f_N\| \leq 1$, hence by Theorem B.9.1, we can derive $\|f_N^{(h)}\| \leq \frac{1}{p^h}$. Next, $c_{m,n,N} \in \mathbb{Z}\ \forall m,n \in \mathbb{N}$. Consequently, by Corollary B.13.30, we have $\log(|f_N|(\frac{1}{p^2})) \leq -(h-1)N^2$ and by Theorem B.9.1, $\log(|f_N^{(h)}|(\frac{1}{p^2})) \leq -(h-1)N^2 + h$. Therefore, by (2) and (4), we obtain

$$(5) \quad (h-1)N^2 - h \leq (q+1)\left(\frac{qsBlN^4}{l^2N^2 - qsN^2} + 12BlN^2 + 2hB\right) + O(\log(N)).$$

Now, by (1) we have $s > (q+1)Bl\Big(12 + \dfrac{qs}{l^2 - qs}\Big)$ and since $h > s$, we can see that (5) is impossible when N is big enough, which ends the proof. □

Similarly:

Theorem B.17.7: *Let a_1, a_2 (resp. $b_1, b_2 \in \mathcal{K}$) be $\mathbb{Q}$-linearly independent and such that $\max_{i=1,2\ j=1,2} |a_ib_j| < 1$ and such that $\frac{a_1}{a_2}$ is algebraic. Then at least one of the numbers $e^{a_ib_j}$ is transcendental over $\mathbb{Q}$.*

B.18. Divisors of analytic functions

In this paragraph, we shall define divisors in $\mathbb{K}$ or in a disk $d(a, R^-)$. We shall then define the divisor of an analytic function and of an ideal. Given a divisor T on $\mathbb{K}$, there is no problem to construct an entire function whose divisor is T. But given a divisor T on a disk $d(a, r^-)$, it is not always possible to find an analytic function (in that disk) whose divisor is T. This is Lazard's problem that we will examine in the Chapter B.19.

Definitions and notations: We call *a divisor in* $\mathbb{K}$ (resp. *a divisor in a disk* $d(a, R^-)$) a mapping T from $\mathbb{K}$ (resp. from $d(a, R^-)$) to $\mathbb{N}$ whose support is countable and has a finite intersection with each disk $d(a, r)$, $\forall r > 0$ (resp. $\forall r \in]0, R[$). Thus, a divisor on $\mathbb{K}$ (resp. of $d(a, R^-)$) is characterized by a sequence $(a_n, q_n)_{n\in\mathbb{N}}$ with $a_n \in \mathbb{K}$, $\lim_{n\to\infty} |a_n| = \infty$ (resp. $a_n \in d(a, R^-)$, $\lim_{n\to\infty} |a_n - a| = R$), $|a_n| \leq |a_{n+1}|$ and $q_n \in \mathbb{N}^*\ \forall n \in \mathbb{N}$. So, we will frequently denote a divisor by the sequence $(a_n, q_n)_{n\in\mathbb{N}}$, which characterizes it.

The set of divisors on $\mathbb{K}$ (resp. on $d(a, R^-)$) is provided with a natural additive law that makes it a semi-group. It is also provided with a natural order relation:

given two divisors T and T', we can set $T \leq T'$ when $T(\alpha) \leq T'(\alpha)\ \forall \alpha \in d(a, R^-)$. Moreover, if T, T' are two divisors such that $T(\alpha) \geq T'(\alpha)\ \forall \alpha \in d(0, R^-)$, we can define the divisor $\frac{T}{T'}$.

Given $f \in \mathcal{A}(\mathbb{K})$ (resp. $f \in \mathcal{A}(d(a, R^-))$), we can define *the divisor of* f, denoted by $\mathcal{D}(f)$ on $\mathbb{K}$ (resp. of $d(a, R^-)$) as $\mathcal{D}(f)(\alpha) = 0$ whenever $f(\alpha) \neq 0$ and $\mathcal{D}(f)(\alpha) = s$ when f has a zero of order s at α.

Similarly, given an ideal I of $\mathcal{A}(\mathbb{K})$ (resp. of $\mathcal{A}(d(a, R^-))$), we will denote by $\mathcal{D}(I)$ the lower bound of the the $\mathcal{D}(f)$ $f \in I$ and $\mathcal{D}(I)$ will be called *the divisor of* I.

Finally, given a divisor $T = (a_n, q_n)_{n\in\mathbb{N}}$, we shall denote by $\overline{T}$ the divisor $(a_n, 1)_{n\in\mathbb{N}}$. Let $T = (a_n, q_n)_{n\in\mathbb{N}}$ be a divisor on $\mathbb{K}$ (resp. of $d(a, R^-)$). For every $r > 0$ (resp. $r \in]0, R[$), we set $|T|(r) = \prod_{|a_j|\leq r} \left(\frac{r}{|a_j|}\right)^{q_j}$. The divisor T on $d(a, R^-)$ is said to be *bounded* if $\lim_{r\to R} |T|(r) < \infty$ and then we put $\|T\| = \lim_{r\to R} |T|(r)$.

The $\mathbb{K}$-algebra $\mathcal{A}(\mathbb{K})$ is provided with the following topology of $\mathbb{K}$-algebra: given $f \in \mathcal{A}(\mathbb{K})$, the neighborhoods of f are the sets $\mathcal{W}(f, r, \epsilon) = \{h \in \{\mathcal{A}(\mathbb{K}) \mid |f - h|(r) \leq \epsilon\}$, with $r > 0$, $\epsilon > 0$. Similarly, given $a \in \mathbb{K}$ and $R > 0$, the $\mathbb{K}$-algebra $\mathcal{A}(d(a, R^-))$ is provided with the following topology of $\mathbb{K}$-algebra: given $f \in \mathcal{A}(d(a, R^-))$, the neighborhoods of f are the sets $\mathcal{W}(f, r, \epsilon) = \{h \in \{\mathcal{A}(d(a, R^-)) \mid |f - h|(r) \leq \epsilon\}$, with $0 < r < R$, $\epsilon > 0$.

Remark: Let $f \in \mathcal{A}(d(a, R^-))$ and let $(a_n, q_n)_{n\in\mathbb{N}} = \mathcal{D}(f)$. Then $\omega_{a_n}(f) = q_n\ \forall n \in \mathbb{N}$ and $\omega_\alpha(f) = 0\ \forall \alpha \in d(a, R^-) \setminus \{a_n \mid n \in \mathbb{N}\}$.

Theorem B.18.1 is immediate.

Theorem B.18.1: *Let $a \in \mathbb{K}, R > 0$. Let $f, g \in \mathcal{A}(\mathbb{K})$ (resp. $f, g \in \mathcal{A}((a, R^-))$) be such that $\mathcal{D}(f) \geq \mathcal{D}(g)$. Then there exists $h \in \mathcal{A}(\mathbb{K})$ (resp. $h \in \mathcal{A}(d(a, R^-))$) such that $f = gh$.*

Proof. Let $T = \mathcal{D}(g) = (a_n, q_n)_{n\in\mathbb{N}}$. Let us fix $r > 0$ (resp. $r \in]0, R[$), let $s \in \mathbb{N}$ be such that $|a_n| \leq r\ \forall n \leq s$ and $|a_n| > r\ \forall n > s$. Let $P_r(x) = \prod_{n=0}^{s} \left(1 - \frac{x}{a_n}\right)^{q_n}$. We can factorize f in the form $P_r\hat{f}$ and similarly, we can factorize g in the form $P_r\hat{g}$, hence $\frac{f}{g} = \frac{\hat{f}}{\hat{g}}$. Since $\hat{g}$ has no zero in $d(0, r)$ it is invertible in $H(d(0, r))$, hence $\frac{f}{g}$ belongs to $H(d(0, r))$. This is true for all $r > 0$ (resp. for all $r \in]0, R[$) and hence $\frac{f}{g}$ belongs to $\mathcal{A}(\mathbb{K})$ (resp. to $\mathcal{A}(d(a, R^-))$). □

Corollary B.18.2: *Let $a \in \mathbb{K}$, $R > 0$. Let I be an ideal of $\mathcal{A}(\mathbb{K})$ (resp. an ideal of $\mathcal{A}(d(a, R^-))$) and suppose that there exists $g \in I$ such that $\mathcal{D}(g) = \mathcal{D}(I)$. Then $I = g\mathcal{A}(\mathbb{K})$ (resp. $I = g\mathcal{A}(d(a, R^-))$).*

As an immediate application of the definitions, by Theorem B.13.26 we have Lemma B.18.3.

Lemma B.18.3: *Let $R \in \mathbb{R}_+^*$ and let $f,\ g \in \mathcal{A}(d(0, R^-)$ be such that $\mathcal{D}(f) \le \mathcal{D}(g)$. Then, given $r, s \in]0, R[$ such that $r < s$ we have $\Psi(f, \log s) - \Psi(f, \log r) \le \Psi(g, \log s) - \Psi(g, \log r)$.*

In the whole field $\mathbb{K}$, given a divisor T, it is always possible to find an entire function admitting T for divisor.

Theorem B.18.4: *Let $T = (\alpha_n, q_n)_{n\in\mathbb{N}}$ be divisor of $\mathbb{K}$. The infinite product $\prod_{n=1}^{\infty}\left(1-\frac{x}{\alpha_n}\right)^{q_n}$ is uniformly convergent in all bounded subsets of $\mathbb{K}$ and defines an entire function $f \in \mathcal{A}(\mathbb{K})$ such that $\mathcal{D}(f) = T$. Moreover, given $g \in \mathcal{A}(\mathbb{K})$ such that $\mathcal{D}(g) = T$, then g is of the form λf.*

Proof. We assume that $|\alpha_n| \le |\alpha_{n+1}|\ \forall n \in \mathbb{N}$. Let us fix $R > 0$ and set $f_m(x) = \prod_{n=1}^{m}(1 - \frac{x}{\alpha_n})^{q_n}$. Consider $N \in \mathbb{N}$ such that $|\alpha_N| > R$ and $m \ge N$. On the one hand, $|f_m|(R) = |f_N|(R)$. Set $M = |f_N|(R)$. On the other hand, we check that

$$|f_{m+1}(x) - f_m(x)| = \left|\prod_{n=1}^{m}\left(1-\frac{x}{\alpha_n}\right)^{q_n}\right| \left|\left(1-\frac{x}{\alpha_{m+1}})^{q_{m+1}}-1\right)\right|$$

$$\le M\left|\sum_{k=1}^{q_{m+1}}(-1)^k\binom{q_{m+1}}{k}\left(\frac{x}{\alpha_{m+1}}\right)^k\right| \le M\frac{R}{|\alpha_{m+1}|}\ \forall x \in d(0, R).$$

Consequently, $|f_{m+1} - f_m|(R) \le M\dfrac{R}{|\alpha_{m+1}|}$, which shows that the sequence $(f_m)_{m\in\mathbb{N}}$ is uniformly converging in $d(0, R)$ to an element of $H(d(0, R))$, hence to a power series. This is true for all $R > 0$, hence the limit f defined in $\mathbb{K}$ belongs to $\mathcal{A}(\mathbb{K})$. Now, for each $m \in \mathbb{N}$, let $r_m = |\alpha_m|$. By construction, the zeros of f_m in $d(0, r_m)$ are the α_n with $1 \le n \le m$, each with multiplicity q_n. And next, we notice that $|(1 - \frac{x}{\alpha_n})^{q_n}| = 1\ \forall n > m,\ \forall x \in d(0, r_m)$. Consequently, the zeros of f in $d(0, r_m)$ are exactly those of f_m. Now, consider $g \in \mathcal{A}(\mathbb{K})$ such that $\mathcal{D}(g) = T$. The function $h \in \mathcal{A}(\mathbb{K})$ such that $f = gh$ has no zero in $\mathbb{K}$ and hence is a constant. □

Corollary B.18.5: *For every divisor T on $\mathbb{K}$, there exists $f \in \mathcal{A}(\mathbb{K})$ such that $\mathcal{D}(f) = T$. Moreover, if $f(0) = 1$, f satisfies $|f|(r) = |T|(r)\ \forall r > 0$.*

Corollary B.18.6: *Let T be a divisor on $\mathbb{K}$, let $g \in \mathcal{A}(\mathbb{K})$ be such that $\mathcal{D}(g) = T$, and let $f \in \mathcal{A}(\mathbb{K})$ be such that $\mathcal{D}(f) \ge T$. Then there exists $h \in \mathcal{A}(\mathbb{K})$ such that $f = gh$.*

Proof. Indeed, let $E = \frac{\mathcal{D}(f)}{T}$ and let $h \in \mathcal{A}(\mathbb{K})$ be such that $\mathcal{D}(h) = \frac{\mathcal{D}(f)}{T}$. Then $\mathcal{D}(f) = \mathcal{D}(gh)$, hence by Theorem B.18.4 $\frac{f}{gh}$ is a constant and we can choose h such that the constant is 1. $\square$

Theorem B.18.7: *Let $f \in \mathcal{A}(\mathbb{K})$ have a divisor of the form (a_n, sq_n) with $s \in \mathbb{N}^*$. Then, there exists $g \in \mathcal{A}(\mathbb{K})$ such that $f = g^s$.*

Proof. By Corollary B.18.5, there exists $h \in \mathcal{A}(\mathbb{K})$ such that $\mathcal{D}(h) = (a_n, q_n)$. Then, $\frac{f}{h^s}$ has no zero and no pole and therefore it is a constant λ. Let $l \in \mathbb{K}$ be such that $l^s = \lambda$ and let $g = lh$. Then $g^s = f$. $\square$

So, by Theorem B.18.4, given a divisor T on $\mathbb{K}$, we can find an entire function whose divisor is just T. It is natural to consider the same problem inside a disk $d(a, r^-)$. Indeed, in $\mathbb{C}$, it is known that the similar problem always admits a solution: in the whole field $\mathbb{C}$ as well as inside an open disk. Actually, in the general context of a complete ultrametric algebraically closed field $\mathbb{K}$, the problem has no solution when $\mathbb{K}$ is not spherically complete.

This problem was first considered by M. Lazard [72] and we will detail the solutions he gave. First, we will construct a function f whose divisor is bigger than the given divisor but narrows it [58].

We shall deal with the problem by showing that given a sequence $(a_n)_{n\in\mathbb{N}}$ such that $|a_n - a| < R$ for all $n \in \mathbb{N}$ and $\lim_{n\to\infty} |a_n - a| = R$, a sequence of integers $(q_n)_{n\in\mathbb{N}}$ and a number $\epsilon > 0$, there exists an analytic function $f \in \mathcal{A}(d(0, R^-))$ that admits each a_n as a zero of order $t_n \geq q_n$ and such that $|f|(r) \leq (1 + \epsilon)\ |T|(r)$. First we need Lemma B.18.8.

Lemma B.18.8: *Let $T = (a_n, q_n)_{n\in\mathbb{N}}$ be a divisor on $d(a, R^-)$ and let $f \in \mathcal{A}(d(a, R^-))$ satisfy $f(0) = 1$, $\mathcal{D}(f) \geq T$ and $|f|(r) = |T|(r)$ $\forall r \in]0, R[$. Then $\mathcal{D}(f) = T$.*

Proof. Since $f(0) = 1$, we may write f in the form $\prod_{j=0}^{\infty} \left(1 - \frac{x}{a_j}\right)^{s_j}$ with $q_j \leq s_j$ $\forall j \in \mathbb{N}$. By hypothesis, we have $q_j \leq s_j$ $\forall j \in \mathbb{N}$. Suppose that $s_k > q_k$ for some index k and let $r_n = |a_n|$, $n \in \mathbb{N}$. Since $|f|(r_k) = |T|(r_k)$, when $r \in]r_k, r_{k+1}[$, we have $|f|(r) = |f|(r_k)\left(\frac{r}{r_k}\right)^{s_k}$, $|T|(r) = |f|(r_k)\left(\frac{r}{r_k}\right)^{q_k}$ and since $|f|(r) = |T|(r)$ $\forall r \in]0, R[$, clearly $s_k = q_k$. $\square$

Definitions and notations: For each divisor E of $\mathbb{K}$, we denote by $\mathcal{T}(E)$ the set of $f \in \mathcal{A}(\mathbb{K})$ such that $E \leq \mathcal{D}(f)$. Similarly, for each divisor E of $d(a, R^-)$, we denote by $\mathcal{T}_R(E)$ the set of $f \in \mathcal{A}(d(a, R^-))$ such that $E \leq \mathcal{D}(f)$.

Theorem B.18.9: *For every divisor E of $\mathbb{K}$, $\mathcal{T}(E)$ is a closed ideal of $\mathcal{A}(\mathbb{K})$. Moreover, $\mathcal{T}$ is a bijection from the set of divisors of $\mathbb{K}$ onto the set of closed ideals of $\mathcal{A}(\mathbb{K})$. Further, given a closed ideal I of $\mathcal{A}(\mathbb{K})$, then $I = \mathcal{T}(\mathcal{D}(I))$.*

Similarly, we have Theorem B.18.10.

Theorem B.18.10: *Let $a \in \mathbb{K}$ and take $R > 0$. For every divisor E of $d(a, R^-)$, $\mathcal{T}_R(E)$ is a closed ideal of $\mathcal{A}(d(a, R^-))$. Moreover $\mathcal{T}_R$ is a bijection from the set of divisors of $d(a, R^-)$ onto the set of closed ideals of $\mathcal{A}(d(a, R^-))$. Further, given a closed ideal I of $\mathcal{A}(d(a, R^-)))$, then $I = \mathcal{T}_R(\mathcal{D}(I))$.*

Proof. (Theorems B.18.9 and B.18.10) Let E be a divisor of $\mathbb{K}$ (resp. of $d(a, R^-)$). First, let us check that $\mathcal{T}(E)$ (resp. $\mathcal{T}_R(E)$) is a closed ideal of $\mathcal{A}(\mathbb{K})$ (resp. of $\mathcal{A}(d(a, R^-)))$. Let $E = (a_n, q_n)_{n\in\mathbb{N}}$ and let $(f_m)_{m\in\mathbb{N}}$ be a sequence of elements of $\mathcal{T}(E)$ (resp. of $\mathcal{T}_R(E)$) converging to a limit f in $\mathcal{A}(\mathbb{K})$ (resp. in $\mathcal{A}(d(a, R^-)))$. For every $n \in \mathbb{N}$, each f_m admits a_n as a zero of order at least q_n, hence by Lemma B.7.1, so does f. Consequently, f belongs to $\mathcal{T}(E)$ (resp. f belongs to $\mathcal{T}_R(E)$).

Now, let us show that $\mathcal{T}$ (resp. $\mathcal{T}_R$) is injective. Let E, F be two distinct divisors of $\mathbb{K}$ (resp. of $d(a, R^-)$). Without loss of generality, we can suppose that E admits a pair (b, s) with $s > 0$ and that F either does not admit any pair (b, m) or admits a pair (b, m) with $m < s$. Let $f \in \mathcal{T}(F)$ (resp. let $f \in \mathcal{T}_R(F)$ and suppose that $\omega_b(f) \geq s$. Then by Lemma B.7.1, f factorizes in the form $(x - b)^{s-m} g$ with $g \in \mathcal{A}(\mathbb{K})$ (resp. $g \in \mathcal{A}(d(a, R^-)))$ and of course g belongs to $\mathcal{T}(F)$ (resp. to $\mathcal{T}_R(F)$). But by construction, g does not belong to $\mathcal{T}(E)$ (resp. to $\mathcal{T}_R(E)$) because $\omega_b(g) < s$. Therefore, $\mathcal{T}(E) \neq \mathcal{T}(F)$ (resp. $\mathcal{T}_R(E) \neq \mathcal{T}_R(F)$). So, $\mathcal{T}$ (resp. $\mathcal{T}_R$) is injective.

Now, let us show that $\mathcal{T}$ (resp. $\mathcal{T}_R$) is injective. Let E, F be two distinct divisors of $\mathbb{K}$ (resp. of $d(a, R^-)$). By Theorem B.18.4, there exists $f \in \mathcal{A}(\mathbb{K})$ (resp. $f \in \mathcal{A}(d(a, R^-)))$ such that $\mathcal{D}(f) = F$, hence $\mathcal{D}(f) \neq E$. Therefore, $\mathcal{D}(f) \notin \mathcal{T}(E)$ and hence $\mathcal{T}(E) \neq \mathcal{T}(F)$. So, $\mathcal{T}$ is injective (resp. $\mathcal{T}_R(E) \neq \mathcal{T}_R(F)$. So, $\mathcal{T}_R$ is injective).

Let us show that it is also surjective. Let I be a closed ideal of $\mathcal{A}(\mathbb{K})$ (resp. of $\mathcal{A}(d(a, R^-)))$ and let $E = \mathcal{D}(I)$. Then E is of the form $(a_n, q_n)_{n\in\mathbb{N}}$ with $|a_n| \leq |a_{n+1}|$ and $\lim\limits_{n\to+\infty} |a_n| = +\infty$ (resp. $\lim\limits_{n\to+\infty} |a_n| = R$), hence there is a unique $s \in \mathbb{N}$ such that $a_n \in d(0, r)\ \forall n \leq s$ and $a_n \notin d(0, r)\ \forall n > s$.

Let $J = \mathcal{T}(E)$ (resp. $J = \mathcal{T}_R(E)$). Then of course, $I \subset J$. Let us show that $J \subset I$. Let $f \in J$ and take $r > 0$. Denoting by P_r the polynomial $\prod\limits_{i=0}^{s} (X - a_i)^{q_i}$ by Theorem B.15.2, $I \cap H(d(0, r)) = P_r(x) H(d(0, r))$. But now all functions $g \in J \cap H(d(0, r))$ also are of the form $P_r(x) h(x)$ with $h \in H(d(0, r))$. Consequently, in $H(d(0, r))$ we can write f in the form $f = \sum\limits_{j=1}^{m} g_j h_j$ with $g_j \in I$ and $h_j \in H(d(0, r))$. Let $\epsilon > 0$ be fixed.

For each $j = 1, \dots, m$, narrowing each h_j by a polynomial ℓ_j in $H(d(0,r))$, we can find $\ell_j \in \mathbb{K}[x]$ such that $|g_j(h_j - \ell_j)|(r) \le \epsilon$. Now, let $\phi_r = \sum_{j=1}^{m} g_j \ell_j$. Then ϕ_r belongs to I and satisfies $|\phi_r - f|(r) \le \epsilon$. This is true for each $r > 0$ and for every $\epsilon > 0$. Consequently, since I is closed, f belongs to I. This finishes proving that $\mathcal{T}$ (rep. $\mathcal{T}_R$) is surjective. Further, we have proven that $I = \mathcal{T}(\mathcal{D}(I))$ (resp. $I = \mathcal{T}_R(\mathcal{D}(I))$). □

Corollary B.18.11: *Every closed ideal of $\mathcal{A}(\mathbb{K})$ is principal.*

Proof. Indeed, consider a closed ideal I and let $E = \mathcal{D}(I)$. By Theorem B.18.9, I is of the form $\mathcal{T}(E)$ with $E = \mathcal{D}(I)$. By Theorem B.18.4, there exists $g \in \mathcal{A}(\mathbb{K})$ such that $\mathcal{D}(g) = E$ and of course, g belongs to I. Hence, $g\mathcal{A}(\mathbb{K}) \subset I$. Now, let $f \in I$. Then $\mathcal{D}(f) \ge E$, hence by Theorem B.18.1, f factorizes in the form gh with $h \in \mathcal{A}(\mathbb{K})$, hence $I = g\mathcal{A}(\mathbb{K})$. □

Theorem B.18.12: *Let $r \in |\mathbb{K}^*|$, let $f \in H(C(0,r))$, and let $P \in \mathbb{K}[x]$ have all its zeros in $d(0,r)$. There exists $g \in H(C(0,r))$ and $L \in \mathbb{K}[x]$ unique such that $f = Pg + L$, $\deg(L) < \deg(P)$, $\Psi(R, \log r) \le \Psi(f, \log r)$, $\Psi(g, \log r) \le \Psi(f, \log r) - \Psi(P, \log r)$. Moreover, if f belongs to $H(d(0,r))$, then so does g.*

Proof. Since $r \in |K|$, without loss of generality we may assume that $r = 1$. Similarly, we may also assume that $\Psi(P, 0) = \Psi(f, 0) = 0$, so P is quasi-monic. Thus, the problem now consists of finding $g \in H(C(0,1))$ and $L \in \mathbb{K}[x]$, each unique, satisfying the statements.

Let $f(x) = \sum_{-\infty}^{\infty} a_m x^m$ and for each $n \in \mathbb{N}$, let $f_n(x) = \sum_{m=-n}^{n} a_m x^m$. For each $n \in \mathbb{N}$, set $u_n = \sup_{|j|_\infty > n} |a_j|$. Then $\lim_{n \to +\infty} a_n = \lim_{n \to -\infty} a_n = 0$. Next, we notice that $x^n P$ is quasi-monic like P. By applying Lemma A.4.3 to $x^n f_n$, we have a Euclidean division of $x^n f_n$ by $x^n P$ in the form $x^n f_n = x^n P g_n + S_n$ with $S_n \in \mathbb{K}[x]$, $\deg(S_n) < n + \deg(P)$, $\Psi(S_n, 0) \le 0$, $\Psi(g_n, 0) \le 0$. Now, by construction $S_n = x^n(f_n - Pg_n)$, hence S_n is of the form $x^n L_n$, with $L_n \in \mathbb{K}[x]$, $\deg(L_n) < \deg(P)$ and $\Psi(L_n, 0) = \Psi(S_n, 0) \le 0$. So, $f_n = Pg_n + L_n$.

Consequently, $f_{n+1} - f_n = P(g_{n+1} - g_n) + L_{n+1} - L_n$. By applying again Lemma A.4.3 to $x^{n+1}(f_{n+1} - f_n)$, we can check that $\Psi(g_{n+1} - g_n, 0) \le \log(u_n)$ and $\Psi(L_{n+1} - L_n, 0) \le \log(u_n)$, hence both sequences $(g_n)_{n \in \mathbb{N}}$, $(L_n)_{n \in \mathbb{N}}$ converge in $H(C(0,1))$ and more precisely, the sequence (L_n) converges to a polynomial L of degree $< \deg(P)$. Moreover, setting $g = \lim_{n \to \infty} g_n$, clearly we have $f = Pg + L$ and $\Psi(g, 0) \le 0$, $\Psi(L, 0) \le 0$, which shows the existence of g and L in the first claim.

Now let us check that they are unique satisfying these relations. Suppose we have $h \in H(C(0,1))$ and $S \in \mathbb{K}[x]$ satisfying the same properties, with particularly

$f = Ph + S$, $\deg(S) < \deg(P)$. Then $P(g-h) = S - L$. Since $\deg(S-L) < \deg(P)$, $S - L$ is an element of $H(C(0,1))$ having strictly less zeros than P in $C(0,1)$, a contradiction, except if $g = h$, hence $L = S$.

Now, assume that f lies in $H(d(0,1))$. Then by Corollary B.12.9, $f - L$ is a quasi-invertible element of $H(d(0,1))$, hence is of the form $Q\phi$ with ϕ invertible in $H(d(0,1))$ and $Q \in \mathbb{K}[x]$, having all its zeros in $d(0,1)$. Hence, $Pg = Q\phi$. This holds in $H(C(0,1))$. But since P has all its zeros in $C(0,1)$, P must divide Q: say $Q = PV$, with $V \in \mathbb{K}[x]$ having all its zeros in $d(0,1)$. So, $Pg = V\phi$, hence $g = V\phi$ and by hypothesis both V, ϕ lie in $H(d(0,1))$, hence so does g, which completes the proof. □

Definitions and notations: Given $r \in |\mathbb{K}^*|$, the division of an element f of $H(C(0,r))$ by a polynomial P having all its zeros in $d(0,r)$, as defined in Theorem B.18.12, will be called *Euclidean division of f by P in $H(C(0,r))$* or *r-Euclidean division of f by P.*

Lemma B.18.13: *Let $(u_n)_{n\in\mathbb{N}}$ be a sequence in $\mathbb{R}_+$ such that $\sum_{n=0}^{\infty} u_n < +\infty$. Let $A = \sum_{n=0}^{\infty} u_n$ and let $B > A$. There exists an increasing sequence $(q_n)_{n\in\mathbb{N}}$ such that $\lim_{n\to\infty} q_n = +\infty$ and such that $\sum_{n=0}^{\infty} q_n u_n \le B$.*

Proof. Let $E = B - A$. For every $n \in \mathbb{N}$, we denote by s_n the smallest integer such that $\sum_{j=s_n}^{\infty} u_j \le 4^{-n}E$. Then for every $j \in \mathbb{N}$ such that $s_n \le j < s_{n+1}$, we set $q_j = 2^n$. We have $\sum_{j=s_n}^{s_{n+1}-1} q_j u_j \le 2^{-n}E$, hence $\sum_{j=0}^{s_{n+1}-1} q_j u_j \le A + E\sum_{k=1}^{n} 2^{-k}$ and finally $\sum_{j=0}^{\infty} q_j u_j \le B$. □

Theorem B.18.14: *Let $T = (a_n, q_n)_{n\in\mathbb{N}}$ be a divisor on the disk $d(a, R^-)$ with $a_n \neq 0$ $\forall n \in \mathbb{N}$ and let $\epsilon > 0$. There exists $f \in \mathcal{A}(d(a, R^-))$ such that $\mathcal{D}(f) \ge T$, $f(a) = 1$ and $|f|(r) \le |T|(r)(1+\epsilon)$ $\forall r \in]0, R[$.*

Proof. Without loss of generality, we can assume $a = 0$. The set $\{|a_j| \quad |j \in \mathbb{N}\}$ is obviously equal to the image of a strictly increasing sequence of limit R that we will denote by $(r_m)_{m\in\mathbb{N}}$. For each $m \in \mathbb{N}$, the set S_m of the a_j lying in $C(0, r_m)$ is of the form $\{a_{h_m}, a_{h_{m+1}}, \ldots, a_{k_m}\}$. We set $B = 1 + \epsilon$ and $P_m = \prod_{j=h_m}^{k_m} \left(1 - \frac{x}{a_j}\right)^{q_j}$.

We can construct a polynomial whose divisor is $(a_n, q_n)_{n<t}$. For every $m \in \mathbb{N}$, we set $\mu_m = \log r_m$ and $\lambda = \log(B)$.

Now, by Lemma B.18.13, there exists an increasing sequence $(t_n)_{n\in\mathbb{N}}$ in $\mathbb{N}$ such that $\lim_{n\to\infty} t_n = +\infty$ and such that

$$(1) \quad \sum_{j=0}^{\infty} t_j(\mu_{j+1} - \mu_j) \leq \lambda.$$

For every $s \in \mathbb{N}$, we put $\tau(s) = \sum_{j=0}^{s} t_j(\mu_{j+1} - \mu_j)$ and $g_s = \prod_{m=0}^{s} P_m$. We notice that $\Psi(g_q, \mu) = \Psi(g_s, \mu)$ whenever $\mu \leq \mu_q$ and $q \leq s$. So, we can define the function ℓ from $]-\infty, \log S[$ into $\mathbb{R}$ as $\ell(\mu) = \lim_{s\to\infty} \Psi(g_s, \mu)$. Then ℓ is an increasing function in μ that satisfies

$$(2) \quad \ell(\mu) = \sum_{j=0}^{k(m)} q_j(\mu - \Psi(a_j)) \text{ whenever } \mu \in [\mu_m, \mu_{m+1}].$$

We will construct a sequence $(f_s)_{s\in\mathbb{N}}$ in $\mathbb{K}[x]$ satisfying the following relations:

$(\alpha_s), (\beta_s), (\gamma_s), (\delta_s)$ for every $s \in \mathbb{N}$ and $(\epsilon_s), (\varphi_s)$ for every $s \in \mathbb{N}^*$.
(α_s) $f_s(0) = 1$,
(β_s) P_j divides f_s for every $j \leq s$,
(γ_s) $\Psi(f_s, \mu_{s+1}) \leq \ell(\mu_{s+1}) + \tau(s)$,
(δ_s) $\Psi(f_s, \mu) \leq \ell(\mu) + \lambda$ whenever $\mu < \log S$,
(ϵ_s) $\Psi(f_s - f_{s-1}, \mu) \leq \ell(\mu) + t_s(\mu - \mu_s) + \lambda$ whenever $\mu \leq \mu_s$,
(φ_s) $\Psi(f_s - f_{s-1}, \mu) \leq \ell(\mu_s) + t_s(\mu - \mu_s) + \tau(s)$ whenever $\mu \in]\mu_s, \log S]$.

We will proceed by induction and will prove that when (α_s), (β_s), (γ_s), (δ_s) are satisfied for $s \in \mathbb{N}$, then we can derive (α_{s+1}), (β_{s+1}), (γ_{s+1}), (δ_{s+1}), (ϵ_{s+1}), (φ_{s+1}).

By taking $f_0 = P_0$, we check that $(\alpha_0), (\beta_0), (\gamma_0), (\delta_0)$ are obviously satisfied. We now suppose already constructed f_m satisfying $(\alpha_m), (\beta_m), (\gamma_m), (\delta_m)$ for every $m = 0, \dots, s$ and $(\epsilon_m), (\varphi_m)$ for every $m = 1, \dots, s$. We will define f_{s+1} satisfying $(\alpha_{s+1}), (\beta_{s+1}), (\gamma_{s+1}), (\delta_{s+1}), (\epsilon_{s+1}), (\varphi_{s+1})$. It is seen that each polynomial P_s has all its zeros in $C(0, r_s)$.

Let R_{s+1} be the rest of the Euclidean division of f_s by P_{s+1} in $H(C(0, r_{s+1}))$. Let $Q_{s+1} = x^{t_{s+1}} g_s$. We have

$$(3) \quad \Psi(Q_{s+1}, \mu) = \mu t_{s+1} + \Psi(g_s, \mu) \text{ whenever } \mu \in \mathbb{R}.$$

We notice that Q_{s+1} admits no zero in $C(0, r_{s+1})$ and then is invertible in $H(C(0, r_{s+1}))$. As a consequence, according to Theorem B.18.12, we can perform the r_{s+1}-Euclidean division of $\frac{R_{s+1}}{Q_{s+1}}$ by P_{s+1} in $H(C(0, r_{s+1}))$. Let V_{s+1} be the rest of this division. Thus, $\frac{R_{s+1}}{Q_{s+1}}$ is of the form $T_{s+1}P_{s+1}+V_{s+1}$ with $T_{s+1} \in H(C(0, r_{s+1}))$ and

(4) $\Psi(Q_{s+1}V_{s+1}, \mu_{s+1}) \le \Psi(R_{s+1}, \mu_{s+1})$.

We have $R_{s+1} = Q_{s+1}(T_{s+1}P_{s+1} + V_{s+1})$. Now we put $f_{s+1} = f_s - Q_{s+1}V_{s+1}$. Of course, f_{s+1} satisfies (α_{s+1}). We will check that f_{s+1} satisfies (β_{s+1}). By definition, each P_j divides g_s for every $j = 0, \dots, s$ and hence it divides Q_{s+1}. Next, by (β_s), P_j also divides f_s. Consequently, P_j divides f_{s+1} for every $j = 0, \dots, s$. Moreover, P_{s+1} divides both $f_s - R_{s+1}$ and $Q_{s+1}T_{s+1}P_{s+1}$. Hence, it also divides f_{s+1} and thereby (β_{s+1}) is satisfied.

Now we will prove (φ_{s+1}). By (4) R_{s+1} satisfies $\Psi(R_{s+1}, \mu_{s+1}) \le \Psi(f_s, \mu_{s+1})$. Hence, by relation (γ_s), we have

(5) $\Psi(R_{s+1}, \mu_{s+1}) \le \ell(\mu_{s+1}) + \tau(s) = \Psi(g_{s+1}, \mu_{s+1}) + \tau(s)$.

Since $\deg(R_{s+1}) < \deg(P_{s+1}) < \deg(g_{s+1})$ and since all zeros of g_{s+1} lie in $d(0, r_{s+1})$, g_{s+1} has more zeros than R_{s+1} in $d(0, r_{s+1})$ and therefore, by Theorem B.13.26 we have

$$\Psi(R_{s+1}, \mu) - \Psi(R_{s+1}, \mu_{s+1}) \le \Psi(g_{s+1}, \mu) - \Psi(g_{s+1}, \mu_{s+1})$$

whenever $\mu \in]\mu_{s+1}, \log S]$ and therefore by (5) we obtain

(6) $\Psi(R_{s+1}, \mu) \le \Psi(g_{s+1}, \mu) + \tau(s)$ whenever $\mu \in]\mu_{s+1}, \log S]$.

Since $\frac{R_{s+1}}{Q_{s+1}} = T_{s+1}P_{s+1} + V_{s+1}$, by Theorem B.18.12 we have $\Psi(Q_{s+1}V_{s+1}, \mu_{s+1})) \le \Psi(R_{s+1}, \mu_{s+1})$ and hence by (5), $\Psi(Q_{s+1}V_{s+1}, \mu_{s+1}) \le \ell(\mu_{s+1}) + \tau(s)$. But $\ell(\mu_{s+1}) = \Psi(g_{s+1}, \mu_{s+1})$, hence by (6) we obtain

(7) $\Psi(Q_{s+1}V_{s+1}, \mu_{s+1}) \le \Psi(g_{s+1}, \mu_{s+1}) + \tau(s)$.

We notice that $\deg(Q_{s+1}V_{s+1}) < \deg(g_{s+1}) + t_{s+1}$ and that all zeros of g_{s+1} lie in $d(0, r_{s+1})$. Hence, by Theorem B.13.26 and by (3) and (7), we have

(8) $\Psi(Q_{s+1}V_{s+1}, \mu) \le \Psi(g_{s+1}, \mu) + t_{s+1}(\mu - \mu_{s+1}) + \tau(s)$ for every $\mu \in]\mu_{s+1}, \log S]$.

Actually, by definition of f_{s+1}, we have $\Psi(Q_{s+1}V_{s+1}, \mu) = \Psi(f_{s+1} - f_s, \mu)$ and $\Psi(g_{s+1}, \mu) \le \ell(\mu)$ for every $\mu \in]\mu_{s+1}, \log S]$, hence by (7) we have proved φ_{s+1}. We will deduce (ϵ_{s+1}).

In particular when $\mu = \mu_{s+2}$, we obtain

$$\Psi(Q_{s+1}V_{s+1}, \mu_{s+2}) \le \ell(\mu_{s+2}) + t_{s+1}(\mu_{s+2} - \mu_{s+1}) + \tau(s).$$

But we notice that $t_{s+1}(\mu_{s+2} - \mu_{s+1}) + \tau(s) = \tau(s + 1)$, hence

(9) $\Psi(Q_{s+1}V_{s+1}, \mu_{s+2}) \le \ell(\mu_{s+2}) + \tau(s + 1)$.

And, by (8) and (1) we obtain (10) $\Psi(f_{s+1} - f_s, \mu) \leq \ell(\mu) + \lambda$ whenever $\mu \in]\mu_{s+1}, \log S]$.

Now we take $\mu \leq \mu_{s+1}$. It is seen that

$$\Psi(Q_{s+1}, \mu) - \Psi(Q_{s+1}, \mu_{s+1}) = \ell(\mu) - \ell(\mu_{s+1}) + t_{s+1}(\mu - \mu_{s+1}).$$

Therefore we have

$$\Psi(Q_{s+1}V_{s+1}, \mu) \leq \Psi(Q_{s+1}V_{s+1}, \mu_{s+1}) + \ell(\mu) - \ell(\mu_{s+1}) + t_{s+1}(\mu - \mu_{s+1}).$$

But by (8) we have $\Psi(Q_{s+1}V_{s+1}, \mu_{s+1}) \leq \ell(\mu_{s+1}) + \lambda$, hence we obtain $\Psi(Q_{s+1}V_{s+1}, \mu) \leq \ell(\mu) + \lambda + t_{s+1}(\mu - \mu_{s+1})$ whenever $\mu \leq \mu_{s+1}$, and this is (ϵ_{s+1}). In particular, we have $\Psi(f_{s+1} - f_s, \mu) \leq \ell(\mu) + \lambda$ whenever $\mu \leq \mu_{s+1}$ and therefore by (δ_s) we obtain (δ_{s+1}).

Now we will show (γ_{s+1}). Obviously, we have

(10) $\Psi(f_0, \mu_{s+2}) \leq \ell(\mu_{s+2})$.

Next, by relations $(\varphi_m)_{1 \leq m \leq s+1}$ for every $m \in \mathbb{N}^*$, we have

$$\Psi(f_m - f_{m-1}, \mu_{s+2}) \leq \ell(\mu_m) + [t_m(\mu_{s+2} - \mu_m) + \tau(m)].$$

But as the sequence $(t_m)_{m \in \mathbb{N}}$ is increasing, it is seen that

$$\tau(m) + t_m(\mu_{s+2} - \mu_m) \leq \sum_{j=0}^{s+1} t_j(\mu_{j+1} - \mu_j) = \tau(s+1).$$

Obviously, $\ell(\mu_{s+2}) \geq \ell(\mu_m)$, hence we obtain $\Psi(f_m - f_{m-1}, \mu_{s+2}) \leq \ell(\mu_{s+2}) + \tau(s+1)$ whenever $m = 1, \ldots, s+1$. Finally by (10), f_{s+1} satisfies $\Psi(f_{s+1}, \mu_{s+2}) \leq \ell(\mu_{s+2}) + \tau(s+1)$ and this is (γ_{s+1}).

We notice that (ϵ_s) and (φ_s) are not used to prove

$$(\alpha_{s+1}),\ (\beta_{s+1}),\ (\gamma_{s+1}),\ (\delta_{s+1}),\ (\epsilon_{s+1}),\ (\varphi_{s+1}).$$

Consequently, (ϵ_1) and (φ_1) are clearly proven by (α_0), (β_0), (γ_0), (δ_0), and therefore we are now done with the recurrence. Therefore, we can now construct the sequence $(f_s)_{s \in \mathbb{N}}$ satisfying (α_s), (β_s), (γ_s), (δ_s), (ϵ_s), *and* (φ_s). By relation (ϵ_s), the sequence is easily seen to converge in each algebra $H(d(0,u))$ whenever $u \in]0, S[$. Indeed, given $u \in]0, S[$ and $N \in \mathbb{N}$ such that $\mu_N < \log u$, by (ϵ_{s+1}) we have

$$\log(\|f_{s+1} - f_s\|_{d(0,u)}) = \Psi(f_{s+1} - f_s, \log u) \leq \ell(\log u) - t_s(\log u - \mu_{s+1}) + \lambda,$$

hence $\log(\|f_{s+1} - f_s\|_{d(0,u)}) \leq \ell(\log u) - t_s(\log u - \mu_N) + \lambda$, whenever $s > N$. As $\lim\limits_{s \to +\infty} t_s = +\infty$, it is seen that $\lim\limits_{s \to \infty} \|f_{s+1} - f_s\|_{d(0,u)} = 0$.

Let f be the function defined in $d(0, R^-)$ as the limit of the sequence $(f_s)_{s \in \mathbb{N}}$ in each disk $d(0,u)$. Obviously, as an element of $H(d(0,u))$ for every $u \in]0, S[$, f belongs to $\mathcal{A}(d(0, R^-))$. By relation (α_s), f satisfies $f(0) = 1$.

We will check $|f|(r) \leq B|T|(r)$. Let $u \in]0, R[$ be such that $\mu_N \leq \log u \leq \mu_{N+1}$. We have $\log \|f\|_{d(0,u)} = \Psi(f, \log u)$.

When s is big enough, $\Psi(f_s, \log u)$ is clearly equal to $\Psi(f, \log u)$, hence f satisfies $\log \|f\|_{d(0,u)} = \Psi(f_s, \log u) \leq \ell(\log u) + \lambda$. Hence, by (2) we obtain $|f|(r) \leq B|T|(r)$. Now we just have to check that every a_j is a zero of f of order $z_j \geq q_j$. Let m be such that $h_m \geq j$. For every $s \geq m$, $(1 - \frac{x}{a_j})^{q_j}$ divides f_s in $H(d(0,u))$ (for every $u \in]0, S]$), hence by Lemma B.7.1 $(1 - \frac{x}{a_j})^{q_j}$ divides f in $H(d(0,u))$ and this finishes the proof of Theorem B.18.14. □

We can obtain a small improvement of Theorem B.18.14.

Theorem B.18.15: *Let $T = (a_n, q_n)_{n \in \mathbb{N}}$ be a divisor on the disk $d(a, R^-)$ with $a_n \neq 0$ $\forall n \in \mathbb{N}$, let $\epsilon > 0$ and $\rho \in]0, R[$. There exists $g \in \mathcal{A}(d(a, R^-))$ such that $\mathcal{D}(g) \geq T$, $g(a) = 1$ and $|g|(r) \leq |T|(r)(1 + \epsilon)$ $\forall r \in]0, R[$ and $\mathcal{D}(g)(\alpha) = T(\alpha)$ $\forall \alpha \in d(a, \rho)$.*

Proof. By Theorem B.18.14, we have a function $f \in \mathcal{A}(d(a, R^-))$ such that $\mathcal{D}(f) \geq T$, $f(a) = 1$ and $|f|(r) \leq |T|(r)(1 + \epsilon)$ $\forall r \in]0, R[$. Now, we can construct a polynomial $P(x)$ such that $P(0) = 1$ and admitting in $d(a, R^-)$ a divisor $\mathcal{D}(P)$ satisfying $\mathcal{D}(P)(\alpha) = \frac{\mathcal{D}(f)(\alpha)}{T(\alpha)}$ $\forall \alpha \in d(a, \rho)$ and $\mathcal{D}(P)(\alpha) = 0$ $\forall \alpha \in d(a, R^-) \setminus d(a, \rho)$. Then the function $g = \frac{f}{P}$ satisfies $\mathcal{D}(g)(\alpha) = T(\alpha)$ $\forall \alpha \in d(a, \rho)$, $T \leq \mathcal{D}(g) \leq \mathcal{D}(f)$ and hence $|g|(r) \leq |f|(r) \leq |T|(r)(1 + \epsilon)$ $\forall r \in]0, R[$. □

Remark: Here we may notice that $H(d(0, R^-))$ is much smaller than $\mathcal{A}_b(d(0, R^-))$. Indeed, by Theorem B.18.15, there exist functions $f \in \mathcal{A}_b(d(0, R^-))$ having infinitely many zeros in $d(0, R^-)$. But by Theorem B.12.8, any element of $H(d(0, R^-))$ is quasi-invertible and hence has finitely many zeros.

B.19. Michel Lazard's problem

This chapter is aimed at studying the following problem mentioned in Chapter B.18 and first considered by M. Lazard in a tremendous work [72]. Let T be a divisor on a disk $d(a, R^-)$. Does a function $f \in \mathcal{A}(d(a, R^-))$ exist such that $\mathcal{D}(f) = T$? The answer depends on whether or not $\mathbb{K}$ is spherically complete.

Theorems B.19.1 and B.19.4 were first proven in [72]. Proofs are long and much technical. Here, we shall try to give an easier presentation of the proofs, which is due to Labib Haddad. More precisely, when $\mathbb{K}$ is spherically complete, Theorem B.19.4 shows the following result, as it was done in [72]: let $T = (a_n, q_n)$ be a divisor on $d(a, R^-)$, with $|a_n| \leq |a_{n+1}|$ and $|a_{u(m)}| < |a_{u(m)+1}|$. For each $m \in \mathbb{N}$, let $P_m(x) = \prod_{j=u(m)+1}^{u(m+1)} (x - a_j)$ and let $(Q_m) \in \mathbb{K}[x]$ be such that $|Q_m|(\rho_m) \leq |T|(\rho_m)$.

Then there exists a function $f \in \mathcal{A}(d(a, R^-))$ such that P_m divides $f - Q_m$. Hence, in particular, given a divisor T on $d(a, R^-)$, there exists a functions f analytic in $d(a, R^-)$, whose divisor is T.

Theorem B.19.1: *Let $\mathbb{K}$ be not spherically complete and let $(D_n)_{n\in\mathbb{N}}$ be a decreasing sequence of disks $d(u_n, \rho_n)$ such that $\bigcap_{n=0}^{\infty} D_n = \emptyset$. Let $R = \dfrac{1}{\lim_{n\to\infty} \mathrm{diam}(D_n)}$. There exists sequences $(c_n)_{n\in\mathbb{N}}$ of $d(0, R^-)$ such that $\lim_{n\to\infty} |c_n| = R$ and such that no function $f \in \mathcal{A}(d(0, R^-))$ admits for the divisor $T = (c_n, 1)_{n\in\mathbb{N}}$.*

Proof. Without loss of generality, we may obviously assume that $R > 1$, hence $\frac{1}{R} < 1 < R$. Consequently, we may assume that $D_0 \subset d(0, R^-)$. For each $n \in \mathbb{N}$, we set $\rho_n = \mathrm{diam}(D_n)$, hence $\rho_n > \rho_{n+1}$. For every $n \in \mathbb{N}$, we can take $\alpha_n \in D_n \setminus D_{n+1}$. Let $\beta_n = \alpha_{n+1} - \alpha_n$, $n \in \mathbb{N}$, hence

(1) $\rho_{n+1} < |\beta_n| \le \rho_n$.

Consider the divisor $T = (\frac{1}{\beta_n}, 1)_{n\in\mathbb{N}}$ and suppose that there exists $f \in \mathcal{A}(d(0, R^-))$ whose divisor is exactly T. Without loss of generality, we can assume that $f(x) = \prod_{n\in\mathbb{N}}(1 - \beta_n x)$. Then $f(x)$ is a series of the form $1 + \sum_{n=1}^{\infty} a_n x^n$. We will show that $\alpha_1 - a_1 \in D_n \ \forall n \ge 1$.

Let us fix $n \in \mathbb{N}$. We can check that $\alpha_n - \alpha_1 = \sum_{j=1}^{n-1} \beta_j$, hence $\alpha_n - (\alpha_1 - a_1) = a_1 + \sum_{j=1}^{n-1} \beta_j$. Since $\alpha_n \in D_n$, then $\alpha_1 - a_1$ lies in D_n if and only if $|a_1 + \sum_{j=1}^{n-1} \beta_j| \le \rho_n$.

For all $n \in \mathbb{N}$, we set $t_n = |(\beta_n)^{-1}|$. By (1) the sequence (t_n) is strictly increasing and satisfies

(2) $|a_n| = \prod_{j=1}^{n} t_j \ \forall n \ge 1$.

Particularly, $a_n \neq 0 \ \forall n \ge 1$. For each $s \ge 1$, we put $f_s = \sum_{n=0}^{s} a_n x^n$. Then $\deg(f_s) = s$. By (2) we see that the s zeros of f_s are distinct and are of the form γ_j^s, $1 \le j \le s$, with $|\gamma_j^s| = t_j$. Thus, f_s is of the form $\prod_{j=1}^{s}(1 - \beta_{s,j} x)$, with $|\beta_{s,j}| = |\beta_j| = t_j^{-1}$. By identification of coefficients, we obtain $a_1 = -\sum_{j=1}^{s} \beta_{s,j}$. Consequently, when

$n \leq s$, we have

(3) $\displaystyle\sum_{j=1}^{s} \beta_j - \beta_{s,j} = a_1 + \sum_{j=1}^{s} \beta_j.$

Let us fix $n \in \mathbb{N}^*$. By Theorem B.15.3, we can see that for each $j = 1, \dots, n$, we have $\lim_{s \to \infty} \beta_{s,j} = \beta_j$. Consequently, when s is big enough, we can see that $\left|\sum_{j=1}^{n-1} \beta_j - \sum_{j=1}^{s} \beta_{s,j}\right| \leq \rho_n$. Therefore by (3), we have $\left|a_1 + \sum_{j=1}^{n-1} \beta_j\right| \leq \rho_n$.

Consequently, $\alpha_1 - a_1$ lies in D_n. This is true for every $n \in \mathbb{N}$, a contradiction to the hypothesis $\bigcap_{n=0}^{\infty} D_n = \emptyset$. □

In order to prove Theorem B.19.4, we must introduce a set of notations that will hold throughout the chapter.

Definitions and notations: We consider a divisor T on $d(0, R^-)$ of the form $(a_{i,m,}, q_{i,m})_{i \leq u_m, m \in \mathbb{N}}$, where the points $a_{i,m}$ lie in the circle $C(0, \rho_m)$ with $0 < \rho_m < \rho_{m+1}$ $\forall m \in \mathbb{N}$ and $\lim_{m \to +\infty} \rho_m = R$. We denote by $(P_m)_{m \in \mathbb{N}}$ the polynomial $P_m = \prod_{i=1}^{u_m} \left(1 - \frac{x}{a_{i,m}}\right)^{q_{i,m}}$ whose zeros by definition, belong to the circle $C(0, \rho_m)$ and for each $m \in \mathbb{N}$, we set $d_m = \deg(P_m)$.

We denote by $(Q_m)_{m \in \mathbb{N}}$ a sequence of $\mathbb{K}[x]$ satisfying $|Q_m|(\rho_m) \leq |T|(\rho_m)$, $\deg(Q_m) < d_m$ $\forall m \in \mathbb{N}$. We notice that $|T|(r) = \prod_{m \geq 1}^{s-1} \left(\frac{r}{\rho_m}\right)^{d_m}$ whenever $r \in [\rho_{s-1}, \rho_s]$. Given $q \in \mathbb{N}$, $s \in \mathbb{N}$, $r \in]0, R[$, we set

$\zeta(q, s, r) = |T|(r)\left(\frac{r}{R}\right)^q \min\left(1, \left(\frac{r}{\rho_s}\right)\right)$, i.e.,

$\zeta(q, s, r) = |T|(r)\left(\frac{r}{R}\right)^q \left(\frac{r}{\rho_s}\right)$ $\forall r \in]0, \rho_s]$,

$\zeta(q, s, r) = |T|(r)\left(\frac{r}{R}\right)^q$ $\forall r \in [\rho_s, R[$.

Particularly, we notice that $\zeta(q, s, r) \leq |T|(r)$ $\forall r < R$. Now, given $n \in \mathbb{N}$, we set $\tau(q, s, n) = \inf_{r < R} \left(\frac{\zeta(q, s, r)}{r^n}\right)$.

Recall that the Euclidean division in $H(C(0, r))$ is defined in Chapter B.18. We will denote by $\Sigma(q, s)$ the subset of the $h \in \mathcal{A}(d(0, R^-))$ satisfying

α) $h(0) = 1$
β) $|h|(r) \leq |T|(r)$ $\forall r < R$

γ) P_m divides $h - Q_m$ in $\mathcal{A}(d(0, R^-))$ $\forall m = 0, \dots, s$

δ) the rest X_m of the Euclidean division of h by P_m in $H(C(0, \rho_m))$ satisfies $|X_m - Q_m|(\rho_m) \le \zeta(q, m, \rho_m) = |T|(\rho_m)\Big(\frac{\rho_m}{R}\Big)^q$ $\forall m \ge 0$.

Remark: By definition, q and n being fixed, the sequence in s: $(\tau(q, s, n))_{s \in \mathbb{N}}$ is decreasing.

In the proof of Theorem B.19.4, we will use Lemmas B.19.2 and B.19.3. Lemma B.19.2 is immediate.

Lemma B.19.2: *$\Sigma(q+1, s) \subset \Sigma(q, s)$ and $\Sigma(q, s)$ is a closed subset of $\mathcal{A}(d(0, R^-))$.*

Lemma B.19.3: *Let $q \in \mathbb{N}$, $s \in \mathbb{N}^*$, $f \in \Sigma(q, s-1)$. There exists $g \in \Sigma(q, s)$ such that $|g - f|(r) \le \zeta(q, s, r)$ $\forall r < R$.*

Proof. Let $h = \prod_{m=0}^{s-1} P_m$ and $u = x^{q+1}h$. Then P_m and x have no zero in $C(0, \rho_s)$, hence they are invertible in $H(C(0, \rho_s))$ and hence so is u. For each $m \in \mathbb{N}$, we denote by R_m the rest of the Euclidean division of f by P_m in $H(C(0, \rho_m))$. Since $f \in \Sigma(q, s-1)$, by γ) above, P_m divides $f - Q_m$ for every $m \le s-1$, hence $R_m = Q_m$ $\forall m = 0, \dots, s-1$.

Consider now the Euclidean division of $(Q_s - R_s)u^{-1}$ by P_s in $H(C(0, \rho_s))$: $(Q_s - R_s)u^{-1} = EP_s + S$, with $\deg(S) < d_s$, $E \in H(C(0, \rho_s))$ and $|S(\rho_s)| \le |(Q_s - R_s)u^{-1}|(\rho_s)$, hence

(1) $|Su|(\rho_s) \le |(Q_s - R_s)|(\rho_s)$.

We then take $g = f + Su$, hence $g = f + Shx^{q+1}$. We shall show that g belongs to $\Sigma(q, s)$ and that $|g - f|(r) \le \zeta(q, s, r)$ $\forall r < R$. We notice that g belongs to $\mathcal{A}(d(0, R^-))$ and that $g(0) = f(0) = 1$.

Next, by hypothesis, P_s divides $f - R_s$ and by construction divides $Su + R_s - Q_s$. But $g - Q_s = f + Su - Q_s = f - R_s + Su + R_s - Q_s \in \mathcal{A}(d(0, R^-))$, hence P_s divides $g - Q_s$ in $\mathcal{A}(d(0, R^-))$.

Now, by hypothesis, f lies in $\Sigma(q, s-1)$ hence particularly $|f|(r) \le |T|(r)$ and we have

(2) $|Q_m - R_m|(\rho_m) \le |T|(\rho_m)\Big(\frac{\rho_m}{R}\Big)^q$ $\forall m \ge 1$.

Thus, by (1) and (2) we have

(3) $|g - f|(\rho_s) \le |T|(\rho_s)\Big(\frac{\rho_s}{R}\Big)^q$.

But since $|S|$ is an increasing function in r, we have $|S|(r) \leq |S|(\rho_s)$ whenever $r \leq \rho_s$. On the other hand, $|h|(r) = |T|(r)\ \forall r \leq \rho_s$. And, as we saw,

$$|g - f|(\rho_s) = |Su|(\rho_s) = |Shx^{q+1}|(\rho_s) = |S|(\rho_s)|T|(\rho_s)(\rho_s)^{q+1} \leq |T|(\rho_s)\Big(\frac{\rho_s}{R}\Big)^q.$$

Consequently, $|S|(r) \leq |S|(\rho_s) \leq \dfrac{1}{R^q \rho_s}\ \forall r \leq \rho_s$ and hence

$$(4) \qquad |g - f|(r) = |Shx^{q+1}|(r) = |S|(r)|h|(r)r^{q+1} \leq |T|(r)\Big(\frac{\rho_s}{R}\Big)^q\Big(\frac{r}{\rho_s}\Big)\forall r \leq \rho_s.$$

Now, when $r \geq \rho_s$, we have

$$|T|(r) = \prod_{m\geq 1} |P_m|(r) = |h|(r) \prod_{m \geq s} |P_m|(r) \geq |h|(r)|P_s|(r) = |h|(r)\Big(\frac{r}{\rho_s}\Big)^{d_s}.$$

Next, Sx^{q+1} is of the form $\lambda \prod_{j=1}^m (x - x_j)$ and is a polynomial of degree $m \leq q + d_s$, hence

$$\frac{|x^{q+1}S|(r)}{|x^{q+1}S|(\rho_s)} = \prod_{j=1}^m \frac{|x - x_j|(r)}{|x - x_j|(\rho_s)} \leq \Big(\frac{r}{\rho_s}\Big)^{q+d_s}.$$

Thus,

$$\begin{aligned}|g - f|(r) &= |Shx^{q+1}|(r) = |Sx^{q+1}|(r)|h|(r) \leq |Sx^{q+1}|(\rho_s)\Big(\frac{r}{\rho_s}\Big)^{q+d_s}\Big(\frac{\rho_s}{r}\Big)^{d_s}|T|(r)\\ &= |T|(r)\Big(\frac{r}{\rho_s}\Big)^q|Sx^{q+1}|(\rho_s) = |T|(r)\Big(\frac{r}{\rho_s}\Big)^q(\rho_s)^{q+1}|S|(\rho_s)\\ &\leq |T|(r)r^q(\rho_s)\frac{1}{R^q\rho_s} = |T|(r)\Big(\frac{r}{R}\Big)^q.\end{aligned}$$

And finally, with (4) we obtain $|g - f|(r) \leq \zeta(q, s, r)\ \forall r \in]0, R[$. □

Theorem B.19.4: *Suppose* $\mathbb{K}$ *is spherically complete. Assume that* $|Q_m|(\rho_m) \leq |T|(\rho_m)\ \forall m \in \mathbb{N}$. *Let* $R \in]0, +\infty[$. *There exists* $f \in \mathcal{A}(d(0, R^-))$ *satisfying*

(i) $f(0) = 1$,
(ii) $|f|(r) \leq |T|(r)\ \forall r < R$,
(iii) P_m *divides* $f - Q_m$ *in* $\mathcal{A}(d(0, R^-))$.

Proof. We mean to construct a sequence of functions $(f_q)_{q\in\mathbb{N}}$, which belong to $\mathcal{A}(d(0, R^-))$, converging in $\mathcal{A}(d(0, R^-))$ to a function f satisfying the claim.

We first fix $q \in \mathbb{N}$ and take $f_q \in \Sigma(q, 0)$. Let $f_q(x) = \sum_{n=0}^{\infty} a_n x^n$. We will construct a sequence $(g_{q,s})_{s\in\mathbb{N}}$ satisfying $g_{q,s} \in \Sigma(q, s)$ and $|g_{q,s} - g_{q,s-1}|(r) \leq \zeta(q, s, r)$, with $g_{q,0} = f_q$. Suppose $g_{q,j}$ has been already constructed for $j = 0, \ldots, s - 1$. By Lemma B.19.3, there exists $h \in \Sigma(q, s)$ such that $|h - g_{q,s-1}|(r) \leq \zeta(q, s, r)$. So, we can set $g_{q,s} = h$ and the sequence is then defined by induction for all $s \in \mathbb{N}$.

Now, for each $s \in \mathbb{N}$ we set $g_{q,s}(x) = \sum_{n=0}^{\infty} b_{q,s,n}x^n$. Since by construction the sequence $g_{q,s}$ satisfies

(1) $|g_{q,s} - g_{q,s-1}|(r) \leq \zeta(q, s, r) \forall r < R$,

then for each fixed $n \in \mathbb{N}$, the sequence $(b_{q,s,n})_{s\in\mathbb{N}}$ satisfies $|b_{q,s,n} - b_{q,s-1,n}| \leq \tau(q, s, n)$. Thus, for each fixed $n \in \mathbb{N}$, we consider the sequence of disks $(D_{s,n})_{s\in\mathbb{N}}$ defined as $D_{q,s,n} = d(b_{q,s,n}, \tau(q, s, n))$. Since the sequence $(\tau(q, s, n))_{s\in\mathbb{N}}$ is decreasing and since $|b_{q,s,n} - b_{q,s-1,n}| \leq \tau(q, s, n)$, the sequence of disks $(D_{q,s,n})_{s\in\mathbb{N}}$ is decreasing with respect to the inclusion. Consequently, since $\mathbb{K}$ is spherically complete, for each $n \in \mathbb{N}$, there exists $a_{q+1,n} \in \bigcap_{s=0}^{\infty} D_{q,s,n}$. Particularly, since $\tau(q, s, n) = 0 \, \forall n \leq q$, we notice that $b_{q,s,0} = b_{q,s-1,0} = 1$ because $g_{q,s-1} \in \Sigma(q, s-1)$. Consequently, $a_{q+1,0} = 1$.

Now, we will show that f_{q+1} belongs to $\Sigma(q+1, 0)$. Let $f_{q+1}(x) = \sum_{n=0}^{\infty} a_{q+1,n}x^n$. Since $a_{q+1,0} = 1$, f_{q+1} satisfies relation α). Next, by construction, we have $|a_{q+1,n} - b_{q,s-1,n}| \leq \tau(q, s, n) \leq \tau(q, 1, n) \, \forall n \in \mathbb{N}$, hence obviously $|a_{q+1,n} - a_{q,n}| \leq \tau(q, 1, n) \, \forall n \in \mathbb{N}$. Consequently, $|a_{q+1,n} - a_{q,n}|r^n \leq \zeta(q, 1, r) \, \forall r < R$, hence

(2) $|f_{q+1} - f_q|(r) \leq \zeta(q, 1, r) \leq |T|(r)\left(\frac{r}{R}\right)^q \, \forall r < R$.

Now since, by hypothesis, $|f_q|(r) \leq |T|(r)$, by (2) we can see that $|f_{q+1}|(r) \leq |T|(r)$ and therefore f_{q+1} satisfies relation β). Since γ) is trivial when $s = 0$, it only remains to show that f_{q+1} satisfies δ).

For each $m \in \mathbb{N}$, let $S_{m,q+1}$ be the rest of the Euclidean division of f_{q+1} by P_m in $H(C(0, \rho_m))$. For each $s \geq m$, since P_m divides $g_{q,s} - Q_m$, the rest of the Euclidean division of $g_{q,s} - f_{q+1}$ by P_m in $H(C(0, \rho_m))$ is equal to $Q_m - S_{m,q+1}$. Consequently, by (1) we have

(3) $|Q_m - S_{m,q+1}|(\rho_m) \leq |g_{q,s} - f_{q+1}|(\rho_m) \leq \zeta(q, s, \rho_m)$

and hence

(4) $|Q_m - S_{m,q+1}|(\rho_m) \leq |T|(\rho_m)\left(\frac{\rho_m}{R}\right)^q\left(\min\left(1, \frac{\rho_m}{\rho_s}\right)\right) \forall s \geq m$.

Now, since $\lim_{s\to\infty} \rho_s = R$, by (4) we have $|Q_m - S_{m,q+1}|(\rho_m) \leq |T|(\rho_m)\left(\frac{\rho_m}{R}\right)^{q+1}$. This finishes showing that δ) is satisfied by $g_{q+1,s}$ and therefore $g_{q+1,s}$ belongs to $\Sigma(q + 1, 0)$. This true for all s, hence by Lemma B.19.2 f_{q+1} also belongs to $\Sigma(q + 1, 0)$.

Thus, we have constructed a sequence $(f_q)_{q\in\mathbb{N}}$ of $\mathcal{A}(d(0, R^-))$ satisfying $f_q \in \Sigma(q, 0) \forall q \in \mathbb{N}$. By (2) we can see that the sequence $(f_q)_{q\in\mathbb{N}}$ converges in all $H(d(0, \rho))$, for every $\rho < R$, to a limit f that thereby belongs to $H(d(0, \rho))$

for all $\rho < R$. Consequently, the function f belongs to $\mathcal{A}(d(0,R^-))$. Moreover, since $\Sigma(q,0)$ is closed, f belongs to $\Sigma(q,0)$ for every $q \in \mathbb{N}$. Consequently, by relation $\delta)$ true for every q, the rest X_m of the Euclidean division of f_q by P_m in $H(C(0,\rho_m))$ satisfies $|X_m - Q_m|(\rho_m) \leq |T|(\rho_m)\Big(\frac{\rho_m}{R}\Big)^q$ for every $q \in \mathbb{N}$, hence $X_m = Q_m$. So, P_m divides $f - Q_m$ for every $m \in \mathbb{N}$. And by construction, f satisfies $\alpha)$ and $\beta)$, which completes the proof. □

Corollary B.19.5: *Suppose $\mathbb{K}$ is spherically complete. Let $(a_n)_{n\in\mathbb{N}}$ be a sequence of $d(0,R^-)$ such that $|a_n| \leq |a_{n+1}|$ $\forall n \in \mathbb{N}$ and $\lim_{n\to+\infty} |a_n| = R$ and let $(b_n)_{n\in\mathbb{N}}$ be a sequence of $\mathbb{K}$. There exists $f \in \mathcal{A}(d(a,R^-))$ such that $f(a_n) = b_n$ $\forall n \in \mathbb{N}$.*

Proof. Indeed we can define a sequence of integers $(s_n)_{n\in\mathbb{N}}$ such that the divisor $T = (a_n, s_n)_{n\in\mathbb{N}}$ satisfies $|T|(|a_n|) \geq |b_n|$ $\forall n \in \mathbb{N}$. □

Theorem B.19.6: *Suppose $\mathbb{K}$ is spherically complete. Let T be a divisor on $d(a,R^-)$. There exists $f \in \mathcal{A}(d(a,R^-))$ such that $\mathcal{D}(f) = T$.*

Proof. Without loss of generality, we may obviously assume $a = 0$. Take $Q_m = 0$ $\forall m \in \mathbb{N}$. By Theorem B.19.4, there exists $f \in \mathcal{A}(d(0,R^-))$ such that

(i) $f(0) = 1$,
(ii) $|f|(r) \leq |T|(r)$ $\forall r < R$,
(iii) P_m divides f in $\mathcal{A}(d(0,R^-))$.

By (iii), clearly $\mathcal{D}(f) \geq T$. Thus, we only have to check that $\mathcal{D}(f) \leq T$. Indeed, for all $s \in \mathbb{N}$ we have

$$|T|(\rho_s) = \prod_{j=1}^{s}\prod_{i=1}^{u_m}\left|\left(1 - \frac{x}{a_{i,m}}\right)^{q_{i,m}}\right|(\rho_s) = \prod_{j=1}^{s}\prod_{i=1}^{u_m}\left(\frac{\rho_s}{\rho_m}\right)^{q_{i,m}}.$$

Now, suppose that $T \neq \mathcal{D}(f)$. Then there exists $\alpha \in d(0,R^-)$ such that $\omega_\alpha(f) > T(\alpha)$. Let s be such that $\rho_s > |\alpha|$. Since $f(0) = 1$, we have

$$|f|(\rho_s) \geq \frac{\rho_s}{|\alpha|}\prod_{j=1}^{s}\prod_{i=1}^{u_m}\left|\left(1 - \frac{x}{a_{i,m}}\right)^{q_{i,m}}\right|(\rho_s) > \prod_{j=1}^{s}\prod_{i=1}^{u_m}\left(\frac{\rho_s}{\rho_m}\right)^{q_{i,m}} = |T|(\rho_s),$$

a contradiction to (iii). □

Similarly to $\mathcal{A}(\mathbb{K})$, the algebra $\mathcal{A}(d(a,R^-))$ is provided with the natural topology of uniform convergence on each disk $d(0,r)$ whenever $0 < r < R$. Such a topology makes $\mathcal{A}(d(a,R^-))$ a topological $\mathbb{K}$-algebra.

In Chapter B.18, we showed that in $\mathcal{A}(\mathbb{K})$ every closed ideal is principal. Here, following the same methods, provided that $\mathbb{K}$ is spherically complete, we can prove similar results with algebras $\mathcal{A}(d(a,R^-))$.

Theorem B.19.7: *Suppose $\mathbb{K}$ is spherically complete. All closed ideals of $\mathcal{A}(d(a, R^-))$ are principal.*

Proof. Let I be a closed ideal of $\mathcal{A}(d(a, R^-))$ and let $E = \mathcal{D}(I)$. By Theorem B.18.10, we have $I = \mathcal{T}_R(E)$. Now, by Theorem B.19.6 there exists $g \in \mathcal{A}(\mathbb{K})$ such that $\mathcal{D}(g) = E$ and of course g belongs to $\mathcal{T}_R(E)$, hence to I. Consequently, $g\mathcal{A}(d(a, R^-)) \subset I$. Conversely, by Corollary B.18.2, we have $I = g\mathcal{A}(d(a, R^-))$. □

B.20. Motzkin factorization and roots of analytic functions

Throughout this chapter, D is a closed infraconnected set and f belongs to $H(D)$.

The idea of factorizing semi-invertible analytic elements into a product of singular factors is a remarkable idea according to E. Motzkin [75]. This factorization has tight links with the Mittag–Leffler series, as it was shown in [58, 68].

Lemma B.20.1: *Let $T = d(a, r^-)$, with $a \in \mathbb{K}$ and $r > 0$, let $E = \mathbb{K} \setminus T$ and take $b \in T$. Let $g \in H(E)$ be invertible in $H(E)$. Then there exist $\lambda \in \mathbb{K}$, $q \in \mathbb{Z}$ and $h \in H(E)$ invertible in $H(E)$, satisfying $\|h - 1\|_E < 1$, $\lim_{|x| \to +\infty} h(x) = 1$ and $g(x) = \lambda(x - b)^q h(x)$. Moreover, λ, q, h, are respectively unique, satisfying those relations. Further, both λ, q do not depend on b in T and $\frac{g'}{g}$ belongs to $H_0(E)$.*

Proof. Without loss of generality we may obviously assume $a = 0$. As g is invertible, if g belongs to $H_0(E)$, then $\frac{1}{g}$ does not. So, we may clearly assume that g does not belong to $H_0(E)$. By Theorem B.2.5, g is of the form $\widetilde{g} + \widehat{g}$, with $\widetilde{g} \in \mathbb{K}[x]$, $\widetilde{g} \neq 0$, and $\widehat{g} \in H_0(E)$. Let $q = \deg(\widetilde{g})$ and let λ be its coefficient of degree q. Now we put $h(x) = \dfrac{g(x)}{\lambda(x - b)^q}$. By definition both λ, q do not depend on b in T. Hence, we may also assume $b = 0$. Clearly, h satisfies $\lim_{|x| \to +\infty} h(x) = 1$. Since $H(E)$ is a $\mathbb{K}$-algebra and since g is invertible, h is invertible in $H(E)$. In particular, we notice that h is bounded and admits no zero in E. Now we check that $\|h - 1\|_E < 1$. Let $s = \frac{1}{r}$, let $A = d(0, s)$, and let $\phi(u) = h(\frac{1}{u})$ whenever $u \in d(0, s)$, $u \neq 0$. Then ϕ belongs to $H(d(0, s) \setminus \{0\})$. But since h is bounded in E, ϕ is bounded in $d(0, s) \setminus \{0\}$. Moreover, the condition $\lim_{|x| \to +\infty} h(x) = 1$ shows that $\lim_{x \to 0} \phi(x) = 1$, hence ϕ belongs to $H(d(0, s))$.

Thus $\phi(u)$ is of the form $\sum_{n=0}^{\infty} a_n u^n$ with $a_0 = 1$ and hence, by Theorem B.13.7, we have $|a_n| s^n < 1 \ \forall n > 0$. Let $\epsilon = \sup\{|a_n| s^n \ | n > 0\}$. Then we have $\|\phi - 1\|_{d(0,s)} = \|h - 1\|_E = \epsilon$. Now, h, q, λ are easily seen to be unique. Indeed, let $g(x) = \alpha x^t l(x)$ with l invertible in $H(E)$, satisfying $\lim_{|x| \to +\infty} l(x) = 1$. Then we have $1 = x^{q-t}\dfrac{\lambda h(x)}{\alpha l(x)}$. Consequently, considering the limit when $|x|$ tends to $+\infty$, we have $q = t$, $\lambda = \alpha$ and therefore $h = l$. Finally, we check that $\dfrac{g'}{g}$ belongs to $H_0(E)$. Indeed, $\dfrac{g'}{g} = \dfrac{q}{x-b} + \dfrac{h'}{h}$. Obviously, $\dfrac{q}{x-b}$ belongs to $H_0(E)$. Since $\lim\limits_{|x|\to\infty} |h(x)| = 1$, it is seen that $h(x)$ is of the form $1 + \sum\limits_{n=1}^{\infty} \dfrac{a_n}{x^n}$ with $\lim\limits_{n\to\infty} |\dfrac{a_n}{s^n}| = 0$ and therefore h' is an element of $H(E)$ such that $\lim\limits_{|x|\to\infty} |h'(x)| = 0$. As a consequence, $\dfrac{h'}{h}$ belongs to $H_0(E)$. Hence, so does $\dfrac{g'}{g}$ and this ends the proof of Lemma B.20.1. □

Definitions and notations: Let $E = \mathbb{K} \setminus d(a, r^-)$ with $a \in \mathbb{K}$ and $r > 0$. Let $f \in H(E)$ be invertible in $H(E)$ and let $\lambda(x-a)^q h(x)$ be the factorization given in Lemma B.20.1. The integer q will be named *the index of* f associated to $d(a, r^-)$ and will be denoted by $m(f, d(a, r^-))$. If $\lambda = 1$, the element f will be called *a pure factor associated to* $d(a, r^-)$. Let $\mathcal{G}^T$ be the group of invertible elements of $H(\mathbb{K} \setminus T)$.

Corollary B.20.2 is an immediate.

Corollary B.20.2: *Let $T = d(a, r^-)$. The set of pure factors associated to T is a sub-multiplicative group of the group $\mathcal{G}^T$. Further, every element of $\mathcal{G}^T$ is of the form λh with h a pure factor associated to T and $\lambda \in \mathbb{K}^*$.*

Lemma B.20.3: *Let $T = d(a, r^-)$, let $E = \mathbb{K} \setminus T$ with $a \in \mathbb{K}$, and let f be a pure factor associated to T such that $\|f - 1\|_E < 1$. Then $m(f, T) = 0$.*

Proof. Without loss of generality, we can assume $a = 0$. Let $q = m(f, T)$. By Lemma B.20.1, there exists a unique element h invertible in $H(E)$ such that $f = x^q h$ and $\lim\limits_{|x|\to+\infty} |h(x)| = 1$. Therefore, by Theorem B.5.6, $h(x)$ is of the form $1 + \sum\limits_{n=1}^{+\infty} \dfrac{a_n}{x^n}$, hence $\|h\|_E < 1$. So, if $q > 0$, then f is unbounded, a contradiction. Next, by Corollary B.20.2, $\dfrac{1}{f}$ is a pure factor satisfying again the hypothesis of Corollary B.20.2, hence the hypothesis $q < 0$ gets to a contradiction again. □

Definitions and notations: Let f belong to $H(D)$. Let T be a hole of D and let h be a pure factor associated to T. If $\dfrac{f}{h}$ belongs to $H(D \cup T)$ and has no zero inside T, h is called *Motzkin factor of f in the hole T*.

Theorem B.20.4: *Let T be a hole of D and let f have a Motzkin factor h in T. Then h is unique. Further, if T is not a f-hole, h is the polynomial of the zeros of f inside T. Moreover, if E is another infraconnected set included in D admitting T as a hole and if g denotes the restriction of f to E, then g admits a Motzkin factor in the hole T as an element of $H(E)$ and this Motzkin factor is equal to h.*

Proof. Let f have another Motzkin factor l in T, let $F = \frac{f}{h}$ and let $G = \frac{f}{l}$. Since G has no zeros inside T, by Theorem B.15.9 there exists a closed bounded infraconnected set D' satisfying $T \subset D' \subset (D \cup T)$, $T \neq D'$, such that G is invertible in $H(D')$. Hence, in $H(D')$ we have $\frac{F}{G} = \frac{l}{h}$ and hence $\frac{l}{h}$ belongs to $H(D')$. Since $T \neq D'$ it is seen that $D' \cap (\mathbb{K} \setminus T)$ is an infraconnected closed bounded set included in D that admits T as a hole. Moreover, we have $D' \cup (\mathbb{K} \setminus T) = \mathbb{K}$. Therefore, by Theorem B.6.10, we see that $\frac{l}{h}$ belongs to $H(\mathbb{K})$ and hence is a polynomial P. Since $\frac{F}{G}$ belongs to $H(D')$ and has no zeros in T, it is seen that $m(h,T) = m(l,T)$, so we have $\lim_{|x| \to \infty} \frac{h(x)}{l(x)} = 1$. Hence, $P = 1$ and this proves that h is unique.

Now we assume that T is not a f-hole, hence f belongs to $H(D \cup T)$. Let Q be the polynomial of the zeros of f inside T. Then by Corollary B.7.6, $\frac{f}{Q}$ belongs to $H(D \cup T)$ and has no zeros inside T. Since its Motzkin factor h is unique, we have $h = Q$. The last statement about g is obvious because $\frac{g}{h}$ clearly belongs to $H(E \cup T)$ and has no zero inside T. This ends the proof of Theorem B.20.4. $\square$

Definitions and notations: We will call *the f-supersequence of D* the sequence of the holes $(T_n)_{n \in I}$ such that either T_n is a f-hole or f belongs to $H(D \cup T)$ and has at least one zero inside T_n. If f admits a Motzkin factor h in a hole T, it will be denoted by f^T and $m(h,T)$ will be called *the Motzkin index of f in T*. For every hole that does not belong to the f-supersequence, we put $f^T = 1$.

Lemma B.20.5 is immediate.

Lemma B.20.5: *Let $D \in$ Alg, let T be a hole of D, and let f, $g \in H(D)$ admitting Motzkin factors in T. Then $(fg)^T = f^T g^T$ and $m(fg,T) = m(f,T) + m(g,T)$. Moreover, if f is invertible in $H(D)$, then $(f^{-1})^T = (f^T)^{-1}$ and $m(f^{-1},T) = -m(f,T)$.*

Lemma B.20.6: *Let $f \in H(D)$ and let $(T_n)_{n \in \mathbb{N}}$ be the f-supersequence. Suppose that for each $n \in \mathbb{N}$, f admits Motzkin factors in T_n. Then there exists $N \in \mathbb{N}$ such that $m(f^{T_n}, T_n) = 0$ whenever $n > N$. Moreover, if $D \in$ Alg, the product $\Big(\prod_{n=1}^{t} f^{T_n}\Big)\Big(\prod_{n=t+1}^{\infty} f^{T_n}\Big)$ does not depend on t whenever $t \geq N$.*

Proof. Indeed, there exists $N \in \mathbb{N}$ such that we have $\|f^{T_n} - 1\|_D < 1$ whenever $n \geq N$ and therefore, by Corollary B.20.2, $m(f^{T_n}, T_n) = 0$. Now in $H_b(D)$, we have

$$\Big(\prod_{n=N+1}^{\infty} f^{T_n}\Big) = \Big(\prod_{n=N+1}^{t} f^{T_n}\Big)\Big(\prod_{n=t+1}^{\infty} f^{T_n}\Big).$$

But then, if D belongs to Alg, we have

$$\begin{aligned}\Big(\prod_{n=1}^{t} f^{T_n}\Big)\Big(\prod_{n=t+1}^{\infty} f^{T_n}\Big) &= \Big(\prod_{n=1}^{N} f^{T_n}\Big)\Big(\prod_{n=N+1}^{t} f^{T_n}\Big)\Big(\prod_{n=t+1}^{\infty} f^{T_n}\Big)\\ &= \Big(\prod_{n=1}^{N} f^{T_n}\Big)\Big(\prod_{n=N+1}^{\infty} f^{T_n}\Big).\end{aligned}$$

□

Definitions and notations: Let $(T_n)_{n\in I}$ be the f-supersequence of D with I a subset of $\mathbb{N}$, which is either finite or equal to $\mathbb{N}$.

If I is finite, f will be said to have *a finite Motzkin factorization* if it factorizes in $H(D)$ in the form $\Big(f^0 \prod_{n\in I} f^{T_n}\Big)$ with f^0 an element of $H(\widetilde{D})$ whose zeros belong to D and for each $n \in \mathbb{N}$, f^{T_n} a Motzkin factor in T_n.

If I is infinite and equal to $\mathbb{N}$, f will be said to have *an infinite Motzkin factorization* if it admits a sequence of Motzkin factors f^{T_n} satisfying $\lim\limits_{n\to\infty} f^{T_n} - 1 = 0$ such that f factorizes in $H(D)$ in the form $\Big(f^0 \prod_{n=1}^{t} f^{T_n}\Big)\Big(\prod_{n=t+1}^{\infty} f^{T_n}\Big)$, with f^0 an element of $H(\widetilde{D})$ whose zeros belong to D. In both cases, f^0 will be called *the principal factor of f.*

Corollary B.20.7: *Let D be bounded and let f have an infinite Motzkin factorization with a f-supersequence $(T_n)_{n\in\mathbb{N}}$. Then we have $f = f^0\Big(\prod_{n=1}^{\infty} f^{T_n}\Big)$.*

Corollary B.20.8: *Let f have an infinite Motzkin factorization with a f-supersequence $(T_n)_{n\in\mathbb{N}}$ such that $m(f^{T_n}, T_n) = 0$ for all $n > 0$. Then we have $f = f^0\Big(\prod_{n=1}^{\infty} f^{T_n}\Big)$.*

Remark 1: Let $f \in H(D)$ be unbounded and have Motzkin factorization of the form $\Big(f^0 \prod_{n=1}^{N} f^{T_n}\Big)\Big(\prod_{n=N+1}^{\infty} f^{T_n}\Big)$. One cannot claim that the product $\Big(f^0 \prod_{n=1}^{\infty} f^{T_n}\Big)$ converges in $H(D)$, even if D is closed and belongs to Alg. Indeed, let $r \in]0,1[$, let $(a_n)_{n\in\mathbb{N}}$ be a sequence in $d(0,1)$ such that $|a_n - a_m| = 1$ whenever $n \neq m$ and $a_1 = 0$.

For every $n \in \mathbb{N}^*$, we put $T_n = d(a_n, r^-)$ and $E = \mathbb{K} \setminus \left(\bigcup_{n=1}^{\infty} T_n\right)$. The holes of E are the T_n. Let $(\lambda_n)_{n\geq 2}$ be a sequence in $d(0, r^-)$ such that $\lim_{n\to\infty} \lambda_n = 0$. For every $n \geq 2$, we put $g_n = 1 + \dfrac{\lambda_n}{x - a_n}$. The sequence $(g_n)_{n\geq 2}$ is seen to satisfy $\|g_n - 1\|_E \leq \dfrac{|\lambda_n|}{\rho} < 1$ and therefore we have $\lim_{n\to\infty} \|g_n - 1\|_E = 0$. Hence, the product $h = \prod_{n=2}^{\infty} g_n$ obviously converges in $H(E)$.

Since E clearly belongs to Alg, we see that $x^2 h$ belongs to $H(E)$ and is invertible in $H(E)$. Now, f clearly has Motzkin factorization with $f^{T_n} = g_n$ for every $n \geq 2$, $f^{T_1} = x^2$, and $f^0 = 1$. However, we will check that the sequence $(f_n)_{n\in\mathbb{N}^*}$ defined by $f_n = x^2 \prod_{j=2}^{n} g_j$ does not converge in $H(E)$. Indeed, we have $f_{n+1}(x) - f_n(x) = x^2\left(\prod_{j=2}^{n} g_j(x)\right)(g_{n+1}(x) - 1)$. For every $x \in \mathbb{K} \setminus d(0,1)$, we have $\left|x^2 \prod_{j=2}^{n} g_j(x)\right|\Big| = |x^2|$ and $|g_{n+1}(x) - 1| = \left|\dfrac{\lambda_{n+1}}{x}\right|$, hence $|f_{n+1}(x) - f_n(x)| = |x||\lambda_{n+1}|$. Thus, $f_{n+1} - f_n$ is not bounded in $H(E)$ and therefore the sequence $(f_n)_{n\in\mathbb{N}^*}$ does not converge in $H(E)$. According to Theorem 4 in [72] , the product $\prod_{n=1}^{\infty} f_n$ should converge to $x^2 h$ in $H(E)$. Here, we see that this is not true in the general case. Actually, the proof given in [72] only shows the simple convergence of the sequence (f_n) and the uniform convergence on bounded subsets of D.

By Lemma B.20.5, Lemma B.20.9 is immediate.

Lemma B.20.9: *Let $D \in$ Alg, let f, $g \in H(D)$ have Motzkin factorization. Then so does fg. Moreover, we have $(fg)^0 = f^0 g^0$. Further, if f is invertible, f^{-1} also has Motzkin factorization and it satisfies $(f^{-1})^0 = (f^0)^{-1}$.*

Corollary B.20.10: (K. Boussaf) *Let $D \in$ Alg, let f have an infinite Motzkin factorization of the form $\left(f^0 \prod_{n=1}^{t} f^{T_n}\right)\left(\prod_{n=t+1}^{\infty} f^{T_n}\right)$. Let $N \in \mathbb{N}$ be such that $m(f^{T_n}, T_n) = 0$ for all $n > N$. Then we have*

$$f = f^0\left(\prod_{n=1}^{N} f^{T_n}\right)\left(\prod_{n=N+1}^{\infty} f^{T_n}\right).$$

Proposition B.20.11 : *Let $f \in H(D)$ satisfy $\|f - 1\|_D < 1$ and have Motzkin factorization of the form $f^0\left(\prod_{n=1}^{\infty} f^{T_n}\right)$ with $\left(T_n\right)_{n\in\mathbb{N}^*}$ the f-supersequence of D. Then for each $n \geq 1$, we have $m(f^{T_n}, T_n) = 0$.*

Proof. For every $n \in \mathbb{N}^*$, we put $q_n = m(f^{T_n}, T_n)$. By Lemma B.20.6, we may assume the $(T_n)_{n\in\mathbb{N}^*}$ ranged in such a way that $q_n \neq 0$ for $n \leq N$, whereas $q_n = 0$ whenever $n > N$. When $n \leq N$, f^{T_n} is of the form $(x - \alpha_n)^{q_n}(1 + \omega_n)$ with $\omega_n \in H_0(\mathbb{K} \setminus T_n)$, $\| \omega_n \|_{\mathbb{K}\setminus T_n} < 1$ and $\alpha_n \in T_n$. When $n > N$, f^{T_n} is just in the form $(1 + \omega_n)$ with $\omega_n \in H_0(\mathbb{K} \setminus T_n)$ and $\|\omega_n\|_{\mathbb{K}\setminus T_n} < 1$. On the other hand, since f has no zero in D, obviously f^0 has no zero in D and therefore it has no zero in $\widetilde{D}$. Hence, by Theorem B.13.7, f^0 is of the form $A(1 + \omega_0(x))$ with $\omega_0 \in H(\widetilde{D})$, $\|\omega_0\|_D < 1$. Let $h(x) = A\prod_{n=1}^{N}(x - \alpha_n)^{q_n}$. We see that f factorizes in the form $h\Big(\prod_{n=0}^{\infty}(1 + \omega_n)\Big)$. Since $\|\omega_n\|_D < 1$ for every $n \in \mathbb{N}$ and since $\lim_{n\to\infty} \omega_n = 0$, it is seen that h satisfies (1) $\quad \|h - 1\|_D < 1$ as does f. Let us suppose $q_1 \neq 0$. We may obviously assume $\alpha_1 = 0$. Let $T_1 = d(0, r^-)$. Thus, in T_1, h admits 0 as a zero of order q_1 if $q_1 > 0$ (resp. a pole of order $-q_1$ if $q_1 < 0$) and has neither any zero nor any pole different from 0. Anyway, when $x \in T_1$, we have (2) $\quad |h(x)| = B|x^{q_1}|$ with $B = |A|\prod_{n=2}^{N}|\alpha_n|^{q_n}$.
We will show that (2) contradicts (1), except if $q_1 = 0$.

Suppose $q_1 > 0$. In T_1, $h(x)$ is a series of the form $\sum_{n=q_1}^{+\infty} c_n x^n$ and $h - 1 = \Big(\sum_{n=q_1}^{+\infty} c_n x^n\Big) - 1$, hence $1 \leq \|h - 1\|_{T_1} \leq \|h - 1\|_D$, which contradicts (1).

Now suppose $q_1 < 0$. By definition, h is obviously invertible in $R(D)$. Hence, we put $F = \frac{1}{h}$ and we see that F satisfies $\|F - 1\|_D < 1$ and admits 0 as a unique zero in T_1, whereas it has no pole in T_1. Hence, the same process lets us get to the same contradiction and finishes showing that $q_1 = 0$ and similarly, $q_n = 0$ for every $n \geq 1$. □

Proposition B.20.12: *Let $f \in H(D)$ be invertible in $H(D)$ and have Motzkin factorization and let $a \in D$. Then f satisfies $\left\|\frac{f}{f(a)} - 1\right\|_D < 1$ if and only if for every hole T of the f-supersequence of D we have $m(f, T) = 0$.*

Proof. Without loss of generality, we may obviously assume $f(a) = 1$. By Proposition B.20.11, we already know that if f satisfies $\|f - 1\|_D < 1$, then for every hole of the f-supersequence, we have $m(f, T) = 0$. Now we suppose that for every hole T of the f-supersequence we have $m(f, T) = 0$ and we will prove that $\|f - 1\|_D < 1$. Indeed, by Lemma B.20.3, for each hole of the f-supersequence, we have $\|f^T - 1\|_D < 1$. Moreover, since f is invertible, f^0 must also be invertible, hence by Theorem B.14.1 it is of the form $(1+\psi(x))$, with $\|\psi\|_D < 1$. Then $\|f-1\|_D < 1$. □

We will show that all semi-invertible elements have Motzkin factorization, step after step and first we consider rational functions.

Proposition B.20.13: *Let $f \in R(D)$. Then f admits Motzkin factorization.*

Proof. The f-supersequence is obviously finite. Let $T_1, \dots, T_s$ be this f-supersequence. We can obviously factorize f in a unique way, in the form $\prod_{j=1}^{s}\left(\frac{h_j}{l_j}\right)$, whereas for each $j = 1, \dots, s$, both h_j, l_j are monic polynomials whose zeros lie in T_j. Thus, we can check that $\frac{h_j}{l_j}$ is the Motzkin factor f^{T_j} of f in the hole T_j. Therefore, putting $f^0 = \frac{f}{\prod_{j=1}^{s}\left(\frac{h_j}{l_j}\right)}$, we have the Motzkin factorization

$$f = f^0 \prod_{j=1}^{s} f^{T_j}.$$ □

Proposition B.20.14: *Let $\phi \in H(D)$ satisfy $\|\phi - 1\|_D < 1$. Then ϕ admits Moztkin factorization $\phi^0\Big(\prod_{n=1}^{\infty}\phi^{T_n}\Big)$ with $(T_n)_{n\in\mathbb{N}^*}$ the f-supersequence. For every ϕ-hole T, we have $\|\phi^T - 1\|_D = \|\overline{\overline{\phi_T}}\|_D$. Moreover, ϕ^0 satisfies $\|\phi^0 - 1\|_D = \|\overline{\overline{\phi_0}} - 1\|_D$.*

Proof. First we suppose $\phi \in R(D)$. Then by Proposition B.20.13, ϕ admits Motzkin factorization. Now by Proposition B.20.12, for each $n > 0$ we have $m(\phi, T_n) = 0$ and therefore, ϕ^{T_n} is of the form $1 + \omega_n$ with $\|\omega_n\|_D < 1$ whenever $n > 0$, whereas $\phi^0 = 1 + \omega_0$ with $\|\omega_0\|_D < 1$. Hence, we see that $\overline{\overline{\phi_{T_n}}} = \overline{\overline{\Big(\omega_n \prod_{\substack{j\neq n\\ j\in\mathbb{N}}}(1+\omega_j)\Big)_{T_n}}}$. Clearly, $\prod_{\substack{j\neq n\\ j\in\mathbb{N}}}(1+\omega_j)$ is of the form $1 + \phi_n$ with $\|\phi_n\|_D < 1$, hence $\|(\omega_n\phi_n)\|_D < \|\omega_n\|_D$ and we obtain

(1) $\|\overline{\overline{(\omega_n\phi_n)_{T_n}}}\|_D \leq \|\omega_n\phi_n\|_D < \|\omega_n\|_D$.

But ω_n is clearly equal to $\overline{\overline{(\phi^{T_n})_{T_n}}}$ and then we have

(2) $\|\overline{\overline{(\omega_n)_{T_n}}}\|_D = \|\omega_n\|_D > \|\omega_n\phi_n\|_D \geq \|\overline{\overline{(\omega_n\phi_n)_{T_n}}}\|_D$.

Moreover, $\overline{\overline{(\omega_n + \omega_n\phi_n)_{T_n}}} = \overline{\overline{(\omega_n)_{T_n}}} + \overline{\overline{(\omega_n\phi_n)_{T_n}}}$, hence by (1) and (2) we have $\|\overline{\overline{(\omega_n(1+\phi_n))_{T_n}}}\|_D = \|\omega_n\|_D$ and finally

(3) $\|\overline{\overline{\phi}}_{T_n}\|_D = \|\overline{\overline{(\omega_n(1+\phi_n))}}_{T_n}\|_D = \|\omega_n\|_D = \|\phi^{T_n} - 1\|_D$.

In the same way, we put $\prod_{n=1}^{\infty}(1+\omega_n) = 1+\psi$ with

(4) $\|\psi\|_D < 1$.

It is seen that ψ belongs to $H_0(\mathbb{K} \setminus (\bigcup_{n=1}^{\infty} T_n))$. Hence Theorem B.6.1, when applied to ψ, shows that

(5) $\overline{\overline{\psi}}_0 = 0$.

Next, we have $\phi = (1+\omega_0)(1+\psi) = 1+\omega_0+\psi+\omega_0\psi$ hence $\overline{\overline{\phi}}_0 = 1 + \overline{\overline{(\omega_0)}}_0 + \overline{\overline{\phi_0}} + \overline{\overline{(\omega_0\psi)_0}}$. By definition $\omega_0 \in H(\widetilde{D})$, hence $\omega_0 = \overline{\overline{(\omega_0)}}_0$ and then by (5) we have $\overline{\overline{\phi}}_0 = 1+\omega_0+\overline{\overline{(\omega_0\psi)_0}}$. But by (4) it is seen that $\|\overline{\overline{(\omega_0\psi)_0}}\|_D < \|\omega_0\|_D$, hence finally we obtain $\|\overline{\overline{\phi_0}} - 1\|_D = \|\omega_0\|_D = \|\phi^0 - 1\|_D$. Thus, we have proven the inequalities satisfied by the ϕ^{T_n} and by ϕ^0 when ϕ belongs to $R(D)$.

Now we consider the general case when $\phi \in H(D)$. Let $(f_m)_{m\in\mathbb{N}}$ be a sequence in $R(D)$ such that $\lim_{m\to\infty} \|\phi - f_m\|_D = 0$. Let $\varepsilon \in]0,1[$ and let $N \in \mathbb{N}$ be such that $\|f_m - \phi\|_D \le \varepsilon$ whenever $m \ge N$. Let T be a hole of the ϕ-supersequence. We will show that the sequence $((f_m)^T)_{m\in\mathbb{N}}$ converges in $H(D)$ and that this convergence is uniform with respect to the ϕ-supersequence. We fix $m \ge N$. It is seen that $\|f_m - 1\|_D < 1$ and then by Lemma B.20.1 and Proposition B.20.12 we have $\|(f_m)^T - 1\|_D < 1$ and in particular $\|(f_m)^T\|_D = 1$. Moreover, we remember that in $H(\mathbb{K} \setminus T)$, the norm $\| \,.\, \|_D$ is multiplicative and actually equal to ${}_D\varphi_T$. Now let $s \ge N$. We have $\|(f_m)^T - (f_s)^T\|_D = \|\frac{(f_m)^T}{(f_s)^T} - 1\|_D$. But by Lemma B.20.5 we have $\frac{(f_m)^T}{(f_s)^T} = \left(\frac{f_m}{f_s}\right)^T$ and then by (3), in $R(D)$, we have $\left\|\left(\frac{f_m}{f_s}\right)^T - 1\right\|_D = \left\|\overline{\overline{\left(\frac{f_m}{f_s}\right)}}_T\right\|_D$. Finally, by Theorem B.6.1, we obtain

(6) $\|(f_m)^T - (f_s)^T\|_D \le \varepsilon$.

Relation (6) does not depend on the hole T and it shows that, for each fixed $n \in \mathbb{N}^*$, the sequence $((f_m)^{T_n})_{m\in\mathbb{N}}$ is a Cauchy sequence, which converges in $H(\mathbb{K} \setminus T_n)$, to an element whose index is equal to 0 and this convergence is uniform with respect to n. For each $n \in \mathbb{N}^*$, we put $\phi_n = \lim_{m\to\infty}(f_m)^{T_n}$. Then it is seen that $\prod_{n=1}^{\infty}\phi_n = \lim_{m\to\infty}\prod_{n=1}^{\infty}(f_m)^{T_n}$. As a consequence, the sequence $(f_m)^0$ is also convergent in $H(D)$ and actually in $H_b(\widetilde{D})$. Let ϕ_0 be its limit. Then we have the factorization $\phi = \prod_{n=0}^{\infty} \phi_n$. We recognize the Motzkin factorization for ϕ. Obviously, for each fixed $n > 0$, the equality satisfied by the $(f_m)^{T_n}$ holds for ϕ^{T_n} and shows that $\|\overline{\overline{(\phi)_{T_n}}}\|_D = \|(\phi)^{T_n} - 1\|_D$. In the same way, the equality satisfied by the $(f_m)^0$ shows that $\|\overline{\overline{(\phi)_0}} - 1\|_D = \|(\phi)^0 - 1\|_D$. This ends the proof of Proposition B.20.14. □

Theorem B.20.15 is given in [48] (see also [46, 47]).

Theorem B.20.15: *Let $a \in D$. Let $\phi \in H_b(D)$ be such that $|\phi(a)| \neq 0$. The statements* (i), (ii), and (iii) *are equivalent.*

(i) $\|\phi - \phi(a)\|_D < |\phi(a)|$.

(ii) *For every hole T, we have $\|\overline{\overline{\phi_T}}\|_D < |\phi(a)|$ and $\|\overline{\overline{\phi_0}} - \overline{\overline{\phi_0}}(a)\|_D < |\phi(a)|$.*

(iii) *ϕ is invertible, admits a Motzkin factorization and for every hole T, ϕ^T satisfies $\|\phi^T - 1\|_D < 1$ and ϕ^0 satisfies $\|\phi^0 - \phi^0(a)\|_D < |\phi(a)|$.*

Further, if statements i), ii), iii) *are satisfied then we have*

(u) $m(\phi, T) = 0$ *for every hole T.*

(v) $\|\overline{\overline{\phi}}_T\|_D = \|\phi^T - 1\|_D|\phi(a)|$ *for every hole T.*

(w) $\|\phi^0 - \phi^0(a)\|_D = \|\overline{\overline{\phi}}_0 - \overline{\overline{\phi}}_0(a)\|_D$.

Proof. Without loss of generality we may obviously assume $|\phi(a)| = 1$ and

(1) $|\phi(a) - 1| < 1$.

Let $(T_m)_{m\in I}$ be the ϕ-supersequence of D. We notice that when (i) is satisfied, ϕ is obviously invertible.

First we suppose (i) is satisfied and will show that so is (ii). By Theorem B.6.1, we have

(2) $\|\overline{\overline{(\phi - \phi(a))_{T_m}}}\|_D \leq \|\phi - \phi(a)\|_D$.

But it is seen that $\overline{\overline{(\phi - \phi(a))_{T_m}}} = \overline{\overline{\phi_{T_m}}}$. Hence, by (2) we have

(3) $\|\overline{\overline{\phi}}_{T_m}\|_D \leq \|\phi - \phi(a)\|_D < 1$, whenever $m \in I$.

In the same way, $\overline{\overline{(\phi - \phi(a))_0}} = \overline{\overline{\phi}}_0 - \phi(a)$ and then by Theorem B.6.1 we have

(4) $\|\overline{\overline{\phi}}_0 - \phi(a)\|_D < 1$.

Besides by (3) we see that $\|\sum_{m=1}^{\infty} \overline{\overline{\phi}}_{T_m}\|_D < 1$, hence $\|\phi - \overline{\overline{\phi}}_0\|_D = \|\sum_{m=1}^{\infty} \overline{\overline{\phi}}_{T_m}\|_D < 1$ and therefore $|\phi(a) - \overline{\overline{\phi}}_0(a)| < 1$, hence by (4) we see that

(5) $\|\overline{\overline{\phi}}_0 - \overline{\overline{\phi}}_0(a)\|_D < 1$.

Finally by (3) and (5), statement (ii) is clearly proven.

Now we will show that each of the statements (ii) and (iii) separately implies (i). We suppose (ii) satisfied. Hence, we have

(6) $\|\sum_{m\in I}\overline{\overline{\phi}}_{T_m}\|_D < 1.$

If D is bounded, by statement (ii) and by (6) we obtain (i). Now let D be not bounded. Then $\overline{\overline{\phi}}_0$ is a constant λ. Hence, ϕ is in the form $\lambda + \sum_{m=1}^{\infty}\overline{\overline{\phi}}_{T_m}$ with $\|\overline{\overline{\phi}}_{T_m}\|_D < 1$ whenever $m \geq 1$, hence

(7) $\|\sum_{m=1}^{\infty}\overline{\overline{\phi}}_{T_m}\|_D < 1.$

Now we have $\phi - \phi(a) = \sum_{m=0}^{\infty}\overline{\overline{\phi}}_{T_m} - \overline{\overline{\phi}}_{T_m}(a) = \sum_{m=1}^{\infty}(\overline{\overline{\phi}}_{T_m} - \overline{\overline{\phi}}_{T_m}(a))$. By (7) we see that $\|\sum_{m=1}^{\infty}(\overline{\overline{\phi}}_{T_m} - \overline{\overline{\phi}}_{T_m}(a))\|_D < 1$, hence finally (i) $\|\phi - \phi(a)\|_D < 1$.

We now suppose (iii) is satisfied. Hence, we have

(8) $\|\phi^{T_m} - 1\|_D < 1$ for all $m \in I$.

If D is bounded we have $\|\phi^0 - \phi^0(a)\|_D < 1$, hence by (8) we directly have (i). If D is not bounded, then ϕ^0 is a constant B such that $\phi(a) = B\prod_{m\in I}\phi^{T_m}(a)$, hence by (8) and (1) we see that $|B-1| < 1$, hence by (8) we obtain (i) again.

Thus, (i) is implied as well by (ii) as by (iii). Obviously by (1), (i) implies $\|\phi - 1\|_D < 1$ and therefore we may apply Proposition B.20.14. Next, we suppose that either (ii) or (iii) is satisfied. Hence, so is (i) and so are (u) and (v) by Proposition B.20.14.

Finally, we will show (w) and at the same time we will finish proving the equivalence between (ii) and (iii). Let $\psi = (\overline{\overline{\phi_0}}(a))^{-1}\phi$. We may apply Proposition B.20.14 to ψ and we have

(9) $\|\overline{\overline{\psi}}_0 - 1\|_D = \|\psi^0 - 1\|_D.$

But we have,

(10) $\|\psi^0 - 1\|_D \geq \|\psi^0 - \psi^0(a)\|_D = \|\phi^0 - \phi^0(a)\|_D$

(11) $\|\overline{\overline{\phi}}_0 - \overline{\overline{\phi}}_0(a)\|_D = \|\overline{\overline{\psi}}_0 - \overline{\overline{\psi}}_0(a)\|_D = \|\overline{\overline{\psi}}_0 - 1\|_D.$

Hence by (9), (10), and (11) we obtain

(12) $\|\phi^0 - \phi^0(a)\|_D \leq \|\overline{\overline{\phi}}_0 - \overline{\overline{\phi}}_0(a)\|_D.$

Now let $\gamma = \phi^0(a)$ and let $\chi = \gamma^{-1}\phi$. By (1) and (7), we see that $|\gamma - 1| < 1$, hence we may apply Proposition B.20.14 to χ and we have

(13) $\|\chi^0 - 1\|_D = \|\overline{\overline{\phi}}_0 - \overline{\overline{\phi}}_0(a)\|_D,$

whereas $\|\overline{\overline{\phi}}_0 - \overline{\overline{\phi}}_0(a)\|_D = \|\overline{\overline{\chi}}_0 - \overline{\overline{\chi}}_0(a)\|_D \leq \|\overline{\overline{\chi}}_0 - 1\|_D$ and $\|\chi^0 - 1\|_D = \|\chi^0 - \chi^0(a)\|_D = \|\phi^0 - \phi^0(a)\|_D$. Hence, by (13) we see that $\|\overline{\overline{\phi}}_0 - \overline{\overline{\phi}}_0(a)\|_D \leq \|\phi^0 - \phi^0(a)\|_D$ and therefore by (12) we obtain (w). This finishes proving the equivalence between (ii) and (iii) and ends the proof of Theorem B.20.15. □

Remark 2: If D is not bounded, as ϕ is bounded, both ϕ^0, $\overline{\overline{\phi_0}}$ are constant and therefore, the statements $\|\overline{\overline{\phi_0}} - \overline{\overline{\phi_0}}(a)\|_D < |\phi(a)|$ and $\|\phi^0 - \phi^0(a)\|_D < |\phi(a)|$ are automatically satisfied. Statement (ii) is then equivalent to

(ii') *For every hole T we have $\|\overline{\overline{\phi_T}}\|_D < |\phi(a)|$*

and statement (iii) is equivalent to

(iii') *ϕ is invertible and for every hole T, ϕ^T satisfies $\|\phi^T - 1\|_D < 1$.*

Theorem B.20.16: *Let $D \in$ Alg. Then f has Motzkin factorization if and only if it is semi-invertible.*

Proof. Without loss of generality we may assume the f-supersequence to be infinite. We denote it by $(T_n)_{n\in\mathbb{N}^*}$. Let f have Motzkin factorization

$$f^0\Big(\prod_{n=1}^{t} f^{T_n}\Big)\Big(\prod_{n=t+1}^{\infty} f^{T_n}\Big),$$

where the product $\Big(\prod_{n=t+1}^{\infty} f^{T_n}\Big)$ converges in $H(D)$. By definition, f^0 is semi-invertible in $H(\widetilde{D})$, hence in $H(D)$. Moreover, $\Big(\prod_{n=1}^{t} f^{T_n}\Big)\Big(\prod_{n=t+1}^{\infty} f^{T_n}\Big)$ is clearly invertible in $H(D)$. So f is semi-invertible.

Now, we suppose f to be semi-invertible and will show it to have Motzkin factorization. By Lemma B.20.9, we may clearly suppose that f is invertible without loss of generality.

First, we suppose that there exists M in $\mathbb{R}_+^*$ satisfying

(1) $M \leq |f(x)|$, whenever $x \in D$.

Let $h \in R(D)$ satisfy

(2) $\|f - h\|_D < \dfrac{M}{2}$,

and let $h^0 \prod_{n=1}^{N} h^{T_n}$ be the Motzkin factorization of h. For every $n = 1, \dots, N$, let $q_n = m(h, T_n)$, let $a_n \in T_n$, let $h_n = (x - a_n)^{-q_n} h^{T_n}$, and let $h_0 = h^0$. Let $u(x) = \prod_{n=1}^{N} (x - a_n)^{q_n}$ and let $l(x) = h^0 \prod_{n=1}^{N} h_n$. By (1) and (2), it is seen that h has no zero in D. Let $a \in D$. Then, by Theorem B.14.1, h^0 satisfies $\|h^0 - h^0(a)\|_D < |h^0(a)|$ and of course for every $n > 0$, h_n satisfy $\|h_n - 1\|_D < 1$. Hence, we have

(3) $\|l - l(a)\|_D < |l(a)|$. Let $b = |l(a)|$.

In particular, we have $|l(x)| = b$ whenever $x \in D$. Moreover, we notice that we have

(4) $\dfrac{M}{|l(a)|} \le |u(x)|$.

Let $F = \dfrac{f}{u}$. Then F does belong to $H_b(D)$. By (3) and (4), we check that $|F(x) - l(x)| < \dfrac{b}{2}$ and therefore by (3) again, we have $|F(x)| = b$ and $\|F - F(a)\|_D \le \dfrac{|F(a)|}{2}$. Now we can apply Theorem B.20.15 to F and then F has Motzkin factorization $F^0 \prod_{n=1}^{\infty} F^{T_n}$, with $m(F, T_n) = 0$ whenever $n > 0$. As a consequence f also has Motzkin factorization $\left(f^0 \prod_{n=1}^{N} f^{T_n}\right)\left(\prod_{n=N+1}^{\infty} f^{T_n}\right)$ with $f^0 = F^0$ and for each $n = 1, \dots, N$, $f^{T_n} = (x - a_n)^{q_n} F^{T_n}$ and finally for each $n > N$, $f^{T_n} = F^{T_n}$.

Now we suppose that $\inf\{|f(x)| \mid x \in D\} = 0$. Since D is closed and since f is invertible, we see that D is unbounded and that the element $G = \dfrac{1}{f}$ is not bounded in D. Hence, by Corollary B.2.7 there exists $q \in \mathbb{N}^*$ such that $\dfrac{x^{-q}}{f}$ has a non-zero limit when $|x|$ tends to $+\infty$, $(x \in D)$. Then it is easily seen that there exists $c > 0$ such that $|G(x)| \ge c$ for all $x \in D$. Indeed, on the one hand there exists r such that $|G(x)| \ge 1$ for all $x \in D \setminus d(0, r)$ and on the other hand f is bounded in $D \cap d(0, r)$, hence we can find $c \in]0, 1[$ such that $|G(x)| \ge c$ whenever $x \in D \setminus d(0, r)$. Thus, G admits Motzkin factorization and then by Lemma B.20.9 so does f. This ends the proof of the theorem. □

Remark 3: If a closed set B does not belong to Alg, there are counter-examples of invertible elements F, which admit certain Motzkin factor F^T such that $\dfrac{F}{F^T}$ does not belong to $H(B)$ (and obviously does not belong to $H(B \cup T)$). Indeed, don't let B belong to Alg. Since by hypothesis B is closed . We know that $\widetilde{B} \setminus \overline{B}$ is not bounded. Hence by Lemma B.20.6, there exists a quasi-minorated element $f \in H_b(B)$ satisfying

(1) $\lim_{\substack{|x| \to \infty \\ x \in B}} f(x) = 0$

and such that xf does not belong to $H(B)$. Since $f \in H_b(B)$, we can take it such that $\|f\|_B < 1$. Without loss of generality, we may assume that 0 belongs to a hole of B. Let $T = d(a, r^-)$ be another hole of B and let $F = \dfrac{x(1+f)}{(x-a)}$. Then it is seen that F belongs to $H_b(B)$ and is invertible in $H_b(B)$ because both $\dfrac{x}{(x-a)}$, $1+f$ are invertible in $H_b(B)$. Hence, F admits Motzkin factorization. In particular, we see that $F^T = \dfrac{1}{(x-a)}$. However, we check that $(x-a)F$ does not belong to $H(B)$ because $(x-a)F = x(1+f)$ and by hypothesis, xf does not belong to $H(B)$.

In the same way, let $G = \dfrac{1}{F}$. Since F is invertible in $H_b(B)$, so is G. But then we see that $\dfrac{1}{x-a}G$ does belong to $H_b(B)$ and has no zero in B, but obviously its inverse does not belong to $H(B)$. Therefore, $\dfrac{1}{x-a}G$ is not semi-invertible in $H(B)$. Thus, there exist invertible elements h, g in $H(B)$ such that hg is not semi-invertible, although it belongs to $H(B)$. This contradicts Theorem 1 in [72], which states that $\dfrac{f}{f^T}$ extends to an element of $H(D \cup T)$.

Theorem B.20.17: (K. Boussaf) *Let D belong to* Alg *and let $T = d(a, r^-)$ be a hole of D. Then f admits a Motzkin factor in the hole T if and only if ${}_D\varphi_{a,r}(f) \neq 0$.*

Proof. On the one hand, we suppose that f admits a Motzkin factor in the hole T. Let $f = gf^T$. Since g belongs to $H(D \cup T)$ and has no zero in T, of course, by Theorem B.13.16 we have ${}_D\varphi_{a,r}(g) \neq 0$. Next, as an invertible element of $H(D)$, it is seen that ${}_D\varphi_{a,r}(f^T) \neq 0$. Hence, ${}_D\varphi_{a,r}(f) \neq 0$.

On the other hand, we suppose ${}_D\varphi_{a,r}(f) \neq 0$. Let $\mathcal{F}$ be the circular filter of center a, of diameter r, and let $M = {}_D\varphi_{a,r}(f)$. There do exist $a_1, \dots, a_q \in d(a, r)$ and s, t satisfying $s < r < t$, such that $|f(x)| \geq M$ whenever $x \in D \bigcap \Big(\bigcap_{j=1}^{q} \Gamma(a_j, s, t) \Big)$.

Let $F = D \bigcap \Big(\bigcap_{j=1}^{q} \Gamma(a_j, s, t) \Big)$. Then T is clearly a hole of F. Next, the restriction g of f to F is invertible in $H(F)$ and therefore, by Theorem B.20.16, it admits a Motzkin factor g^T in the hole T. But then, $\dfrac{f}{g^T}$ belongs to $H(D)$ and to $H(F \cup T)$. Let $E = F \cup T$. Clearly, a hole of $D \cap E$ is either a hole of D included in E, or a hole of E. Hence, D and E are infraconnected sets that satisfy the hypothesis of Theorem B.6.10 and then $\dfrac{f}{g^T}$ belongs to $H(D \cup F) = H(D \cup T)$. Finally, as g^T is the Motzkin factor of g in T, $\dfrac{g}{g^T}$ has no zero inside T. This ends the proof. $\square$

Theorem B.20.18: *Let $D \in$ Alg and let $\mathcal{G}$ be the multiplicative group of the invertible elements in $H(D)$. Let $\mathcal{T}$ be the set of the holes of D. Let $\mathcal{G}^0$ be the* *subgroup of the elements invertible in* $H(\widetilde{D})$. *Let* $\mathcal{H} = \mathcal{G}^0 \prod_{T \in \mathcal{T}} \mathcal{G}^T$. *The product* $\mathcal{H}$ *is a direct product and is dense in* $\mathcal{G}$.

Proof. The product is direct because for each element, Motzkin factorization is unique. Thus, $\mathcal{H}$ is the set of the invertible elements whose Motzkin factorization is finite. Since every element of $\mathcal{G}$ has Motzkin factorization, it obviously belongs to the closure of $\mathcal{H}$. □

Thanks to the Motzkin factorization, the question on whether the n-th root of an analytic element is an analytic element appears to be linked to the number of zeros of each Motzkin factor. Several of these results were given in [58].

Theorem B.20.19: *Let* D *be a closed bounded infraconnected set and let* $f \in H(D)$ *be semi-invertible. Let* T *be a hole of* D. *We assume* f^s *to have continuation to an element of* $H(D \cup T)$ *for some* $s \in \mathbb{N}^*$. *Then the number of the zeros of* f^s *inside* T *is a multiple of* s *(taking multiplicities into account). Moreover, if* f *does not belong to* $H(D \cup T)$, *then the number of the zeros of* f^s *inside* T *is different from* 0.

Proof. By Lemma B.20.9, we have $(f^s)^T = (f^T)^s$. But as f^s belongs to $H(D \cup T)$, by Theorem B.20.4 $(f^s)^T$ is the polynomial of the zeros of f^s inside T. Let $Q = (f^s)^T$. Then we have $\deg(Q) = m((f^s), T) = sm(f, T)$. So s divides $\deg(Q)$. Now assume $\deg(Q) = 0$. We have $Q = 1$, $m(f, T) = 0$ and therefore $(f^T)^s = 1$ and $\lim_{|x| \to \infty} f^T(x) = 1$. Thus, f^T is just the constant 1 and therefore f belongs to $H(D \cup T)$, which ends the proof of Theorem B.20.19. □

In particular, Theorem B.20.19 applies to open disks.

Theorem B.20.20: *Let* $r \in |\mathbb{K}|$ *and let* $f \in H(d(0, r^-)) \setminus H(d(0, r))$ *satisfy* $f^s \in H(d(0, r))$. *Then* f *has continuation to an analytic element in a set* D *of the form* $d(0, r) \setminus \left(\bigcup_{i=1}^{t} d(a_i, r^-) \right)$ *with* $|a_i| = r = |a_i - a_j|$ *whenever* $i \neq j$, *such that for each* $i = 1, \ldots, t$ *the number of zeros of* f^s *in* $d(a_i, r^-)$ *is a multiple of* s *different from* 0.

Proof. The Mittag–Leffler series of f in $d(0, r^-)$ is of the form $\sum_{n=0}^{\infty} f_n$ with $f_0 = \overline{\overline{f_0}} \in H(d(0, r))$, and for every $n > 0$, $f_n = \overline{\overline{f_{T_n}}}$, with $T_n = d(a_n, r^-)$ and $|a_n - a_j| = |a_n| = r$ whenever $n \neq j$. Since $f \notin H(d(0, r))$, at least one of the f_n is different from 0. Let l be an integer such that $f_l \neq 0$. Now since f^s belongs to $H(d(0, r))$, by Theorem B.20.19, f^s has a number of zeros inside T_l, which is different from 0 and a multiple of s. Since any element of $H(d(0, r))$ has finitely many zeros in $d(0, r)$, we see that there are finitely many integers l such that $f_l \neq 0$. Let I be the finite set of the $l \in \mathbb{N}^*$ such that $f_l \neq 0$ and let $D = d(0,r) \setminus \Big(\bigcup_{l\in I} d(a_n, r^-)\Big)$. Then by definition, f belongs to $H(D)$ and for every $l \in I$, the number of zeros of f^s in T_n is different from 0 and a multiple of s. This ends the proof of Theorem B.20.20. □

Corollary B.20.21: *Let f be a power series whose radius of convergence is r though f does not belong to $H(d(0,r))$. If for some $s \in \mathbb{N}^*$, f^s has a radius of convergence r' strictly superior to r and if f^s has strictly less than s zeros inside $C(0,r)$ (taking multiplicities into account), then f does not belong to $H(d(0,r^-))$.*

Proof. Since $r' > r$ obviously we have $s > 1$. We assume that f belongs to $H(d(0,r^-))$ and therefore r must belong to $|\mathbb{K}|$. Since $r' > r$, f^s belongs to $H(d(0,r))$ and then by Theorem B.20.20, its number of zeros inside $C(0,r)$ is different from 0 and is a multiple of s, which contradicts the hypothesis. Hence, finally f does not belong to $H(d(0,r^-))$. □

We have now got to recall the definition of the function $\sqrt[q]{u}$ when $u \in d(1,1^-)$.

Definitions and notations: Henceforth and up to the end of the chapter, we suppose that $\mathbb{K}$ has characteristic zero and residue characteristic p. Let $q \in \mathbb{N}^*$. Let $(1+x)^q = 1 + qx + \sum_{j=2}^{q} b_j x^j$. So, $|q| = 1$ and $|b_j| \leq 1$ whenever $j = 2,\dots,q$. Recall that r_k was defined in Chapter A.6.

Theorem B.20.22: *Let $q \in \mathbb{N}^*$. If q is prime to p, the mapping $g_q(x) = (1+x)^q$ is injective in $d(1,1^-)$ and maps $d(1,1^-)$ onto $d(1,1^-)$. If $p \neq 0$ and if $q = p$, the mapping $g_p(x) = (1+x)^p$ is injective in $d(1,r_1^-)$ and maps $d(1,(r_1)^-)$ onto $d(1,(r_2)^-)$.*

Proof. Suppose first q prime to p. Since $|q| = 1 \geq |b_j|\ \forall j \geq 2$, by Corollaries B.13.10 and B.14.14, the mapping g_q defines a bijection from $d(0,1^-)$ onto $d(1,1^-)$.

Suppose now $p \neq 0$ and take $q = p$. By Theorems B.16.3 and B.16.5, inside $d(0,(r_1)^-)$ we can write $g_p(x) = exp(pLog(1+x))$. This way, we notice that when $x \in d(0,(r_1)^-)$, g_p is injective and that we have $|Log(1+x)| = |x|$ and $|exp(x)-1| = |x|$, hence

$$|(1+x)^p - 1| = |exp(pLog(1+x)) - 1| = \frac{x}{p}.$$

Consequently, the image of $d(0,r_1^-)$ by g_p is the disk $d\Big(1,\Big(\frac{r_1}{p}\Big)^-\Big) = d(1,(r_2)^-)$. □

Definitions and notations: Suppose $q \in \mathbb{N}^*$ prime to p. The mapping η_q defined in $d(1,1^-)$ by $\eta_q(u) = u^q$ is a bijection from $d(1,1^-)$ onto $d(1,1^-)$. We denote by

$\sqrt[q]{u}$ the inverse mapping from $d(1,1^-)$ onto $d(1,1^-)$ and we put $\phi_q(x) = \sqrt[q]{1+x}$ whenever $x \in d(0,1^-)$.

Suppose now $p \neq 0$ and take $q = p$. The mapping η_p defined in $d(1,(r_1)^-)$ by $\eta_p(u) = u^p$ is a bijection from $d(1,(r_1)^-)$ onto $d(1,(r_2)^-)$. So, we can denote by $\sqrt[p]{u}$ that inverse mapping from $d(1,(r_2)^-)$ onto $d(1,(r_1)^-)$ and we put $\phi_p(x) = \sqrt[p]{1+x}$ whenever $x \in d(0,(r_2)^-)$.

Theorem B.20.23: *Let $q \in \mathbb{N}^*$. If q is prime to p, ϕ_q belongs to $\mathcal{A}_b(d(0,1^-))$ but does not belong to $H(d(0,1^-))$. Next, suppose now $p > 0$. Then ϕ_p belongs to $\mathcal{A}_b(d(0,r_1^-))$ but does not belong to $H(d(0,r_1^-))$.*

Proof. Suppose first that q is prime to p. By construction and by Corollary B.14.15, ϕ_q belongs to $\mathcal{A}_b(d(0,1^-))$. Suppose that ϕ_q belongs to $H(d(0,1))$. Then it must satisfy $(\phi_q(-1))^q = 0$, hence $\phi_q(-1) = 0$ and therefore $\phi_q(x)^q$ admits a zero of order q at -1. But this contradicts the identity $(\phi_q(x))^q = 1+x$ and therefore $\phi_q(x)$ does not belong to $H(d(0,1))$. Finally, since $(\phi_q'x))^q$ has a unique zero in $C(0,1)$, by Corollary B.20.21 we see that ϕ_q does not belong to $H(d(0,1^-))$.

Suppose now $p > 0$. By Theorem B.20.22, the function $f(x) = (1+x)^p$ is strictly injective inside $d(0,r_1^-)$ and maps $d(0,r_1^-)$ onto itself. So by Corollary B.14.15, it admits an inverse mapping ϕ_p defined inside $d(0,r_1^-)$ that belongs to $\mathcal{A}_b(d(0,r_1^-))$.

Let us show that ϕ_p does not belong to $H(d(0,r_1^-))$. Indeed, suppose ϕ_p belongs to $H(d(0,r_1^-))$. Consider the Mittag–Leffler decomposition of ϕ_p on the infraconnected set $d(0,r_1^-)$. It is of the form $\sum_{n=0}^{\infty} g_n$ with $g_0 \in H(d(0,r_1)$ and $g_n \in H_0(\mathbb{K} \setminus d(a_n,r_n^-))$ with $a_n \in C(0,r_1)$. Now, by Theorem A.7.4, we can consider a complete algebraically closed extension $\widehat{\mathbb{K}}$ of $\mathbb{K}$ whose residue class field is not countable.

Let χ be the residue class field of $\mathbb{K}$. Thus, we can find a class G of $\widehat{C}(0,r_1)$ that has an empty intersection with $\bigcup_{n=1}^{\infty} d(a_n,r_1^-)$ and then ϕ_p has continuation to an element of $H(d(0,r_1^-)\cup G$. Let $c \in G$. Since $|c| = r_1$, the function $h(x) = (1+x)^p - c$ satisfies $\nu^-(h,\log(r_1) = 0$, $\nu^+(h,\log(r_1) > 1$, hence h admits a zero $a \in \widehat{C}(0,r_1)$. Then of course a does not belong to $\mathbb{K}$. Now, let ζ be a p-th root of 1 different from 1 and let $t = \zeta(1+a)$. By Corollary A.6.7, we have $|\zeta - 1| = r_1$, hence t is of the form $1+b$ with $b \in \widehat{C}(0,r_1)$. We then have $(1+a)^p = (1+b)^p$. Set $E = \widehat{d}(a,r_1^-)$, $F = \widehat{d}(b,r_1^-)$, $D' = d(0,r_1^-) \cup E \cup F$, and $D'' = d(0,r_1^-) \cup G$. Since the image of $d(0,r_1^-)$ by the function f is $d(0,r_1^-)$ we can check that for each $u \in \widehat{C}(0,r_1)$, the image of $d(u,r_1^-)$ by f is $d(1,r_1^-)$. Consequently, both images of E and F by f are equal to G. Now, since f belongs to $\widehat{H}(D')$, the image of D' by f is D'' and ϕ_p belongs to $\widehat{H}(D'')$. Consequently, by Corollary B.3.3 $\phi_p \circ f$ belongs to $\widehat{H}(D')$ and we have $\phi_p \circ f(x) = 1+x \ \forall x \in d(0,r_1^-)$. But since D' has no pierced filter, by Corollary B.12.4 it is an analytic set. Consequently, the equality

$\phi_p \circ f(x) = 1 + x \ \forall x \in d(0, r_1^-)$ holds in all D', a contradiction since f is not injective in D'. That finishes showing that ϕ_p does not belong to $H(d(0, r_1^-))$. □

Remark 4: Let q be prime to p. Since ϕ_q belongs to $\mathcal{A}_b(d(0, 1^-))$, obviously ϕ_q belongs to $H(d(0, r))$, whenever $r \in]0, 1[$. Now, let E be a closed bounded set in $\mathbb{K}$ and let $h \in H(E)$ satisfy $\|h\|_E < 1$. Then by Corollary B.3.3, $\phi_q \circ h$ belongs to $H(E)$. In other words, if $g \in H(E)$ and if $\|g - 1\|_E < 1$, then $\sqrt[q]{g}$ also belongs to $H(E)$.

Remark 5: Theorems B.20.19 and B.20.20 couldn't be significantly improved as this example shows. Let q be an integer prime to p, let $a, b \in d(1, 1^-)$, with $a \neq b$, and let $P(x) = (a - x)^{q-1}(b - x)$. It is easily seen that $|P(x) - 1| < 1$ whenever $x \in d(0, 1^-)$ and then we can consider $f(x) = \sqrt[q]{P(x)}$. We will show that $f \in H(d(0, 1^-)) \setminus H(d(0, 1))$. Indeed we have

$$\begin{aligned} f(x) &= \sqrt[q]{(a-x)^q\Big(\frac{b-x}{a-x}\Big)} = (a-x)\sqrt[q]{1 + \Big(\frac{b-a}{a-x}\Big)} \\ &= (a-x)\sum_{n=0}^{\infty}\binom{\frac{1}{q}}{n}\Big(\frac{b-a}{a-x}\Big)^n = a - x + \frac{1}{q}(b-a) + \sum_{j=1}^{\infty}\binom{\frac{1}{q}}{j+1}\frac{(b-a)^{j+1}}{(a-x)^j}. \end{aligned}$$

This is just a Mittag–Leffler series of the form $f_0 + f_1 \in H(d(0, 1^-))$, with

$$f_0 = -x + a + \frac{1}{q}(b-a) \in H(d(0,1)),$$

$$f_1 = -\sum_{j=1}^{\infty}\binom{\frac{1}{q}}{j+1}\frac{(a-b)^{j+1}}{(x-a)^j} \in H_0(\mathbb{K} \setminus d(1, 1^-)).$$

Thus, we see that f belongs to $H(d(0, 1^-))$ and more precisely $f \in H(\mathbb{K} \setminus d(1, 1^-))$, but $f \notin H(d(0, 1))$. Actually, f^q has exactly q zeros in $d(1, 1^-)$.

B.21. Order of growth for entire functions

Here, we mean to introduce and study the notion of order of growth of an entire function on $\mathbb{K}$ in relation with the distribution of zeros in disks and in relation with the question whether an entire function can be divided by its derivative inside the algebra of entire functions. Results were published in [20].

Definitions and notations: Let $f \in \mathcal{A}(\mathbb{K})$. Similarly to the definition known on complex entire functions, $\limsup_{r \to +\infty} \dfrac{\log(\log(|f|(r)))}{\log(r)}$ is called *the order of growth of f* or *the growth order of f* in brief and is denoted by $\rho(f)$. We say that f has *finite order* if $\rho(f) < +\infty$. In this chapter and in Chapters B.22 and B.23, for simplicity in certain calculations, we put $\theta = e$, i.e., we denote by log the Neperian logarithm function. However, we can check that the definitions do not depend on the basis $b > 1$ of the logarithm.

Theorem B.21.1 is easily proven.

Theorem B.21.1: *Let $f,\ g \in \mathcal{A}(\mathbb{K})$. Then*

(i) *if $c(|f|(r))^{\alpha} \geq |g|(r)$ with α and $c > 0$, when r is big enough, then $\rho(f) \geq \rho(g)$,*
(ii) *$\rho(f+g) \leq \max(\rho(f), \rho(g))$ and if $\rho(g) < \rho(f)$, then $\rho(f+g) = \rho(f)$,*
(iii) *$\rho(fg) = \max(\rho(f), \rho(g))$.*

Proof. Suppose $c(|f|(r))^{\alpha} \geq |g|(r)$ with α and $c > 0$ when r is big enough. Then we check that $\left(\limsup_{r\to+\infty} \dfrac{\log(\log(c(|f|(r))^{\alpha}))}{\log(r)}\right) \geq \limsup_{r\to+\infty} \dfrac{\log(\log(|g|(r)))}{\log(r)}$, hence $\rho(f) \geq \rho(g)$.

Next, we have $|f+g|(r) \leq \max(|f|(r), |g|(r))$, hence

$$\frac{\log(\log(|f+g|(r)))}{\log(r)} \leq \max\left(\frac{\log(\log(|f|(r)))}{\log(r)}, \frac{\log(\log(|g|(r)))}{\log(r)}\right)$$

and hence $\rho(f+g) \leq \max(\rho(f), \rho(g))$.

Now, suppose $\rho(f) > \rho(g)$. Then when r is big enough, we have $|f|(r) > |g|(r)$, hence $|f+g|(r) = |f|(r)$ and therefore $\rho(f+g) = \rho(f)$.

Let us now show that $\rho(fg) = \max(\rho(f), \rho(g))$. Since $|h|(r)$ tends to $+\infty$ with r for every $h \in \mathcal{A}(\mathbb{K})$ and since, by Corollary B.5.9, $|\ .\ |(r)$ is an absolute value on $\mathcal{A}(\mathbb{K})$, we have

$$\max(|f|(r), |g|(r)) \leq |fg|(r) = |f|(r)|g|(r) \leq (\max(|f|(r), |g|(r)))^2,$$

hence

$$\log(\max(|f|(r), |g|(r))) \leq \log(|fg|(r)) \leq \log(\max(|f|(r), |g|(r))) + \log(2)$$

and therefore we can easily conclude that $\rho(fg) = \max(\rho(f), \rho(g))$. □

Corollary B.21.2: *Let $f,\ g \in \mathcal{A}(\mathbb{K})$. Then $\rho(f^n) = \rho(f)\ \forall n \in \mathbb{N}^*$.*

Definitions and notations: Given a number $t \geq 0$, we will denote by $\mathcal{A}(\mathbb{K}, t)$ the $\mathbb{K}$-algebra of entire functions of order inferior or equal to t and we put $\mathcal{A}^*(\mathbb{K}) = \bigcup_{t>0} \mathcal{A}(\mathbb{K}, t)$.

Corollary B.21.3: *For any $t \geq 0$, $\mathcal{A}(\mathbb{K}, t)$ is a $\mathbb{K}$-subalgebra of $\mathcal{A}(\mathbb{K})$ and so is $\mathcal{A}^0(\mathbb{K})$.*

Corollary B.21.4: *Consider the differential equation*

$$(\mathcal{E})\ f^{(n)} + a_{n-1}(x)f^{(n-1)}(x) + \cdots + a_0(x)f(x) = 0$$

with $a_j \in \mathcal{A}^*(D)$, $j = 0, \ldots, n-1$ and $\rho(a_j) < \rho(a_0)\ \forall j = 1, \ldots, n-1$. Then every non-trivial solution f of $(\mathcal{E})$ satisfies $\rho(f) \geq \rho(a_0)$.

Theorem B.21.5: *Let $f \in \mathcal{A}(\mathbb{K})$ and let $P \in \mathbb{K}[x]$. Then $\rho(P \circ f) = \rho(f)$ and $\rho(f \circ P) = \deg(P)\rho(f)$.*

Proof. Let $n = \deg(P)$. For r big enough, we have

$$\begin{aligned}\log(\log(|f|(r))) &\le \log(\log(|P \circ f|(r))) \le \log((n+1)\log(|f|(r))) \\ &= \log(n+1) + \log(\log(|f|(r))).\end{aligned}$$

Consequently,

$$\begin{aligned}\limsup_{r \to +\infty}\left(\frac{\log(\log(|f|(r)))}{\log(r)}\right) &\le \limsup_{r \to +\infty}\left(\frac{\log(\log(|P \circ f|(r)))}{\log(r)}\right) \\ &\le \limsup_{r \to +\infty}\left(\frac{\log(n+1) + \log(\log(|f|(r)))}{\log(r)}\right)\end{aligned}$$

and therefore $\rho(P \circ f) = \rho(f)$.

Next, for r big enough, we have

$$\frac{\log(\log(|f|(r)))}{\log(r)} \le \frac{\log(\log(|f \circ P|(r)))}{\log(r)} = \Big(\frac{\log(\log(|f \circ P|(r)))}{\log(|P|(r))}\Big)\Big(\frac{\log(|P|(r))}{\log(r)}\Big).$$

Now,

$$\limsup_{r \to +\infty}\Big(\frac{\log(\log(|f \circ P|(r)))}{\log(|P|(r))}\Big) = \limsup_{r \to +\infty}\Big(\frac{\log(\log(|f|(r)))}{\log(r)}\Big),$$

because the function h defined in $[0, +\infty[$ as $h(r) = |P|(r)$ is obviously an increasing continuous bijection from $[0, +\infty[$ onto $[|P(0)|, +\infty[$. On the other hand, it is obviously seen that $\limsup_{r \to +\infty}\Big(\frac{\log(|P|(r))}{\log(r)}\Big) = n$. Consequently,

$$\limsup_{r \to +\infty}\Big(\frac{\log(\log(|f \circ P|(r)))}{\log(|P|(r))}\Big) = n \limsup_{r \to +\infty}\Big(\frac{\log(\log(|f|(r)))}{\log(r)}\Big),$$

and hence $\rho(f \circ P) = n\rho(f)$. □

Theorem B.21.6: *Let f, $g \in \mathcal{A}(\mathbb{K})$ be transcendental. Then $\rho(f \circ g) \ge \max(\rho(f), \rho(g))$. If $\rho(f) \neq 0$, then $\rho(f \circ g) = +\infty$.*

Proof. Let $f(x) = \sum_{j=0}^{\infty} a_n x^n$ and $g(x) = \sum_{j=0}^{\infty} b_n x^n$. Since g is transcendental, for every $n \in \mathbb{N}$, there exists r_n such that $\zeta(r_n, g) \ge n$. Then $|g|(r) \ge |b_n| r^n \ \forall r \ge r_n$ and hence, by Theorem B.21.5, we have

$$\rho(f \circ g) \ge n\rho(f). \tag{1}$$

Therefore, $\rho(f \circ g) \ge \rho(f)$.

Now, let $k \in \mathbb{N}$ be such that $a_k \neq 0$ and let s_0 be such that $\zeta(s_0, f) \ge k$. Then, $|f|(r) \ge |a_k| r^k \ \forall r \ge s_0$, hence $|f \circ g|(r) \ge |a_k|(|g|(r))^k \ \forall r \ge s_0$, hence by

Theorems B.21.1 and B.21.5 we have $\rho(f \circ g) \geq \rho(g)$. Next, Relation (1) is true for every $n \in \mathbb{N}$. Suppose now that $\rho(f) \neq 0$. Then by (1) we have $\rho(f \circ g) = +\infty$. □

Definitions and notations: Let $f \in \mathcal{A}(d(0, R^-))$. For each $r \in]0, R[$, we denote by $\zeta(r, f)$ the number of zeros of f in $d(0, r)$, taking multiplicity into account and set $\xi(r, f) = \zeta(r, \frac{1}{f})$.

Theorem B.21.7: *Let $f \in \mathcal{A}(\mathbb{K})$ be not identically zero and such that for some $t \geq 0$, $\limsup_{r \to +\infty} \dfrac{\zeta(r, f)}{r^t}$ is finite. Then $\rho(f) \leq t$.*

Proof. Set $\limsup_{r \to +\infty} \dfrac{\zeta(r, f)}{r^t} = b \in [0, +\infty[$. Let us fix $\epsilon > 0$. We can find $R > 0$ such that $\dfrac{\zeta(r, f)}{r^t} \leq b + \epsilon \ \forall r \geq R$ and hence, by Corollary B.13.30, we have $\dfrac{|f|(r)}{|f|(R)} \leq \left(\dfrac{r}{R}\right)^{\zeta(r,f)} \leq \left(\dfrac{r}{R}\right)^{r^t(b+\epsilon))}$. Therefore, putting $M = |f|(R)$, we have

$$\log(|f|(r)) \leq \log(M) + r^t(b + \epsilon)(\log(r) - \log(R)).$$

Now, when $u > 2$, $v > 2$, we know that $\log(u+v) \leq \log(u) + \log(v)$. Applying that inequality with $u = M$ and $v = r^t(b+\epsilon)(\log(r) - \log(R))$ when $r^t(b+\epsilon)(\log(r) - \log(R)) > 2$, since $\log R \geq 0$, which yields,

$$\begin{aligned}\log(\log(|f|(r))) &\leq \log(\log(M)) + t\log(r) + \log(b+\epsilon) + \log(\log(r) - \log R)) \\ &\leq \log(\log(M)) + t\log(r) + \log(b+\epsilon) + \log(\log(r)).\end{aligned}$$

Consequently,

$$\frac{\log(\log(|f|(r)))}{\log(r)} \leq \frac{\log(\log(M)) + t\log(r) + \log(b+\epsilon) + \log(\log(r))}{\log(r)},$$

and hence we can check that

$$\limsup_{r \to +\infty} \frac{\log(\log(|f|(r)))}{\log(r)} \leq t.$$

□

Theorem B.21.8: *Let $f \in \mathcal{A}(\mathbb{K})$ be not identically zero. If there exists $s \geq 0$ such that*

$$\limsup_{r \to +\infty} \left(\frac{\zeta(f, r)}{r^s}\right) < +\infty,$$

then $\rho(f)$ is the lowest bound of the set of $s \in [0, +\infty[$ such that

$$\limsup_{r \to +\infty} \left(\frac{\zeta(f, r)}{r^s}\right) = 0.$$

Moreover, if $\limsup_{r \to +\infty} \left(\dfrac{\zeta(f, r)}{r^t}\right)$ is a number $b \in]0, +\infty[$, then $\rho(f) = t$.

If there exists no s such that $\limsup_{r \to +\infty} \left(\dfrac{\zeta(f, r)}{r^s}\right) < +\infty$, then $\rho(f) = +\infty$.

Proof. The proof holds in two statements. First we will prove that given $f \in \mathcal{A}(\mathbb{K})$ non-constant and such that for some $t \geq 0$, $\limsup\limits_{r \to +\infty} \dfrac{\zeta(f,r)}{r^t}$ is finite, then $\rho(f) \leq t$.

Set $\limsup\limits_{r \to +\infty} \left(\dfrac{\zeta(f,r)}{r^t}\right) = b \in [0, +\infty[$. Let us fix $\epsilon > 0$. We can find $R > 1$ such that $|f|(R) > e^2$ and $\dfrac{\zeta(f,r)}{r^t} \leq b + \epsilon \ \forall r \geq R$ and hence, by Corollary B.13.30, we have $\dfrac{|f|(r)}{|f|(R)} \leq \left(\dfrac{r}{R}\right)^{\zeta(f,r)} \leq \left(\dfrac{r}{R}\right)^{r^t(b+\epsilon))}$. Therefore, since $R > 1$, we have

$$\log(|f|(r)) \leq \log(|f|(R)) + r^t(b+\epsilon)(\log(r)).$$

Now, when $u > 2$, $v > 2$, we check that $\log(u+v) \leq \log(u) + \log(v)$. Applying that inequality with $u = \log(|f|(R))$ and $v = r^t(b+\epsilon)(\log(r))$ when $r^t(b+\epsilon)(\log(r)) > 2$, that yields

$$\log(\log(|f|(r))) \leq \log(\log(|f|(R))) + t\log(r) + \log(b+\epsilon) + \log(\log(r)).$$

Consequently,

$$\frac{\log(\log(|f|(r)))}{\log(r)} \leq \frac{\log(\log(|f|(R))) + t\log(r) + \log(b+\epsilon) + \log(\log(r))}{\log(r)},$$

and hence we can check that

$$\limsup_{r \to +\infty} \frac{\log(\log(|f|(r)))}{\log(r)} \leq t,$$

which proves the first claim.

Second, we will prove that given $f \in \mathcal{A}(\mathbb{K})$ not identically zero and such that for some $t \geq 0$, we have $\limsup\limits_{r \to +\infty} \dfrac{\zeta(f,r)}{r^t} > 0$, then $\rho(f) \geq t$.

By hypotheses, there exists a sequence $(r_n)_{n \in \mathbb{N}}$ such that $\lim_{n \to +\infty} r_n = +\infty$ and such that $\lim\limits_{n \to +\infty} \dfrac{\zeta(f,r_n)}{r_n^t} > 0$. Thus, there exists $b > 0$ such that $\lim\limits_{n \to +\infty} \dfrac{\zeta(f,r_n)}{r_n^t} \geq b$. We can assume that $|f|(r_0) \geq 1$, hence by Corollary B.13.27, $|f|(r_n) \geq 1 \ \forall n$. Let $\lambda \in]1, +\infty[$. By Corollary B.13.30, we have

$$\frac{|f|(\lambda r_n)}{|f|(r_n)} \geq (\lambda)^{\zeta(f,r_n)} \geq (\lambda)^{[b(r_n)^t]},$$

hence

$$\log(|f|(\lambda r_n) \geq \log(|f|(r_n)) + b(r_n)^t \log(\lambda).$$

Since $|f|(r_n) \geq 1$, we have $\log(\log(|f|(\lambda r_n))) \geq \log(b\log(\lambda)) + t\log(r_n)$, therefore

$$\frac{\log(\log(|f|(\lambda r_n)))}{\log(r_n)} \geq t + \frac{\log(b\log(\lambda))}{\log(r_n)} \quad \forall n \in \mathbb{N},$$

and hence

$$\limsup_{r\to+\infty} \frac{\log(\log(|f|(r)))}{\log(r)} \geq t,$$

which ends the proof of the second claim.

Definitions and notations: Let $t \in [0, +\infty[$ and let $f \in \mathcal{A}(\mathbb{K})$ of order t. We set $\psi(f) = \limsup\limits_{r\to+\infty} \frac{\zeta(r, f)}{r^t}$ and call $\psi(f)$ the *cotype of growth* of f, or just the *cotype* of f in brief. □

Lemma B.21.9: *Let $f,\ g \in \mathcal{A}(\mathbb{K})$. Then* $\max\big(\zeta(r, f), \zeta(r, g)\big) \leq \zeta(r, fg) = \zeta(r, f) + \zeta(r, g)$.

Proof. Let $r' > r$ be such that both $\zeta(u, f)$ and $\zeta(u, g)$ are constant in $[r, r']$. Then, when $\mu = \log(u)$, by Corollary B.10.3 we have

$$\zeta(u, f) = \frac{d\Psi(f, \mu)}{d\mu},\ \zeta(u, g) = \frac{d\Psi(g, \mu)}{d\mu}\ \forall \mu \in]\log(r), \log(r')[.$$

But since $\Psi(fg, \mu) = \Psi(f, \mu) + \Psi(g, \mu)$, the inequalities

$$\max\big(\zeta(r, f), \zeta(r, g)\big) \leq \zeta(r, fg) = \zeta(r, f) + \zeta(r, g)$$

are clear. □

Theorem B.21.10: *Let $f,\ g \in \mathcal{A}^*(\mathbb{K})$. Then $\psi(fg) \leq \psi(f) + \psi(g)$. Moreover, if $\rho(f) \geq \rho(g)$, then $\psi(f) \leq \psi(fg)$. If $\rho(f) = \rho(g)$, then* $\max\big(\psi(f), \psi(g)\big) \leq \psi(fg)$.

Proof. Set $s = \rho(f)$, $t = \rho(g)$, and suppose $s \geq t$. By Theorem B.21.1, we have $\rho(f.g) = \rho(f) = s$. By Lemma B.21.9, for each $r > 0$, we have $\max\big(\zeta(r, f), \zeta(r, g)\big) \leq \zeta(r, f.g) = \zeta(r, f) + \zeta(r, g)$. Consequently,

$$\limsup_{r\to+\infty} \frac{\zeta(r, f.g)}{r^s} \leq \limsup_{r\to+\infty} \frac{\zeta(r, f)}{r^s} + \limsup_{r\to+\infty} \frac{\zeta(r, g)}{r^s} \leq \limsup_{r\to+\infty} \frac{\zeta(r, f)}{r^s} + \limsup_{r\to+\infty} \frac{\zeta(r, g)}{r^t} = \psi(f) + \psi(g).$$

Moreover, assuming again $s \geq t$, then

$$\psi(f) = \limsup_{r\to+\infty} \frac{\zeta(r, f)}{r^s} \leq \limsup_{r\to+\infty} \frac{\zeta(r, fg)}{r^s} = \limsup_{r\to+\infty} \frac{\zeta(r, fg)}{r^{\rho(fg)}} = \psi(fg).$$

Consequently, if $\rho(f) = \rho(g)$, then $\max\big(\psi(f), \psi(g)\big) \leq \psi(fg)$. □

Theorem B.21.11: *Let* $f(x) = \sum_{n=0}^{+\infty} a_n x^n \in \mathcal{A}(\mathbb{K})$. *Then*

$$\rho(f) = \limsup_{n \to +\infty} \Big(\frac{n \log(n)}{-\log |a_n|}\Big).$$

Proof. We will follow a similar way as this of [86] when $\rho(f) < +\infty$.

Let $t = \rho(f)$ and suppose first that $t < +\infty$. Let $\alpha = \limsup_{n \to +\infty} \frac{n \log(n)}{-\log |a_n|}$. Take $s > t$. For all $n \in \mathbb{N}$, we have $|a_n| r^n \leq |f|(r)$ and therefore $|a_n| r^n \leq e^{(r^s)}$ and hence $|a_n| \leq r^{-n} e^{(r^s)}$, i.e.,

$$\log |a_n| \leq r^s - n \log(r) \tag{1}$$

when r is big enough.

Now, choose $r = \left(\frac{n}{s}\right)^{\frac{1}{s}}$. So, we have $\log |a_n| \leq \frac{n}{s} - \frac{n}{s} \log(\frac{n}{s})$, i.e., $-\log(|a_n|) \geq -\frac{n}{s} + \frac{n}{s} \log\left(\frac{n}{s}\right)$.

Consequently, when n is big enough we have

$$\frac{n \log n}{(-\log |a_n|)} \leq \frac{n \log n}{\frac{n}{s} \log(\frac{n}{s}) - \frac{n}{s}} \leq s + O(1).$$

Therefore, we have $\alpha \leq s$ and since this is true for each $s > t$, that shows that $\alpha \leq t$.

Now, take $\beta > \alpha$ so that $\frac{n \log n}{(-\log |a_n|)} < \beta$ for n big enough. Then, when n is big enough, we have $n \log(n) \leq \beta(-\log |a_n|)$, hence $n^{\frac{n}{\beta}} \leq \frac{1}{|a_n|}$ and hence $|a_n| \leq \frac{1}{n^{\frac{n}{\beta}}}$. Consequently, $|a_n| r^n \leq \frac{r^n}{n^{\frac{n}{\beta}}}$. Now, for r big enough, $|f|(r) = \sup_{n \in \mathbb{N}} |a_n| r^n \leq \sup_{n \in \mathbb{N}} \frac{r^n}{n^{\frac{n}{\beta}}}$.

Putting $\varphi(n) = \frac{n}{\beta}$ and $R = \frac{r}{\beta}$, we have

$$|f|(r^{\frac{1}{\beta}}) \leq \sup_{n \in \mathbb{N}} \frac{r^{\varphi(n)}}{n^{\varphi(n)}} \leq \sup_{x > 0} \frac{R^x}{x^x}.$$

Now we check that the maximum on $[0, +\infty[$ of the function $g(x) = \frac{R^x}{x^x}$ is reached when $x = \frac{R}{e}$ and hence is $e^{\frac{R}{e}} = e^{\frac{r}{\beta e}}$. Therefore, we have $|f|(r^{\frac{1}{\beta}}) \leq e^{\frac{r}{\beta e}}$. Putting now $u = r^{\frac{1}{\beta}}$, we can derive $|f|(u) \leq e^{\frac{u^\beta}{\beta e}}$, hence

$$\log(\log(|f|(u))) \leq \beta \log(u) - \log(e\beta).$$

Consequently,

$$\limsup_{r \to +\infty} \frac{\log(\log(|f|(r)))}{\log(r)} \leq \beta.$$

So we have $t \leq \beta$ and since this is true for all $\beta > \alpha$, we have proven that $t \leq \alpha$, which ends the proof when $t < +\infty$.

Suppose now that $t = +\infty$ and suppose that $\limsup_{n\to+\infty} \frac{n\log n}{(-\log|a_n|)} < +\infty$. Let us take $s \in \mathbb{N}$ such that

$$\frac{n\log n}{(-\log|a_n|)} < s \ \forall n \in \mathbb{N}. \tag{2}$$

By Theorem B.21.8, we have $\limsup_{r\to+\infty} \frac{\zeta(f,r)}{r^s} = +\infty$. So, we can take a sequence $(r_m)_{m\in\mathbb{N}}$ such that

$$\lim_{m\to+\infty} \frac{\zeta(f,r_m)}{(r_m)^s} = +\infty. \tag{3}$$

For simplicity, set $u_m = \zeta(f,r_m)$, $m \in \mathbb{N}$. By (2), for m big enough we have

$$u_m \log(u_m) < s(-\log(|a_{u_m}|) = s\log\left(\frac{1}{|a_{u_m}|}\right),$$

hence

$$\frac{1}{(u_m)^{u_m}} > |a_{u_m}|^s,$$

therefore

$$|a_{u_m}|^s (r_m)^{su_m} < \frac{(r_m)^{su_m}}{(u_m)^{u_m}},$$

i.e.,

$$(|f|(r_m))^s < \left(\frac{(r_m)^s}{u_m}\right)^{u_m}.$$

But by Corollary B.13.27, we have $\lim_{r\to+\infty} |f|(r_m) = +\infty$, hence $(r_m)^s > u_m$ when m is big enough and therefore $\limsup_{m\to+\infty} \frac{\zeta(f,r_m)}{(r_m)^s} \leq 1$, a contradiction to (3). Consequently, (2) is impossible and therefore

$$\limsup_{n\to+\infty}\left(\frac{n\log(n)}{-\log|a_n|}\right) = +\infty = \rho(f).$$

□

Remark: Of course, polynomials have a growth order equal to 0. On $\mathbb{K}$ as on $\mathbb{C}$ we can easily construct transcendental entire functions of order 0 or of order ∞.

Example 1: Suppose that for each $r > 0$, we have $\zeta(r,f) \in [r^t \log r, r^t \log r + 1]$. Then of course, for every $s > t$, we have $\limsup_{r\to+\infty} \frac{\zeta(r,f)}{r^s} = 0$ and $\limsup_{r\to+\infty} \frac{\zeta(r,f)}{r^t} = +\infty$, so there exists no $t > 0$ such that $\frac{\zeta(r,f)}{r^t}$ have non-zero superior limit $b < +\infty$. Consequently, $\rho(f) = +\infty$.

Example 2: Let $(a_n)_{n\in\mathbb{N}}$ be a sequence in $\mathbb{K}$ such that $-\log|a_n| \in [n(\log n)^2, n(\log n)^2+1]$. Then clearly, $\lim\limits_{n\to+\infty} \dfrac{\log|a_n|}{n} = -\infty$, hence the function $\sum\limits_{n=0}^{\infty} a_n x^n$ has radius of convergence equal to $+\infty$. On the other hand, $\lim\limits_{n\to+\infty} \dfrac{n\log n}{-\log|a_n|} = 0$, hence $\rho(f) = 0$.

Example 3: Let $(a_n)_{n\in\mathbb{N}}$ be a sequence in $\mathbb{K}$ such that $-\log|a_n| \in [n\sqrt{\log n}, n\sqrt{\log n}+1]$. Then $\lim\limits_{n\to+\infty} \dfrac{\log|a_n|}{n} = -\infty$ again and hence the function $\sum\limits_{n=0}^{\infty} a_n x^n$ has radius of convergence equal to $+\infty$. On the other hand, $\lim\limits_{n\to+\infty} \Big(\dfrac{n\log n}{-\log|a_n|}\Big) = +\infty$ hence $\rho(f) = +\infty$.

Theorem B.21.12: *Let $f \in \mathcal{A}^*(\mathbb{K})$. Then*

$$\rho(f) = \inf\left\{ s \in]0,+\infty[\ \Big|\ \lim_{r\to+\infty} \frac{\log(|f|(r))}{r^s} = 0 \right\}.$$

Proof. Indeed, let $M = \inf\{s \in]0,+\infty[\ |\ \lim_{r\to+\infty} \frac{\log(|f|(r))}{r^s} = 0\}$. First we will prove that $\rho(f) \le M$. Let s be such that $\lim_{r\to+\infty} \frac{\log(|f|(r))}{r^s} = 0$. Let us fix $\epsilon > 0$. For r big enough, we have $\dfrac{\log(|f|(r))}{r^s} \le \epsilon$, hence $\log(|f|(r)) \le \epsilon r^s$, therefore $\log(\log(|f|(r))) \le \log\epsilon + s\log(r)$, hence $\dfrac{\log(\log(|f|(r)))}{\log(r)} \le s + \dfrac{\epsilon}{\log(r)}$. This is true for every $\epsilon > 0$, therefore $\limsup\limits_{r\to+\infty} \dfrac{\log(\log(|f|(r)))}{\log(r)} \le s$, i.e., $\rho(f) \le s$ and hence, $\rho(f) \le M$.

On the other hand, we notice that

$$M = \sup\{s \in]0,+\infty[\ |\ \limsup_{r\to+\infty} \frac{\log(|f|(r))}{r^s} > 0\}.$$

Now, suppose that for some $s > 0$, we have $\limsup\limits_{r\to+\infty} \dfrac{\log(|f|(r))}{r^s} = b > 0$. Let us fix $\epsilon \in]0,b[$. There exists a sequence $(r_n)_{n\in\mathbb{N}}$ such that, when n is big enough, we have $b - \epsilon \le \dfrac{\log(|f|(r_n))}{(r_n)^s} \le b + \epsilon$, hence $s\log(r_n) + \log(b-\epsilon) < \log(\log(|f|(r_n))) <$ $s\log(r_n) + \log(b-\epsilon)$, therefore

$$s + \frac{\log(b-\epsilon)}{\log(r_n)} < \frac{\log(\log(|f|(r_n)))}{\log(r_n)} < s + \frac{\log(b+\epsilon)}{\log(r_n)}.$$

Consequently, $\lim_{n\to+\infty} \dfrac{\log(\log(|f|(r_n)))}{\log(r_n)} = s$ and therefore $\rho(f) \geq s$, hence $\rho(f) \geq M$. Finally, $\rho(f) = M$. □

B.22. Type of growth for entire functions

Definitions and notations: In complex analysis, the *type of growth* is defined for an entire function having a finite order of growth t as $\sigma(f) = \limsup_{r\to+\infty} \dfrac{\log(M_f(r))}{r^t}$, with $t < +\infty$. Of course, the same notion may be defined for $f \in \mathcal{A}(\mathbb{K})$. Here, as in Chapters B.21 and B.23, we put $\theta = e$ and we denote by log the Neperian logarithm. Then, given $f \in \mathcal{A}^*(\mathbb{K})$ of order t, we set $\sigma(f) = \limsup_{r\to+\infty} \dfrac{\log(|f|(r))}{r^t}$.

Moreover, we put $\widetilde{\sigma}(f) = \liminf_{r\to+\infty} \dfrac{\log(|f|r))}{r^t}$.

Theorem B.22.1: *Let $f,\ g \in \mathcal{A}^*(\mathbb{K})$. Then $\sigma(fg) \leq \sigma(f)+\sigma(g)$. If $\rho(f) \geq \rho(g)$, then $\sigma(f) \leq \sigma(fg)$ and if $\rho(f) = \rho(g)$, then $\max(\sigma(f), \sigma(g)) \leq \sigma(fg)$. Moreover, if $\rho(f) = \rho(g)$ and if $c|f|(r) \geq |g|(r)$ with $c > 0$ when r is big enough, then $\sigma(f) \geq \sigma(g)$. If $\rho(f) = \rho(g)$ and $\sigma(f) > \sigma(g)$ then $\rho(f+g) = \rho(f)$ and $\sigma(f+g) = \sigma(f)$. If $\rho(f+g) = \rho(f) \geq \rho(g)$ then $\sigma(f+g) \leq \max(\sigma(f), \sigma(g))$.*

Proof. Let $t = \rho(g)$ and $s = \rho(f)$ and suppose $s \geq t$. When r is big enough, we have $\max(\log(|f|(r)), \log(|g|(r)) \leq \log(|f.g|(r)) = \log(|f|(r)) + \log(|g|(r))$. By Theorem B.21.1, then $\rho(fg) = s$. Therefore,

$$\begin{aligned}\sigma(fg) &= \limsup_{r\to+\infty} \left(\frac{\log(|f.g|(r))}{r^s}\right) \leq \limsup_{r\to+\infty} \left(\frac{\log(|f|(r))}{r^s}\right) + \limsup_{r\to+\infty} \left(\frac{\log(|g|(r))}{r^s}\right) \\ &\leq \limsup_{r\to+\infty} \left(\frac{\log(|f|(r))}{r^s}\right) + \limsup_{r\to+\infty} \left(\frac{\log(|g|(r))}{r^t}\right) = \sigma(f) + \sigma(g).\end{aligned}$$

Now, suppose $s > t$. Then by Theorem B.21.1, $\rho(f+g) = \rho(f) = s$. Consequently,

$$\begin{aligned}\sigma(f+g) &= \limsup_{r\to+\infty} \left(\frac{\log|f+g|(r)}{r^s}\right) \leq \limsup_{r\to+\infty} \left(\frac{\max(\log|f|(r), \log|g|(r))}{r^s}\right) \\ &= \max\left(\limsup_{r\to+\infty} \left(\frac{\log|f|(r)}{r^s}\right), \limsup_{r\to+\infty} \left(\frac{\log|g|(r)}{r^s}\right)\right) \\ &\leq \max\left(\limsup_{r\to+\infty} \left(\frac{\log|f|(r)}{r^s}\right), \limsup_{r\to+\infty} \left(\frac{\log|g|(r)}{r^t}\right)\right) = \max\left(\sigma(f), \sigma(g)\right).\end{aligned}$$

Now, just suppose $s \geq t$. Then

$$\sigma(f) = \limsup_{r\to+\infty} \frac{\log(|f|(r))}{r^s} \leq \limsup_{r\to+\infty} \frac{\log(|fg|(r))}{r^s}.$$

But by Theorem B.21.1, $\rho(fg) = s$, hence $\sigma(f) \leq \sigma(fg)$.

Now, suppose $\rho(f) = \rho(g) = s$. Then

$$\max\Big(\limsup_{r\to+\infty}\Big(\frac{\log(|f|(r))}{r^s}\Big), \limsup_{r\to+\infty}\Big(\frac{\log(|g|(r))}{r^s}\Big)\Big) \leq \limsup_{r\to+\infty}\Big(\frac{\log(|f.g|(r))}{r^s}\Big),$$

because both $|f|(r)$ and $|g|(r)$ tend to $+\infty$ with r. Consequently, $\sigma(fg) \geq \max(\sigma(f), \sigma(g))$.

Suppose now $c|f|(r) \geq |g|(r)$ when r is big enough, then, assuming again that $s = t$, it is obvious that $\sigma(f) \geq \sigma(g)$.

Now, suppose again that $\rho(f) = \rho(g)$ and suppose $\sigma(f) > \sigma(g)$. Let $s = \rho(f)$, $b = \sigma(f)$. Then $b > 0$. Let $(r_n)_{n\in\mathbb{N}}$ be a sequence such that $\lim_{n\to+\infty} r_n = +\infty$ and $\lim\limits_{n\to+\infty} \dfrac{\log(|f|(r_n))}{(r_n)^s} = b$. Since $\sigma(g) < \sigma(f)$, we notice that when n is big enough we have $|g|(r_n) < |f|(r_n)$. Consequently, when n is big enough, we have $|f + g|(r_n) = |f|(r_n)$ and hence

$$\lim_{n\to+\infty} \frac{\log(|f+g|(r_n))}{(r_n)^s} = b. \tag{1}$$

Now, by definition of σ we have $\sigma(f+g) \geq \lim\limits_{n\to+\infty} \dfrac{\log(|f+g|(r_n))}{(r_n)^{\rho(f+g)}}$. By Theorem B.21.1, we have $\rho(f+g) \leq s$, hence

$$\sigma(f+g) \geq \lim_{n\to+\infty} \frac{\log(|f+g|(r_n))}{(r_n)^{\rho(f+g)}} \geq \lim_{n\to+\infty} \frac{\log(|f+g|(r_n))}{(r_n)^s}$$

$$= \lim_{n\to+\infty} \frac{\log(|f|(r_n))}{(r_n)^s} = \sigma(f),$$

therefore by (1), $\sigma(f+g) \geq \sigma(f)$.

Suppose that $\sigma(f+g) > \sigma(f)$. Putting $h = f + g$, we have $f = h - g$ with $\sigma(g) < \sigma(h)$, hence $\sigma(h-g) \geq \sigma(h)$, i.e., $\sigma(f) > \sigma(f+g)$, a contradiction. Consequently, $\sigma(f+g) = \sigma(f)$. Thus, $\limsup\limits_{r\to+\infty} \dfrac{\log(|f+g|(r))}{r^s} = b > 0$. But then, $\limsup\limits_{r\to+\infty} \dfrac{\log(|f+g|(r))}{r^m} = 0\ \forall m > s$. Therefore, by Theorem B.21.12, we have $\rho(f+g) = \rho(f)$.

Finally, suppose now that $\rho(f+g)=\rho(f)\geq\rho(g)$. Then,

$$\sigma(f+g)=\limsup_{r\to+\infty}\frac{\log(|f+g|(r))}{r^s}\leq\max\Big(\limsup_{r\to+\infty}\frac{\log(|f|(r))}{r^s},\limsup_{r\to+\infty}\frac{\log(|g|(r))}{r^s}\Big)$$

$$\leq\max\Big(\limsup_{r\to+\infty}\frac{\log(|f|(r))}{r^s},\limsup_{r\to+\infty}\frac{\log(|g|(r))}{r^t}\Big)=\max(\sigma(f),\sigma(g)).$$

The last statement is derived from the previous ones and from Theorem B.21.1(iii). □

Corollary B.22.2: *Let $f,\ g\in\mathcal{A}(\mathbb{K})$ be such that $\rho(f)\neq\rho(g)$. Then $\sigma(f+g)\leq\max(\sigma(f),\sigma(g))$.*

Proof. Indeed, assuming that $\rho(f)>\rho(g)$, we have $\rho(f+g)=\rho(f)$ and hence the conclusion comes from the last statement of Theorem B.22.1. □

Now we will show that $\sigma(f)$ may be computed by the same formula as on $\mathbb{C}$.

Theorem B.22.3: *Let $f(x)=\sum_{n=0}^{\infty}a_nx^n\in\mathcal{A}(\mathbb{K})$ be such that $0<\rho(f)<+\infty$. Then $\sigma(f)\rho(f)e=\limsup_{n\to+\infty}\big(n\sqrt[n]{|a_n|^{\rho(f)}}\big)$.*

Proof. Let $t=\rho(f)$. First, let us show that $et\sigma(f)\geq\limsup_{n\to+\infty}n|a_n|^{\frac{t}{n}}$. We will follow a similar way as in [86]. Let $u=\sigma(f)$ and let us take $w>\sigma(f)$. For r big enough we have $\log(|f|(r))\leq wr^t$, hence for all $n\in\mathbb{N}$, we can derive

$$|a_n|\leq\frac{|f|(r)}{r^n}\leq\frac{e^{wr^t}}{r^n}. \tag{1}$$

Now, let us take r such that the derivative of the logarithm of the function $\dfrac{e^{wr^t}}{r^n}$ vanishes: we have $wtr^{t-1}-\frac{n}{r}=0$. So, we can choose $r_n=\left(\frac{n}{wt}\right)^{\frac{1}{t}}$ and we can check that

$$|a_n|\leq\frac{e^{\frac{n}{t}}}{\left(\frac{n}{wt}\right)^{\frac{n}{t}}}=\left(\frac{ewt}{n}\right)^{\frac{n}{t}}.$$

Consequently, we have $n|a_n|^{\frac{t}{n}}\leq ewt$, therefore $\limsup_{n\to+\infty}n|a_n|^{\frac{t}{n}}\leq etw$. This is true for all $w>\sigma(f)$ and hence $\limsup_{n\to+\infty}n|a_n|^{\frac{t}{n}}\leq et\sigma(f)$.

Now let us show the reverse inequality. Take $c>\dfrac{1}{et}\limsup_{n\to+\infty}n|a_n|^{\frac{t}{n}}$. When n is big enough we have $|a_n|\leq\left(\frac{ect}{n}\right)^{\frac{n}{t}}$, hence $|a_n|r^n\leq\left(\frac{ect}{n}\right)^{\frac{n}{t}}r^n$ and consequently

$|f|(r) \leq \sup_{n\geq 1}\left(\frac{ect}{n}\right)^{\frac{n}{t}} r^n$. Therefore, $|f|(r) \leq \sup_{x>1} \frac{(ect)^{\frac{x}{t}} r^x}{x^{\frac{x}{t}}}$. Now, set $y = \frac{x}{t}$ and $R = ecr$. Then

$$|f|(r^{\frac{1}{t}}) \leq \sup_{y>0} \frac{(ect)^y r^y}{(ty)^y} = \sup_{y>0}\left(\frac{ecr}{y}\right)^y = \sup_{y>0} \frac{R^y}{y^y} = e^{\frac{R}{e}} = e^{cr}.$$

Thus, we have $|f|(r) \leq e^{cr^t}$ and hence $\limsup_{r\to+\infty} \frac{\log(|f|(r))}{r^t} \leq c$. Therefore $\sigma(f) \leq c$, which ends the proof. □

In the proof of Theorem B.22.5, we will use Lemma B.22.4.

Lemma B.22.4: *Let g, h be the real functions defined in $]0, +\infty[$ as $g(x) = \frac{e^{tx}-1}{x}$ and $h(x) = \frac{1-e^{-tx}}{x}$ with $t > 0$. Then*
(i) $\inf\{|g(x)| \; |x > 0\} = t$.
(ii) $\sup\{|h(x)| \; |x > 0\} = t$.

Definitions and notations: Given $f \in \mathcal{A}^*(\mathbb{K})$, we put $\widetilde{\sigma}(f) = \liminf_{r\to+\infty} \frac{\log(|f|(r))}{r^t}$.

Theorem B.22.5: *Let $f \in \mathcal{A}^*(\mathbb{K})$ be not identically zero. Then*

$$\rho(f)\sigma(f) \leq \psi(f) \leq \rho(f)(e\sigma(f) - \widetilde{\sigma}(f)).$$

Moreover, if $\psi(f) = \lim_{r\to+\infty} \frac{q(f,r)}{r^{\rho(f)}}$ or if $\sigma(f) = \lim_{r\to+\infty} \frac{\log(|f|(r))}{r^{\rho(f)}}$, then $\psi(f) = \rho(f)\sigma(f)$.

Proof. Without loss of generality, we can assume that $f(0) \neq 0$. Let $t = \rho(f)$ and set $\ell = \log(|f(0)|)$. Let $(a_n)_{n\in\mathbb{N}}$ be the sequence of zeros of f with $|a_n| \leq |a_{n+1}|$, $n \in \mathbb{N}$ and for each $n \in \mathbb{N}$, let w_n be the multiplicity order of a_n. For every $r > 0$, let $k(r)$ be the integer such that $|a_n| \leq r \; \forall n \leq k(r)$ and $|a_n| > r \; \forall n > k(r)$. Then by Corollary B.13.27, we have $\log(|f|(r)) = \ell + \sum_{n=0}^{k(r)} w_n(\log(r) - \log(|a_n|))$, hence

$$\sigma(f) = \limsup_{r\to+\infty} \left(\frac{\ell + \sum_{n=0}^{k(r)} w_n(\log(r) - \log(|a_n|))}{r^t}\right).$$

Given $r > 0$, set $c_n = |a_n|$ and let us keep the notations above. Then

$$\sigma(f) = \limsup_{r\to+\infty} \sigma(f,r), \quad \psi(f) = \limsup_{r\to+\infty} \psi(f,r). \tag{1}$$

We will first show the inequality $\rho(f)\sigma(f) \leq \psi(f)$. By the definition of $\sigma(f,r)$ we can derive

$$\sigma(f,r) \leq \sum_{n=0}^{k(re^{-\alpha})} \frac{w_n\big(\log(r) - \log(re^{-\alpha})\big)}{r^t}$$

$$+ \sum_{n=0}^{k(re^{-\alpha})} \frac{w_n\big(\log(re^{-\alpha}) - \log(c_n)\big)}{r^t} + \alpha \sum\nolimits_{k(re^{-\alpha})<n\leq k(r)} \frac{w_n}{r^t},$$

hence

$$\sigma(f,r) \leq \alpha \sum_{n=0}^{k(re^{-\alpha})} \frac{w_n}{r^t} + \sum\nolimits_{n=0}^{k(re^{-\alpha})} \frac{w_n\big(\log(re^{-\alpha}) - \log(c_n)\big)}{r^t}$$

$$+ \alpha \sum_{k(re^{-\alpha})<n\leq k(r)} \frac{w_n}{r^t},$$

therefore

$$\sigma(f,r) \leq \alpha \sum_{n=0}^{k(re^{-\alpha})} \frac{w_n}{r^t} + \sum\nolimits_{n=0}^{k(re^{-\alpha})} \frac{w_n\big(\log(re^{-\alpha}) - \log(c_n)\big)}{r^t}$$

$$+ \alpha \sum_{k(re^{-\alpha})<n\leq k(r)} \frac{w_n}{r^t},$$

hence

$$\sigma(f,r) \leq \alpha \sum_{n=0}^{k(re^{-\alpha})} \frac{w_n}{r^t} + e^{-t\alpha} \sum_{n=0}^{k(re^{-\alpha})} \frac{w_n(\log(re^{-\alpha}) - \log(c_n))}{(re^{-\alpha})^t}$$

$$+ \alpha \sum_{0\leq n\leq k(r)} \frac{w_n}{r^t} - \alpha \sum_{0\leq n\leq k(re^{-\alpha})} \frac{w_n}{r^t},$$

hence

$$\sigma(f,r) \leq e^{-t\alpha} \sum_{n=0}^{k(re^{-\alpha})} \frac{w_n(\log(re^{-\alpha}) - \log(c_n))}{(re^{-\alpha})^t} + \alpha \sum_{0\leq n\leq k(r)} \frac{w_n}{r^t}.$$

Thus, we have

$$\sigma(f,r) \leq e^{-t\alpha}\sigma(f, re^{-\alpha}) + \alpha\psi(f,r).$$

We check that we can pass to superior limits on both sides, so we obtain $\sigma(f) \leq e^{-t\alpha}\sigma(f) + \alpha\psi(f)$ therefore $\sigma(f)\dfrac{(1-e^{-t\alpha})}{\alpha} \leq \psi(f)$. That holds for every $\alpha > 0$, hence by Lemma B.22.4 (ii), we can derive

$$\psi(f) \geq \rho(f)\sigma(f). \tag{2}$$

We will now show the inequality

$$\psi(f) \leq \rho(f)(e\sigma(f) - \widetilde{\sigma}(f)).$$

Let us fix $\alpha > 0$. We can write

$$\sigma(f,r) = \sum_{n=0}^{k(re^{-\alpha})} \frac{w_n(\log(r) - \log(re^{-\alpha}))}{r^t}$$
$$+ \sum_{j=0}^{k(re^{-\alpha})} \frac{w_j(\log(re^{-\alpha}) - \log(c_n))}{r^t} + \sum_{k(re^{-\alpha})<j\leq k(r)} \frac{w_j(\log(r) - \log(c_j))}{r^t},$$

hence

$$\sigma(f,r) \geq \alpha \sum_{n=0}^{k(re^{-\alpha})} \frac{w_n}{r^t} + \sum_{j=0}^{k(re^{-\alpha})} \frac{w_j(\log(re^{-\alpha}) - \log(c_n))}{r^t},$$

hence

$$\sigma(f,r) \geq \alpha e^{-t\alpha} \sum_{n=0}^{k(re^{-\alpha})} \frac{w_n}{(re^{-\alpha})^t} + e^{-t\alpha} \sum_{j=0}^{k(re^{-\alpha})} \frac{w_n(\log(re^{-\alpha}) - \log(c_n))}{(re^{-\alpha})^t},$$

and hence

$$\sigma(f,r) \geq \alpha e^{-t\alpha}\psi(f,re^{-\alpha}) + e^{-t\alpha}\sigma(f,re^{-\alpha}).$$

Therefore, we can derive

$$\alpha e^{-t\alpha}\psi(f) \leq \limsup_{r\to+\infty} \Big(\sigma(f,r) - e^{-t\alpha}\sigma(f,re^{-\alpha}))\Big),$$

and therefore

$$\alpha e^{-t\alpha}\psi(f) \leq \sigma(f) - e^{-t\alpha}\widetilde{\sigma}(f)). \tag{3}$$

That holds for every $\alpha > 0$ and hence, when $t\alpha = 1$, by (3) we obtain $\psi(f) \leq \rho(f)\big(e\sigma(f) - \widetilde{\sigma}(f)\big)$, which is the left-hand inequality of the general conclusion.

Now, suppose that $\sigma(f) = \lim_{r\to+\infty} \dfrac{\log(|f|(r))}{r^t}$. Then by (3) we have $\limsup_{r\to+\infty} \psi(f,r) \leq \sigma(f)\Big(\dfrac{e^{t\alpha}-1}{\alpha}\Big)$ and hence $\psi(f) \leq \sigma(f)\Big(\dfrac{e^{t\alpha}-1}{\alpha}\Big)$. That holds for every $\alpha > 0$ and then, by Lemma B.22.4 (i) we obtain $\psi(f) \leq t\sigma(f)$, i.e., $\psi(f) \leq \rho(f)\sigma(f)$, hence by (2) we have, $\psi(f) = \rho(f)\sigma(f)$.

Now, suppose that

$$\psi(f) = \lim_{r\to+\infty} \sum_{n=0}^{k(r)} \frac{w_n}{r^t} = \lim_{r\to+\infty} \psi(f,r).$$

We can obviously find a sequence $(r_n)_{n\in\mathbb{N}}$ in $]0,+\infty[$ of limit $+\infty$ such that $\sigma(f) = \lim_{n\to+\infty} \sigma(f, r_n e^{-\alpha})$. Then, by (1) we have

$$\sigma(f,r_n) \geq \alpha e^{-t\alpha}\psi\left(f, \frac{r_n}{e^\alpha}\right) + e^{-t\alpha}\sigma\left(f, \frac{r_n}{e^\alpha}\right),$$

hence

$$\limsup_{n\to+\infty} \sigma(f,r_n) \geq \alpha e^{-t\alpha}\psi(f) + e^{-t\alpha}\sigma(f),$$

and hence

$$\sigma(f) \geq \alpha e^{-t\alpha}\psi(f) + e^{-t\alpha}\sigma(f),$$

therefore, $\psi(f) \leq \left(\frac{e^{t\alpha}-1}{\alpha}\right)\sigma(f)$. Finally, by Lemma B.22.4 (i), we have, $\psi(f) \leq \rho(f)\sigma(f)$ and hence by (2), $\psi(f) = \rho(f)\sigma(f)$. □

Remark: The conclusions of Lemma B.22.4 hold for $\psi(f) = \sigma(f) = +\infty$.

We will now present Example 1 where neither $\psi(f)$ nor $\sigma(f)$ are obtained as limits but only as superior limits: we will show that the equality $\psi(f) = \rho(f)\sigma(f)$ holds again.

Example 1: Let $r_n = 2^n$, $n \in \mathbb{N}$ and let $f \in \mathcal{A}(\mathbb{K})$ have exactly 2^n zeros in $C(0,r_n)$ and satisfy $f(0) = 1$. Then $\zeta(r_n, f) = 2^{n+1} - 2$ $\forall n \in \mathbb{N}$. We can see that the function $h(r)$ defined in $[r_n, r_{n+1}[$ by $h(r) = \frac{\zeta(r,f)}{r}$ is decreasing and satisfies $h(r_n) = \frac{2^{n+1}-2}{2^n}$ and $\lim_{r\to r_{n+1}} h(r) = \frac{2^{n+1}-2}{2^{n+1}}$. Consequently, $\sup\left(h(r) \mid r_n \leq r < r_{n+1}\right) = \frac{2^{n+1}-2}{2^n}$. Therefore, $\limsup_{r\to+\infty} h(r) = 2$ and $\liminf_{r\to+\infty} h(r) = 1$. Particularly, by Theorem B.21.8, we have $\rho(f) = 1$ and of course $\psi(f) = 2$.

Now, let us compute $\sigma(f)$ and consider the function in r: $E(r) = \dfrac{\log(|f|(r))}{r}$. When r belongs to $[r_n, r_{n+1}]$, we have

$$E(r) = \frac{(2^{n+1} - 2)\log r - (\log 2)(\sum_{k=1}^{n} 2^k)}{r}$$

and its derivative is $E'(r) = \dfrac{\sum_{k=1}^{n} 2^k(1 + k\log(2)) - \log(r))}{r^2}$. We will need to compute

$$\sum_{k=1}^{n} k2^k = 2(n2^{n+1} - (n+1)2^n + 1). \tag{1}$$

Now, the numerator of $E'(r)$ is $U(r) = \sum_{k=1}^{n} 2^k(1+k\log(2)) - \log(r))$ is decreasing in the interval $[r_n, r_{n+1}]$ and has a unique zero α_n satisfying, by (1),

$$\log(\alpha_n) = \frac{2^n\Big((\log 2)(n - 1 + 2^{-n}) + 2 - 2^{-n+1}\Big)}{2^n - 2},$$

thereby $\log(\alpha_n)$ is of the form $n\log(2) + \epsilon_n$ with $\lim\limits_{n\to+\infty} \epsilon_n = 0$.

Since $E'(r)$ is decreasing in $[r_n, r_{n+1}]$, we can check that $E(r)$ passes by a maximum at α_n and consequently,

$$\sigma(f) = \limsup{}_{n\to+\infty} \frac{E(\alpha_n)}{\alpha_n}.$$

Therefore, $\sigma(f) = 2 = \psi(f)$.

Now, we can check that $\liminf\limits_{r\to+\infty} \dfrac{E(r)}{r} < \sigma(f)$. Indeed, consider

$$\begin{aligned}\frac{E(r_n)}{r_n} &= \frac{(2^{n+1} - 2)(\log r_n) - (\log 2)\sum_{k=1}^{n} k2^k}{r_n} \\ &= \frac{(2^{n+1} - 2)(n\log 2) - (\log 2)\sum_{k=1}^{n} k2^k}{2^n},\end{aligned}$$

hence by (1), we obtain

$$\begin{aligned}E(r_n) &= \frac{(2^{n+1} - 2)(n\log 2) - 2(\log 2)(n2^{n+1} - (n+1)2^n + 1)}{2^n} \\ &= \frac{2(\log 2)(2^n - n - 1)}{2^n},\end{aligned}$$

therefore $\lim\limits_{n\to\infty} E(r_n) = 2\log 2$ and hence $\liminf\limits_{r\to+\infty} E(r) < \sigma(f)$. Thus, in that Example 1, we have $\liminf\limits_{r\to+\infty} E(r) < \sigma(f)$ but however, $\psi(f) = \rho(f)\sigma(f)$.

Therefore, Theorem B.22.5 and Example 1 suggest the following conjecture:

Conjecture: *Let $f \in \mathcal{A}^*(\mathbb{K})$ be such that either $\sigma(f) < +\infty$ or $\psi(f) < +\infty$. Then $\psi(f) = \rho(f)\sigma(f)$.*

Example 2: Infinite type and cotype: Here is an example of $f \in \mathcal{A}(\mathbb{K}, 1)$ such that $\sigma(f) = \psi(f) = +\infty$.

For each $n \in \mathbb{N}$, set $\phi(n) = \sqrt{\log n}$ and let u_n be defined by $\log(u_n) = -\dfrac{n \log n}{1 + \frac{1}{\phi(n)}}$. For simplicity, suppose first that the set of absolute values of $|\mathbb{K}|$ is the whole set $[0, \mathbb{R}[$. We can take a sequence (a_n) of $\mathbb{K}$ such that $|a_n| = u_n \ \forall n \in \mathbb{N}^*$, with $a_0 = 1$. Then $\dfrac{\log |a_n|}{n} = -\dfrac{(\log n)\phi(n)}{\phi(n) + 1}$, hence $\lim\limits_{n \to +\infty} \dfrac{\log |a_n|}{n} = -\infty$, therefore $f \in \mathcal{A}(\mathbb{K})$. Next, $\dfrac{n \log n}{-\log[a_n|} = \dfrac{\phi(n)}{\phi(n) + 1,}$ hence $\lim\limits_{n \to +\infty} \dfrac{n \log n}{-\log[a_n|} = 1$, therefore $\rho(f) = 1$.

Henceforth, $\log(n|a_n|^{\frac{1}{n}}) = \log n + \dfrac{\log |a_n|}{n} = \log n - \dfrac{\phi(n) \log n}{\phi(n) + 1} = \dfrac{\log n}{\phi(n) + 1}$ and hence $\sigma(f) = +\infty$.

Let us now compute $\psi(f)$. Now, for each $n \in \mathbb{N}^*$, take $r_n = \dfrac{u_{n-1}}{u_n}$. We will first check that the sequence $(r_n)_{n \in \mathbb{N}^*}$ is strictly increasing when n is big enough. Indeed, we just have to show that there exists $M \in \mathbb{N}$ such that

$$\log(u_n) - \log(u_{n+1}) > \log(u_{n-1}) - \log(u_n) \ \forall n > M. \tag{1}$$

Let g be the function defined in $]0, +\infty[$ as $g(x) = -\dfrac{x \log x}{1 + \frac{1}{\sqrt{\log x}}}$. Then we can check that g is convex and therefore (1) is proven.

Now, since the sequence $(r_n)_{n \in \mathbb{N}^*}$ obviously tends to $+\infty$, there exists a rank $N \geq M$ such that $r_{n+1} > r_n \ \forall n \geq M$ and $r_M > r_k \ \forall k < N$. Consequently, for each $n > N$, we have $|a_n|r^n > |a_k|r^k \ \forall k \neq n$ and therefore, f admits $n - 1$ zeros inside $d(0, (r_n)^-)$ and a unique zero in $C(0, r_n)$, hence f admits exactly n zeros in $d(0, r_n)$. Consequently, we have

$$\zeta(r_n, f) = n \ \forall n \geq N. \tag{2}$$

Since $\zeta(r, f)$ remains equal to $\zeta(r_n, f)$ for all $r \in [r_n, r_{n+1}[$, by (2) we can derive that

$$\limsup_{r \to +\infty} \frac{\zeta(r, f)}{r} = \limsup_{n \to +\infty} \frac{\zeta(r_n, f)}{r_n}. \tag{3}$$

Now, for $n \geq N$, we have

$$\begin{aligned} \log \frac{\zeta(r_n, f)}{r_n} &= \log(n) - \log(u_{n-1}) + \log(u_n) \\ &= \log(n) - \frac{n \log n}{1 + \frac{1}{\phi(n)}} + \frac{(n-1)\log(n-1)}{1 + \frac{1}{\phi(n-1)}}. \end{aligned}$$

Set $S_n = \dfrac{n\log n}{1+\frac{1}{\phi(n)}} - \dfrac{(n-1)\log(n-1)}{1+\frac{1}{\phi(n-1)}}$. Then

$$\log\left(\frac{\zeta(r_n,f)}{r_n}\right) = \log n - S_n. \tag{4}$$

Now, we have

$$S_n = \frac{\phi(n)\phi(n-1)\big(n\log(n)-(n-1)\log(n-1)\big)+n\log(n)\phi(n)-(n-1)\log(n-1)\phi(n-1)}{(\phi(n)+1)(\phi(n-1)+1)}.$$

Set $A_n = \dfrac{\phi(n)\phi(n-1)\big(n\log(n)-(n-1)\log(n-1)\big)}{(\phi(n)+1)(\phi(n-1)+1)}$ and $B_n = \dfrac{n\log(n)\phi(n)-(n-1)\log(n-1)\phi(n-1)}{(\phi(n)+1)(\phi(n-1)+1)}$. Then $S_n = A_n + B_n$ and both A_n, B_n are positive. By finite increasings theorem applied to the function $g(x) = x\log x$, we have

$$A_n \le \frac{\phi(n)\phi(n-1)(\log n)}{(\phi(n)+1)(\phi(n-1)+1)}. \tag{5}$$

On the other hand, by finite increasings theorem applied to the function $h(x) = x(\log x)^{\frac{3}{2}}$, we have

$$B_n \le \frac{\phi(n)(\log n + \frac{3}{2})}{(\phi(n)+1)(\phi(n-1)+1)}. \tag{6}$$

Then by (1), (5), and (6) we have $\log(\zeta(r_n,f)) - A_n - B_n$

$$\begin{aligned}
&\ge \frac{\log n\Big(\phi(n)+1)(\phi(n-1)+1)-\phi(n)\phi(n-1)-\phi(n)\Big)-\frac{3}{2}\phi(n)}{(\phi(n)+1)(\phi(n-1)+1)}\\
&= \frac{\log(n)\big(\phi(n)+\phi(n-1)+1-\phi(n)\big)-\frac{3\phi(n)}{2}}{(\phi(n)+1)(\phi(n-1)+1)}\\
&= \frac{\log n}{\phi(n)+1}-\frac{3\phi(n)}{2(\phi(n)+1)(\phi(n-1)+1)}.
\end{aligned}$$

Now, since $\phi(n) = \sqrt{\log n}$, it is obvious that

$$\lim_{n\to+\infty} \log(\zeta(r_n,f)) - S_n = +\infty$$

and therefore by (3), (4), and (5), $\psi(f) = +\infty$.

B.23. Growth of the derivative of an entire function

Similarly to the situation in complex entire functions, here we will see that the order and the type of the derivative of an entire function f are respectively equal to those of f. As in Chapters B.21 and B.22, we put $\theta = e$ and denote by log the Neperian logarithm.

Throughout the chapter, $\mathbb{K}$ *is supposed to have characteristic* 0.

Theorem B.23.1: *Let $f \in \mathcal{A}(\mathbb{K})$ be not identically zero. Then $\rho(f) = \rho(f')$.*

Proof. By Theorem B.21.11, $\rho(f') = \limsup_{n\to+\infty} \Big(\frac{n\log(n)}{-\log(|(n+1)a_{n+1}|)}\Big)$. But since $\frac{1}{n} \le |n| \le 1$, we have

$$\log(|a_{n+1}|) - \log(n+1) \le \log(|(n+1)a_{n+1}|) \le \log(|a_{n+1}|),$$

hence

$$-\frac{\log(|a_{n+1}|)}{n\log(n)} \le -\frac{\log(|(n+1)a_{n+1}|)}{n\log(n)} \le -\frac{\log(|a_{n+1}|) - \log(n+1)}{n\log(n)},$$

hence

$$\liminf_{n\to+\infty}\Big(-\frac{\log(|a_{n+1}|)}{n\log(n)}\Big)$$
$$\le \liminf_{n\to+\infty}\Big(-\frac{\log(|(n+1)a_{n+1}|)}{n\log(n)}\Big) \le \liminf_{n\to+\infty}\Big(-\frac{\log(|a_{n+1}|) - \log(n+1)}{n\log(n)}\Big).$$

But since

$$\lim_{n\to+\infty}\frac{\log(n+1)}{n\log(n)} = 0,$$

we have

$$\liminf_{n\to+\infty}\Big(-\frac{\log(|a_{n+1}|)}{n\log(n)}\Big) = \liminf_{n\to+\infty}\Big(-\frac{\log(|a_{n+1}|) + \log(n+1)}{n\log(n)}\Big),$$

therefore

$$\liminf_{n\to+\infty}\Big(\frac{-\log(|a_{n+1}|)}{n\log(n)}\Big) = \liminf_{n\to+\infty}\Big(-\frac{\log((n+1)|a_{n+1}|)}{n\log(n)}\Big).$$

But since all quantities are positive, we can derive

$$\limsup_{n\to+\infty}\frac{n\log n}{-\log(|a_{n+1}|)} = \limsup_{n\to+\infty}\Big(\frac{n\log n}{-\log(|(n+1)a_{n+1}|)}\Big),$$

therefore

$$\limsup_{n\to+\infty}\frac{n\log n}{-\log(|a_{n+1}|)} = \limsup_{n\to+\infty}\Big(\frac{(n+1)\log(n+1)}{-\log(|(n+1)a_{n+1}|)}\Big) = \rho(f),$$

and hence $\rho(f') = \rho(f)$. □

Corollary B.23.2: *Consider the differential equation*

$$(\mathcal{E})\ f^{(n)} + a_{n-1}(x)f^{(n-1)}(x) + \cdots + a_0(x)f(x) = 0$$

with $a_j \in \mathcal{A}^*(\mathbb{K})$, $j = 0, \ldots, n-1$ and $\rho(a_j) < \rho(a_0)\ \forall j = 1, \ldots, n-1$. Then every non-trivial solution f of $(\mathcal{E})$ satisfies $\rho(f) \ge \rho(a_0)$.

Corollary B.23.3: *The derivation on $\mathcal{A}(\mathbb{K})$ restricted to the algebra $\mathcal{A}(\mathbb{K},t)$ (resp. to $\mathcal{A}^*(\mathbb{K})$) provides that algebra with a derivation.*

In complex analysis, it is known that if an entire function f has order $t<+\infty$, then f and f' have same type. We will check that it is the same here.

Theorem B.23.4: *Let $f\in\mathcal{A}(\mathbb{K})$ of order $t\in]0,+\infty[$. Then $\sigma(f)=\sigma(f')$.*

Proof. By Theorem B.22.3, we have

$$\begin{aligned}
e\rho(f')\sigma(f') &= \limsup_{n\to+\infty}\left(n\Big(|n+1||a_{n+1}|\Big)^{\frac{t}{n}}\right)\\
&= \limsup_{n\to+\infty}\left(\left((n+1)\Big(|n+1||a_{n+1}|\Big)^{\frac{t}{n}}\right)^{\frac{n}{n+1}}\left(\frac{n}{n+1}\right)\right)\\
&= \limsup_{n\to+\infty}\left((n+1)\Big(|n+1||a_{n+1}|\Big)^{\frac{t}{n+1}}\right)=e\rho(f)\sigma(f).
\end{aligned}$$

But since $\rho(f)=\rho(f')$ and since $\rho(f)\neq 0$, we can see that $\sigma(f')=\sigma(f)$. □

Theorem B.23.4 shows a way to compare the growth of an entire function f to this of its derivative. Of course, we know that the inequality $|f'|(r)\leq|f|(r)$ holds always. But we don't have an inequality in the other side. However, thanks to Theorem B.23.4, we can get Corollary B.23.5.

Corollary B.23.5: *Let $f\in\mathcal{A}^*(\mathbb{K})$ be not identically zero, of order $t<+\infty$. Given $\epsilon>0$, there exists a sequence of intervals $[r'_n,r''_n]$, with $\lim_{n\to+\infty}r'_n=+\infty$, such that $|f'|(r)\geq|f|(r)e^{-(\epsilon r^t)}$ $\forall r\in\bigcup_{n\in\mathbb{N}}[r'_n,r''_n]$.*

By Theorems B.22.5 and B.23.4, we can now derive Corollary B.23.6.

Corollary B.23.6: *Let $f\in\mathcal{A}^*(\mathbb{K})$ be not identically zero, of order $t<+\infty$. If $\psi(f)=\lim_{r\to+\infty}\frac{\zeta(f,r)}{r^t}$ and if $\psi(f')=\lim_{r\to+\infty}\frac{\zeta(f',r)}{r^t}$, then $\psi(f')=\psi(f)$.*

Remarks: If the conjecture presented in Chapter B.22 is true, then $\psi(f)=\psi(f')$ $\forall f\in\mathcal{A}^*(\mathbb{K})$. Of course, polynomials have a growth order 0. On $\mathbb{K}$ as on $\mathbb{C}$ we can easily construct transcendental entire functions of order 0 or of order ∞.

Example 1: Let $(a_n)_{n\in\mathbb{N}}$ be a sequence in $\mathbb{K}$ such that $-\log|a_n|\in[n(\log n)^2, n(\log n)^2+1]$. Then clearly, $\lim_{n\to+\infty}\frac{\log|a_n|}{n}=-\infty$, hence the function $\sum_{n=0}^{\infty}a_nx^n$ has a radius of convergence equal to $+\infty$. On the other hand, $\lim_{n\to+\infty}\frac{n\log n}{-\log|a_n|}=0$ hence $\rho(f)=0$.

Example 2: Let $(a_n)_{n\in\mathbb{N}}$ be a sequence in $\mathbb{K}$ such that $-\log|a_n| \in [n\sqrt{\log n}, n\sqrt{\log n}+1]$. Then $\lim_{n\to+\infty} \frac{\log|a_n|}{n} = -\infty$ again and hence the function $\sum_{n=0}^{\infty} a_n x^n$ has radius of convergence equal to $+\infty$. On the other hand, $\lim_{n\to+\infty} \left(\frac{n\log n}{-\log|a_n|}\right) = +\infty$ hence $\rho(f) = +\infty$.

Similarly, comparing the number of zeros of f' to this of f inside a disk is very uneasy. Now, we can give some precisions thanks to Theorems B.21.8 and B.23.1.

Theorem B.23.7: *Let $f,\ g \in \mathcal{A}(\mathbb{K})$ be transcendental and of order $t \in [0,+\infty[$. Then for every $\epsilon > 0$,*

$$\limsup_{r\to+\infty} \left(\frac{r^{\epsilon}\zeta(r,g)}{\zeta(r,f)}\right) = +\infty.$$

Proof. Suppose first $t = 0$. The proof then is almost trivial. Indeed, for all $\epsilon > 0$, we have $\lim_{r\to+\infty} \frac{\zeta(r,f)}{r^{\epsilon}} = 0$, hence $\lim_{r\to+\infty} \frac{r^{\epsilon}}{\zeta(r,f)} = +\infty$, therefore $\lim_{r\to+\infty} \frac{r^{\epsilon}\zeta(r,g)}{\zeta(r,f)} = +\infty$.

Now suppose $t > 0$. By Theorem B.21.8, there exists $\lambda > 0$ such that

$$\zeta(r,f) \le \lambda r^t \ \forall r > 1. \tag{1}$$

Let us fix $s \in]0,t[$. By hypothesis, $\rho(g) = \rho(f)$ and hence by Theorem B.21.8, we have $\limsup_{r\to+\infty} \frac{\zeta(r,g)}{r^s} = +\infty$ so, there exists an increasing sequence $(r_n)_{n\in\mathbb{N}}$ of $\mathbb{R}_+$ such that $\lim_{n\to+\infty} r_n = +\infty$ and $\frac{\zeta(r_n,g)}{(r_n)^s} \ge n$. Therefore, by (1), we have

$$\frac{\lambda(r_n)^t\zeta(r_n,g)}{(r_n)^s\zeta(r_n,f)} > \frac{\zeta(r_n,g)}{(r_n)^s} > n,$$

and hence

$$\lambda \lim_{n\to+\infty} \left(\frac{(r_n)^{t-s}\zeta(r_n,g)}{\zeta(r_n,f)}\right) = +\infty.$$

Consequently,

$$\limsup_{r\to+\infty} \left(\frac{(r)^{t-s}\zeta(r,g)}{\zeta(r,f)}\right) = +\infty. \tag{2}$$

Now, since that holds for all $s \in]0,t[$, the statement comes from (2). □

By Theorem B.23.1, we can derive Corollary B.23.8.

Corollary B.23.8: *Let $f \in \mathcal{A}(\mathbb{K})$ be transcendental and of order $t \in [0, +\infty[$. Then for every $\epsilon > 0$, we have* $\limsup_{r\to+\infty}\left(\frac{r^{\epsilon}\zeta(r,f')}{\zeta(r,f)}\right) = +\infty$ *and* $\limsup_{r\to+\infty}\left(\frac{r^{\epsilon}\zeta(r,f)}{\zeta(r,f')}\right) = +\infty$.

B.24. Growth of an analytic function in an open disk

In Chapters B.21, B.22, and B.23, we defined the order of growth and the type of growth for entire functions in $\mathbb{K}$ in a similar way as it is done for complex entire functions and we also defined a cotype of growth strongly linked to the order and the type. In most of the cases, the cotype is the product of the order of growth by the type of growth.

Here, we consider analytic functions in an "open" disk $d(a, R^-)$ that we will denote by E throughout the chapter.

Notations and definitions: Let $f = \sum_{n=0}^{\infty} a_n x^n \in \mathcal{A}(E)$. In order to define a growth order similarly as it was done in the algebra of entire functions in $\mathbb{K}$, we can define in $\mathcal{A}(E)$ a growth order in the following way: given $r \in]0, R[$, as it was done in complex analysis, given an unbounded function $f \in \mathcal{A}(E)$, when r is close enough to R, we put $\rho(f,r) = \frac{\log(\log(|f|(r)))}{-\log(R-r)}$ and $\rho(f) = \limsup_{r\to R^-} \rho(f,r)$, hence $\rho(f) = \limsup_{r\to R^-} \frac{\log(\log(|f|(r)))}{-\log(R-r)}$. Then $\rho(f)$ is called *the order of growth of f*.

On the other hand, for every $r \in]0, R[$, if the set of the $s > 0$ such that $\lim_{r\to R^-} \zeta(f,r)(R-r)^s = 0$ is empty, we put $\theta(f) = +\infty$. Else, we then denote by $\theta(f)$ the lowest bound of the $s > 0$ such that $\lim_{r\to R^-} \zeta(f,r)(R-r)^s = 0$. Similarly, if the set of the $s > 0$ such that $\lim_{r\to R^-} \log(|f|(r))(R-r)^s = 0$ is empty, we put $\lambda(f) = +\infty$. Else, we denote by $\lambda(f)$ the lowest bound of the $s > 0$ such that $\lim_{r\to R^-} \log(|f|(r))(R-r)^s = 0$. And if $0 < \rho(f) < +\infty$, we put $\sigma(f,r) = \log(|f|(r))(R-r)^{\rho(f)}$, $\sigma(f) = \limsup_{r\to R^-} \sigma(f,r)$, $\psi(f,r) = \zeta(f,r)(R-r)^{\rho(f),}$ and $\psi(f) = \limsup_{r\to R^-} \psi(f,r)$. We call $\sigma(f)$ *the type of growth of f* and $\psi(f)$ *the cotype of growth of f*.

Let us recall that, as far as ultrametric entire functions are concerned, the order of growth is equal to the lowest bound of the $s > 0$ such that $\lim_{r\to+\infty} \frac{\log(|f|(r))}{r^s} = 0$ and to the lowest bound of the $s > 0$ such that $\lim_{r\to+\infty} \frac{\zeta(f,r)}{r^s} = 0$. Here, we will try to prove similar results. This paper is aimed at showing relations between these expressions $\rho(f), \sigma(f), and\ \psi(f)$.

Definitions and notations: We will denote by $\mathcal{A}^*(E)$ the set of unbounded functions $f \in \mathcal{A}(E)$ such that $0 < \rho(f) < +\infty$.

Theorems B.24.1 and B.24.3 are easy and don't need any proof.

Theorem B.24.1: *Let $f,\ g \in \mathcal{A}^*(E)$. Then $\rho(f+g) \leq \max(\rho(f), \rho(g))$ and $\rho(fg) = \max(\rho(f), \rho(g))$.*

Corollary B.24.2: *Let $f,\ g \in \mathcal{A}^*(E)$. Then $\rho(f^n) = \rho(f)\ \forall n \in \mathbb{N}^0$. If $\rho(f) > \rho(g)$, then $\rho(f+g) = \rho(f)$.*

Theorem B.24.3: *Let $f \in \mathcal{A}^*(E)$ and let $P \in \mathbb{K}[x]$ be non-constant. Then $\rho(P \circ f) = \rho(f)$.*

Theorem B.24.4: *Let $f,\ g \in \mathcal{A}^*(E)$. Then $\psi(fg) \leq \psi(f) + \psi(g)$. Moreover, if $\rho(f) = \rho(g)$, then $\max(\psi(f), \psi(g)) \leq \psi(fg)$.*

Proof. Set $\rho(f) = s,\ \rho(g) = t$. Without loss of generality, we can assume $s \geq t$. By Theorem 1, we have $\rho(f.g) = \rho(f) = s$. Now, for each $r > 0$, we have $\zeta(f.g, r) = \zeta(f, r) + \zeta(g, r)$, hence

$$\psi(fg) = \limsup_{r \to R^-}(\zeta(f,r) + \zeta(g,r))(R-r)^s \leq \limsup_{r \to R^-} \zeta(f,r)(R-r)^s$$
$$+ \limsup_{r \to R^-} \zeta(g,r)(R-r)^t,$$

hence $\psi(fg) \leq \psi(f) + \psi(g)$. Now, suppose $s = t$. Then

$$\psi(fg) = \limsup_{r \to R^-}(\zeta(f,r) + q(g,r))(R-r)^s \geq \limsup_{r \to R^-} \max(\zeta(f,r), \zeta(g,r))(R-r)^s$$
$$= \max(\psi(f), \psi(g)).$$

□

Remark 1: Let $f \in \mathcal{A}^*(E)$. If $s > \theta(f)$, then by definition, $\lim_{r \to R^-} \zeta(f,r)(R-r)^s = 0$. But if $s < \theta(f)$ then $\limsup_{r \to R^-} \zeta(f,r)(R-r)^s = +\infty$ because if $\limsup_{r \to R^-} \zeta(f,r)(R-r)^s < +\infty$, we can find $s' \in]s, \theta(f)[$ and then we can check that $\lim_{r \to R^-} \zeta(f,r)(R-r)^{s'} = 0$, a contradiction.

Thanks to the classical inequality $|f'|(r) \leq \dfrac{|f|(r)}{r}$ [58], Theorem B.24.5 is then immediate.

Theorem B.24.5: *Suppose* $\mathbb{K}$ *has characteristic* 0. *Let* $f \in \mathcal{A}^*(E)$. *Then* $\rho(f') \leq \rho(f)$.

Remark 2: In a field of characteristic $p \neq 0$, certain analytic functions have a null derivative. This is why we must suppose that $\mathbb{K}$ has characteristic 0 in all statements involving derivatives.

In complex analysis, many estimates were given concerning the growth order of solutions of linear differential equations. Here, by Corollary B.24.2 and Theorem B.24.5 we can immediately obtain Corollary B.24.6.

Corollary B.24.6: *Suppose* $\mathbb{K}$ *has characteristic* 0. *Consider the differential equation*

$$(\mathcal{E})\ f^{(n)} + a_{n-1}(x)f^{(n-1)}(x) + \cdots + a_0(x)f(x) = 0$$

with $a_j \in \mathcal{A}^*(E)$, $j = 0, \dots, n-1$ *and* $\rho(a_j) < \rho(a_0)\ \forall j = 1, \dots, n-1$. *Then every non-trivial solution* f *of* $(\mathcal{E})$ *satisfies* $\rho(f) \geq \rho(a_0)$.

Theorem B.24.7: *Suppose* $\mathbb{K}$ *has residue characteristic* 0. *Then for every* $f \in \mathcal{A}^*(D)$, *we have* $\rho(f') = \rho(f)$, $\theta(f') = \theta(f)$, $\sigma(f') = \sigma(f)$, *and* $\psi(f') = \psi(f)$.

Remark 3: Theorem B.24.7 does not hold in residue characteristic $p > 0$ because there exist functions $f \in \mathcal{A}^*(D)$ such that $\rho(f) > 0$ and that f' is bounded, as shows the following example with $R = 1$: $g(x) = \sum_{m=0}^{\infty} \frac{x^{p^m}}{p^m}$. We can see that $g'(x) = \sum_{n=0}^{\infty} x^{p^m-1}$, hence g' is bounded and therefore $\rho(g') = 0$. However, consider the sequence $(r_m)_{m \in \mathbb{N}}$ defined as $r_m = 1 - \frac{1}{p^m}$. We can check that $|g|(r_m) \geq p^m (r_m)^{p^m}$, hence

$$\log(|g|(r_m)) \geq m + p^m \log(r_m) = m + p^m \log\left(1 - \frac{1}{p^m}\right).$$

When m is big enough, we have $\log\left(1 - \frac{1}{p^m}\right) \geq \frac{-2}{p^m}$, hence

$$\log\left(|g|(r_m)\right) \geq m - p^m \left(\frac{2}{p^m}\right) = m - 2.$$

Therefore, when m is big enough, we have

$$\frac{\log\left(\log(|g|(r_m))\right)}{-\log(r_m)} \geq \frac{\log(m-2)}{-\log(1 - \frac{1}{p^m})} > \frac{\log(m-2)}{\frac{2}{p^m}} = \frac{p^m}{2}\log(m-2).$$

Thus, we have $\rho(g) = +\infty$.

Remark 4: Theorem B.24.7 applies for instance to the complex Levi-Civita field whose residue characteristic is 0 [12].

Theorem B.24.8: *Let $f \in \mathcal{A}^*(E)$. Then $\lambda(f) = \rho(f)$.*

Proof. First we will prove that $\rho(f) \leq \lambda(f)$. Obviously, we can assume that $\lambda(f) < +\infty$. Let s be such that $\lim_{r \to R^-} \log(|f|(r))(R-r)^s = 0$. Let us fix $\epsilon > 0$. For r close enough to R, we have $\log(|f|(r))(R-r)^s \leq \epsilon$, hence $\log(|f|(r)) \leq \frac{\epsilon}{(R-r)^s}$, therefore $\log(\log(|f|(r))) \leq \log \epsilon - s\log(R-r)$, hence

$$\frac{\log(\log(|f|(r)))}{(-\log(R-r))} \leq \frac{\log(\epsilon)}{(-\log(R-r))} + s,$$

and hence

$$\limsup_{r \to R^-} \frac{\log(\log(|f|(r)))}{(-\log(R-r))} \leq s,$$

i.e., $\rho(f) \leq s$. This is true for every s such that $\lim_{r \to R^-} \log(|f|(r))(R-r)^s = 0$ and hence $\rho(f) \leq \lambda(f)$.

On the other hand, we notice that, by definition of $\lambda(f)$, either $\lambda(f) = 0$ and then $\lambda(f) \leq \rho(f)$, or

$$\lambda(f) = \sup\{s \in]0, +\infty[\mid \limsup_{r \to R^-} \log(|f|(r))(R-r)^s > 0\}.$$

Thus, suppose that $\lambda(f) > 0$. Let us take $s \in]0, \lambda(f)[$. We have a number $b > 0$ such that

$$\limsup_{r \to R^-} (\log(|f|(r)(R-r)^s) \geq b > 0.$$

Let us fix $\epsilon \in]0, b[$. There exists a sequence $(r_n)_{n \in \mathbb{N}}$ in $]0, R[$ such that $\lim_{n \to +\infty} r_n = R$ and such that, when n is big enough, we have $b - \epsilon \leq \log(|f|(r_n))(R-r_n)^s$, hence $-s\log(R-r_n) + \log(b-\epsilon) < \log(\log(|f|(r_n)))$, therefore

$$s + \frac{\log(b-\epsilon)}{(-\log(R-r_n))} \leq \frac{\log(\log(|f|(r_n)))}{(-\log(R-r_n))}.$$

Consequently, $\limsup_{n \to +\infty} \frac{\log(\log(|f|(r_n)))}{(-\log(R-r_n))} \geq s$, therefore $\rho(f) \geq s$. But this holds for every $s < \lambda(f)$. Thus, $\rho(f) \geq \lambda(f)$ and finally, $\rho(f) = \lambda(f)$. □

Theorem B.24.9: *Let f, $g \in \mathcal{A}^*(E)$. Then $\sigma(fg) \leq \sigma(f) + \sigma(g)$. If $\rho(f) \geq \rho(g)$, then $\sigma(f) \leq \sigma(fg)$. If $\rho(f) = \rho(g)$, then $\max(\sigma(f), \sigma(g)) \leq \sigma(fg)$.*

If $\rho(f) = \rho(g)$ and $\sigma(f) > \sigma(g)$, then $\sigma(f+g) \geq \sigma(f)$. If $\rho(f+g) = \rho(f) \geq \rho(g)$, then $\sigma(f+g) \leq \max(\sigma(f), \sigma(g))$.

Proof. Let $s = \rho(f)$, $t = \rho(g)$, and suppose $s \geq t$. When r is close enough to R, we have $\max(\log(|f|(r)), \log(|g|(r)) \leq \log(|f.g|(r)) = \log(|f|(r)) + \log(|g|(r))$ and by Theorem 1, we have $\rho(fg) = s$. Therefore,

$$\begin{aligned}\sigma(fg) &= \limsup_{r \to R^-} \big(\log(|f.g|(r))(R-r)^s\big) \\ &\leq \limsup_{r \to R^-} \big(\log(|f|(r))(R-r)^s\big) \\ &\quad + \limsup_{r \to R^-} \big(\log(|g|(r))(R-r)^t\big) = \sigma(f) + \sigma(g).\end{aligned}$$

On the other hand,

$$\sigma(f) = \limsup_{r \to R^-} \log(|f|(r))(R-r)^s \leq \limsup_{r \to +R^-} (\log(|fg|(r)))(R-r)^s.$$

But $\rho(fg) = s$, hence $\sigma(f) \leq \sigma(fg)$. Particularly, if $\rho(f) = \rho(g)$, then $\max(\sigma(f), \sigma(g)) \leq \sigma(fg)$.

Now, suppose again that $\rho(f) = \rho(g) = s$ and suppose $\sigma(f) > \sigma(g)$. Let $s = \rho(f)$, $b = \sigma(f)$. Then $b > 0$. Let $(r_n)_{n \in \mathbb{N}}$ be a sequence such that $\lim_{n \to +\infty} r_n = R$ and $\lim\limits_{n \to +\infty} (\log(|f|(r_n))(R-r_n)^s) = b$. Since $\sigma(g) < \sigma(f)$, we notice that when n is big enough we have $|g|(r_n) < |f|(r_n)$. Consequently, when n is big enough, we have $|f+g|(r_n) = |f|(r_n)$ and hence

$$\lim_{n \to +\infty} (\log(|f+g|(r_n)))(R-r_n)^s) = b. \tag{1}$$

By definition of σ, we have $\sigma(f+g) \geq \lim\limits_{n \to +\infty} (\log(|f+g|(r_n)))(R-r_n)^{\rho(f+g)}$. By Theorem 1, we have $\rho(f+g) \leq s$, hence

$$\begin{aligned}\sigma(f+g) &\geq \lim_{n \to +\infty} (\log(|f+g|(r_n)))(R-r_n)^{\rho(f+g)} \\ &\geq \lim_{n \to +\infty} (\log(|f+g|(r_n)))(R-r_n)^s \\ &= \lim_{n \to +\infty} \log(|f|(r_n))(R-r_n)^s = \sigma(f),\end{aligned}$$

therefore by (1), $\sigma(f+g) \geq \sigma(f)$.

Finally, suppose now that $\rho(f+g) = \rho(f) \geq \rho(g)$. Let $s = \rho(f)$ and $t = \rho(g)$. Then

$$\begin{aligned}\sigma(f+g) &= \limsup_{r \to R^-} (\log(|f+g|(r)))(R-r)^s \\ &\leq \max\Big(\limsup_{r \to R^-} (\log(|f|(r)))(R-r)^s, \limsup_{r \to R^-} (\log(|g|(r)))(R-r)^s\Big) \\ &\leq \max\Big(\limsup_{r \to R^-} (\log(|f|(r)))(R-r)^s, \limsup_{r \to R^-} (\log(|g|(r)))(R-r)^t\Big) \\ &= \max(\sigma(f), \sigma(g)),\end{aligned}$$

which ends the proof. □

Corollary B.24.10: *Let $f,\ g \in \mathcal{A}^*(E)$ be such that $\rho(f) \neq \rho(g)$. Then*

$$\sigma(f+g) \leq \max(\sigma(f), \sigma(g)).$$

Lemma B.24.11: *Let $a \in [1, +\infty[$ and $b \in [0, +\infty[$. Then* $\log(a+b) \leq \log(a) + \log(b+1)$.

Proof. Indeed, since $a \geq 1$, we have $\log(a+b) \leq \log(a(b+1)) = \log(a) + \log(b+1)$. □

Theorem B.24.12: *Let $f \in \mathcal{A}^*(E)$. Then $\theta(f) - 1 \leq \rho(f) \leq \theta(f)$. Moreover, if $0 < \psi(f) < \infty$, then $\rho(f) = \theta(f)$.*

Proof. We will denote by $|\ .\ |_\infty$ the Archimedean absolute value of $\mathbb{R}$. Let us first choose $s > \theta(f)$. Then $\lim\limits_{r \to R^-} \zeta(f,r)(R-r)^s = 0$. Now, since $\lim\limits_{r \to R^-} |f|(r) = +\infty$, we can take $\ell \in]0, R[$ such that $|f|(\ell) > 1$. Then we can take $b > 0$ such that

$$\zeta(f,r) \leq b(R-r)^{-s}\ \forall r \in [\ell, R[.$$

Now, taking $r \in [\ell, R[$, by Theorem B.13.26, we have

$$\log(|f|(r)) \leq \log(|f|(\ell))) + \zeta(f,r)\left(\log\left(\frac{r}{\ell}\right)\right),$$

which leads to

$$\log(|f|(r)) \leq \log(|f|(\ell))) + b(R-r)^{-s}\left(\log\left(\frac{r}{\ell}\right)\right),$$

hence

$$\log(\log(|f|(r))) \leq \log\Big(\log(|f|(\ell))) + b(R-r)^{-s}\left(\log\left(\frac{r}{\ell}\right)\right)\Big)$$

therefore, by Lemma B.24.11, we can derive

$$(1)\qquad \log(\log(|f|(r))) \leq \log(\log(|f|(\ell))) + \log\left(b(R-r)^{-s}\left(\log\left(\frac{r}{\ell}\right)\right) + 1\right).$$

Now, since $s > 0$, there obviously exists $h \in [\ell, R[$ such that $b(R-r)^{-s} \geq 1\ \forall r \in [h, R^-[$, therefore by Lemma B.24.11 again,

$$\log(\log(|f|(r))) \leq \log(\log(|f|(\ell))) + \log\left(b(R-r)^{-s}\left(\log\left(\frac{r}{\ell}\right)\right)\right) + \log(1+1),$$

i.e.,

(2)

$$\log(\log(|f|(r))) \leq \log(\log(|f|(\ell))) + \log(b) - s\log(R-r) + \log\left(\left(\log\left(\frac{r}{\ell}\right)\right) + \log(2)\right).$$

Consequently, by (2), we obtain

$$\frac{\log(\log(|f|(r)))}{-\log(R-r)} \leq \frac{\log(\log(|f|(\ell)))}{-\log(R-r)} + \frac{\log(b)}{-\log(R-r)} + s + \frac{\log(\log(\frac{r}{\ell})) + \log(2)}{-\log(R-r)}.$$

We can check that

$$\lim_{r\to R^-}\frac{\log(\log(|f|(\ell)))+\log(b)}{-\log(R-r)}=\lim_{r\to R^-}\frac{\log(\log(\frac{r}{\ell})+\log(2)}{-\log(R-r)}=0,$$

and hence $\limsup_{r\to R^-}\frac{\log(\log(|f|(r)))}{-\log(R-r)}\leq s$. Consequently, choosing $\epsilon>0$, there exists $u\in[\ell,1[$ such that $\frac{\log(\log(|f|(r)))}{-\log(R-r)}\leq s+\epsilon\ \forall r\in[u,R[$ and hence $\rho(f)\leq s+\epsilon$. But since that holds for every $s>\theta(f)$ and for every $\epsilon>0$, we have $\rho(f)\leq s$ and hence $\rho(f)\leq\theta(f)$.

Let us now show that $\rho(f)\geq\theta(f)-1$. By Theorem B.13.26, we have

$$\log(|f|(r))-\log\left(|f|\left(\frac{r^2}{R}\right)\right)\geq\zeta\left(f,\frac{r^2}{R}\right)\left(\log(r)-\log\left(\frac{r^2}{R}\right)\right)\tag{3}$$
$$=\zeta\left(f,\frac{r^2}{R}\right)(\log(R)-\log(r)).$$

Consider now a number $s<\theta(f)$ and a sequence $(r_n)_{n\in\mathbb{N}}$ of $]0,R[$ such that $\lim_{n\to+\infty}r_n=R$ and such that $\limsup_{n\to+\infty}\zeta(f,r_n)(R-r_n)^s\geq b>0$. Then by (3) we have

$$\log(|f|(r_n))\geq\frac{b(\log(R)-\log(r_n))}{\left(R-\frac{r_n^2}{R}\right)^s}.$$

Consequently,

$$\log(\log(|f|(r_n)))\geq\log(b)+\log(\log(R)-\log(r_n)))-s\big(\log(R-r_n)$$
$$+\log(R+r_n)\big)+2s\log(R),$$

and therefore

$$\frac{\log(\log(|f|(r_n)))}{-\log(R-r_n)}\geq\frac{\log(b)}{-\log(R-r_n)}+\frac{\log(\log(R)-\log(r_n))}{-\log(R-r_n)}$$
$$+s\left(1+\frac{\log(R+r_n)+2\log(R)}{-\log(R-r_n)}\right).$$

Clearly,

$$\lim_{n\to+\infty}\left(\frac{\log(b)}{\log(R-r_n)}\right)=\lim_{n\to+\infty}\frac{\log(R+r_n)+2\log(R)}{\log(R-r_n)}=0,$$

and by elementary reasonings, we can check that

$$\lim_{t\to R^-}\frac{\log(\log(R)-\log(t))}{\log(R-t)}=1,$$

therefore

$$\lim_{n\to+\infty}\frac{\log(\log(R)-\log(r_n))}{\log(R-r_n)}=1.$$

Consequently,

$$\limsup_{n\to+\infty}\frac{\log(\log(|f|(r_n)))}{-\log(R-r_n)}\geq s-1,$$

and therefore

$$\limsup_{r\to R^-}\frac{\log(\log(|f|(r)))}{-\log(R-r)}\geq s-1.$$

That holds for every $s<\theta(f)$ and shows that if $\theta(f)<+\infty$, then $\rho(f)\geq\theta(f)-1$. Next, if $\theta(f)=+\infty$, then we would have $\rho(f)=+\infty$, which is excluded by hypothesis since $f\in\mathcal{A}^*(E)$. Consequently, the inequality $\rho(f)\geq\theta(f)-1$ is established.

Let us now show that $\rho(f)\geq\theta(f)$ when $\psi(f)<+\infty$. Suppose $\theta(f)>\rho(f)$ and let $s\in]\rho(f),\theta(f)[$. Then by Remark 1, we have $\limsup\limits_{r\to R^-}\zeta(f,r)(R-r)^s=+\infty$, but then $\limsup\limits_{r\to R^-}\zeta(f,r)(R-r)^{\rho(f)}=+\infty$, i.e., $\psi(f)=+\infty$, a contradiction. Therefore, $\theta(f)\leq\rho(f)$ and hence whenever $\psi(f)<+\infty$, we have $\theta(f)=\rho(f)$. □

Theorem B.24.12 obviously suggests the following conjecture:

Conjecture: *Let $f\in\mathcal{A}^*(E)$. Then $\rho(f)=\theta(f)$.*

Theorem B.24.13 is much different from the relations concerning ρ, σ, and ψ obtained for entire functions.

Theorem B.24.13: *Let $f\in\mathcal{A}^*(E)$ be such that, $\psi(f)<+\infty$. Then $\sigma(f)=0$.*

Proof. Without loss of generality, we can assume that $f(0)\neq 0$. Let us fix $\epsilon>0$ and let R' be such that $\log(R)-\log(R')=\epsilon$. Let $(a_n)_{n\in\mathbb{N}}$ be the sequence of zeros of f, for each $n\in\mathbb{N}$, let w_n be the order of a_n and let $r_n=|a_n|$. Now, let u be the biggest integer n such that $r_n<R'$ and for each $r>0$, let $m(r)$ be the biggest integer n such that $r_n\leq r$.

Let $A_u=\sum_{n=0}^u w_n$ and let $B_u=\log(|f(0)|)+\sum_{n=0}^u w_n(\log(R')-\log(r_n))$. Let us take $r\in]R',R[$. Now, we can write

$$\frac{\sigma(r,f)}{\psi(r,f)}=\frac{B_u+\sum_{n=u+1}^{m(r)}w_n(\log(r)-\log(r_n))}{A_u+\sum_{n=u+1}^{m(r)}w_n}.$$

But by hypothesis, $\log(r)-\log(r_n)\leq\epsilon\ \forall n\geq u$, hence

$$\frac{\sigma(r,f)}{\psi(r,f)}\leq\frac{B_u+\epsilon\sum_{n=u+1}^{m(r)}w_n}{A_u+\sum_{n=u+1}^{m(r)}w_n}.$$

Let us put $\phi(r) = \sum_{n=u+1}^{m(r)} w_n$. Thus,

$$\frac{\sigma(f,r)}{\psi(f,r)} \leq \frac{B_u + \epsilon\phi(r)}{A_u + \phi(r)}.$$

But since f belongs to $\mathcal{A}^*(D)$, it has infinitely many zeros in D, hence $\phi(r)$ is an increasing unbounded function tending to $+\infty$ when r tends to R. Consequently, it is obvious that

$$\lim_{r\to R} \frac{\sigma(r,f)}{\psi(r,f)} = 0.$$

Therefore, if $\limsup_{r\to R^-} \psi(r,f) < +\infty$, then $\sigma(f) = 0$. □

C. Meromorphic functions and Nevanlinna theory

C.1. Meromorphic functions in $\mathbb{K}$

In this chapter, we will define and examine the basic properties of meromorphic functions: relations with poles of analytic elements, absolute values on fields of meromorphic functions defined by circular filters, value of the derivative on a circular filter, development in a Laurent series in an annulus, and existence of primitives [51–53].

Definitions and notations: We denote by $\mathcal{M}(\mathbb{K})$ the field of fractions of $\mathcal{A}(\mathbb{K})$. The elements of $\mathcal{M}(\mathbb{K})$ are called *meromorphic functions in* $\mathbb{K}$.

In the same way, given $a \in \mathbb{K}$ and $r > 0$, we denote by $\mathcal{M}(d(a, r^-))$ (resp. $\mathcal{M}_b(d(a, r^-))$, resp. $\mathcal{M}_u(d(a, r^-))$) the field of fractions of $\mathcal{A}(d(a, r^-))$ (resp. the field of fractions of $\mathcal{A}_b(d(a, r^-))$, resp. the set $\mathcal{M}(d(a, r^-)) \setminus \mathcal{M}_b(d(a, r^-))$).

Let $b \in \mathbb{K}$ (resp. $b \in d(a, R^-)$) and let $r \in \mathbb{R}_+^*$ (resp. $r \in]0, R[$). The absolute value $\varphi_{b,r}$ defined on $\mathcal{A}(\mathbb{K})$ (resp. on $\mathcal{A}(d(a, R^-))$) has an immediate continuation to $\mathcal{M}(\mathbb{K})$ (resp. to $\mathcal{M}(d(a, R^-))$) that we shall denote again by $\varphi_{b,r}$. In the same way, $\varphi_{0,r}$ will be denoted by $|\ .\ |(r)$ on $\mathcal{M}(\mathbb{K})$ and on $\mathcal{M}(d(0, R^-))$. Similarly, the function $\Psi(\ .\ \mu)$ defined on $\mathcal{A}(\mathbb{K})$ and on $\mathcal{A}(d(0, R^-))$ has an immediate continuation to $\mathcal{M}(\mathbb{K})$ and to $\mathcal{M}(d(0, R^-))$ as $\Psi(\frac{h}{l}, \mu) = \Psi(h, \mu) - \Psi(l, \mu)$, with $h, l \in \mathcal{A}(\mathbb{K})$ (resp. $h, l \in \mathcal{A}(d(0, R^-))$).

Let $f = \frac{h}{l} \in \mathcal{M}(\mathbb{K})$ (resp. $f = \frac{h}{l} \in \mathcal{M}(d(a, R^-))$). For each $\alpha \in \mathbb{K}$ (resp. $\alpha \in d(a, R^-)$), the number $\omega_\alpha(h) - \omega_\alpha(l)$ does not depend on the functions h, l chose to make $f = \frac{h}{l}$. Thus, we can generalize the notation by setting $\omega_\alpha(f) = \omega_\alpha(h) - \omega_\alpha(l)$.

If $\omega_\alpha(f)$ is an integer $q > 0$, α is called *a zero of f of order q*.

If $\omega_\alpha(f)$ is an integer $q < 0$, α is called *a pole of f of order $-q$*.

If $\omega_\alpha(f) \geq 0$, f will be said to be *holomorphic* at α.

Similarly, as for $\mathcal{A}(\mathbb{K})$, given $f \in \mathcal{M}(\mathbb{K})$ (resp. $f \in \mathcal{M}(d(a, R^-))$), we can define the divisor $\mathcal{D}(f)$ on $\mathbb{K}$ (resp. of $d(a, R^-)$) as $\mathcal{D}(f)(\alpha) = 0$ whenever $f(\alpha) \neq 0$ and $\mathcal{D}(f)(\alpha) = s$ when f has a zero of order s at α.

Lemma C.1.1: *Let $r, R \in \mathbb{R}_+^*$ with $0 < r < R$ and let $f \in \mathcal{M}(d(a, R^-))$. Then f has finitely many poles $a_1, \dots, a_q$ in $d(a, r^-)$. Let $E = d(a, r) \setminus \{a_1, \dots, a_q\}$. Then f belongs to $H(E)$. If there exists $s \in \mathbb{N}^*$ such that f^s is a constant, then so is f.*

Proof. Without loss of generality, we can assume $a = 0$. Let $f = \frac{h}{l}$ with $h,\ l \in \mathcal{A}(d(0, R))$. Since l belongs to $\mathcal{A}(d(0, R))$, by Corollary B.13.19 l has finitely many zeros in $d(0, r)$, hence f has finitely many poles $a_1, \dots, a_q$ in $d(0, r)$. Suppose first f is of the form $\frac{1}{l}$ with $l \in \mathcal{A}(d(0, R^-))$. By Corollary B.5.16, l factorizes in the form $P(x)u(x)$ with $P \in \mathbb{K}[x]$ a polynomial whose zeros in $d(a, r)$ are $a_1, \dots, a_q$ and $u \in \mathcal{A}(d(0, R^-))$ is invertible in $H(d(0, r))$. On the other hand, $\frac{1}{P(x)}$ obviously belongs to $R(E)$. And by Proposition B.8.3, E belongs to Alg. Consequently, l is invertible in $H(E)$. Consider now the general case $f = \frac{h}{l}$ with $h,\ l \in \mathcal{A}(d(0, R))$. Then both h, $\frac{1}{l}$ belong to $H(E)$), hence by Proposition B.8.3, so does f.

Suppose now that f^s is a constant. Since $\mathbb{K}$ is algebraically closed and since $\mathcal{M}(d(a, R^-))$ is a field extension of $\mathbb{K}$, f belongs to $\mathbb{K}$. □

Corollary C.1.2: *Let $f \in \mathcal{M}(d(a, R^-))$, let $r \in]0, R[$, let $\alpha_j, 1 \leq j \leq q$ be the poles of f in $d(a, r)$, let $\rho \in]0, \min_{i \neq j} |\alpha_i - \alpha_j|[$, and for each $j = 1, \dots, q$, let $\rho_j \in]0, \rho[$, let $T_j = d(\alpha_j, \rho_j^-)$. Let $D = d(a, r) \setminus \left(\bigcup_{j=1}^q T_j \right)$. Then f belongs to $H(D)$.*

Lemma C.1.3: *Let $f \in \mathcal{M}(\mathbb{K})$. There exists $h \in \mathcal{A}(\mathbb{K})$ such that $\mathcal{D}(h) = \mathcal{D}(f)$ and then the function $l = \frac{h}{f}$ belongs to $\mathcal{A}(\mathbb{K})$. Then $\mathcal{D}(\frac{1}{f}) = \mathcal{D}(l)$ and we can write f in the form $\frac{h}{l}$ with $h,\ l \in \mathcal{A}(\mathbb{K})$, having no common zero.*

Proof. Indeed, by Theorem B.18.4 there exists $h \in \mathcal{A}(\mathbb{K})$ such that $\mathcal{D}(h) = \mathcal{D}(f)$ and hence conclusion follows. □

Remark: Let $f \in \mathcal{M}(d(a, R^-))$, let $r \in]0, R[$, and let $\alpha_j,\ 1 \leq j \leq n$ be the poles of f in $d(a, r)$, of respective order q_j. By Lemma C.1.1, f belongs to $H(d(a, r) \setminus \{\alpha_1, \dots, \alpha_n\})$. Now, according to the definition of poles for analytic elements (see Chapter B.2), f also admits each α_j as a pole of order q_j, considered as an element of $H(d(a, r) \setminus \{\alpha_1, \dots, \alpha_n\})$.

By Theorem B.18.4, we have already seen that if $f \in \mathcal{M}(\mathbb{K})$ has no zero and no pole in $\mathbb{K}$, then it is a constant. Here, we can generalize that with functions inside a disk.

Theorem C.1.4: *Let $f \in \mathcal{M}(\mathbb{K})$ (resp. $f \in \mathcal{M}(d(a, R^-))$) have no pole in $\mathbb{K}$ (resp. in $d(a, R^-)$). Then f belongs to $\mathcal{A}(\mathbb{K})$ (resp. to $\mathcal{A}(d(a, R^-))$).*

Proof. Suppose $f \in \mathcal{M}(\mathbb{K})$ has no pole in $\mathbb{K}$. By Lemma C.1.3, we can write f in the form $\frac{h}{l}$ with $\mathcal{D}(f) = \mathcal{D}(h)$. Since f has no pole in $\mathbb{K}$, l has no zero and hence is a constant, which ends the proof when f belongs to $\mathcal{M}(\mathbb{K})$.

Suppose now that f belongs to $\mathcal{M}(d(a, R^-))$ and has no pole in $d(a, R^-)$. By Proposition B.5.12, it is sufficient to show that for each $\rho \in]0, R[$, f belongs to $H(d(a, \rho))$. Let $f = \frac{h}{l}$, with $h,\ l \in \mathcal{A}(d(a, R^-))$. By Proposition B.5.12, both h, l belong to $H(a, \rho))$.

By hypothesis, each zero α of l is a zero of h such that $\omega_\alpha(h) \geq \omega_\alpha(l)$. Let P be the polynomial admitting for zeros; the zeros of l inside $d(a, \rho)$ with the same multiplicity and no other zero. Then P divides h and l in $\mathcal{A}(d(a, R^-))$, say $h = P\phi$, $l = P\psi$. So, ψ is a power series that has no zero in $d(a, \rho)$, hence by Theorem B.7.9, it is an invertible element of $H(d(a, \rho))$, which ends the proof. □

Corollary C.1.5: *Let $f,\ g \in \mathcal{A}(\mathbb{K})$ (resp. $f,\ g \in \mathcal{A}(d(a, R^-))$) be such that $\mathcal{D}(g) \leq \mathcal{D}(f)$. There exists $h \in \mathcal{A}(\mathbb{K})$ (resp. $h \in \mathcal{A}(d(a, R^-))$ such that $f = gh$.*

Proof. Indeed, $\frac{f}{g}$ belongs to $\mathcal{M}(\mathbb{K})$ (resp. to $\mathcal{M}(d(a, R^-))$) and has no pole. □

Corollary C.1.6: *Let $f \in \mathcal{M}(\mathbb{K})$ (resp. $f \in \mathcal{M}(d(a, R^-))$) have no zero and no pole in $\mathbb{K}$ (resp. in $d(a, R^-)$). Then it is a constant (resp. an invertible element of $\mathcal{A}_b(d(a, R^-))$).*

Corollary C.1.7: *Let $f,\ g \in \mathcal{A}(\mathbb{K})$ (resp. $f,\ g \in \mathcal{A}(d(a, R^-))$) satisfy $\mathcal{D}(f) = \mathcal{D}(g)$. Then $\frac{f}{g}$ belongs to $\mathbb{K}$ (resp. is invertible in $\mathcal{A}_b(d(a, R^-))$).*

Corollary C.1.8: *Let $f \in \mathcal{A}(\mathbb{K})$ be such that $\mathcal{D}(f) = (a_n, q_n)_{n \in \mathbb{N}}$ with $a_n \neq 0$ $\forall n \in \mathbb{N}$, $\lim_{n \to +\infty} |a_n| = +\infty$. Then $f(x)$ is of the form $\lambda \prod_{n=0}^{\infty} \left(1 - \frac{x}{a_n}\right)^{q_n}$ with $\lambda \in \mathbb{K}$.*

By Theorem B.19.6, Lemma C.1.9 and Corollary C.1.10 are immediate.

Lemma C.1.9: *Let $\mathbb{K}$ be spherically complete, let $a \in \mathbb{K}$, $r \in \mathbb{R}_+^*$, let $\mathcal{B}$, $\mathcal{C}$ be divisors on $d(a, R^-)$. There exists $f \in \mathcal{M}(d(a, R^-))$ such that $\mathcal{D}(f) = \mathcal{B}$ and $\mathcal{D}(\frac{1}{f}) = \mathcal{C}$.*

Corollary C.1.10: *Let $\mathbb{K}$ be spherically complete, let $a \in \mathbb{K}$, $r \in \mathbb{R}_+^*$ and let $f \in \mathcal{M}(d(a, R^-))$. There exists $g,\ h \in \mathcal{A}(d(a, R^-))$, having no common zero, such that $f = \frac{g}{h}$.*

Lemma C.1.11: *Let $a \in \mathbb{K}$, $r \in \mathbb{R}_+^*$ and let $f \in \mathcal{A}(d(a, R^-))$. If $\mathbb{K}$ is spherically complete, there exists $h \in \mathcal{A}(d(a, R^-))$ such that $\mathcal{D}(h) = \mathcal{D}(f)$ and then the function*

$l = \frac{h}{f}$ *belongs to* $\mathcal{A}(d(a, R^-))$. *We have* $\mathcal{D}(\frac{1}{f}) = \mathcal{D}(l)$ *and we can write* f *in the form* $\frac{h}{l}$ *with* $h,\ l \in \mathcal{A}(d(a, R^-))$, *having no common zero.*

Remark: If $\mathbb{K}$ is not spherically complete, in the general case, as shown in Theorem B.19.1, we cannot find an analytic function $h \in \mathcal{A}(d(a, R^-))$ such that $\mathcal{D}(h) = \mathcal{D}(f)$. Consequently, in a field such as $\mathbb{C}_p$, we can't write f in the form $f = \frac{h}{l}$ with $h,\ l \in \mathcal{A}(d(a, R^-))$, having no common zero (this gap was forgotten in several works).

However, by Theorem A.7.4 we can take an algebraically closed spherically complete extension $\widehat{\mathbb{K}}$ of $\mathbb{K}$ and consider f as an analytic function on the disk $\widehat{d}(a, R^-)$ in the field $\widehat{\mathbb{K}}$: then f may be written in the form $f = \frac{\widehat{h}}{\widehat{l}}$ with $\widehat{h},\ \widehat{l} \in \mathcal{A}(\widehat{d}(a, R^-))$, with $\widehat{h},\ \widehat{l}$ having no common zero.

Theorem C.1.12: *Let* $r \in \mathbb{R}_+^*$, *let* $f(x) = \mathcal{M}(\mathbb{K})$ *(resp.* $f \in \mathcal{M}(d(a, r^-))$*), let* S *be the set of zeros and poles of* f *in* $\mathbb{K}$ *(resp. in* $d(a, r^-)$*). Let* t *be the g.c.d. of* $\{\omega_\alpha(f) \mid \alpha \in S\}$ *and let* $n \in \mathbb{N}^*$. *If there exists* $g \in \mathcal{M}(\mathbb{K})$ *(resp.* $g \in \mathcal{M}(d(a, r^-))$*) such that* $g^q = f$, *then* q *divides* t. *Conversely, if* q *divides* t, *then there exists* $g \in \mathcal{M}(\mathbb{K})$ *such that* $g^q = f$ *(resp. if* p *is prime to* q *and if* q *divides* t *then there exists* $g \in \mathcal{M}(d(a, R^-))$ *such that* $g^q = f$*).*

Proof. If there exists $g \in \mathcal{M}(d(a, r^-))$ (resp. $g \in \mathcal{M}(\mathbb{K})$) such that $g^q = f$, then of course, $\omega_\alpha(g)$ divides $\omega_\alpha(f)$ for every $\alpha \in S$ and hence it divides t. Now suppose q divides t and set $t = lq$. For each $\alpha \in S$, $\omega_\alpha(f)$ is of the form $ts_\alpha = qls_\alpha$.

Suppose first $f \in \mathcal{M}(\mathbb{K})$. By Lemma C.1.3, in $\mathcal{M}(\mathbb{K})$ there exists $g \in \mathcal{M}(\mathbb{K})$ admitting each zero α of f as a zero of order ls_α and each pole α of f as a pole of order $-ls_\alpha$. Then $\frac{f}{g^q}$ has no zero and no pole in $\mathbb{K}$, hence, by Corollary C.1.6, it is a constant λ. Let v be a $q-th$ root of λ. Then $f = (vg)^q$.

Now suppose that q is prime to p and suppose $f \in \mathcal{M}(d(a, R^-))$. Suppose first that $\mathbb{K}$ is spherically complete. By Lemma C.1.9, there exists $g \in \mathcal{M}(d(a, R^-))$ admitting each zero α of f as a zero of order ls_α and each pole α of f as a pole of order $-ls_\alpha$. Then $\frac{f}{g^q}$ has no zero and no pole in $d(a, R^-)$, hence it belongs to $\mathcal{A}(d(a, r^-))$. But since it has no zero, by Corollary B.14.2 it satisfies $|h(x) - h(a)| < |h(a)|\ \forall x \in d(a, r^-)$. Let $\psi(x) = \frac{h(x)}{h(a)}$. Then we have $|\psi(x) - 1| < 1\ \forall x \in d(a, r^-)$ and then, since q is prime to p, by Theorem B.20.23 we can apply the function $\sqrt[q]{.}$ to $\psi(x)$ in order to get a function $\sqrt[q]{\psi(x)} \in \mathcal{A}(d(a, r^-))$. Now, let v be a q-th root of $h(a)$. We have $f(x) = h(a)\psi(x)(g(x))^q = \left(v \sqrt[n]{\psi(x)} g(x)\right)^q$, which ends the proof when $\mathbb{K}$ is spherically complete.

Consider now the general case, when $\mathbb{K}$ is no longer supposed to be spherically complete. Let $\widehat{\mathbb{K}}$ be a spherically complete algebraically closed extension of $\mathbb{K}$. The function f has continuation to a function $\widehat{f}$, which belongs to $\mathcal{A}(\widehat{d}(\alpha, R^-))$ and hence there exists a function $g \in \mathcal{A}(\widehat{d}(\alpha, R^-))$ such that $g^q = f$. Then by Lemma B.5.5, g is a power series that has all coefficients in $\mathbb{K}$ and hence belongs to $\mathcal{A}(d(a, R^-))$. $\square$

Corollary C.1.13: *Let $f(x) \in \mathcal{M}(\mathbb{K})$, let S be the set of zeros and poles of f in $\mathbb{K}$, and let t be the* g.c.d. *of $\{\omega_\alpha(f) \mid \alpha \in S\}$. Then t is the greatest of the integers n such that there exists $g \in \mathcal{M}(\mathbb{K})$ satisfying $g^n = f$.*

Theorem C.1.14: *Let $f \in \mathcal{M}(\mathbb{K})$ (resp. $f \in \mathcal{M}(d(a, R^-))$, resp $f \in \mathcal{M}(D)$) be constant inside a disk included in $\mathbb{K}$ (resp. in $d(a, R^-)$, resp. in D). Then f is constant in $\mathbb{K}$ (resp. in $d(a, R^-)$, resp. in D).*

Proof. For a non-identically zero meromorphic function, the zeros and the poles of f are isolated. Consequently, if $f(x)$ is equal to a constant inside a disk, it is constant in the set of definition. $\square$

Definitions and notations: Given $f \in \mathcal{M}(\mathbb{K})$ (resp. $f \in \mathcal{M}(d(a, R^-))$, resp. $f \in \mathcal{M}(D)$), we will call *divisor of the poles of f on $\mathbb{K}$ (resp. on $d(a, R^-)$)* the divisor of $\frac{1}{f}$ on $\mathbb{K}$ (resp. on $d(a, R^-)$, resp. on D).

Lemma C.1.15: *Let $f \in \mathcal{M}(\mathbb{K}) \setminus \mathcal{A}(\mathbb{K})$ (resp. $f \in \mathcal{M}(d(0, R^-)) \setminus \mathcal{A}(d(0, R^-))$) and suppose 0 is not a pole of f. Let r be the minimal distance of the poles of f to 0. Then f belongs to $\mathcal{A}(d(0, r^-))$ and its radius of convergence is r.*

Proof. Consider the divisor T of the poles of f on $d(0, R^-)$. If $f \in \mathcal{M}(\mathbb{K}) \setminus \mathcal{A}(\mathbb{K})$, there is no problem to write f in the form $\frac{h}{l}$ with $h,\ l \in \mathcal{A}(\mathbb{K})$ where l has no zero in $d(0, r^-)$. Consequently, by Corollary B.14.2, the restriction of l to $d(0, r^-)$ is invertible in $\mathcal{A}(d(0, r^-))$. Therefore, $\frac{h}{l}$ belongs to $\mathcal{A}(d(0, r^-))$ and hence its radius of convergence is $\geq r$. Conversely, since f has a pole in $C(0, r)$, it is not equal to a power series in x in $d(0, r)$ and hence, the radius of convergence is r.

Now suppose $f \in \mathcal{M}(d(0, R^-)) \setminus \mathcal{A}(d(0, R^-))$. By Theorem B.18.15, we can find a function $l \in \mathcal{A}(d(0, R^-))$ such that $\mathcal{D}(l) \geq T$ and such that none of the zeros of l lie in $d(0, r^-)$. Next, we set $h = fl$ and see that h has no pole in $d(0, R^-)$. So, in both cases, we have made f in the form $\frac{h}{l}$ with $h,\ l \in \mathcal{A}(\mathbb{K})$ where l has no zero in $d(0, r^-)$. The proof is then similar to the case $f \in \mathcal{A}(\mathbb{K})$. $\square$

Corollary C.1.16: *Let $f \in \mathcal{M}(\mathbb{K}) \setminus \mathcal{A}(\mathbb{K})$ (resp. $f \in \mathcal{M}(d(0, R^-)) \setminus \mathcal{A}(d(0, R^-))$). If 0 is not a pole, $f(x)$ has a development in a power series whose radius of convergence is the minimal distance of poles of f to 0. If 0 is a pole of order q of f, then*

$f(x)$ has a development in a Laurent series $\sum_{k=-q}^{\infty} a_k x^k$ with $a_{-q} \neq 0$ and the radius of convergence of the series $\sum_{k=0}^{\infty} a_k x^k$ is equal to the minimal distance of non-zero poles of f to 0.

Theorem C.1.17: *Let $f \in \mathcal{M}(d(0, R^-))$ have no pole in an annulus $\Gamma(0, r, s)$ with $s < R$. Then $f(x)$ is equal to a Laurent series $\sum_{-\infty}^{+\infty} a_n x^n$ converging in all $\Gamma(0, r, s)$. For each $\mu \in [\log r, \log s]$, if f has q zeros and t poles in $d(0, \theta^\mu)$ taking multiplicity into account, one has $\nu^+(f, \mu) = q - t$ and if f has q' zeros and t' poles in $d(0, (\theta^\mu)^-)$, one has $\nu^-(f, \mu) = q' - t'$. Then the functions in $\rho : \nu^+(f, \log \rho)$, $\nu^-(f, \log \rho)$, and $|f|(\rho)$ are increasing. Let $k = \nu^+(f, \log r)$. Then $|f|(\rho) \geq |a_k| \rho^k \ \forall \rho \in [r, s]$.*

Proof. Since f belongs to $\mathcal{M}(d(0, R^-))$ and since $s < R$, f has finitely many zeros and poles in $d(0, s)$, hence we can write it $\frac{h}{l}$ with h, $l \in \mathcal{A}(d(0, R^-))$ having no common zero in $d(0, s)$. Since f has no pole in $\Gamma(0, r, s)$, l has no zero in $\Gamma(0, r, s)$. Let $\rho = \theta^\mu$. By Corollary B.10.3, we have $\nu^+(f, \mu) = \nu^+(h, \mu) - \nu^+(l, \mu) = q - t$. Similarly, in $d(0, \rho^-)$ we have $\nu^-(f, \mu) = \nu^-(h, \mu) - \nu^-(l, \mu) = q' - t'$.

We can write $f(x)$ in the form $\frac{h(x)}{Q(x)}$ with $h \in \mathcal{A}(d(0, s^-)$ and $Q \in \mathbb{K}[x]$. Then Q has no zero in $\Gamma(0, r, s)$ and hence $\nu^+(Q, \mu)$ is constant in $[\log r, \log s[$. On the other hand, $\nu^+(h, \mu)$ is increasing, hence so is $\nu^+(f, \mu)$. Consequently, the function $|f|(\rho)$ is increasing. Therefore, $|f|(\rho) \geq |a_k| \rho^k \ \forall \rho \in [r, s]$. □

Corollary C.1.18: *Let $f \in \mathcal{M}(d(0, R^-))$ have no pole in $\Gamma(0, r, s)$, with $0 < r < s < R$ and let $q = \nu^-(f, \log s)$, $k = \nu^+(f, \log r)$. Then*

$$\left(\frac{s}{r}\right)^k \leq \frac{|f|(s)}{|f|(r)} \leq \left(\frac{s}{r}\right)^q.$$

Corollary C.1.19: *Let $f \in \mathcal{M}(\mathbb{K}) \setminus \mathbb{K}(x)$ have finitely many poles. For every $q \in \mathbb{N}$, f satisfies $\lim_{r \to \infty} \frac{|f|(r)}{r^q} = +\infty$.*

Proof. Let $f = \frac{h}{Q}$ with Q a monic polynomial and $h \in \mathcal{A}(\mathbb{K})$. Since $f \notin \mathbb{K}(x)$, h does not lie in $\mathbb{K}[x]$, hence it has infinitely many zeros and therefore infinitely many terms $a_n \neq 0$ when $n > 0$. □

We will also need Lemma C.1.20 in the future.

Lemma C.1.20: *Let $f \in \mathcal{M}(\mathbb{K})$ be transcendental and have finitely many poles and let P be a polynomial. There exists $s > 0$ such that $|f + P|(r) = |f|(r)\ \forall r \geq s$ and then f has the same number of zeros as $f + P$ in $d(0, r)$.*

Proof. Let $R \in \mathbb{R}_+^*$ be such that all poles of f and all zeros of P lie in $d(0, R)$, let q be the number of poles of f and let $t = \deg(P)$. Then $f(x)$ is of the form $\dfrac{g(x)}{\prod_{j=1}^{q}(x - b_j)}$ with $g(x)$ of the form $\sum_{n=0}^{\infty} a_n x^n$. Now, when $r > R$, we have $|f|(r) = \dfrac{|g|(r)}{r^q}$ and $|P|(r)| = \dfrac{|P|(R)}{R^t} r^t$. Consequently, by Theorem B.13.22, $|f|(r)$ gets bigger than $|P|(r)$ when r is big enough and hence there exists $s > R$ such that $|P|(r) < |f|(r)\ \forall r > s$. But then, by Corollary B.13.23, we have $\nu(f, \log r) = \nu(f + P, \log r)\ \forall r > s$ and hence, f and $f + P$ have the same number of zeros in $d(0, r)$. □

Definitions and notations: Let $f \in \mathcal{M}(\mathbb{K})$ (resp. $f \in \mathcal{M}_u(d(a, R^-))$) and let $b \in \mathbb{K}$. Then b will be said to be *an exceptional value for f* if $f - b$ has no zero in $\mathbb{K}$ (resp. in $d(a, R^-)$) and b will be said to be *a pseudo-exceptional value for f* if $\lim_{r \to \infty} |f - b|(r) = 0$ (resp. $\lim_{r \to R^-} |f - b|(r) = 0$). Moreover, if $f \in \mathcal{M}(\mathbb{K}) \setminus \mathbb{K}(x)$ (resp. if $f \in \mathcal{M}_u(d(a, R^-))$), b will be said to be *a quasi-exceptional value for f* if $f - b$ has finitely many zeros in $\mathbb{K}$ (resp. in $d(a, R^-)$).

Theorem C.1.21: *Let $f \in \mathcal{M}(\mathbb{K}) \setminus \mathbb{K}$ (resp. $f \in \mathcal{M}_u(d(a, R^-))$). If b is an exceptional value for f, then it is a pseudo-exceptional value for f. Let $f \in \mathcal{M}(\mathbb{K}) \setminus \mathbb{K}(x)$ (resp. $f \in \mathcal{M}_u(d(a, R^-))$). If b is a quasi-exceptional value for f, then it is a pseudo-exceptional value for f.*

Proof. Without loss of generality we may assume that $a = b = 0$. Suppose first that $f \in \mathcal{M}(\mathbb{K}) \setminus \mathbb{K}$ and that 0 is an exceptional value for f. So, $\dfrac{1}{f}$ has no pole in $\mathbb{K}$ (resp. in $d(0, R^-)$), hence it is a function $h \in \mathcal{A}(\mathbb{K}) \setminus \mathbb{K}$ (resp. $h \in \mathcal{A}_u(d(0, R^-))$) so that $f = \dfrac{1}{h}$. Then, by Corollary B.5.18 (resp. by Theorem B.5.20) we have $\lim_{r \to +\infty} |h|(r) = +\infty$ (resp. $\lim_{r \to R^-} |h|(r) = +\infty$).

Suppose now that $f \in \mathcal{M}(\mathbb{K}) \setminus \mathbb{K}(x)$ and that 0 is a quasi-exceptional value for f. Then f is of the form $\dfrac{P(x)}{h(x)}$ with $P \in \mathbb{K}[x]$ and $h \in \mathcal{A}(\mathbb{K}) \setminus \mathbb{K}(x)$. By Corollary B.5.7, we have $\lim_{r \to +\infty} \dfrac{|P|(r)}{|h|(r)} = 0$, which proves the claim when $f \in \mathcal{M}(\mathbb{K}) \setminus \mathbb{K}(x)$.

Next, suppose that $f \in \mathcal{M}_u(d(0, R^-))$ admits 0 as a quasi-exceptional value. Then f is of the form $\dfrac{P(x)}{h(x)}$ with $P \in \mathbb{K}[x]$ and $h \in \mathcal{A}_u(d(0, R^-))$. So P is bounded in $d(0, R^-)$, hence of course $\lim\limits_{r \to +\infty} \dfrac{|P|(r)}{|h|(r)} = 0$, which ends the proof. $\square$

Theorem C.1.22: *Let $f \in \mathcal{M}(\mathbb{K}) \setminus \mathbb{K}$ (resp. $f \in \mathcal{M}_u(d(a, R^-))$). Then f admits at most one pseudo-exceptional value. Moreover, if $f \in \mathcal{A}(\mathbb{K}) \setminus \mathbb{K}$ (resp. $f \in \mathcal{A}(d(a, R^-))$), then f has no pseudo-exceptional value.*

Proof. Suppose that b is a pseudo-exceptional value for f. Without loss of generality we may assume that $a = b = 0$. Let $t \in \mathbb{K}^*$. Since $\lim\limits_{r \to +\infty} |f|(r) = 0$ (resp. $\lim\limits_{r \to R^-} |f|(r) = 0$), it is obvious that $\lim\limits_{r \to +\infty} |f - t|(r) = |t|$ (resp. $\lim\limits_{r \to R^-} |f - t|(r) = |t|$), so t is not a pseudo-exceptional value for f.

Now, suppose $f \in \mathcal{A}(\mathbb{K}) \setminus \mathbb{K}$. Since $\lim\limits_{r \to +\infty} |f|(r) = +\infty$, of course 0 is not a pseudo-exceptional value of f. Finally, suppose $\mathcal{A}(d(0, R^-))$). Then if $f \in \mathcal{A}_u(d(0, R^-))$, we have $\lim\limits_{r \to R} |f|(r) = +\infty$, hence 0 is not a pseudo-exceptional value of f. And if $f \in \mathcal{A}_b(d(0, R^-))$, we have $\lim\limits_{r \to R} |f|(r) = \|f\|_{d(0, R^-)}$, which is not 0, hence 0 is not a pseudo-exceptional value of f either. $\square$

Corollary C.1.23: *Let $f \in \mathcal{M}(\mathbb{K}) \setminus \mathbb{K}$ (resp. $f \in \mathcal{M}_u(d(a, R^-))$). Then f admits at most one exceptional value. Moreover, if $f \in \mathcal{M}(\mathbb{K}) \setminus \mathbb{K}(x)$ (resp. $f \in \mathcal{M}_u(d(a, R^-))$), then f admits at most one quasi-exceptional value. Further, if $f \in \mathcal{A}(\mathbb{K}) \setminus \mathbb{K}$ (resp. if $f \in \mathcal{A}_u(d(a, R^-))$), then f admits no exceptional value. And if $f \in \mathcal{A}(\mathbb{K}) \setminus \mathbb{K}[x]$ (resp. if $f \in \mathcal{A}_u(d(a, R^-))$), then f admits no quasi-exceptional value.*

C.2. Residues of meromorphic functions

Throughout this chapter, D is infraconnected, T is a hole of D, and V is a disk of the form $d(a, r)$ or $d(a, r^-)$, included in $\widetilde{D}$, such that $\widetilde{V} \cap D \neq \emptyset$.

Definitions and notations: Let $f \in \mathcal{M}(\mathbb{K})$ (resp. $f \in \mathcal{M}(d(0, R^-))$ have a pole α of order q and let $f(x) = \sum\limits_{k=-q}^{-1} a_k (x - \alpha)^k + h(x)$ with $a_{-q} \neq 0$ and $h \in \mathcal{M}(\mathbb{K})$ (resp. $f \in \mathcal{M}(d(0, R^-))$ and h holomorphic at α. Accordingly, to previous notations for analytic elements in Chapters B.2 and B.6, the coefficient a_{-1} is called *residue of f at α* and denoted by $\text{res}(f, \alpha)$.

We can now compare residues on a hole defined for analytic elements and residues at a point, we just defined for a meromorphic function.

Theorem C.2.1: *Let $a \in \mathbb{K}$, let $R \in \mathbb{R}_+^*$, let $f \in \mathcal{M}(d(a, R^-))$, and let $r \in]0, R[$. Let α_j, $1 \leq j \leq q$ be the poles of f in $d(a, r)$, let $\rho \in]0, \min_{i \neq j} |\alpha_i - \alpha_j|[$ and for each $j = 1, \dots, q$, let $\rho_j \in]0, \rho[$, let $T_j = d(\alpha_j, \rho_j^-)$. Let $D = d(\alpha, r) \setminus \left(\bigcup_{j=1}^q T_j\right)$. Then f belongs to $H(D)$ and* $\text{res}(f, \alpha_j) = \text{res}(f, T_j)$, $j = 1, \dots, q$.

Proof. By Corollary C.1.2, f belongs to $H(D)$. On the other hand, assuming that α_j is a pole of order s_j, by Corollary C.1.16, $f(x)$ has a development at α_j in a Laurent series $\sum_{m=-s_j}^{\infty} b_{m,j}(x - \alpha_j)^m$. Consequently, by Theorem B.6.1, the Mittag–Lefflerterm of f on T_j with respect to the infraconnected set D is $\sum_{m=-s_j}^{-1} b_{m,j}(x - \alpha_j)^m$. Then $\text{res}(f, T_j) = b_{-1,j} = \text{res}(f, \alpha_j)$, which ends the proof. □

Corollary C.2.2: *Let $f \in H_b(D)$ be meromorphic in $T = d(b, r^-)$ and admit only one pole b inside T. Let q be the multiplicity order of b. Then the Mittag–Leffler term of f associated to T is of the form $\sum_{j=1}^{q} \frac{a_j}{(x - b)^j}$, with $a_q \neq 0$ and also is of the form $\frac{P}{(x - a_j)^q}$ where P is a polynomial of degree $s < q$. Moreover, it does not depend on r when r tends to 0.*

Definitions and notations: Let $f \in \mathcal{M}(d(a, R^-))$ and let b be a pole of order t of f, let $r > 0$ be such that $d(b, r^-)$ contains no pole of f other than b and let $\frac{P(x)}{(x - \alpha)^t}$ be the Mittag– Leffler term of f associated to $d(b, r^-)$. Then $\frac{P(x)}{(x - b)^t}$ will be called *the singular part of f at b*.

An element $f \in H(D)$ will be said to be *meromorphic in V* if there exists finitely many points $(a_i)_{(1 \leq i \leq n)}$ in V such that f has continuation to an element of $H((D \cup V) \setminus \{a_i |\ 1 \leq i \leq n\})$.

Let f be meromorphic in V and belong to $H((D \cup V) \setminus \{a_i |\ 1 \leq i \leq n\})$. For each $i = 1, \dots, n$, if $f \notin H((D \cup V) \setminus \{a_h\ |h \neq i\})$, then by Corollary B.2.9 and Theorem B.2.10, a_i is a pole of f as an element of $H((D \cup V) \setminus \{a_j | 1 \leq j \leq n\})$. Let q_i be its order. Then a_i will be called *a pole of f of order q_i in V*. The polynomial $P(x) = \prod_{i=1}^{n} (x - a_i)^{q_i}$ will be called *the polynomial of the poles of f in V*.

Lemma C.2.3: *Let D be bounded or belong to* Alg *and let $f \in H(D)$. If f is meromorphic in T, the polynomial of its poles P in T satisfies $Pf \in H(D \cup T)$.*

Proof. Indeed, let $D' = (D \cup T) \setminus \{a_1, \dots, a_n\}$. If D is bounded, then so is D' and therefore by Theorem B.2.4, Pf belongs to $H(D')$. But by construction Pf is bounded at each point a_i and therefore by Corollary B.2.6, Pf belongs to $H(D \cup T)$.

Now, suppose $D \in \text{Alg}$. Then by Theorem B.8.9, D' belongs to Alg and therefore Pf belongs to $H(D')$ so we have the same conclusion. □

Lemma C.2.4: *Let D be bounded (resp. let $D \in \text{Alg}$) and let f be invertible in $H(D)$. Then f is meromorphic in T if and only if so is $\frac{1}{f}$.*

Proof. First we suppose f meromorphic in T. Let P be the polynomial of its poles in T and let $g(x) = f(x)P(x)$. Since D is bounded (resp. belongs to Alg) by Lemma C.2.3, g belongs to $H(D \cup T)$. Let Q be the polynomial of the zeros of g in T. Since f has no zero in D, Q actually is the polynomial of the zeros of g in $D \cup T$ and then g is of the form $Q(x)h(x)$ with h an element of $H(D \cup T)$ that has no zero in T. Hence, we have $\frac{1}{h} = \frac{1}{f}\frac{Q}{P}$. If $D \in \text{Alg}$, $\frac{1}{h}$ obviously belongs to $H(D)$. If D is bounded, we have $\frac{Q}{P} \in R_b(D)$ and then by Theorem B.2.4, $\frac{1}{h}$ belongs to $H(D)$. Thus, $\frac{1}{h}$ belongs to $H(D)$ anyway. Now, by Lemma B.15.1, h is invertible in $H(D \cup T)$. Hence, f factorizes in the form $\frac{P}{Q}h$ and then $\frac{1}{f} = \frac{1}{Q}\frac{P}{h}$. But $\frac{P}{h}$ belongs to $H(D \cup T)$ and therefore $\frac{1}{f}$ is meromorphic in T and admits Q as the polynomial of its poles in T. We may obviously apply the same reasoning to $\frac{1}{f}$ and this shows the converse. □

Theorem C.2.5: *Let $F \in H(D)$ be meromorphic in T and satisfy $\|F - 1\|_D < 1$. Then F has as many poles as many zeros in T.*

Proof. Without loss of generality, we may obviously assume that D is bounded because the hypothesis remains true in any set $D \cap d(0, R)$. We may also assume that $T = d(0, r^-)$. Let P (resp. Q) be the polynomial of the zeros (resp. the poles) of F in T. Then by Lemma C.2.3, F factorizes in the form $\frac{f}{Q}$ with $f \in H(D \cup T)$. Now, the zeros of f inside T are just those of F, hence f factorizes in the form Pg with $g \in H(D \cup T)$, g having no zero in T. Let $h = \frac{P}{Q}$. Since g has no zero in T, by Theorem B.13.14, $|g(x)|$ is equal to a constant inside T. Let $s \in]0, r[$ be such that all zeros of P and of Q lie in $d(0, s)$. Then obviously F belongs to $H(\Gamma(0, s, r))$. Now by hypothesis there exists $\lambda > 0$ such that $\Psi(F, \mu) \leq \lambda$ for all $\mu \geq \log r$. Hence, by continuity, there exists $\lambda' > 0$ and s' in $]s, r[$ such that $\Psi(F, \mu) \leq \lambda'$ for all $\mu \geq \log s'$. Thus, there exists $b \in]0, 1[$ and t in $]s', r[$ such that $|F(x) - 1| \leq b$ for all $x \in \Gamma(0, t, r)$ and then, $|h(x)|$ is constant in $\Gamma(0, t, r)$. Hence, we have $\frac{d}{d\mu}\Psi(h, \mu) = 0$ for all $\mu \in [\log t, \log r]$. Since h has neither any zero nor any pole in $C(0, t)$, by Corollary A.3.17, h has as many zeros as many poles in $d(0, t)$ and therefore in $d(0, r)$ and this ends the proof. □

It is useful to consider again elements meromorphic at a point.

Lemma C.2.6: *Let $a \in \overline{D}$ and let $f \in H(D)$ be meromorphic but not holomorphic at a. Then a is a pole of f.*

Proof. By hypothesis, there exists $r > 0$ such that f belongs to $H\big(D \cup \big(d(a,r) \setminus \{a\}\big)\big)$. Suppose that a is not a pole of f. Then by Theorem 11.10, in [58], f belongs to $H(D \cup d(a,r))$ and then f is holomorphic at a. □

Corollary C.2.7: *Let $a \in \overline{D}$ and let $f \in H(D)$. Then f is meromorphic at a and admits a as a pole of order q if and only if there exists a disk $d(a,r)$ included in $\overline{D}$ and an element $h \in H(d(a,r))$ such that $f(x)(x-a)^q = h(x)$ whenever $x \in d(a,r) \setminus \{a\}$ and $h(a) \neq 0$.*

Corollary C.2.8: *Let D satisfies Condition (B) and let $f \in H(D)$. For every $a \in \overline{D}$, f is meromorphic at a. For every $a \in D$, f is holomorphic at a.*

Remark: Let $a \in \overline{D} \setminus \overset{\circ}{D}$ and let f admits a as a pole of order q. This does not imply that f is meromorphic at a. Indeed, by [43, 49] we know that there exist infraconnected sets E with a point $a \in E \setminus \overset{\circ}{E}$ and elements $h \in H(E)$ such that $\lim\limits_{|x-a| \to 0} h(x) = 0$ and such that $\limsup\limits_{|x-a| \to 0} \left|\dfrac{h(x)}{x-a}\right| = +\infty$. Let $E' = E \setminus \{a\}$ and let $g = \dfrac{1+h}{x-a}$. It is easily seen that a is a pole of order 1 for g. But h does not belong to any space $H(d(a,r))$, whenever $r > 0$ (because if it did, it should factorize in $H(d(a,r))$ in the form $(x-a)\ell(x)$, with $\ell \in H(d(a,r))$). Thus, it is seen that $(x-a)g$ does not belong to $H(E)$ and therefore g is not meromorphic at a.

Concerning the derivation, Theorem C.2.9 is easy and follows the classical rules.

Theorem C.2.9: *Let $f \in \mathcal{M}(\mathbb{K})$ (resp. $f \in \mathcal{M}(d(a,R^-))$). For each $\alpha \in \mathbb{K}$ (resp. $\alpha \in d(a,R^-)$) such that f is holomorphic at α, f has a derivative $f'(\alpha)$ at α. Further, given a point $\beta \in \mathbb{K}$ (resp. $\beta \in d(a,R^-)$) and the Laurent development of f at β : $\sum\limits_{k=-q}^{\infty} a_k(x-\beta)^k$ with $a_{-q} \neq 0$, the development of f' at β is*

$$\sum_{k=-q}^{0} ka_k(x-\beta)^{k-1} + \sum_{k=1}^{\infty} ka_k(x-\beta)^{k-1}.$$

Proof. Suppose first f is holomorphic at α. By Theorem C.1.4, $f(x)$ is equal to a power series $\sum\limits_{k=0}^{\infty} a_k(x-\alpha)^k$ converging inside a disk $d(\alpha,r^-)$ where r is the minimal distance from α to the various poles. Then by Theorem B.9.1, we know that f has a derivative whose development is obtained by deriving term by term.

Suppose now that β is a pole of order q and let $r \in]0,R[$ be strictly inferior to the minimal distance from β to the other poles (with just $r < R$ if β is the unique pole of f). By Lemma C.1.1, for every $\rho \in]0, r[$, f belongs to $H(d(\beta, r) \setminus d(\beta, \rho^-))$ and the Laurent development of f at β is its Mittag– Leffler development as an element of $H(d(\beta, r) \setminus d(\beta, \rho^-))$: the Mittag– Leffler term associated to the hole $d(\beta, \rho^-)$ is $\sum_{k=-q}^{-1} a_k(x-\beta)^k$ with $a_{-q} \neq 0$ and the term associated to $d(\beta, r)$ is $\sum_{k=0}^{\infty} a_k(x-\beta)^k$ with $a_{-q} \neq 0$. Consequently, by Theorem B.9.19, the derivative has a Mittag– Leffler development at β: $\sum_{k=-q}^{-1} ka_k(x-\beta)^{k-1} + \sum_{k=1}^{\infty} ka_k(x-\beta)^{k-1}$. This is true for all $r \in]0, R[$ strictly inferior to the minimal distance from β to the other poles and for every $\rho \in]0, r[$, which ends the proof. $\square$

Theorem C.2.10 is an improvement of the classical upper bound f' in function of f. That is due to J.P. Bézivin [8].

Theorem C.2.10: *For each $n \in \mathbb{N}$ and for all $r \in]0, R[$, we have*

$$|f^{(n)}|(r) \leq |n!| \frac{|f|(r)}{r^n}.$$

Moreover, given $r \in]0, R[$ such that $\nu^+(f, \log r) = \nu^-(f, \log r)$, if the residue characteristic p does not divide $\nu(f, \log r)$, then $\nu(f', \log r) = \nu(f, \log r) - 1$ and

$$|f'|(r) = \frac{|f|(r)}{r}.$$

Proof. When $f \in \mathcal{A}(d(0, R^-))$, this was shown at Corollary B.9.16. Now, consider the general case and set $f = \frac{U}{V}$ with $U, V \in \mathcal{A}(d(0, R^-))$. The stated inequality is trivial when $q = 1$. So, we assume it holds for $q \leq n-1$ and consider $f^{(n)}$. Writing $U = V\left(\frac{U}{V}\right)$, by Leibniz theorem we have

$$U^{(n)} = \sum_{q=0}^{n} \binom{n}{q} V^{(n-q)} \left(\frac{U}{V}\right)^{(q)}$$

and hence

$$V\left(\frac{U}{V}\right)^{(n)} = U^{(n)} - \sum_{q=0}^{n-1} \binom{n}{q} V^{(n-q)} \left(\frac{U}{V}\right)^{(q)}.$$

Now, by Corollary B.9.16 we have

$$|U^{(n)}|(R) \leq |n!| \frac{|U|(R)}{R^n} \tag{1}$$

and for each $q \leq n-1$, we have $|V^{(n-q)}|(R) \leq |(n-q)!| \frac{|V|(R)}{R^{n-q}}$

and $\left|\left(\frac{U}{V}\right)^{(q)}\right|(R) \leq |q!|\frac{|U|(R)}{|V|(R)R^q}$. Consequently,

$$\left|\left(\frac{U}{V}\right)^{(q)}\right|(R)\left|\left(\frac{U}{V}\right)^{(n-q)}\right|(R) \leq |((n-q)!)q!|\frac{|U|(R)}{|V|(R)R^n}$$

and then we can derive.

$$\left|\binom{n}{q}\right|\left|\left(\frac{U}{V}\right)^{(q)}\right|(R)\left|\left(\frac{U}{V}\right)^{(n-q)}\right|(R) \leq |n!|\frac{|U|(R)}{|V|(R)R^n}. \tag{2}$$

So, by (1) and (2) the first conclusion holds for $q = n$.

Suppose now that $\nu^+(f, \log r) = \nu^-(f, \log r)$ and that the residue characteristic of $\mathbb{K}$ does not divide $\nu(f, \log r)$. Without loss of generality, we may assume that f has no pole in $C(0, r)$ because all conclusions hold by continuity. In $C(0, r)$, $f(x)$ is equal to a power series $\sum_{-\infty}^{+\infty} a_n x^n$. Set $q = \nu(f, -\log(r))$. Then $|f|(r) = |a_q|r^q$ and $|f'|(r) = |q||a_q|r^{q-1} = |a_q|r^{q-1}$, which ends the proof. □

It seems obvious that the condition for a meromorphic function to admit primitives is that all residues are null. This is stated by Theorem C.2.11 but the proof is not this immediate [22].

Let us remark that the topology of uniform convergence in all disks of $\mathbb{K}$ (resp. of all disk included inside an open disk $d(a, R^-)$) is obviously defined on the algebra $\mathcal{M}(\mathbb{K})$ (resp. $\mathcal{M}(d(a, R^-))$) and that $\mathcal{M}(\mathbb{K})$ (resp. $\mathcal{M}(d(a, R^-))$) is complete for that topology.

We are now able to solve the problem of Bezout rings $\mathcal{A}(\mathbb{K})$ and $\mathcal{A}(d(a, R^-))$. Theorem C.2.11 is a Mittag–Leffler theorem similar to this known in complex analysis.

Theorem C.2.11: *Let $(a_m, q_m)_{m\in\mathbb{N}}$ be a divisor of $\mathbb{K}$ (resp. of $d(a, R^-)$), $a \in \mathbb{K}$, $R > 0$) and for every $n \in \mathbb{N}$, let $Q_m \in \mathbb{K}[x]$ be of degree $< q_m$. There exists $f \in \mathcal{M}(\mathbb{K})$ (resp. $f \in \mathcal{M}(d(a, R^-))$) admitting for poles each a_m of order q_m and no other pole and such that its singular part is $\frac{Q_m}{(x - a_m)^{q_m}}$.*

Proof. The proof is similar to that in the complex case. Without loss of generality, we can suppose that $|a_m| \leq |a_{m+1}|$. Let $(t(n))_{n\in\mathbb{N}}$ be the strictly increasing sequence such that $a_{t(n)}| < |a_{t(n+1)}|$ and let $r_n = |a_{t(n)}|$ $n \in \mathbb{N}$.

For each $m \in \mathbb{N}$, set $S_m(x) = \frac{Q_m(x)}{(x - a_m)^{q_m}}$. Now, for each $n \in \mathbb{N}$, we can set $f_n = \sum_{m\in L_n} S_m(x)$. So, by construction, f_n belongs to $H(d(0, r_{n-1}))$, hence there exists $P_n \in \mathbb{K}[x]$ such that $\|f_n - P_n\|_{d(0;r_{n-1})} \leq \left(\frac{1}{2}\right)^n$. Consequently, the sequence $(f_n - P_n)_{n\in\mathbb{N}}$ converges to 0 with respect to the topology of $\mathcal{M}(\mathbb{K})$ (resp. of

$\mathcal{M}(d(a, R^-)))$. Set $f(x) = f_1(x) + \sum_{n=2}^{\infty}(f_n(x) - P_n(x))$. By construction, f belongs to $\mathcal{M}(d(0, r^-))$ $\forall r > 0$, hence f belongs to $\mathcal{M}(\mathbb{K})$ (resp. to $\mathcal{M}(d(a, R^-)))$. Moreover, the poles of f are the points a_m, $m \in \mathbb{N}$. Let us take $q \geq 1$ and $\rho > 0$ such that $|a_m - a_q| \geq \rho$ $\forall m \neq q$. Then $f - S_q$ belongs to $H(d(a, \rho))$, so S_q is the singular part of f at S_q. □

Now, we can give an easy proof of Theorem C.2.12 already proven in [22] in a more complicated way.

Theorem C.2.12: *$\mathbb{K}$ is supposed to have characteristic 0. A function $f \in \mathcal{M}((\mathbb{K})$ (resp. $f \in \mathcal{M}((d(a, R^-))$, $a \in \mathbb{K}$, $R > 0$) admits primitives in $\mathcal{M}(\mathbb{K})$ (resp. in $\mathcal{M}((d(a, R^-)))$ if and only if all residues of f are null.*

Proof. Let a be a pole of f. According to the Laurent series of f at a, if f admits primitives, then f has no residue different from zero at a because the function $\frac{1}{x-a}$ has no primitive in $\mathcal{M}(d(a, r))$ (whenever $r > 0$). Now let $(a_m)_{m \in \mathbb{N}}$ be the sequence of poles of f, each of respective degree q_m and suppose that $\text{res}(f, a_m) = 0$. Since $\text{res}_{a_m}(f) = 0$, the singular part of f at a_m is of the form $\dfrac{Q_m(x - a_m))}{(x - a_m)^{q_m}}$ with $q_m \geq 2$ and $Q_m(X)$ is a polynomial of degree $\leq q_m - 2$. Consequently, the singular part of f at a_m admits a primitive of the form $\dfrac{P_m(x - a_m)}{(x - a_m)^{q_m - 1}}$ with $\deg(P_m(X)) \leq q_m - 2$. Then by Theorem C.2.11, there exists $G \in \mathcal{M}(\mathbb{K})$ (resp. $G \in \mathcal{M}(d(a, R^-)))$ admitting the a_m for poles with respective singular part $\dfrac{P_m}{(x - a_m)^{q_m - 1}}$ and no other pole. By construction, for each $m \in \mathbb{N}$, the singular part of G' at a_m is $\dfrac{Q_m(x - a_m))}{(x - a_m)^{q_m}}$, hence $G' - f$ has no pole at a_m and hence has no pole in $\mathbb{K}$ (resp. in $d(a, R^-)$). Consequently, $G' - f$ belongs to $\mathcal{A}(\mathbb{K})$ (resp. to $\mathcal{A}(d(a, R^-)))$. But then by Corollary B.9.7, $G' - f$ admits a primitive $L \in \mathcal{A}(\mathbb{K})$ (resp. $L \in \mathcal{A}(d(a, R^-)))$ and hence the function $F = G - L$ is a primitive of f that belongs to $L \in \mathcal{M}(\mathbb{K})$ (resp. to $L \in \mathcal{M}(d(a, R^-)))$. □

Corollary C.2.13: *The field $\mathbb{K}$ is supposed to have characteristic 0. Let $f \in \mathcal{M}((\mathbb{K})$ (resp. $f \in \mathcal{M}((d(a, R^-))$, $a \in \mathbb{K}$, $R > 0$). Then f' belongs to $\mathbb{K}(x)$ if and only if so does f.*

Proof. If f belongs to $\mathbb{K}(x)$, of course, so does f'. Now, suppose f' belongs to $\mathbb{K}(x)$. We can write it in the form $\sum_{j=1}^{q} \dfrac{b_j}{(x - a_j)^{q_j}}$. And by Theorem C.2.11, we have $q_j \geq 2$ $\forall j = 1, \ldots, q$. Consequently, since $\mathbb{K}$ has characteristic 0, $f(x)$ is of the form $-\sum_{j=1}^{q} \dfrac{b_j}{q_j(x - a_j)^{q_j - 1}} + c$ with $c \in \mathbb{K}$ and hence f belongs to $\mathbb{K}(x)$. □

C.3. Meromorphic functions out of a hole

Definitions and notations: We fix $R > 0$ and denote by I the interval $[R, +\infty[$. Throughout the chapter, we denote by S the disk $d(0, R^-)$ and put $D = \mathbb{K} \setminus S$.

We denote by $H_0(D)$ the $\mathbb{K}$-subvector space of the $f \in H(D)$ such that $\lim_{|x|\to+\infty} f(x) = 0$.

By classical properties of analytic elements, we know that given a circle $C(a, R)$ and an element f of $H(C(a, R))$, i.e., a Laurent series $f(x) = \sum_{-\infty}^{+\infty} c_n(x-a)^n$ converging whenever $|x| = r$, then $|f(x)|$ is equal to $\sup_{n\in\mathbb{Z}} |c_n|r^n$ in all classes of the circle $C(a, r)$ except maybe infinitely many. When $a = 0$, we put $|f|(r) = \sup_{n\in\mathbb{Z}} |c_n|r^n$. Then $|f|(r)$ is a multiplicative norm on $H(C(0, r))$.

We denote by $\mathcal{A}(D)$ the $\mathbb{K}$-algebra of Laurent series converging in D and by $\mathcal{A}^c(D)$ the set of $f \in \mathcal{A}(D)$ having infinitely many zeros in D. Similarly, we will denote by $\mathcal{M}(D)$ the field of fractions of $\mathcal{A}(D)$ that we will call *field of meromorphic functions in* D and we denote by $\mathcal{M}^c(D)$ the set of functions $f \in \mathcal{M}(D)$, which have infinitely many zeros or poles in D.

Similarly, as we did in $\mathbb{K}$ and inside a disk, here we define a pseudo-exceptional value and a quasi-exceptional value in D. Given a meromorphic function $f \in \mathcal{M}(D))$, a value $b \in \mathbb{K}$ is called *a pseudo-exceptional value for* f. If $\lim_{|x|\to+\infty} f(x) = 0$, it is called *a quasi-exceptional value for* f if $f - b$ has finitely many zeros in D and it is called *an exceptional value for* f if has no zero in in $\mathbb{K}$ (resp. in $d(a, R^-)$, resp. in D).

Proposition C.3.1: *Let $f \in \mathcal{M}^C(D)$ and let b be a quasi-exceptional value. Then b is a pseudo-exceptional value.*

Proof. Without loss of generality, we may assume that $b = 0$. Therefore, we can write f in the form $\frac{P}{h}$ with P a polynomial whose zeros lie in D and $h \in \mathcal{A}^c(D)$. On the other hand, $h(x)$ is a Laurent series $\sum_{-\infty}^{+\infty} a_n x^n$ converging in all D, having infinitely many zeros, hence infinitely many coefficients a_n with $n > 0$, are different from zero, therefore one sees that $\lim_{|x|\to+\infty} \frac{|P|(r)|}{|h|(r)} = 0$, and hence $\lim_{|x|\to+\infty} \frac{|P(x)|)|}{|h(x)|} = 0$. □

Proposition C.3.2: *Let $f \in \mathcal{M}(D)$. If f has infinitely many zeros in D (resp. infinitely many poles in D), the set of zeros (resp. the set of poles) is a sequence*

$(\alpha_n)_{n\in\mathbb{N}}$ *such that* $\lim_{n\to+\infty} |\alpha_n| = +\infty$. *If* f *has no zero in* D, *then it is of the form* $\sum_{-\infty}^{+\infty} a_n x^n$ *with* $|a_q|r^q > |a_n|r^n\ \forall n \in \mathbb{Z},\ n \neq q, \forall r \geq R$.

Proof. Suppose first $f \in \mathcal{A}(D)$. For each $L > R$, f belongs to $H(Delta(0, R, L))$ and by Corollary B.12.11, it is quasi invertible in $H(Delta(0, R, L))$, hence it has finitely many zeros in $Delta(0, R, L)$, for every $L > R$. Consequently, if f has infinitely many zeros in D, the zeros form a sequence $(\alpha_n)_{n\in\mathbb{N}}$ such that $\lim_{n\to+\infty} |\alpha_n| = +\infty$. Suppose now $f \in \mathcal{M}(D)$. Then f is of the form $\frac{g}{h}$ with $g,\ h \in f \in \mathcal{A}(D)$. If f has infinitely many zeros, so does g, and each zero of f is a zero of g, hence the zeros of f form a sequence $(\alpha_n)_{n\in\mathbb{N}}$ such that $\lim_{n\to+\infty} |\alpha_n| = +\infty$. Similarly, if f has infinitely many poles, the h has infinitely many zeros, and each pole of f is a zero of h, hence the poles of f form a sequence $(\beta_n)_{n\in\mathbb{N}}$ such that $\lim_{n\to+\infty} |\beta_n| = +\infty$. □

Theorem C.3.3: *Let* $f \in \mathcal{M}(D)$ *have no zero and no pole in* D. *Then* $f(x)$ *is of the form* $\sum_{-\infty}^{q} a_n x^n$ *with* $|a_q|r^q > |a_n|r^n\ \forall n < q,\ \forall r \geq R$, and $|f(x)| = |a_q|r^q\ \forall x \in D$.

Proof. For every $r \geq R$, f belong to $H(C(0, r))$ and by Theorem B.13.1, we have $\nu^+(f, \log(r)) = \nu^-(f, \log(r))$. Consequently, by continuity, $\nu(f, \log(r)$ is a constant q in $\log(R), +\infty[$. It is then clear that $|f(x)| = |a_q|r^q\ \forall x \in D$. □

Theorem C.3.4: *Let* $f \in \mathcal{M}(D)$ *have at least infinitely many zeros or infinitely many poles in* D. *Then* f *admits at most one pseudo-exceptional value.*

Proof. Suppose that $f \in \mathcal{M}^c(D)$ has two distinct pseudo-exceptional values a and b. Without loss of generality, we can assume that $a = 0$ and hence f is of the form $\frac{\phi}{\psi}$ with *phi* and $\psi \in \mathcal{A}(D)$, ψ admitting infinitely many zeros and satisfying $\lim_{r\to+\infty} \frac{\phi|(r)}{|psi|(r)} = 0$. Then $f - b = \frac{\phi - b\psi}{\psi}$. But when r is big enough, we have $|\phi - b\psi(r)| = |b\psi|(r)$, therefore $P - b\psi(r)$ does not admit 0 as a pseudo-exceptional value, a contradiction. □

Corollary C.3.5: *Let* $f \in \mathcal{M}(D)$ *have at least infinitely many zeros or infinitely many poles in* D. *Then* f *admits at most one exceptional value.*

Definitions and notations: Let $f \in H(D)$ have no zero in D, $f(x) = \sum_{-\infty}^{q} a_n x^n$ with $|a_q|R^q > |a_n|R^n\ \forall n < q$ and $a_q = 1$. Then f is a Motzkin factor associated to S and the integer q is called *the Motzkin index of f* and will be denoted by $m(f, S)$.

Theorem C.3.6: *Let $f \in \mathcal{M}(D)$. We can write f in a unique way in the form $f^S f^0$ with $f^S \in H(D)$ a Motzkin factor associated to S and $f^0 \in \mathcal{M}(\mathbb{K})$, having no zero and no pole in S.*

Proof. Suppose first $f \in \mathcal{A}(D)$ and take $V > R$. Then as a quasi-invertible element of $H(\Delta(0, R, V))$, by Theorem B.20.16, f admits a factorization in the form $f^S f^0$ where f^S is a Motzkin factor and f^0 belongs to $H(d(0, V))$ and has no zero in S. Moreover, by Lemma B.20.6, f^S does not depend on V. Consequently, since f^S is obviously invertible in $\mathcal{A}(D)$, we can factorize $f \in \mathcal{A}(D)$ in the form $f^S f^0$ where f^0 belongs to $\mathcal{A}(\mathbb{K})$ and has no zero in S.

Consider now the general case: $f = \dfrac{g}{h}$ with $g,\ h \in \mathcal{A}(D)$. Then we can write $g = g^S g^0,\ h = h^S h^0$, hence $f = \left(\dfrac{g^S}{h^S}\right)\left(\dfrac{g^0}{h^0}\right)$. Then we can check that this is the factorization announced in the statement: $f^S = \dfrac{g^S}{h^S}$ and $f^0 = \dfrac{g^0}{h^0}$. □

Lemma C.3.7 is immediate.

Lemma C.3.7: *The set of Motzkin factors associated to S makes a multiplicative group. Let $f,\ g \in \mathcal{M}(D)$. Then $(fg)^S = (f^S)(g^S)$, $\left(\dfrac{1}{f}\right)^S = \dfrac{1}{f^S}$, $(fg)^0 = (f^0)(g^0)$, $\left(\dfrac{1}{f}\right)^0 = \dfrac{1}{f^0}$ and $m(fg, S) = m(f, S) + m(g, S)$, $m\left(\dfrac{1}{f}, S\right) = -m(f, S)$.*

Definitions and notations: We will denote by $\mathcal{M}^*(D)$ the set of $f \in \mathcal{M}(D)$ such that $f^0 \notin \mathbb{K}(x)$, i.e., the set of f admitting at least infinitely many zeros in D or infinitely many poles in D. Similarly, we will denote by $\mathcal{A}^*(D)$ the set of $f \in \mathcal{A}(D)$ such that $f^0 \notin \mathbb{K}[x]$, i.e., the set of f admitting infinitely many zeros in D. Next, we set $\mathcal{M}^0(D) = \mathcal{M}(D) \setminus \mathcal{M}^*(D)$ and $\mathcal{A}^*(D) = \mathcal{A}(D) \setminus \mathcal{A}^*(D)$.

C.4. Nevanlinna theory in $\mathbb{K}$ and in an open disk

Throughout the next chapters, the field $\mathbb{K}$ is supposed to have characteristic 0.

The Nevanlinna theory was made by Rolf Nevanlinna on complex functions [64, 77]. It consists of defining counting functions of zeros and poles of a meromorphic function f and giving an upper bound for multiple zeros and poles of various functions $f - b$, $b \in \mathbb{C}$.

A similar theory for functions in a p-adic field was constructed and correctly proved by A. Boutabaa [29] in the field $\mathbb{K}$, after some previous work by Ha Huy Khoai [63]. In [31] the theory was extended to functions in $\mathcal{M}(d(0, R^-))$ by taking into account Lazard's problem. A new extension to functions out of a hole was made in Chapter C.6.

Definitions and notations: Recall that given three functions ϕ, ψ, ζ defined in an interval $J =]a, +\infty[$ (resp. $J =]a, R[$), with values in $[0, +\infty[$, we shall write $\phi(r) \leq \psi(r) + O(\zeta(r))$ if there exists a constant $b \in \mathbb{R}$ such that $\phi(r) \leq \psi(r) + b\zeta(r)$. We shall write $\phi(r) = \psi(r) + O(\zeta(r))$ if $|\psi(r) - \phi(r)|$ is bounded by a function of the form $b\zeta(r)$.

Similarly, we shall write $\phi(r) \leq \psi(r) + o(\zeta(r))$ if there exists a function h from $J =]a, +\infty[$ (resp. from $J =]a, R[$) to $\mathbb{R}$ such that $\lim_{r\to+\infty} \frac{h(r)}{\zeta(r)} = 0$ (resp. $\lim_{r\to R} \frac{h(r)}{\zeta(r)} = 0$) and such that $\phi(r) \leq \psi(r) + h(r)$. And we shall write $\phi(r) = \psi(r) + o(\zeta(r))$ if there exists a function h from $J =]a, +\infty[$ (resp. from $J =]a, R[$) to $\mathbb{R}$ such that $\lim_{r\to+\infty} \frac{h(r)}{\zeta(r)} = 0$ (resp. $\lim_{r\to R} \frac{h(r)}{\zeta(r)} = 0$) and such that $\phi(r) = \psi(r) + h(r)$.

Throughout the next paragraphs, we will denote by I the interval $[t, +\infty[$ and by J an interval of the form $[t, R[$ with $t > 0$.

We have to introduce the counting function of zeros and poles of f, counting or not multiplicity. Here we will choose a presentation that avoids assuming that all functions we consider admit no zero and no pole at the origin.

Definitions and notations: We denote by $Z(r, f)$ the counting function of zeros of f in $d(0, r)$ in the following way.

Let (a_n), $1 \leq n \leq \sigma(r)$ be the finite sequence of zeros of f such that $0 < |a_n| \leq r$, of respective order s_n.

We set $Z(r, f) = \max(\omega_0(f), 0) \log r + \sum_{n=1}^{\sigma(r)} s_n(\log r - \log |a_n|)$ and so, $Z(r, f)$ is called *the counting function of zeros of f in $d(0, r)$, counting multiplicity.*

In order to define the counting function of zeros of f without multiplicity, we put $\overline{\omega_0}(f) = 0$ if $\omega_0(f) \leq 0$ and $\overline{\omega_0}(f) = 1$ if $\omega_0(f) \geq 1$.

Now, we denote by $\overline{Z}(r, f)$ the counting function of zeros of f without multiplicity: $\overline{Z}(r, f) = \overline{\omega_0}(f) \log r + \sum_{n=1}^{\sigma(r)} (\log r - \log |a_n|)$ and so, $\overline{Z}(r, f)$ is called *the counting function of zeros of f in $d(0, r)$ ignoring multiplicity.*

In the same way, considering the finite sequence (b_n), $1 \leq n \leq \tau(r)$ of poles of f such that $0 < |b_n| \leq r$, with respective multiplicity order t_n, we put $N(r, f) = \max(-\omega_0(f), 0) \log r + \sum_{n=1}^{\tau(r)} t_n(\log r - \log |b_n|)$ and then $N(r, f)$ is called *the counting function of the poles of f, counting multiplicity.*

Next, in order to define the counting function of poles of f without multiplicity, we put $\overline{\overline{\omega_0}}(f) = 0$ if $\omega_0(f) \geq 0$ and $\overline{\overline{\omega_0}}(f) = 1$ if $\omega_0(f) \leq -1$ and we set $\overline{N}(r, f) = \overline{\overline{\omega_0}}(f) \log r + \sum_{n=1}^{\tau(r)} (\log r - \log |b_n|)$ and then $\overline{N}(r, f)$ is called *the counting function of the poles of f, ignoring multiplicity.*

Now we can define the the Nevanlinna function $T(r, f)$ in I or J as $T(r, f) = \max(Z(r, f), N(r, f))$ and the function $T(r, f)$ is called *characteristic function of f or Nevanlinna function of f.*

Finally, if S is a subset of $\mathbb{K}$ we will denote by $Z_0^S(r, f')$ the counting function of zeros of f', excluding those which are zeros of $f - a$ for any $a \in S$.

Remark: If we change the origin, the functions Z, N, T are not changed, up to an additive constant.

By Corollary B.13.2, Lemma C.4.1 is easy.

Lemma C.4.1: *Let $\widehat{\mathbb{K}}$ be a complete algebraically closed extension of $\mathbb{K}$ whose absolute value extends that of $\mathbb{K}$ and let $f \in \mathcal{M}(\mathbb{K})$ (resp. let $f \in \mathcal{M}(d(0, R^-))$). Let $\widehat{d}(0, R) = \{x \in \widehat{\mathbb{K}} \mid |x| < R\}$. The meromorphic function $\widehat{f}$ defined by f in $\widehat{d}(0, R)$ has the same Nevanlinna functions as f.*

In a p-adic field such as $\mathbb{K}$, the first Main theorem is almost immediate and is an immediate consequence of Corollary B.13.27.

Theorem C.4.2: *Let $f \in \mathcal{M}(\mathbb{K})$ (resp. $f \in \mathcal{M}(d(0, R^-))$) have no zero and no pole at 0. Then $\log(|f|(r)) = \Psi(f, \log r) = \log(|f(0)|) + Z(r, f) - N(r, f)$.*

Proof. We can write $f(x) = \dfrac{h}{l}$ with $h,\ l \in \mathcal{A}(\mathbb{K})$ (resp. $h,\ l \in \mathcal{A}(d(0, R^-))$) such that $l(0)h(0) \neq 0$. By Corollary B.13.27, we have $\log(|h|(r)) = \log(|f(0)|) + Z(r, h)$, $\log(|l|(r)) = \log(|l(0)|) + N(r, l)$, so the conclusion is obvious. □

Theorem C.4.3 is now immediate.

Theorem C.4.3: *Let $f,\ g \in \mathcal{M}(\mathbb{K})$ (resp. $f,\ g \in \mathcal{M}(d(0, R^-))$). Then $Z(r, fg) \leq Z(r, f) + Z(r, g)$, $N(r, fg) \leq N(r, f) + N(r, g)$, $T(r, fg) \leq T(r, f) + T(r, g)$, $T(r, f + g) \leq T(r, f) + T(r, g) + O(1)$, $T(r, cf) = T(r, f)\ \forall c \in \mathbb{K}^*$, $T(r, \frac{1}{f}) = T(r, f))$, $T(r, \frac{f}{g}) \leq T(r, f)) + T(r, g)$.*

Suppose now $f, g \in \mathcal{A}(\mathbb{K})$ (resp. $f,\ g \in \mathcal{A}(d(0, R^-))$). Then $Z(r, fg) = Z(r, f) + Z(r, g)$, $T(r, f) = Z(r, f))$, $T(r, fg) = T(r, f) + T(r, g) + O(1)$, and $T(r, f + g) \leq \max(T(r, f), T(r, g))$. Moreover, if $\lim_{r\to+\infty} T(r, f) - T(r, g) = +\infty$, then $T(r, f + g) = T(r, f)$ when r is big enough.

Lemma C.4.4: *Let $\alpha_1, \dots, \alpha_n \in \mathbb{K}$ be pairwise distinct, let $P(u) = \prod_{i=1}^{n}(u - \alpha_i)$ and let $f \in \mathcal{M}(d(0, R^-))$. Then $Z(r, P(f)) = \sum_{i=1}^{n} Z(r, f - \alpha_i)$ and $\overline{Z}(r, P(f)) = \sum_{i=1}^{n} \overline{Z}(r, f - \alpha_i)$.*

Lemma C.4.5: *Let $f \in \mathcal{M}(\mathbb{K})$. Then f belongs to $\mathbb{K}(x)$ if and only if $T(r, f) = O(\log r)$.*

Proof. If f belongs to $\mathbb{K}(x)$, one can write it $\frac{P(x)}{Q(x)}$ with $P,\ Q \in \mathbb{K}[x]$ having no common zeros, hence $Z(r, f) = Z(r, P)$ and $N(r, f) = Z(r, Q)$ and hence $T(r, f) = O(\log r)$.

Now suppose that $f \notin \mathbb{K}(x)$. Suppose for instance that f has infinitely many zeros (a_n) of respective order q_n. Then let us fix $s \in \mathbb{N}$ and let $t \in \mathbb{N}$ be $> s+1$. For r big enough, we have $Z(r, f) > \sum_{n=1}^{t}(\log r - \Psi(a_n)) > s \log r$, hence $Z(r, f)$ is not $O(\log r)$. Similarly, if f has infinitely many poles we get to the same conclusion. □

Applying Lemma C.4.1 and C.2.10 to $\frac{f'}{f}$, up to a change of origin, we can derive Corollary C.4.6.

Corollary C.4.6: *Let $f \in \mathcal{M}(\mathbb{K})$ (resp. $f \in \mathcal{M}(d(0, R^-))$). Then*

$$Z\left(r, \frac{f'}{f}\right) - N\left(r, \frac{f'}{f}\right) \leq -\log r + O(1).$$

Theorem C.4.7: *Let $f \in \mathcal{A}(\mathbb{K})$ (resp. $f \in \mathcal{A}(d(0, R^-))$) and let $b \in \mathbb{K}$. Then $Z(r, f) = Z(r, f - b) + O(1)$ $r \in I$ (resp. $r \in J$).*

Proof. Let $f(x) = \sum_{n=0}^{\infty} a_n x^n$ and let $\rho \in \mathbb{R}_+^*$ (resp. $\rho \in]0, R[$) be such that $\nu^+(f, \log \rho) > 0$ and $\nu^+(f - b, \log \rho) > 0$. Then we have $\nu^+(f, \mu) = \nu^+(f - b, \mu)\ \forall \mu > \log \rho$ (resp. $\forall \mu \in]\log \rho, \log R[$). Consequently, on each circle $C(0, r)$ such that $r > \rho$ (resp. $r \in]\rho, R[$), f and $f - b$ have the same number of zeros, taking multiplicity into account. Let (a_n) be the sequence of zeros of f, with respective multiplicity q_n, with $|a_n| \leq |a_{n+1}|$, $n \in \mathbb{N}^*$, and $|a_n| > \rho$ if and only if $n \geq t$.

Similarly, let (b_n) be the sequence of zeros of $f - b$, with respective multiplicity s_n, with $|b_n| \leq |b_{n+1}|, n \in \mathbb{N}^*$, and $|b_n| > \rho$ if and only if $n \geq u$. Since f and $f - b$ have the same number of zeros in $d(0, \rho)$, we also have

$$\sum_{n=1}^{t} q_n = \sum_{n=1}^{u} s_n. \tag{1}$$

Consequently, for all $r > \rho$ (resp. $r \in]\rho, R[$), we have

$$\sum_{\substack{n\geq t,\\ |a_n|\leq r}} q_n(\log r - \Psi(a_n)) = \sum_{\substack{n\geq u\\ |b_n|\leq r}} s_n(\log r - \Psi(b_n)).$$

Now, suppose that both $f(0)$, $f(0) - b$ are not 0. Then

$$Z(r, f) = \sum_{|a_n|\leq r} q_n(\log r - \Psi(a_n)),\ Z(r, f - b) = \sum_{|b_n|\leq r} s_n(\log r - \Psi(b_n)).$$

Therefore, $Z(r, f) - Z(r, f - b)$ is reduced to

$$\sum_{\substack{|a_n|\leq r,\\ |a_n|\leq \rho}} (\log r - \Psi(a_n)) - \sum_{\substack{|b_n|\leq r,\\ |b_n|\leq \rho}} (\log r - \Psi(b_n))$$

$$= \sum_{n=1}^{t} q_n(\log r - \Psi(a_n)) - \sum_{n=1}^{u} s_n(\log r - \Psi(b_n))$$

that this is a constant by (1), for $r > \rho$ (resp. for $r \in]0, R[$).

And now, suppose that 0 is a zero of order q_1 of f. Then,

$$Z(r, f) = q_1 \log r + \sum_{n\geq 2, |a_n|\leq r} q_n(\log r - \Psi(a_n))$$

and therefore $Z(r, f) - Z(r, f - b)$ is reduced to

$$q_1 \log r + \sum_{\substack{n\geq 2,\\ |a_n|\leq \rho}} q_n(\log r - \Psi(a_n)) - \sum_{|b_n|\leq \rho} s_n(\log r - \Psi(b_n))$$

$$= q_1 \log r + \sum_{n=2}^{t} q_n(\log r - \Psi(a_n)) - \sum_{n=1}^{u} s_n(\log r - \Psi(b_n))$$

and we check that this is a constant. Again thanks to (1).

Similarly, if $f(0) = b$, then f and $f - b$ playing the same role, we have the same conclusion. $\square$

Theorem C.4.8 (First Main Fundamental Theorem): *Let $f,\ g \in \mathcal{M}(\mathbb{K})$ (resp. let $f,\ g \in \mathcal{M}(d(0, R^-))$). Then $T(r, f + b) = T(r, f) + O(1)$. Let h be a Moebius function. Then $T(r, f) = T(r, h \circ f) + O(1)$. Let $P(X) \in \mathbb{K}[X]$. Then $T(r, P(f)) = \deg(P)T(r, f) + O(1)$ and $T(r, f'P(f) \geq T(r, P(f))$.*

Suppose now $f, g \in \mathcal{A}(\mathbb{K})$ (resp. $f,\ g \in \mathcal{A}(d(0, R^-))$). Then $Z(r, fg) = Z(r, f) + Z(r, g)$, $T(r, f) = Z(r, f))$, $T(r, fg) = T(r, f) + T(r, g) + O(1)$, and $T(r, f + g) \leq \max(T(r, f), T(r, g))$. Moreover, if $\lim_{r\to+\infty} T(r, f) - T(r, g) = +\infty$ then $T(r, f+g) = T(r, f)$ when r is big enough.

Proof. $T(r, f+b) \le T(r, f) + O(1) \le T(r, f+b) + O(1)$, hence $T(r, f+b) = T(r, f) + O(1)$ $\forall b \in \mathbb{K}$. Now, consider $T(r, f+g)$ when $f,\ g \in \mathcal{A}(\mathbb{K})$ (resp. if $f,\ g \in \mathcal{A}(d(0, R^-)))$. We have $T(r, f+g) = Z(r, f+g) = (\Psi(f+g, \log r) + O(1) \le \max(\Psi(f, \log r), \Psi(g, \log r)) = \max(T(r, f) + O(1), T(r, f) + O(1))$.

Let $h(X) = \dfrac{aX+b}{cX+d}$ be a Moebius function and let $g(x) = h \circ f(x)$. We can write $h(X) = \dfrac{a}{c} + \dfrac{\lambda}{cX+d}$ with $\lambda = d\Big(1 - \dfrac{a}{c}\Big)$. Then

$$T(r, g) = T\left(r, \frac{\lambda}{cf(x)+d}\right) + O(1) = T(r, cf(x)+d) + O(1) = T(r, f(x)) + O(1).$$

Now, let $P(X) = \prod_{k=1}^{q}(X - a_k) \in \mathbb{K}[x]$ be a polynomial of degree q and let $F(x) = P(f(x))$. Then $T(r, f - a_k) = T(r, f) + O(1)$ $\forall k = 1, \dots, q$ and hence $T(r, F) = qT(r, f) + O(1)$. Moreover, zeros of $P(f)$ are not poles of f' and poles of f' are poles of f and hence are not zeros of $P(f)$. Consequently, $N(r, f'P(f) = N(r, P(f)) + N(r, f') = N(r, P(f)) + N(r, f) + \overline{N}(r, f)$, $Z(r, f'P(f)) = Z(r, P(f) + Z(r, f')$. Therefore, $T(r, f'P(f) \ge T(r, P(f))$.

It now only remains to prove that $T(r, P(f)) = qT(r, f) + O(1)$. Let $P(X) = \prod_{j=1}^{q}(X - a_j)$. It is immediate to check that $Z(r, P(f)) = \sum_{j=1}^{q} Z(r, f - a_j) = qZ(r, f) + O(1)$ and that $N(r, P(f)) = qN(r, f)$. Therefore, $T(r, P(f)) = qT(r, f) + O(1)$. □

Theorem C.4.9: *Let $f \in \mathcal{M}(\mathbb{K})$ (resp. $f \in \mathcal{M}(d(0, R^-)))$. There exists $\phi,\ \psi \in \mathcal{A}(\mathbb{K})$ (resp. $\phi,\ \psi \in \mathcal{A}(d(0, R^-)))$ such that $f = \dfrac{\phi}{\psi}$ and* $\max(T(r, \phi), T(r, \psi)) \le T(r, f) + O(1)$, *$r \in I$ (resp. $(r \in J)$).*

Proof. Let $V_1 = \mathcal{D}(f)$ and let $V_2 = \mathcal{D}\Big(\dfrac{1}{f}\Big)$. Suppose first $f \in \mathcal{M}(\mathbb{K})$. By Theorem B.18.4 there exists $\phi,\ \psi \in \mathcal{A}(\mathbb{K})$ such that $\mathcal{D}(f) = \mathcal{D}(\phi)$, $\mathcal{D}\Big(\dfrac{1}{f}\Big) = \mathcal{D}(\psi)$. Consequently, $Z(r, f) = Z(r, \phi), N(r, f) = Z(r, \psi)$ and the claim is immediate. Now, suppose $f \in \mathcal{M}(d(0, R^-))$. By Theorem B.18.14 there exists $\phi \in \mathcal{A}(\mathbb{K})$ such that $\mathcal{D}(f) \le \mathcal{D}(\phi)$ and such that $|\mathcal{D}(\phi)|(r) \le |V_1|(r) + 1$, $r \in J$, hence $Z(r, \phi) \le Z(r, f) + 1$, $r \in J$. Let $\psi = \dfrac{\phi}{f}$. Then ψ lies in $\mathcal{A}(d(0, R^-))$ because $\mathcal{D}(f) \le \mathcal{D}(\phi)$. And $\mathcal{D}(\psi) = \mathcal{D}\Big(\dfrac{1}{f}\Big) + \mathcal{D}(\phi) - \mathcal{D}(f)$. Consequently, $|\mathcal{D}(\psi)|(r) \le |\mathcal{D}(\dfrac{1}{f})|(r) + 1$. But $T(r, \phi) = Z(r, \phi) + O(1) = \log(|\mathcal{D}(\phi)|(r)) + o(1)$ and $T(r, \psi) = Z(r, \psi) + O(1) = \log(|\mathcal{D}(\psi)|(r)) + O(1)$. Therefore, $\max(T(r, \phi), T(r, \psi)) \le \max(Z(r, f), N(r, f)) + O(1)$, $r \in I$ (resp. $(r \in J)$). □

Theorem C.4.10: *Let $f \in \mathcal{M}(d(0, R^-))$. Then f belongs to $\mathcal{M}_b(d(0, R^-))$ if and only if $T(r, f)$ is bounded in $[0, R[$.*

Proof. Suppose first $f \in \mathcal{A}(d(0,R^-))$. Without loss of generality, we can obviously suppose that $f(0) \neq 0$. By Theorem C.4.2, we have $\log|f|(r) = \log(|f(0)|) + Z(r,f)$. And $|f|(r) = \sup\{|f(x)| \mid x \in d(0,r^-)\}$, so the claim is clear. Now, consider the general case. Suppose $T(r,f)$ is not bounded, so either $Z(r,f)$ or $N(r,f)$ is not bounded. Let $f = \dfrac{\phi}{\psi}$ with $\phi,\ \psi \in \mathcal{A}(d(0,R^-))$. If $Z(r,f)$ is not bounded, then $\phi \notin \mathcal{A}_b(d(0,R^-))$. If $N(r,f)$ is not bounded, then $\psi \notin \mathcal{A}_b(d(0,R^-))$. Thus, f cannot be put in the form $\dfrac{\phi}{\psi}$ with $\phi,\ \psi \in \mathcal{A}_b(d(0,R^-))$ and therefore $f \notin \mathcal{A}_b(d(0,R^-))$.

Conversely, if $f \in \mathcal{A}_b(d(0,R^-))$, then it is of the form $\dfrac{\phi}{\psi}$ with $\phi,\ \psi \in \mathcal{A}_b(d(0,R^-))$, hence both $Z(r,\phi), Z(r,\psi)$ are bounded. But since $Z(r,f) \leq Z(r,\phi)$ and $N(r,f) \leq Z(r,\psi)$, $T(r,f)$ is clearly bounded in $[0,R[$. □

Corollary C.4.11: *Let $f \in \mathcal{M}_u(d(a,R^-))$ and let $h \in \mathcal{M}_b(d(a,R^-))$, $h \neq 0$. Then fh belongs to $\mathcal{M}_u(d(a,R^-))$.*

By Theorems C.4.8 and C.4.10, we can also derive Corollary C.4.12.

Corollary C.4.12: *Let $f \in \mathcal{M}(d(a,R^-))$ and let $P \in \mathbb{K}[x]$. Then $P(f)$ belongs to $\mathcal{M}_b(d(a,R^-))$ if and only if so does f.*

Lemma C.4.13 is classical and easily checked.

Lemma C.4.13: *Let $\alpha_1, \dots, \alpha_q \in \mathbb{K}$ be pairwise distinct, let $S = \{\alpha_1, \dots, \alpha_q\}$ and let $P(x) = \prod_{j=1}^{q}(x - \alpha_j)$. Let $f \in \mathcal{M}(\mathbb{K})$ (resp. $f \in \mathcal{M}(d(0,R^-))$). Then*

$$\sum_{j=1}^{n} Z(r, f - \alpha_j) = Z(r, P(f)), \quad \sum_{j=1}^{n} \overline{Z}(r, f - \alpha_j) = \overline{Z}(r, P(f))\ \forall r \in I$$

(resp. $\forall r \in J$). Moreover, assuming that $\mathbb{K}$ is of characteristic 0, we have $\sum_{j=1}^{n}\Big(Z(r,f-\alpha_j) - \overline{Z}(r,f-\alpha_j)\Big) = Z(r,f') - Z_0^S(r,f')\ \forall r \in I$ *(resp. $\forall r \in J$).*

Theorem C.4.14: *We assume that $\mathbb{K}$ is of characteristic 0. Let $f \in \mathcal{M}(\mathbb{K})$ (resp. $f \in \mathcal{M}(d(0,R^-))$). Then $Z(r,f') - N(r,f') \leq Z(r,f) - N(r,f) - \log r + O(1)$, $r \in I$ (resp. $r \in J$). Moreover, $N(r,f^{(k)}) = N(r,f) + k\overline{N}(r,f) + O(1)$, $r \in I$ and $Z(r,f^{(k)}) \leq Z(r,f) + k\overline{N}(r,f) - k\log r + O(1)$, $r \in I$ (resp. $r \in J$).*

Proof. Without loss of generality, we can assume that $f, f', \dots, f^{(k)}$ have no zero and no pole at 0.

The first statement is immediate and just comes from this basic property: if α is a pole of f of order q, then it is a pole of $f^{(k)}$ of order $q+k$. Next, by Theorem C.4.2, we have $Z(r,f)-N(r,f)=\Psi(f,\log r)-\log(|f(0)|)$ and $Z(r,f')-N(r,f')=\Psi(f',\log r)-\log(|f'(0)|)$. But $\Psi(f',\log r)\leq\Psi(f,\log r)-\log r$, hence we obtain $Z(r,f')\leq N(r,f')-N(r,f)+Z(r,f)-\log r+O(1)$. Actually, $N(r,f')-N(r,f)=\overline{N}(r,f)$, hence $N(r,f^{(k)})=N(r,f)+k\overline{N}(r,f)$. Next, $Z(r,f')\leq Z(r,f)+\overline{N}(r,f)-\log r+O(1)$. Now, suppose the second statement has been proved for $k\leq t$. Thus, we have $Z(r,f^{(t+1)})\leq Z(r,f^{(t)})+\overline{N}(r,f^{(t)})-\log r+O(1)$. But as we just noticed, $\overline{N}(r,f^{(t)})=\overline{N}(r,f)$, hence $Z(r,f^{(t+1)})\leq Z(r,f)+t\overline{N}(r,f^{(t)})+\overline{N}(r,f^{(t)})-(t+1)\log r+O(1)$. □

Corollary C.4.15: *We assume that* $\mathbb{K}$ *is of characteristic* 0. *Let* $f\in\mathcal{M}(\mathbb{K})$ *(resp.* $f\in\mathcal{M}(d(0,R^-))$*). Then* $T(r,f^{(k)})\leq(k+1)T(r,f)+O(1)$ $(r\in I)$ *(resp.* $r\in J$*).*

Theorem C.4.16: *We assume that* $\mathbb{K}$ *is of characteristic* 0. *Let* $f\in\mathcal{M}(\mathbb{K})$ *(resp.* $f\in\mathcal{M}(d(0,R^-))$*). Then,* $T(r,f)-Z(r,f)\leq T(r,f')-Z(r,f')+O(1)$. *Further, given* $\alpha\in\mathcal{M}(d(0,R^-))$, *we have* $T(r,\alpha f)-Z(r,\alpha f)\leq T(r,f)-Z(r,f)+T(r,\alpha)$.

Proof. By Theorem C.4.14, the first statement is immediate. Let us check the last one. On the one hand, $T(r,f)-Z(r,f)=\max(Z(r,f),N(r,f))-Z(r,f)=\max(0,N(r,f)-Z(r,f))$ $r<R$. On the other hand,

$$\begin{aligned}T(r,f')-Z(r,f')&=\max(Z(r,f'),N(r,f'))-Z(r,f')=\max(0,N(r,f')-Z(r,f'))\\&=\max(0,N(r,f)+\overline{N}(r,f)-Z(r,f'))\ r<R.\end{aligned}$$

But by Theorem C.4.14, $-Z(r,f')\geq-Z(r,f)-\overline{Z}(r,f)+\log(r)+O(1)$ $r<R$, hence $T(r,f')-Z(r,f')\geq\max(0,N(r,f)-Z(r,f)+\log(r))+O(1)\geq T(r,f)-Z(r,f)+O(1)$, $r<R$.

Now, let $\alpha\in\mathcal{M}(d(0,R^-))$. Suppose $N(r,\alpha f)\geq Z(r,\alpha f)$, $r<R$. Then $T(r,\alpha f)-Z(r,\alpha f)=N(r,\alpha f)-Z(r,\alpha f)$ $r<R$. We can write α in the form $\frac{\beta(x)}{\lambda(x)}$ with β, $\lambda\in H(d(0,r))$, β,λ having no common zero. Next, we can write λ in the form $\lambda_1\lambda_2$, where each zero of λ_1 is not a zero of f and each zero of λ_2 is a zero of f. Then we can check that $N(r,\alpha f)=N(r,f)+Z(r,\lambda_1)$ and $Z(r,\alpha f\geq Z(r,f)-Z(r,\lambda_2)$. Consequently, $N(r,\alpha f)-Z(r,\alpha f)\leq N(r,f)+Z(r,\lambda_1)-(Z(r,f)-Z(r,\lambda_2))=N(r,f)-Z(r,f)+Z(r,\lambda_1)+Z(r,\lambda_2)=N(r,f)-Z(r,f)+Z(r,\lambda)\leq N(r,f)-Z(r,f)+T(r,\lambda)$ $r<R$.

Suppose now that $N(r,\alpha f)\leq Z(r,\alpha f)$. We can do a symmetric reasoning with the zeros of β. □

Lemma C.4.17 is an immediate consequence of Corollary B.13.27 and Theorem C.2.10.

Lemma C.4.17: *Let $\mathbb{K}$ be of characteristic 0. Let $f \in \mathcal{M}(\mathbb{K})$ (resp. $f \in \mathcal{M}(d(0,R^-))$) and let $G = \frac{f'}{f}$. Then, G satisfies $Z(r,G) \leq N(r,G) - \log r + O(1)$ $r \in I$ (resp. ($r \in J$).*

Proof. Without loss of generality we can assume that 0 is neither a pole of f nor a zero for ff'. By Theorem C.2.10, G satisfies $\Psi(G, \log r) \leq -\log r$. On the other hand, by Theorem C.4.2 we have $\Psi(G, \log r) = \log|G(0)| + Z(r,G) - N(r,G)$. Consequently, we obtain $\log|G(0)| + Z(r,G) - N(r,G) \leq \log r$, which proves the claim. □

We can now prove the Second Main theorem under different forms. Lemma C.4.18 is essential and directly leads to the theorems.

Lemma C.4.18: *Let $f \in \mathcal{M}(\mathbb{K})$ (resp. $f \in \mathcal{M}_u(d(0,R^-))$). Suppose that there exists $\xi \in \mathbb{K}$ (resp. $\xi \in \mathcal{M}_b(d(0,R^-))$) and a sequence of intervals $I_n = [u_n, v_n]$ such that $u_n < v_n < u_{n+1}$, $\lim_{n\to+\infty} u_n = +\infty$ (resp. $\lim_{n\to+\infty} u_n = R$) and $\lim_{n\to+\infty}\Big(\inf_{r\in I_n} T(r,f) - Z(r,f-\xi)\Big) = +\infty$ (resp. $\lim_{n\to+\infty}\Big(\inf_{r\in I_n} T(r,f) - Z(r,f-\xi)\Big) = +\infty$).*

Let $\tau \in \mathbb{K}$ (resp. let $\tau \in \mathcal{M}_b(d(0,R^-))$), $\tau \neq \xi$. Then $Z(r,f-\tau) = T(r,f) + O(1)$ $\forall r \in I_n$ when n is big enough.

Proof. We know that the Nevanlinna functions of a meromorphic function f are the same in $\mathbb{K}$ and in an algebraically closed complete extension of $\mathbb{K}$ whose absolute value extends that of $\mathbb{K}$. Consequently, without loss of generality, we can suppose that $\mathbb{K}$ is spherically complete because we know that such a field does admit a spherically complete algebraically closed extension whose absolute value expands that of $\mathbb{K}$. If f belongs to $\mathcal{M}(\mathbb{K})$, we can obviously set it in the form $\frac{g}{h}$ where g, h belong to $\mathcal{A}(\mathbb{K})$ and have no common zero. Next, since $\mathbb{K}$ is supposed to be spherically complete, if f belongs to $\mathcal{M}(d(0,R^-))$ we can also set it in the form $\frac{g}{h}$ where g, h belong to $\mathcal{A}(d(0,R^-))$ and have no common zero. Consequently, we have $T(r,f) = \max(Z(r,g), Z(r,h))$.

When ξ is a constant we can obviously suppose that $\xi = 0$. Suppose now $\xi \in \mathcal{M}_u(d(0,R^-))$. Then $f - \xi$ also belongs to $\mathcal{M}_u(d(0,R^-))$ and $\tau - \xi$ belongs to $\mathcal{M}_b(d(0,R^-))$. Consequently, in both cases, we can assume $\xi = 0$ to prove the claim. Next, up to a change of origin, we can also assume that none of the functions we consider have a pole or a zero at the origin.

Now, we have $\lim_{n\to+\infty}\Big(\inf_{r\in I_n} T(r,f) - Z(r,f)\Big) = +\infty$, i.e.,

$$\lim_{n\to+\infty}\Big(\inf_{r\in I_n} (Z(r,h) - Z(r,g))\Big) = +\infty. \tag{1}$$

Particularly, we notice that $T(r,f) = Z(r,h) + O(1)$ whenever $r \in I_n$ when n is big enough.

Consider now $Z(r, f - \tau) = Z(r, g - \tau h)$. But by (1) we can see that $|g|(r) < |\tau| h|(r)$ $\forall r \in I_n$ when n is big enough. Hence, $Z(r, g - \tau h) = Z(r, \tau h)$ $\forall r \in I_n$ when n is big enough. Hence, $Z(r, \tau h) = Z(r, h)$ $\forall r \in I_n$ when n is big enough. Therefore, $Z(r, f - \tau) = Z(r, h) + O(1) = T(r, f) + O(1)$, $\forall r \in I_n$ when n is big enough. So the claim is proven when τ is a constant.

Suppose now that $f \in \mathcal{M}(d(0, R^-))$ and $\tau \in \mathcal{M}_b(d(0, R^-))$. By Theorem C.1.10, we can write τ in the form $\dfrac{\phi}{\psi}$ where $\phi, \psi \in \mathcal{A}_b((d(0, R^-))$ have no common zero. Consider $Z(r, f - \tau h) = Z(r, \dfrac{\psi g - \phi h}{\psi h})$. Since g and h have no common zero and since both ϕ, ψ are bounded, we have $Z(r, \dfrac{\psi g - \phi h}{\psi h}) = Z(r, \psi g - \phi h) + O(1)$. By (1), in I_n we have $|\psi g|(r) < |\phi h|(r)$ when n is big enough and since $|\ .\ |(r)$ is an absolute value, $|\psi g - \phi h|(r) = |\phi h|(r)$ in I_n when n is big enough. Therefore, we have $Z(r, \psi g - \phi h) = Z(r, \phi h) = Z(r, h) + O(1)$ in I_n when n is big enough. Consequently, $Z(r, f - \tau) = Z(r, h) + O(1) = T(r, h) + O(1) = T(r, f) + O(1)$ $\forall r \in I_n$ when n is big enough. That finishes proving Lemma C.4.18. □

Theorems C.4.19 and C.4.21 may be found in a very different form in [85].

Theorem C.4.19: *Let $f \in \mathcal{M}(\mathbb{K})$ and let $a_1, \dots, a_q \in \mathbb{K}$ be distinct. Then*

$$(q-1)T(r,f) \le \max_{1 \le k \le q} \Big(\sum_{j=1, j \neq k}^{q} Z(r, f - a_j) \Big) + O(1).$$

Corollary C.4.20: *Let $f \in \mathcal{M}(\mathbb{K})$ and let $a_1, \dots, a_q \in \mathbb{K}$ be distinct. Then $(q-1)T(r,f) \le \sum_{j=1}^{q} Z(r, f - a_j) + O(1)$.*

Theorem C.4.21: *Let $f \in \mathcal{M}(d(0, R^-))$ and let $\tau_1, \dots, \tau_q \in \mathcal{M}_b(d(0, R^-))$ be distinct. Then*

$$(q-1)T(r,f) \le \max_{1 \le k \le q} \Big(\sum_{j=1, j \neq k}^{q} Z(r, f - \tau_j) \Big) + O(1).$$

Corollary C.4.22: *Let $f \in \mathcal{M}(d(0, R^-))$ and let $\tau_1, \dots, \tau_q \in \mathcal{M}_b(d(0, R^-))$ be distinct. Then $(q-1)T(r,f) \le \sum_{j=1}^{q} Z(r, f - \tau_j) + O(1)$.*

Proof. Suppose Theorem C.4.19 (resp. Theorem C.4.21) is wrong. In order to make a unique proof for the two theorems, in Theorem C.4.19 we set $\tau_j = a_j$. Thus, there exists $f \in \mathcal{M}(\mathbb{K})$ (resp. $f \in \mathcal{M}(d(0, R^-))$) and $\tau_1, \dots, \tau_q \in \mathbb{K}$ (resp. $\tau_1, \dots, \tau_q \in \mathcal{M}_b(d(0, R^-))$) such that $(q-1)T(r,f) - \max_{1 \le k \le q} \Big(\sum_{j=1, j \neq k}^{q} Z(r, f - \tau_j) \Big)$ admits no superior bound in $]0, +\infty[$. So, there exists a sequence of intervals $J_s = [w_s, y_s]$

such that $w_s < y_s < w_{s+1}$, $\lim_{s\to+\infty} w_s = +\infty$ (resp. $\lim_{s\to+\infty} w_s = R$) and two distinct indices $m \le q$ and $t \le q$ such that

$$\lim_{s\to+\infty} \inf_{r\in J_s} \big(T(r,f) - Z(r, f-\tau_m)\big) = +\infty$$

and

$$\lim_{s\to+\infty} \inf_{r\in J_s} \big(T(r,f) - Z(r, f-\tau_t)\big) = +\infty.$$

But by Lemma C.4.18, this is impossible. This ends the proof of Theorems C.4.19 and C.4.21. □

Remark: Theorem C.4.19 does not hold in complex analysis. Indeed, let f be a meromorphic function in $\mathbb{C}$ omitting two values a and b, such as $f(x) = \dfrac{e^x}{e^x - 1}$. Then $Z(r, f-a) + Z(r, f-b) = 0$.

Theorem C.4.23 *Let $\alpha_1, \dots, \alpha_q \in \mathbb{K}$, with $q \ge 2$, let $S = \{\alpha_1, \dots, \alpha_q\}$, and let $f \in \mathcal{M}(\mathbb{K})$ (resp. $f \in \mathcal{M}(d(0, R^-))$). Then $(q-1)T(r,f) \le \sum_{j=1}^{q} \overline{Z}(r, f-\alpha_j) + Z(r, f') - Z_0^S(r, f') + O(1)$ $\forall r \in I$ (resp. $\forall r \in J$).*

Moreover, if f belongs to $f \in \mathcal{A}(\mathbb{K})$ (resp. $\mathcal{A}(d(0, R^-))$), then $qT(r,f) \le \sum_{j=1}^{q} \overline{Z}(r, f-\alpha_j) + Z(r, f') - Z_0^S(r, f') + O(1)$ $\forall r \in I$ (resp. $\forall r \in J$).

Theorem C.4.24 (Second Main Theorem): *Let $\alpha_1, \dots, \alpha_q \in \mathbb{K}$, with $q \ge 2$, let $S = \{\alpha_1, \dots, \alpha_q\}$ and let $f \in \mathcal{M}(\mathbb{K})$ (resp. $f \in \mathcal{M}_u(d(0, R^-))$). Then $(q-1)T(r,f) \le \sum_{j=1}^{q} \overline{Z}(r, f-\alpha_j) + \overline{N}(r,f) - Z_0^S(r, f') - \log r + O(1)$ $\forall r \in I$ (resp. $\forall r \in J$).*

Proof. By Theorem C.4.23 (resp. Theorem C.4.24), there exists a constant $B > 0$ and for each $r > 0$ (resp. for each $r \in]0, R[$), there exists $k(r) \in \mathbb{N}$, $k(r) \le q$, such that $(q-1)T(r,f) \le \sum_{j=1, j\ne k(r)}^{q} Z(r, f-a_j) + B$ i.e. $(q-1)T(r,f) \le \sum_{j=1}^{q} Z(r, f-a_j) - Z(r, a_{k(r)} + B$. Now, $\sum_{j=1}^{q} Z(r, f-a_j) = \sum_{j=1}^{q} \overline{Z}(r, f-a_j) + Z(r, f') - Z_0^S(r, f') - \log r$. Consequently, $(q-1)T(r,f) \le \sum_{j=1}^{q} \overline{Z}(r, f-a_j) + Z(r, f') - Z_0^S(r, f') - Z(r, f-a_{k(r)}) + B$ and this proves the first claim of Theorem C.4.23. Particularly, if $f \in \mathcal{A}(\mathbb{K})$ (resp. if $f \in \mathcal{A}(d(0, R^-))$), then we have $Z(r, f-a_j) = T(r, f-a_j) = T(r,f) + O(1)$ $\forall j = 1, \dots, q$, hence $Z(r, f-a_{k(r)}) = T(r,f) + O(1)$ and therefore $qT(r,f) \le \sum_{j=1}^{q} \overline{Z}(r, f-a_j) + Z(r, f') - Z_0^S(r, f') + O(1)$, which ends the proof of Theorem C.4.23.

Henceforth, by Theorem C.4.14, there exists a constant $c_j > 0$ such that $Z(r, f') \le Z(r, f-a_j) = N(r, f-a_j) - \log r + c_j$. Let $c = \max(c_1, \dots, c_q)$. Then

$Z(r,f') - Z_0^S(r,f') - Z(r, f - a_{k(r)} \leq \overline{N}(r, f - a_{k(r)}) + c - \log r = \overline{N}(r,f) + c - \log r$.
Consequently,

$$\sum_{j=1}^{q} Z(r, f - a_j) = \sum_{j=1}^{q} \overline{Z}(r, f - a_j) + \overline{N}(r,f) - \log r + O(1).$$

That finishes the proof of Theorem C.4.24. □

Remark: In Theorem C.4.21, in the hypothesis $f \in \mathcal{M}(d(0, R^-))$, the term $-\log r$ has no veritable meaning since r is bounded.

Corollary C.4.25: *Let $\alpha_1, \dots, \alpha_q \in \mathbb{K}$, with $q \geq 2$, let $S = \{\alpha_1, \dots, \alpha_q\}$ and let $f \in \mathcal{M}(\mathbb{K})$ (resp. $f \in \mathcal{M}(d(0, R^-))$). Then $\sum_{j=1}^{q} \big(Z(r, f - \alpha_j) - \overline{Z}(r, f - \alpha_j)\big) \leq T(r,f) + \overline{N}(r,f) - Z_0^S(r,f') - \log r + O(1) \quad \forall r \in I$ (resp. $\forall r \in J$).*

C.5. Nevanlinna theory out of a hole

Here we mean to construct a Nevanlinna theory for meromorphic functions in the complement of an open disk. Thanks to the use of specific properties of the Analytic Elements on infraconnected subsets of $\mathbb{K}$ already examined in Chapter C.3.

Definitions and notations: Throughout the chapter and in the next, we will conserve the notations introduced in Chapter C.3. Particularly, we denote by S the disk $d(0, R^-)$ and put $D = K|S$. Recall that we denote by $H_0(D)$ the $\mathbb{K}$-vector space of analytic elements f in D such that $\lim_{|x| \to \infty} f(x) = 0$. The definitions of $\mathcal{A}(D)$, $\mathcal{A}_u(D)$, $\mathcal{M}(D)$, $\mathcal{M}_u(D)$ are those given in Chapter C.3.

Given $f \in \mathcal{M}(D)$, for $r > R$, here we will denote by $Z_R(r,f)$ the counting function of zeros of f between R and r, i.e., if $\alpha_1, \dots, \alpha_m$ are the distinct zeros of f in $\Delta(0, R, r)$, with respective multiplicity u_j, $1 \leq j \leq m$, then $Z_R(r,f) = \sum_{j=1}^{m} u_j(\log(r) - \log(|\alpha_j|))$. Similarly, we denote by $N_R(r,f)$ the counting function of poles of f between R and r, i.e., if $\beta_1, \dots, \beta_n$ are the distinct poles of f in $\Delta(0, R, r)$, with respective multiplicity v_j, $1 \leq j \leq m$, then $N_R(r,f) = \sum_{j=1}^{n} v_j(\log(r) - \log(|\beta_j|))$. Finally, we put $T_R(r,f) = \max\big(Z_R(r,f), N_R(r,f)\big)$.

Next, we denote by $\overline{Z}_R(r,f)$ the counting function of zeros without counting multiplicity: if $\alpha_1, \dots, \alpha_m$ are the distinct zeros of f in $\Delta(0, R, r)$, then we put $\overline{Z}_R(r,f) = \sum_{j=1}^{m} \log(r) - \log(|\alpha_j|)$.

Similarly, we denote by $\overline{N}_R(r, f)$ the counting function of poles without counting multiplicity: if $\beta_1, \dots, \beta_n$ are the distinct poles of f in $\Delta(0, R, r)$, then we put $\overline{N}_R(r, f) = \sum_{j=1}^{n} \log(r) - \log(|\beta_j|)$.

Finally, putting $W = \{a_1, \dots, a_q\}$, we denote by $Z_R^W(r, f')$ the counting function of zeros of f' on points x where $f(x) \notin W$.

Throughout the chapter, we denote by $| \,.\, |_\infty$ the Archimedean absolute value of $\mathbb{R}$. Given two functions defined in an interval $I = [b, +\infty[$, we will write $\phi(r) = \psi(r) + O(\log(r))$ (resp. $\phi(r) \leq \psi(r) + O(\log(r))$) if there exists a constant $B > 0$ such that $|\phi(r) - \psi(r)|_\infty \leq B\log(r),\ r \in I$ (resp. $\phi(r) - \psi(r) \leq B\log(r),\ r \in I$).

We will write $\phi(r) = o(\psi(r)),\ r \in I$ if $\lim_{r\to+\infty} \dfrac{\phi(r)}{\psi(r)} = 0$.

Theorem C.5.1: *Let $f \in \mathcal{M}(D)$. Then $\log(|f|(r)) - \log(|f|(R)) = Z_R(r, f) - N_R(r, f) + m(f, S)(\log r - \log R)$ $(r \in I)$.*

Proof. By Theorem C.3.6, we have $f = f^S f^0$. Since f^S has no zero and no pole in D, by Theorem C.3.3 it satisfies $|f^S|(r)) = r^{m(f,S)}\ \forall r \in I$, hence $\log(|f^S|(r)) - \log(|f^S|(R)) = m(f, S)(\log r - \log R)$ $(r \in I)$. Next, since f^0 has no zero and no pole in S, we have $\log(|f^0|(r)) - \log(|f^0|(R)) = Z_R(r, f^0) - N_R(r, f^0)$ $(r \in I)$, therefore the statement is clear. □

Corollary C.5.2: *Let $f \in \mathcal{M}(D)$. Then $T_R(r, f)$ is identically zero if and only if f is a Motzkin factor.*

Corollary C.5.3: *Let $f \in \mathcal{A}(D)$ and let $\phi \in H_0(D)$. Then $Z_R(r, f + \phi) = Z_R(r, f) + O(\log(r))$ $(r \in I)$.*

Proof. Indeed, since ϕ is bounded and tends to zero at infinite, we have $\log|f|(r) = \log|f + \phi|(r)$ when r is big enough. □

Corollary C.5.4: *Let $f,\ g \in \mathcal{A}(D)$ satisfy $\log(|f|(r)) \leq \log(|g|(r))\ \forall r \geq R$ $(r \in I)$. Then $Z_R(r, f) \leq Z_R(r, g) + (m(g, S) - m(f, S))(\log(r) - \log(R))$, $(r \in I)$.*

Theorem C.5.5: *Let $f \in \mathcal{A}(D)$. Then $Z_R(r, f') \leq Z_R(r, f) + O(\log(r))$ $(r \in I)$.*

Proof. Indeed, by Theorem B.9.2 we have $|f'|(r) \leq \dfrac{|f|(r)}{r}$. Therefore, the conclusion comes from Theorem C.5.1. □

We can now characterize the set $\mathcal{M}^*(D)$:

Theorem C.5.6: *Let $f \in \mathcal{M}(D)$. The following three statements are equivalent:*

(i) $\lim_{r\to+\infty} \dfrac{T_R(r, f)}{\log(r)} = +\infty$ $(r \in I)$,

(ii) $\dfrac{T_R(r,f)}{\log(r)}$ *is unbounded, and*
(iii) f *belongs to* $\mathcal{M}^*(D)$.

Proof. Consider an increasing sequence $(u_n)_{n\in\mathbb{N}}$ in $\mathbb{R}_+$ such that $\lim_{n\to+\infty} u_n = +\infty$ and let $(k_n)_{n\in\mathbb{N}}$ be a sequence of $\mathbb{N}^*$. Clearly, we have

$$\lim_{r\to+\infty} \frac{\sum_{u_n\le r} k_n(\log(r)-\log(u_n))}{\log(r)} = +\infty.$$

Consequently, if a function $f \in \mathcal{M}^*(D)$ has infinitely many zeros (resp. infinitely many poles in D), then $\lim_{n\to+\infty} \dfrac{Z_R(r,f)}{\log(r)} = +\infty$ (resp. $\lim_{n\to+\infty} \dfrac{N_R(r,f)}{\log(r)} = +\infty$), hence in both cases, $\lim_{n\to+\infty} \dfrac{T_R(r,f)}{\log(r)} = +\infty$. Conversely, if f has finitely many zeros and finitely many poles in D, then we check that $\lim_{n\to+\infty} \dfrac{T_R(r,f)}{\log(r)} < +\infty$. Thus, the equivalence of the three statements is clear. □

Operations on $\mathcal{M}(D)$ work almost like for meromorphic functions in the whole field.

Theorem C.5.7: *Let* $f,\ g \in \mathcal{M}(D)$. *Then for every* $b \in \mathbb{K}$, *we have* $T_R(r,f+b) = T_R(r,f) + O(\log(r))$, $(r \in I)$ $T_R(r,f.g) \le T_R(r,f) + T_R(r,g) + O(\log(r))$ $(r \in I)$, $T_R\Big(r,\dfrac{1}{f}\Big) = T_R(r,f))$, $T_R(r,f+g) \le T_R(r,f) + T_R(r,g) + O(\log(r))$ $(r \in I)$ *and* $T_R(r,f^n) = nT_R(r,f)$.

Let h *be a Moebius function. Then* $T_R(r,h\circ f) = T_R(r,f) + O(\log(r))$ $(r \in I)$.

Moreover, if both f *and* g *belong to* $\mathcal{A}(D)$, *then*

$$T_R(r,f+g) \le \max(T_R(r,f),T_R(r,g)) + O(\log(r))\ (r \in I)$$

and $T_R(r,fg) = T_R(r,f) + T_R(r,g)$, $(r \in I)$. *Particularly, if* $f \in \mathcal{A}^*(D)$, *then* $T_R(r,f+b) = T_R(r,f) + O(1)$ $(r \in I)$. *Given a polynomial* $P(X) \in \mathbb{K}[X]$, *then* $T_R(r,P\circ f) = qT_R(r,f) + O(\log(r))$.

Proof. Suppose first $f,\ g \in \mathcal{A}(D)$. It is immediate to check that $T_R(r,fg) = Z_R(fg) = Z_R(f) + Z_R(r,g) = T_R(r,f) + T_R(r,g)$, that $T_R(r,f^n) = nT_R(r,f)$ and that $T_R(r,\frac{1}{f}) = T_R(r,f)$.

Then $T_R(r,f+g) = Z_R(r,f+g) = \log(|f+g|(r)) - m(f+g,S)(\log(r) - \log(R))$, $(r \in I)$. But $\log(|f+g|(r)) \le \max\big(\log(|f|(r)),\log(|g|(r))\big)$, hence

$$\begin{aligned} Z_R(r,f+g) \le \max\big(&Z_R(r,f) + m(f,S)(\log(r) - \log(R), Z_R(r,g) \\ &+ m(g,S)(\log(r) - \log(R)\big), \end{aligned}$$

and hence $T_R(r,f+g) \le \max\big(T_R(r,f),T_R(r,g)\big) + O(\log(r))$.

Particularly, given $b \in \mathbb{K}$, we have $T_R(r, f+b) \leq T_R(r, f)+O(\log(r) \leq T_R(r, f)+O(\log(r))$, hence $T_R(r, f+b) = T_R(r, f) + O(\log(r)$.

Now, given a polynomial of degree q, we have $Z_R(r, P\circ f) = qZ_R(r, f)+O(\log(r))$ and $N_R(r, P \circ f) = qN_R(r, f)$, hence $T_R(r, P \circ f) = qT_R(r, f)$.

Now, suppose $f \in \mathcal{A}^*(D)$. Then $f(x)$ is a Laurent series $\sum_{-\infty}^{+\infty} a_n x^n$ convergent in all D such that $\lim_{r\to+\infty} |f|(r) = +\infty$. Let $b \in \mathbb{K}$ and take V be such that $|f|(r) > |b|\ \forall r \geq V$. Then for every $r > V$, $|f|(r)$ is of the form $|a_k|r^k$ with $k > 0$, $|a_n|r^n < |a_k|r^k\ \forall n > k$ and the number of zeros of f in $\Delta(0, R, r)$ is $k-m(f, S)$. Next, $f-b$ is of the form $\sum_{-\infty}^{+\infty} c_n x^n$ with $c_n = a_n\ \forall n \neq 0$ and $c_0 = a_0-b$. Consequently, $f - b$ has the same number of zeros in $\Delta(0, R, r)$ and in each circle $C(0, r)$ for $r > V$. Therefore, $T_R(r, f) = T_R(r, f - b)$ when r is big enough.

Next, consider the general case: $f,\ g \in \mathcal{M}(D)$. First it is immediate to check that $T_R(r, fg) \leq T_R(r, f)+T_R(r, g)$. Similarly, for $T_R(r, \frac{1}{f})$. By definition, we have $Z_R\Big(r, \frac{1}{f}\Big) = N_R(r, f)$ and $N_R\Big(r, \frac{1}{f}\Big) = Z_R(r, f)$, hence $T_R\Big(r, \frac{1}{f}\Big) = T_R(r, f)$.

Now consider $T_R(r, f+g)$ in the general case: $f,\ g \in \mathcal{M}(D)$. By Theorem C.3.6, we can write

$$f + g = f^S\Big(\frac{f_1^0}{f_2^0}\Big) + g^S\Big(g_1^0,\ g_2^0\Big),$$

hence

$$T_R(r, f + g) = T_R\Big(r, \frac{f^S f_1^0 g_2^0 + g^S g_1^0 f_2^0}{f_2 g_2}\Big)$$

with $f_1^0,\ f_2^0,\ g_1^0,\ g_2^0 \in \mathcal{A}(\mathbb{K})$, having no zero in T and $f^S,\ g^S$ Motzkin factors associated to S. Then $Z_R(r, f^S f_1^0 g_2^0) = Z_R(r, f_1^0) + Z_R(r, g_2^0), Z_R(r, g^S g_1^0 f_2^0) = Z_R(r, g_1^0) + Z_R(r, f_2^0)$, hence, by what we just saw, $Z_R(r, f^S f_1^0 g_2^0 + g^S g_1^0 f_2^0) \leq \max\big(T_R(r, f), T_R(r, g)\big) + O(\log(r))$. And obviously, $Z_R(r, f_2 g_2) \leq T_R(r, f) + T_R(r, g)$. So we obtain in the general case $T_R(r, f+g) \leq T_R(f)+T_R(r, g)+O(\log(r))$.

Finally, consider a Moebius function h. Then $h \circ f(x)$ is of the form $C + \dfrac{e}{\alpha f(x) + \beta}$ and thereby, $T_R(r, h \circ f) = T_R(r, f) + O(\log(r))$. □

Corollary C.5.8: *Let $f,\ g \in \mathcal{M}^0(D)$. Then $T_R\Big(r, \frac{f}{g}\Big) \geq T_R(r, f) - T_R(r, g)$ $(r \in I)$ and $T_R\Big(r, \frac{f}{g}\Big) \geq T_R(r, g) - T_R(r, f)$ $(r \in I)$.*

By Theorems C.5.6 and C.5.7, we have this immediate corollary:

Corollary C.5.9: *$\mathcal{M}^0(D)$ is a subfield of $\mathcal{M}(D)$.*

Theorem C.5.10: *Every $f \in \mathcal{M}^*(D)$ is transcendental over $\mathcal{M}^0(D)$.*

Proof. Consider a polynomial $P(Y) = \sum_{j=0}^{n} a_j Y^j \in \mathcal{M}^0(D)[Y]$ with $a_n = 1$. Let $f \in \mathcal{M}^*(D)$ and suppose that $P(f) = 0$. Then $f^n = -\sum_{j=0}^{n-1} a_j f^j$. Set $\Xi = \sum_{j=0}^{n-1} a_j f^j$ and $f = f^0 \frac{g}{h}$ with $g,\ h \in \mathcal{A}(D)$ having no zero in S. Then $\Xi = \frac{\sum_{j=0}^{n-1} a_j g^j h^{n-1-j}}{h^{n-1}}$.

Since $\sum_{j=0}^{n-1} a_j g^j h^{n-1-j}$ belongs to $\mathcal{A}(D)$, by Theorem C.5.7 we have

$$T_R\left(r, \sum_{j=0}^{n-1} a_j g^j h^{n-1-j}\right) \leq (n-1)T_R(r,f) + O(\log(r)),\ (r \in I)$$

and of course $T_R(r, h^{n-1}) \leq (n-1)T_R(r,f)$, $(r \in I)$. Consequently

$$T_R(r, \Xi) \leq (n-1)T_R(r,f) + O(\log(r),\ (r \in I).$$

But on the other hand, by Theorem C.5.7, $T_R(r, f^n) = nT_R(r,f)$. Therefore we should have $nT_R(r,f) \leq (n-1)T_R(r,f) + O(\log(r),\ (r \in I)$, which is impossible by Theorem C.5.6 because f belongs to $\mathcal{M}^*(D)$. Consequently, the equality $P(f) = 0$ is impossible, which proves that f is transcendental over $\mathcal{M}^0(D)$. □

Theorem C.5.11: *Let $f \in \mathcal{M}(D)$. Then $N_R(r, f^{(k)}) = N_R(r,f) + k\overline{N}_R(r,f)$, $(r \in I)$ and $Z_R(r, f^{(k)}) \leq Z_R(r,f) + k\overline{N}_R(r,f) + O(\log(r))$, $(r \in I)$.*

Proof. The inequality $N_R(r, f^{(k)}) = N_R(r,f) + k\overline{N}_R(r,f) + O(1)$, $r \in I$ is obvious. Next consider f in the form $\frac{g}{h}$ with $g,\ h \in \mathcal{A}(\mathbb{K})$. Recall that we can write h in the form $\overline{h}\widetilde{h}$ with $\overline{h}$ and $\widetilde{h}$ in $\mathcal{A}(\mathbb{K})$, each zero of $\overline{h}$ being of order one and all zeros of h being a zero of $\overline{h}$. So, h' is of the form $\widetilde{h}\widehat{h}$ where $\widehat{h}$ belong to $\mathcal{A}(\mathbb{K})$ and none of the zeros of $\widehat{h}$ is a zero of h. Then f' is of the form $\frac{g'\overline{h} - g\widehat{h}}{h\overline{h}}$. So, $Z_R(r, f') \leq Z_R(r, g'\overline{h} - g\widehat{h})$ and hence, by Theorem C.5.7,

$$(1) \qquad Z_R(r, f') \leq \max(Z_R(r, g'\overline{h}), Z_R(r, g\widehat{h}).$$

On one hand, by Theorem C.5.5, $Z_R(r, g') \leq Z_R(r, g) + O(\log r)$ and by Corollary C.5.4, we have $Z_R(r, g') \leq Z_R(r, f) + O(\log(r)$. Obviously, $Z_R(r, \overline{h}) \leq \overline{Z}_R(r, h) = \overline{N}_R(r, \ell) = \overline{N}_R(r,f)$, hence $Z_R(r, (r, g'\overline{h}) \leq Z_R(r,f) + \overline{N}_R(r,f)$.

Now, let us estimate $Z_R(r,\widehat{h})$. Since $\log(|h'|(r)) \leq \log(|h|(r)) - \log r$, we have $Z_R(r,h') \leq Z_R(r,h)+O(\log(r))$. But since $h' = \widehat{\widetilde{h}h}$, we have $Z_R(r,\widehat{h}) = Z_R(r,h') - Z_R(r,\widetilde{h}) \leq Z_R(r,h) - Z_R(r,\widetilde{h}) + O(\log(r)) = Z_R(r,\overline{h}) + O(\log(r)) = \overline{N}_R(r,f) + O(\log(r))$. Consequently,

$$Z_R(r,g\widehat{h}) \leq Z_R(r,g) + \overline{N}_R(r,f) + O(\log(r)) = Z_R(r,f) + \overline{N}_R(r,f) + O(\log(r)).$$

Thus, by (1) we have proven the claim when $k = 1$ and then it is immediately derived by induction on k. □

Lemma C.5.12 will be necessary in the proof of Theorem C.5.13.

Lemma C.5.12: *Let $f \in \mathcal{M}(D)$. Suppose that there exists $\xi \in \mathbb{K}$ and a sequence of intervals $J_n = [u_n, v_n]$ such that $u_n < v_n < u_{n+1}$, $\lim_{n\to+\infty} u_n = +\infty$, and* $\lim_{n\to+\infty}\left[\inf_{r\in J_n} \dfrac{T_R(r,f) - Z_R(r,f-\xi)}{\log(r)}\right] = +\infty.$

Let $\tau \in \mathbb{K}$ $\tau \neq \xi$. Then $Z_R(r,f-\tau) = T_R(r,f) + O(\log(r)))$ $\forall r \in J_n$ when n is big enough.

Proof. Without loss of generality, we can obviously suppose that $\xi = 0$. By Theorem C.3.6, f is of the form $f^S f^0$ and f^0 is of the form $\dfrac{g}{h}$ with g, $h \in \mathcal{A}(D)$, having no zero in S. Set $w = f^S$. Thus we have

$$\lim_{n\to+\infty}\left[\inf_{r\in J_n} \frac{Z_R(r,h) - Z_R(r,g)}{\log(r)}\right] = +\infty.$$

Consequently, by Theorem C.5.1,

$$\lim_{n\to+\infty}\left[\inf_{r\in J_n} \frac{\log(|h|(r) - \log(|g|(r)}{\log(r)}\right] = +\infty. \tag{1}$$

Consider now $f - \tau$. We have $f - \tau = \dfrac{wg - \tau h}{h}$, hence

$$\log(|f|(r)) = \log\big(|wg - \tau h|(r) - \log(|h|(r)).$$

But by (1), we have $\log(|\tau h|(r)) > \log(|wg|(r))$ because $\log(|w|(r) = O(\log(r))$, therefore $\log\big(|wg - \tau h|(r)\big) = \log(|\tau h|(r))$ $\forall r \in J_n$ when n is big enough and hence

$$\lim_{n\to+\infty}\left[\sup_{r\in J_n} \frac{\log(|\tau h - wg|(r) - \log(|h|(r)}{\log(r)}\right] = 0. \tag{2}$$

Consequently, by (2) and by Theorem C.5.1,

$$\lim_{n\to+\infty}\left[\sup_{r\in J_n} \frac{Z_R(r,\tau h - wg) - Z_R(r,h)}{\log(r)}\right] = 0,$$

i.e.,

$$\lim_{n\to+\infty}\left[\sup_{r\in J_n} \frac{Z_R(r,f-\tau) - T_R(r,f)}{\log(r)}\right] = 0,$$

which proves the claim. □

The Nevanlinna Second Main theorem is based on Theorem C.5.13.

Theorem C.5.13: *Let $f \in \mathcal{M}(D)$ and let $a_1, \dots, a_q \in \mathbb{K}$ be distinct. Then*

$$(q-1)T_R(r,f) \le \max_{1\le k\le q}\Big(\sum_{j=1, j\neq k}^{q} Z_R(r, f-a_j)\Big) + O(\log(r))\ (r \in I).$$

Proof. Suppose Theorem C.5.13 is wrong. Thus, there exists $f \in \mathcal{M}(D)$ and $a_1, \dots, a_q \in \mathbb{K}$ such that $(q-1)T_R(r,f) - \max_{1\le k\le q}\Big(\sum_{j=1, j\neq k}^{q} Z_R(r, f-a_j)\Big)$ admits no superior bound in $]0, +\infty[$. So, there exists a sequence of intervals $J_s = [w_s, y_s]$ such that $w_s < y_s < w_{s+1}$, $\lim_{s\to+\infty} w_s = +\infty$ and two distinct indices $m \le q$ and $t \le q$ such that

$$\lim_{s\to+\infty}\left[\inf_{r\in J_s}\frac{(T_R(r,f) - Z_R(r, f-a_m))}{\log(r)}\right] = +\infty$$

and

$$\lim_{s\to+\infty}\left[\inf_{r\in J_s}\frac{(T_R(r,f) - Z_R(r, f-a_t))}{\log(r)}\right] = +\infty.$$

But by Lemma C.5.12, that is impossible. □

We can now state and prove the Second Main Theorem for $\mathcal{M}(D)$.

Theorem C.5.14: *Let $f \in \mathcal{M}(D)$, let $\alpha_1, \dots, \alpha_q \in \mathbb{K}$, with $q \ge 2$, and let $W = \{\alpha_1, \dots, \alpha_q\}$. Then $(q-1)T_R(r,f) \le \sum_{j=1}^{q} \overline{Z}_R(r, f-\alpha_j) + Z_R(r, f') - Z_R^W(r, f') + O(\log(r)) \quad (r \in I)$.*

Moreover, if f belongs to $\mathcal{A}(D)$, then $qT_R(r,f) \le \sum_{j=1}^{q} \overline{Z}_R(r, f-\alpha_j) + Z_R(r, f') - Z_R^W(r, f') + O(\log(r)) \quad (r \in I)$.

Theorem C.5.15 (Second Main Theorem): *Let $f \in \mathcal{M}(D)$, let $\alpha_1, \dots, \alpha_q \in \mathbb{K}$, with $q \ge 2$ and let $W = \{\alpha_1, \dots, \alpha_q\}$. Then*

$$(q-1)T_R(r,f) \le \sum_{j=1}^{q} \overline{Z}_R(r, f-\alpha_j) + \overline{N}_R(r,f) - Z_R^W(r, f') + O(\log(r)) \quad (r \in I).$$

Proof. (Theorems C.5.14 and C.5.15) By Theorem C.5.13 there exists a constant $B > 0$ and for each $r > R$ there exists $k(r) \in \mathbb{N}$, $k(r) \leq q$, such that

$$(q-1)T_R(r,f) \leq \sum_{j=1, j\neq k(r)}^{q} Z_R(r, f-a_j) + B\log(r),$$

i.e., $(q-1)T_R(r,f) \leq \sum_{j=1}^{q} Z_R(r, f-a_j) - Z_R(r, a_{k(r)} + O(\log(r))$. Now,

$$\sum_{j=1}^{q} Z_R(r, f-a_j) = \sum_{j=1}^{q} \overline{Z}_R(r, f-a_j) + Z_R(r, f') - Z_R^W(r, f') + B\log(r).$$

Consequently,

$$(q-1)T_R(r,f) \leq \sum_{j=1}^{q} \overline{Z}_R(r, f-a_j,,,,) + Z_R(r, f') - Z_R^W(r, f')$$
$$- Z_R(r, f-a_{k(r)}) + O(\log(r))$$

and this proves the first claim of Theorem C.5.14.

Particularly, if $f \in \mathcal{A}(D)$ then we have $Z_R(r, f-a_j) = T_R(r, f-a_j) = T_R(r,f) + O(\log(r))\ \forall j = 1, \ldots, q$, hence $Z_R(r, f-a_{k(r)}) = T_R(r,f) + O(\log(r))$ and therefore

$$qT_R(r,f) \leq \sum_{j=1}^{q} \overline{Z}_R(r, f-a_j) + Z_R(r, f') - Z_R^W(r, f') + O(\log(r)),$$

which ends the proof of Theorem C.5.14.

Consider now the situation in Theorem C.5.15. By Theorem C.5.11, for each $j = 1, \ldots, q$, there exists a constant $B_j > 0$ such that $Z_R(r, f') \leq Z_R(r, f-a_j) + \overline{N}_R(r, f-a_j) + B_j\log(r))$. Consequently, there exists a constant $C > 0$ such that $Z_R(r, f') \leq Z_R(r, f-a_{k(r)}) + \overline{N}_R(r, f-a_{k(r)}) + C\log(r)\ \forall r > R$.

Therefore, by Relation (9) that remains true in Theorem C.5.15, we can derive

$$(q-1)T_R(r,f) \leq \sum_{j=1}^{q} \overline{Z}_R(r, f-\alpha_j) + \overline{N}_R(r,f) - Z_R^W(r, f') + O(\log(r)) \quad \forall r \in I.$$

□

Corollary C.5.16: *Let $f \in \mathcal{M}(\mathbb{K})$ and let $a_1, \ldots, a_q \in \mathbb{K}$ be distinct. Then $(q-1)T_R(r,f) \leq \sum_{j=1}^{q} Z_R(r, f-a_j) + O(\log(r))$ $(r \in I)$.*

Corollary C.5.17: *Let $f \in \mathcal{M}(D)$, let $\alpha_1, \ldots, \alpha_q \in \mathbb{K}$, with $q \geq 2$ and let $W = \{\alpha_1, \ldots, \alpha_q\}$. Then*

$$\sum_{j=1}^{q} \left(Z_R(r, f-\alpha_j) - \overline{Z}_R(r, f-\alpha_j)\right) \leq T_R(r,f) + \overline{N}_R(r,f) - Z_R^W(r, f')$$
$$+ O(\log(r))\ (r \in I).$$

C.6. Immediate applications of the Nevanlinna theory

Definitions and notations: As in Chapter C.6, we denote by D the set $\mathbb{K} \setminus d(0, R^-)$ with R a positive number. The definitions of $\mathcal{A}(D)$, $\mathcal{A}_u(D)$, $\mathcal{M}(D)$, $\mathcal{M}_u(D)$ are those given in Chapter C.3.

As immediate applications of the Second Main theorem, we can notice following Theorems C.6.1, C.6.2, C.6.3, and C.6.4.

Theorem C.6.1: *Let $a_1, a_2 \in \mathbb{K}$ $(a_1 \neq a_2)$ and let $f,\ g \in \mathcal{A}(\mathbb{K})$ satisfy $f^{-1}(\{a_i\}) = g^{-1}(\{a_i\})$ $(i = 1, 2)$. Then $f = g$.*

Remark: Theorem C.6.1 does not hold in complex analysis. Indeed, let $f(z) = e^z$, $g(z) = e^{-z}$, let $a_1 = 1$, $a_2 = -1$. Then $f^{-1}(\{a_i\}) = g^{-1}(\{a_i\})$ $(i = 1, 2)$, though $f \neq g$.

Theorem C.6.2: *Let $a_1,\ a_2,\ a_3 \in \mathbb{K}$ $(a_i \neq a_j\ \forall i \neq j)$ and let $f,\ g \in \mathcal{A}_u(d(a, R^-))$ (resp. $f,\ g \in \mathcal{A}_u(D)$) satisfy $f^{-1}(\{a_i\}) = g^{-1}(\{a_i\})$ $(i = 1, 2, 3)$. Then $f = g$.*

Theorem C.6.3: *Let $a_1,\ a_2,\ a_3,\ a_4 \in \mathbb{K}$ $(a_i \neq a_j\ \forall i \neq j)$ and let $f,\ g \in \mathcal{M}(\mathbb{K})$ satisfy $f^{-1}(\{a_i\}) = g^{-1}(\{a_i\})$ $(i = 1,\ 2,\ 3,\ 4)$. Then $f = g$.*

Theorem C.6.4: *Let $a_1,\ a_2,\ a_3,\ a_4,\ a_5 \in \mathbb{K}$ $(a_i \neq a_j\ \forall i \neq j)$ and let $f,\ g \in \mathcal{M}_u(d(a, R^-)))$ (resp. $f, g \in \mathcal{M}_u(D)$ satisfy $f^{-1}(\{a_i\}) = g^{-1}(\{a_i\})$ $(i = 1, 2, 3, 4, 5)$. Then $f = g$.*

Remark: Let $f(x) = \dfrac{x}{3x-1}$, $g(x) = \dfrac{x^2}{x^2+2x-1}$. Let $a_0 = 0$, $a_1 = 1$, $a_2 = \dfrac{1}{2}$. Then we can check that $f^{-1}(\{a_i\}) = g^{-1}(\{a_i\})$, $i = 1,\ 2,\ 3$. So, Theorem C.6.3 is sharp.

Proof. (Theorems C.6.1, C.6.2, C.6.3, C.6.4) Let $I =]0, +\infty[$ in Theorems C.6.1, and C.6.3 and let $I =]0, R[$ in Theorems C.6.2 and C.6.4. For each $j = 1, \dots, n$, let S_j be the set of all zeros of $f - a_j$ (without taking multiplicities into account). Since $a_i \neq a_j\ \forall i \neq j$, we have $S_i \cap S_j = \emptyset\ \forall i \neq j$. Next, we notice that $f(x) = a_j$ implies $f(x) - g(x) = 0$. Consequently, we check that

(1) $\sum_{j=1}^{n} \overline{Z}(r, f - a_j) \leq \overline{Z}(r, f - g)$.

Suppose first that f and g either belong to $\mathcal{A}(K)$ or belong to $\mathcal{A}(d(0, R^-))$.

By applying Theorem C.4.24 to f we obtain

$$(n-1)T(r,f) \le \sum_{j=1}^{n} \overline{Z}(r, f-a_j) + \overline{N}(r,f) - \log(r) + O(1)$$
$$\le n\overline{Z}(r, f-g) + \overline{N}(r,f) - \log(r) + O(1) \ (r \in I),$$

hence by (1),

$$(n-1)T(r,f) \le T(r, f-g) + \overline{N}(r,f) - \log(r) + O(1) \ (r \in I),$$

and finally

$$(n-1)T(r,f) \le T(r, f-g) + N(r,f) - \log(r) + O(1) \ (r \in I).$$

Similarly,

$$(n-1)T(r,g) \le T(r, f-g) + N(r,g) - \log(r) + O(1) \ (r \in I),$$

therefore we obtain

(2) $(n-1)\max(T(r,f), T(r,g)) \le T(r, f-g)) + \max(N(r,f), N(r,g)) - \log(r) + O(1) \ (r \in I)$.

Assume we are in the hypothesis of Theorem C.6.1. We have $N(r,f) = N(r,g) = 0$ and by Theorem C.4.3, $T(r, f-g) \le \max(T(r,f), T(r,g)) + O(1)$. Consequently, by (2),

$$(n-1)\max(T(r,f), T(r,g)) \le \max(T(r,f), T(r,g)) - \log(r) + O(1) \ (r \in I).$$

Since r is not bounded, we can see that the inequality does not hold with $n = 2$, when r goes to $+\infty$.

Now, assume the hypothesis of Theorem C.6.2. Again, we have $N(r,f) = N(r,g) = 0$ and by Theorem C.4.3, $T(r, f-g) \le \max(T(r,f), T(r,g)) + O(1)$, hence by (2),

$$(n-1)\max(T(r,f), T(r,g)) \le \max(T(r,f), T(r,g)) + O(1) \ (r \in I).$$

Since f, g are unbounded, by Theorem C.4.10, so are $T(r,f)$, $T(r,g)$ in intervals $]r_0, R[$, hence the inequality does not hold with $n = 3$.

Suppose now that f and g belong to $\mathcal{A}_u(D)$. We then obtain

$$(n-1)T_R(r,f) \le T_R(r, f-g) + O(\log(r))$$
$$(r > R) \le \max(T_R(r,f), T_R(r,g)) + O(\log(r)$$

and similarly

$$(n-1)T_R(r,g) \le T_R(r, f-g) + O(\log(r))$$
$$(r > R) \le \max(T_R(r,f), T_R(r,g)) + O(\log(r),$$

therefore

$$(n-1)\max(R_R(r,f),T(r,g)) \leq \max(T_R(r,f),T_R(r,g)) + O(\log(r),$$

and hence $n \leq 2$, which proves the conclusion whenever $n \geq 3$.

Assume now the hypothesis of Theorem C.6.3. Since

$$\max(N(r,f),N(r,g)) \leq \max(T(r,f),T(r,g)),$$

by (2) and Theorem C.4.3 we have

$$(n-1)\max(T(r,f),T(r,g)) \leq 3\max(T(r,f),T(r,g)) - \log(r) + O(1)\ (r \in I).$$

Since r is not bounded, the inequality does not hold with $n = 4$, when r goes to ∞.

Finally, assume we are in the hypothesis of Theorem C.6.4. Suppose first that f and g belong to $\mathcal{M}_u(d(0,R^-))$. By (2) and Theorem C.4.3, we have

$$(n-1)\max(T(r,f),T(r,g)) \leq 3\max(T(r,f),T(r,g)) + O(1)\ (r \in I).$$

Since $T(r,f)$, $T(r,g)$ are not bounded, the inequality does not hold with $n = 5$.

And now, suppose that f and g belong to $\mathcal{M}_u(D)$. We then obtain

$$(n-1)T_R(r,f) \leq T_R(r,f-g)+O(\log(r))\ (r > R) \leq 3(T_R(r,f)+T_R(r,g))+O(\log(r)$$

and similarly

$$(n-1)T_R(r,g) \leq T_R(r,f-g)+O(\log(r))\ (r > R) \leq 3(T_R(r,f)+T_R(r,g))+O(\log(r),$$

therefore

$$(n-1)(T_R(r,f)+T_R(r,g)) \leq 3(T_R(r,f)+T_R(r,g)) + O(\log(r),$$

and hence $n \leq 4$, which proves the conclusion whenever $n \geq 5$. That finishes the proof of Theorem C.6.4. □

Definitions and notations: Let $f \in \mathcal{M}(\mathbb{K})$. The function f will be called *a function of uniqueness* (resp. *a function of strong uniqueness*) for a family $\mathcal{F}$ of functions defined in a suitable subset of $\mathbb{K}$ if given any two functions $f,\ g \in \mathcal{F}$ satisfying $h \circ f = h \circ g$ (resp. $h \circ f = b(h \circ g)$ with $b \in \mathbb{K}^*$), then f and g are identical.

Similarly, we will consider the same question in the purely algebraic context. Let E be an algebraically closed field, let $h \in E(x)$ and let $\mathcal{F}$ be a subset of $E(x)$. Then h will be called *a function of uniqueness for* $\mathcal{F}$ (resp. *a function of strong uniqueness for* $\mathcal{F}$) if given any two functions $f,\ g \in \mathcal{F}$ satisfying $h \circ f = h \circ g$ (resp. $h \circ f = b(h \circ g)$, with $b \in E^*$), f and g are identical.

Particularly in each case, if h is a polynomial, it will be called *a polynomial of uniqueness for the family* $\mathcal{F}$ (resp. *a polynomial of strong uniqueness for the family* $\mathcal{F}$).

In Theorem C.6.6, we will need the basic Lemma C.6.5 [59].

Lemma C.6.5: *Let E be an algebraically closed field of characteristic 0 and let $P(x) = (n-1)^2(x^n - 1) - n(n-2)(x^{n-1} - 1)^2 \in E[x]$. Then P admits 1 as a zero of order 4 and all other zeros u_j $(1 \le j \le 2n-6)$ are simple.*

Theorem C.6.6: *Let*

$$Q(x) = b\Big((n+2)(n+1)x^{n+3} - 2(n+3)(n+1)x^{n+2} + (n+3)(n+2)x^{n+1}\Big)$$

with $b \in \mathbb{K}^*$. Let $R \in]0, +\infty[$. Then Q is a polynomial of uniqueness for $\mathcal{M}(\mathbb{K})$ for every $n \ge 2$ and is a polynomial of uniqueness for $\mathcal{M}_u(d(0, R^-))$ for every $n \ge 3$.

Proof. Suppose $f,\ g \in \mathcal{M}(\mathbb{K})$ (resp. $f,\ g \in \mathcal{M}_u(d(0, R^-))$, resp. $f,\ g \in \mathcal{M}_u(D)$) and suppose that $Q(f) = Q(g)$. Let $h = \dfrac{f}{g}$. We can derive

$$\begin{aligned}(n+2)(n+1)(h^{n+3} - 1)g^2 - 2(n+3)(n+1)(h^{n+2} - 1)g \\ + (n+3)(n+2)(h^{n+1} - 1) = 0.\end{aligned}$$

If h is a constant, it is 1, a contradiction. So, we suppose h is not a constant. If g lies in $\mathcal{M}(\mathbb{K})$, so does h. Now, if g belongs to $\mathcal{M}_u(d(0, R^-))$ or to $\mathcal{M}_u(D)$ so does h, respectively. Indeed, suppose that $h \in \mathcal{M}_b(d(0, R^-))$. Then clearly we have $T(r, (n+2)(n+1)(h^n + 2)g^2) \ge 2T(r, g) + O(1)$, whereas $T(r, -2(n+3)(n+1)(h^{n+2} - 1)g + (n+3)(n+2)((h^{n+1} - 1)) \le T(r, g) + O(1)$, a contradiction. Similarly, if $h \in \mathcal{M}_b(D)$, we have the same contradiction.

Let $P(x) = (n+2)^2(x^{n+3}) - 1) - (n+3)(n+1)(x^{n+2} - 1)^2 \in \mathbb{K}[x]$. By Lemma C.6.5, P admits 1 as a zero of order 4 and all other zeros u_j $(1 \le j \le 2n)$ are simple. By change of variable, we can obviously assume that $h - u_j$ has no zero and no pole at 0. Consequently, we check that

$$\left(g - \left(\frac{n+3}{n+2}\right)\left(\frac{h^{n+2} - 1}{h^{n+3} - 1}\right)\right)^2 = \frac{(n+3)(h-1)^4 \prod_{j=1}^{2n}(h - u_j)}{(n+2)^2(n+1)(h^{n+3} - 1)^2}.$$

Since $\dfrac{(n+3)(h-1)^4 \prod_{j=1}^{2n}(h - u_j)}{(n+2)^2(n - +1)(h^{n+3} - 1)^2}$ is equal to a square, clearly each zero of $h - u_j, (1 \le j \le 2n)$ has order at least 2. Let $J =]0, +\infty[$ (resp. $J =]0, R[$, resp. $J = [R, +\infty[$). Consequently,

$$\sum_{j=1}^{2n} \overline{Z}(r, h - u_j) \le \frac{1}{2} \sum_{j=1}^{2n} Z(r, h - u_j) \le \frac{1}{2}(2n)T(r, h) + O(1)\ (r \in J).$$

Suppose first that f and g belong to $\mathcal{M}(\mathbb{K})$ or to $\mathcal{M}(d(0,R^-))$. Then, applying Theorem C.4.24 to h at the points u_j $(1 \leq j \leq 2n)$, we obtain

$$\begin{aligned}(2n-1)T(r,h) &\leq \sum_{j=1}^{2n} \overline{Z}(r,h-u_j) + \overline{N}(r,h) - \log(r) + O(1)\\ &\leq \frac{1}{2}\sum_{j=1}^{2n} Z(r,h-u_j) + \overline{N}(r,h) + O(1)\\ &\leq \frac{1}{2}(2n)T(r,h) + \overline{N}(r,h) - \log(r) + O(1) \ (r \in J),\end{aligned}$$

and therefore $(2n-1)T(r,h) \leq nT(r,h) + T(r,h) - \log(r) + O(1)$. If f, g belong to $\mathcal{M}(\mathbb{K})$, we conclude that $n \leq 1$. And if f, g belong to $\mathcal{M}_u(d(0,R^-))$, we conclude that $n \leq 2$.

Suppose now that f and g belong to $\mathcal{M}(D)$. Then we can apply Theorem C.5.15 and we have $(2n-1)T(r,h) \leq nT(r,h)+T(r,h)+O(\log(r))$, therefore we have again $n \leq 2$, which ends the proof. □

Corollary C.6.7: *Let $P(x) \in \mathbb{K}[x]$ have a derivative of the form $c(x-a)^n(x-b)^2$. Then P is a polynomial of uniqueness for $\mathcal{M}(\mathbb{K})$ $\forall n \geq 2$ and is a polynomial of uniqueness for $\mathcal{M}(d(\alpha,R^-))$ and for $\mathcal{M}(D)$ for every $n \geq 3$.*

Theorem C.6.8: *Let $Q(x) = x^{n+1} - x^n$. Let $a \in \mathbb{K}$ and $R \in]0,+\infty[$. Then Q is a polynomial of uniqueness for $\mathcal{A}(\mathbb{K})$ and for $\mathcal{A}_u(d(0,R^-))$ for every $n \geq 2$.*

Proof. Let f, $g \in \mathcal{A}(\mathbb{K}))$ (resp. f, $g \in \mathcal{A}_u(d(0,R^-))$, (resp. f, $g \in \mathcal{A}_u(D)$. Suppose $f^n(f-1) = g^n(g-1)$. Let $h = \frac{f}{g}$ and suppose h is not the constant 1. Then we have $g = \left(\frac{h^n-1}{h^{n+1}-1}\right)$. Consequently, if h belongs to $\mathbb{K}$ (resp. to $\mathcal{M}_b(d(0,R^-))$, resp. to $\mathcal{M}_b(D)$), so does g, a contradiction. Thus, h belongs to $\mathcal{M}(\mathbb{K}) \setminus \mathbb{K}$ (resp. to $\mathcal{M}_u(d(0,R^-))$, resp. to $\mathcal{M}_u(D)$).

Now, since $n \geq 2$, by Corollary C.1.23 h has to take at least one of the $(n+1)$-th roots of 1 other than 1 and such a $(n+1)$-th root of 1 cannot be a n-th root of 1. Consequently, g admits a pole, a contradiction. Therefore, h is identically equal to 1 and hence $f = g$. □

Now, we must examine the situation in $\mathcal{M}(D)$ in order to obtain a result similar to Theorem C.6.8.

Definitions and notations: We will denote by $\mathcal{A}^c(D)$ the set of functions $f \in \mathcal{A}(D)$ having infinitely many zeros in D and by $\mathcal{M}^c(D)$ the set of functions $f \in \mathcal{M}(D)$ admitting at least either infinitely zeros or infinitely many poles.

Theorem C.6.9: *Let $Q(x) = x^{n+1} - x^n$. Let $a \in \mathbb{K}$ and $R \in]0, +\infty[$. Then Q is a polynomial of uniqueness for $\mathcal{A}^c(D)$ for every $n \geq 2$.*

Proof. Let $f,\ g \in \mathcal{A}^c(D)$. Suppose $f^n(f-1) = g^n(g-1)$. Let $h = \dfrac{f}{g}$ and suppose h is not the constant 1. Then we have $g = \left(\dfrac{h^n - 1}{h^{n+1} - 1}\right)$. Consequently, if h does not belong to $\mathcal{M}^c(D)$, neither does g, a contradiction. Thus, h belongs to $\mathcal{M}^c(D)$).

Now, since $n \geq 2$, by Theorem C.3.4, h has to take at least one of the $(n+1)$-th roots of 1 other than 1 and such a $(n+1)$-th root of 1 cannot be a n-th root of 1. Consequently, g admits a pole, a contradiction. Therefore, h is identically equal to 1 and hence $f = g$. □

We will examine particular cases where curves are defined by their equations so that, for most of them, the p-adic Nevanlinna theory lets us find easy proofs. Most of the results come from [32].

Definitions and notations: Let $F(x,y) \in \mathbb{K}[x,y]$. A point (a,b) of the algebraic curve of equation $F(x,y) = 0$ is called a *singular point* if $\dfrac{\partial F}{\partial x} = \dfrac{\partial F}{\partial y} = 0$. An algebraic curve is said to be *degenerate* if it admits a singular point. An algebraic curve of degree 2 (resp. 3) is called *a conic* (resp. *an elliptic curve*).

Remark: The p-adic functions sin and cos are bounded inside $d(0, (p^{-\frac{1}{p-1}})^-)$ when the residue characteristic is p (resp. inside $d(0, 1^-)$ when the residue characteristic is 0) and satisfy $\sin^2 x + \cos^2 x = 1$.

Throughout the chapter, we will denote by D an infinite bounded set included in a disk $d(a, r)$, for some $r < R$.

Remark: Let $P,\ Q \in \mathbb{K}[x]$. A point $(\alpha,\ \beta) \in \mathbb{K}^2$ is a singular point of the curve of equation $P(x) = Q(y)$ if and only if $P(\alpha) = Q(\beta)$ and $P'(\alpha) = Q'(\beta) = 0$.

C.7. Branched values

In complex functions theory, a notion closely linked to Picard's exceptional values was introduced: the notion of "perfectly branched value" [37]. Here, we shall consider the same notion on $\mathcal{M}(\mathbb{K})$, on $\mathcal{M}(d(a, R^-))$, and on $\mathcal{M}(D)$. Most of the results come from [28, 54, 56].

Definitions and notations: Let $f \in \mathcal{M}(\mathbb{K})$ (resp. let $f \in \mathcal{M}(d(a, R^-))$, resp. let $f \in \mathcal{M}(D)$) and let $b \in \mathbb{K}$. The value b is said to be *a perfectly branched value for* f if all zeros of $f - b$ are multiple zeros, except finitely many. And in the present book, b will be said to be *a totally branched value for f* if all zeros of $f - b$ are multiple zeros, without exception. Similarly, ∞ will be called *a perfectly branched value for f* if all poles of f are multiple but finitely many and it will be called *a totally branched value for f* if all poles of f are multiple, without exception.

In $\mathbb{C}$, it is known that a transcendental meromorphic function admits at most 4 perfectly branched values and an entire function admits at most 2 perfectly branched values. As explained by K. S. Charak in [37], these numbers, respectively, 4 and 2, are sharp. The Weierstrass function $\mathcal{P}$ has 4 totally branched values (considering ∞ as a value) and of course, sine and cosine functions admit 2 totally branched values: 1 and -1.

Here, we will do a similar study on p-adic functions and obtain sometimes certain better results. Particularly, an entire function admits at most one perfectly branched value.

Lemma C.7.1 is immediate.

Lemma C.7.1: *Let $f \in \mathcal{M}(\mathbb{K})$ (resp. let $f \in \mathcal{M}_u(d(0, R^-))$, resp. let $f \in \mathcal{M}^c(D)$), admitting a perfectly branched value $b \neq 0$. Then $\frac{1}{f}$ admits $\frac{1}{b}$ as a perfectly branched value. If f admits 0 as a perfectly branched value, $\frac{1}{f}$ admits ∞ as a perfectly branched value. If f admits ∞ as a perfectly branched value, $\frac{1}{f}$ admits 0 as a perfectly branched value.*

We have an immediate application of the definition with meromorphic functions whose denominator is a small function with respect to the numerator or vice versa.

Theorem C.7.2: *Let $f,\ g \in \mathcal{A}(\mathbb{K}) \setminus \mathbb{K}[x]$ (resp. $f,\ g \in \mathcal{A}_u(d(0, R^-))$, resp. $f,\ g \in \mathcal{A}^c(D)$) be such that $\limsup\limits_{r \to +\infty} \frac{T(r, f)}{T(r, g)} > 2$ (resp. $\limsup\limits_{r \to R^-} \frac{T(r, f)}{T(r, g)} > 2$, resp. $\limsup\limits_{r \to +\infty} \frac{T_R(r, f)}{T_R(r, g)} > 2$). Then both $\frac{f}{g}$ and $\frac{g}{f}$ have at most two perfectly branched values.*

Proof. Set $\phi = \frac{f}{g}$. Without loss of generality, we can suppose that f and g have no common zero. Indeed, suppose first that $f,\ g \in \mathcal{A}(\mathbb{K}) \setminus \mathbb{K}[x]$ or $f,\ g \in \mathcal{A}^c(D)$. By Lemma C.1.3 we can write f and g in the form $f = \widetilde{f}.h$ and $g = \widetilde{g}.h$, where $\widetilde{f}$ and $\widetilde{g}$ have no common zero and then $Z(r, f) = Z(r, \widetilde{f}) + Z(r, h)$, $Z(r, g) = Z(r, \widetilde{g}) + Z(r, h)$ and so much the more, we have $\frac{T(r, \widetilde{f})}{T(r, \widetilde{g})} > 2$.

Now, if $f,\ g \in \mathcal{A}_u(d(0,r)^-))$, we can place ourselves in an algebraically closed spherically complete extension to obtain the same conclusion because the Nevanlinna functions are the same in such an extension. Therefore, we assume that f and g have no common zero.

Suppose first that f and g belong to $\mathcal{A}(\mathbb{K}) \setminus \mathbb{K}[x]$ or $\mathcal{A}_u(d(0,R^-))$. Since $f,\ g$ have no common zero, we have $Z(r,\phi) = Z(r,f)$ and $N(r,\phi) = Z(r,g)$, hence $T(r,\phi) = \max(Z(r,f), Z(r,g)) + O(1)$.

Now, by hypothesis, there exists $\lambda < \dfrac{1}{2}$ and a sequence $(r_n)_{n\in\mathbb{N}}$ such that $\lim\limits_{n\to+\infty} r_n = +\infty$ (resp. $\lim\limits_{n\to+\infty} r_n = R$) and such that

$$T(r_n, g) \le \lambda T(r_n, f)\ \forall n \in \mathbb{N}. \tag{1}$$

Suppose that ϕ has 3 perfectly branched values $b_j,\ j = 1,2,3$. Applying Theorem C.4.24, we have

$$2T(r,\phi) \le \sum_{j=1}^{3} \overline{Z}(r,\phi - b_j) + \overline{N}(r,\phi) - \log r + O(1). \tag{2}$$

But here, for each $j = 1,\ 2,\ 3$, we notice that $\overline{Z}(r,\phi - b_j) \le \dfrac{Z(r,\phi - b_j)}{2} + q_j\log(r)$ with $q_j \in \mathbb{N}$ and $Z(r,\phi - b_j) = Z(r, f - b_j g) \le \max(T(r,f), T(r,g))$. But since $T(r_n,f) > T(r_n,g)$, we have $T(r_n,\phi - b_j) \le T(r_n,f) + O(1)$, hence $\overline{Z}(r,\phi - b_j) \le \frac{T(r_n,f)}{2} + q_j\log(r_n) + O(1)$. Now, putting $q = q_1 + q_2 + q_3$, by (2) we obtain

$$2T(r_n,f) \le \frac{3T(r_n,f)}{2} + T(r_n,g) + 2q\log(r_n) + O(1),$$

hence

$$T(r_n,f) \le 2T(r_n,g) + q\log(r_n) + O(1),$$

a contradiction to (1).

Similarly, if f and g belong to $\mathcal{A}^c(D)$, we can make the same reasoning by replacing T by T_R, Z by Z_R, and N by N_R. □

Theorem C.7.3: *Let $f,\ g \in \mathcal{A}(\mathbb{K}))$ (resp. let $f,\ g \in \mathcal{A}(d(0,R^-)))$ be such that $+\infty > \rho(f) > \rho(g)$. Then*

$$\liminf_{r\to R^-} \frac{T(r,g)}{T(r,f)} = 0.$$

Proof. Suppose first $f,\ g \in \mathcal{A}(\mathbb{K}))$. Let $\gamma = \dfrac{\rho(g)}{\rho(f)}$ and let $(r_n)_{n\in\mathbb{N}}$ be a sequence in $]0,+\infty[$ such that $\lim\limits_{n\to+\infty} r_n = +\infty$ and $\lim\limits_{n\to+\infty} \dfrac{\log(\log(|f|(r_n)))}{-\log(R)} = \rho(f)$.

By hypothesis, we have

$$\lim_{n\to+\infty} \frac{\log(\log(|g|(r_n)))}{\log(\log(|f|(r_n)))} \leq \gamma,$$

hence

$$\lim_{n\to+\infty} \frac{\log(T(r_n,g))}{\log(T(r_n,f))} \leq \gamma.$$

Take $\beta \in]\gamma, 1[$. Then when n is big enough, we can get

$$\frac{T(r_n,g)}{T(r_n,f)} \leq (T(r_n,f))^{\beta-1}.$$

But since $\beta < 1$ and since, by Lemma G, $\lim_{n\to\infty} T(r_n,f) = +\infty$, one sees that $\lim_{n\to\infty} (T(r_n,f))^{\beta-1} = 0$, which ends the proof when f and g belong to $\mathcal{A}(\mathbb{K})$.

Next, when f and g belong to $\mathcal{A}(d(0,R^-))$, we can make the same proof with a sequence (r_n) of limit R. □

Corollary C.7.4: *Let $f,\ g \in \mathcal{A}(\mathbb{K})$ (resp. let $f,\ g \in \mathcal{A}_u(d(0,R^-))$) be such that $\rho(f) \neq \rho(g)$. Then both $\frac{f}{g}$ and $\frac{g}{f}$ have at most two perfectly branched values.*

Corollary C.7.5: *Let $f \in \mathcal{A}_u(d(0,R^-))$ and let $f \in \mathcal{A}_b(d(0,R^-))$. Then both $\frac{f}{g}$ and $\frac{g}{f}$ have at most two perfectly branched values.*

The proof of the next theorems will require several basic lemmas.

Lemma C.7.6. *Let $(\alpha_i)_{1\leq i\leq t}$, $(\beta_i)_{1\leq i\leq t}$ be two finite sequences of $\mathbb{K}$ such that $|\alpha_i| < R$, $|\beta_i| < R$ $\forall i = 1,\dots,t$. Let $\Theta(x) = \prod_{i=1}^{t} \left(\frac{1-\frac{\beta_i}{x}}{1-\frac{\alpha_i}{x}}\right)$. Then the function $\sqrt{\Theta(x)}$ is defined and belongs to $\mathcal{A}(\mathbb{K}\setminus d(0,4R))$. Moreover, if $p \neq 2$, it belongs to $\mathcal{A}(\mathbb{K}\setminus d(0,R))$.*

Proof. By Theorem B.20.23, there exists a unique function $\ell \in \mathcal{A}(d(1,(\frac{1}{4})^-))$ with value in $d(1,1^-)$ such that $(\ell(u))^2 = u$ $\forall u \in d(1,(\frac{1}{4})^-)$. Moreover, if $p \neq 2$, then ℓ has continuation to a function $\overline{\ell} \in \mathcal{A}(d(1,1^-))$ with value in $d(1,1^-)$ again and such that $(\overline{\ell}(u))^2 = u$ $\forall u \in d(1,1)$. Here, we put $u = \prod_{i=1}^{t} \left(\frac{1-\frac{\beta_i}{x}}{1-\frac{\alpha_i}{x}}\right)$. □

Lemma C.7.7. *Let $g,\ h \in \mathcal{A}(\mathbb{K})$ with $\frac{g}{h}$ transcendental and let $\Theta(x) = \prod_{i=1}^{t} \left(\frac{1-\frac{\beta_i}{x}}{1-\frac{\alpha_i}{x}}\right)$ with $|\alpha_i| < R$ and $|\beta_i| < R$ $\forall i = 1,\dots,t$. Then the function*

$g(x)^2 - h(x)^2\Theta(x)$ belongs to $\mathcal{A}(\mathbb{K} \setminus d(0, 4R))$ and satisfies $\lim_{r\to+\infty} \frac{|g^2 - h^2\Theta|(r)}{r^m} = +\infty$ $\forall m \in \mathbb{N}$.

Proof. We first notice that $g^2 - h^2\Theta$ obviously belongs to $\mathcal{A}(\mathbb{K} \setminus d(0, R))$. Let us fix $m \in \mathbb{N}$. By Lemma C.7.6, $\sqrt{\Theta}$ is defined in $\mathbb{K}\setminus d(0, R)$ and belongs to $\mathcal{A}(\mathbb{K}\setminus d(0, 4R))$. Let $\ell = \sqrt{\Theta}$. Then can write $g^2 - h^2\Theta = (g - h\ell)(g + h\ell)$.

Since $\frac{g}{h}$ is transcendental, $g^2 - h^2\Theta$ is not identically zero. So, by Theorem B.13.22 there exists $a > 0$ and $q \in \mathbb{Z}$ such that $|g - h\ell|(r) \geq ar^q$ $\forall r > R$.

Suppose first h is transcendental. Since h is entire and since $|\ell|(r) = 1$ $\forall r > R$, by Theorem B.13.22, we have

$$\forall m \in \mathbb{N}, \lim_{r\to\infty} \frac{|h\ell|(r)}{r^m} = +\infty. \tag{1}$$

Consequently $|h\ell|(r) > |g - h\ell|(r)$ and hence $|h\ell|(r) = |g + h\ell|(r)$. Thus,

$$|g - h\ell|(r)|g + h\ell|(r) \geq ar^q|h\ell|(r).$$

Then the conclusion comes from (1).

Suppose now h is not transcendental, hence it is a polynomial and then g is transcendental. Consequently, when r is big enough, we have $|g - h\ell|(r) = |g|(r) = |g+h\ell|(r)$ and hence $|g^2 - h^2\Theta|(r) = (|g|(r))^2$, which yields the same conclusion. □

Theorem C.7.8: *Let $f \in \mathcal{M}(\mathbb{K})$ be transcendental (resp. let $f \in \mathcal{M}_u(d(a, R^-))$). Then f has at most four perfectly branched values. Moreover, any function $g \in \mathcal{M}(\mathbb{K})$ has at most three totally branched values.*

Remark. Let $f \in \mathcal{M}(\mathbb{K})$. If $a \in \mathbb{K}$ is a perfectly branched value for f, then for every $b \in \mathbb{K}$, $a + b$ is a perfectly branched value for $f + b$. Moreover, if $a \neq 0$, then $\frac{1}{a}$ is a perfectly branched value for $\frac{1}{f}$. So, we are going to construct a function $f \in \mathcal{M}(\mathbb{K})$ admitting three distinct totally branched values. Let $\ell = \prod_{j=1}^{\infty}(1 - \frac{x}{a_j}) \in \mathcal{A}(\mathbb{K})$ with $\lim_{j\to+\infty} |a_j| = +\infty$ and $a_j \neq a_k$ $\forall j \neq k$. Let $u = \prod_{k=1}^{\infty}(1 - \frac{x}{a_{2k}})$ and let $w = \prod_{k=1}^{\infty}(1 - \frac{x}{a_{2k-1}})$. So, both u and w belong to $\mathcal{A}(\mathbb{K})$ and satisfy $uw = \ell$. Now, let $\phi = \frac{u^2 + w^2}{2}$ and $\psi = \frac{w^2 - u^2}{2}$. Then $\phi^2 - \psi^2 = \ell^2$. Now, let $g = \left(\frac{\phi}{\ell}\right)^2$. Note that g admits 0 and 1 as totally branched values. Consequently, $g + 1$ admits 1 and 2 as totally branched values and hence the function $f = \frac{1}{g+1}$ admits 1 and $\frac{1}{2}$ as totally branched values. But, on the other hand, all poles of g are multiple, hence so are those of $g+1$. Consequently, f also admits 0 as a totally branched value. Thus, Theorem C.7.8 is sharp as far as totally branched values are concerned for meromorphic functions. One can only ask whether there exist meromorphic functions admitting 4 perfectly branched values where some of them are not totally branched values.

Theorem C.7.9: *Let $f \in \mathcal{M}(\mathbb{K})$ be transcendental and have finitely many poles. Then f has at most one perfectly branched value.*

Corollary C.7.10: *Let $f \in \mathcal{A}(\mathbb{K})$ be transcendental. Then f has at most one perfectly branched value.*

Remark. However, a polynomial can admit two values looking like "perfectly branched values." Yet, the definition of a perfectly branched value does not really apply to a polynomial or a rational function.

Example: Let $P(x) = x^3 - x^2 + \frac{4}{27}$. Then 0 and $-\frac{4}{27}$ are two perfectly branched values that are not totally branched. Indeed, on the one hand $-\frac{4}{27}$ is perfectly but not totally branched since $P(x) - \frac{4}{27} = x^2(x-1)$. On the other hand, we can check that $P(x) = \left(x - \frac{2}{3}\right)^2 \left(x + \frac{1}{3}\right)$.

Theorem C.7.11: *Let $f \in \mathcal{M}_u(d(a, R^-))$ have finitely many poles. Then f has at most two perfectly branched values.*

Corollary C.7.12: *Let $f \in \mathcal{A}_u(d(a, R^-))$. Then f has at most two perfectly branched values.*

Corollary C.7.13: *Let $a_i \in \mathbb{K}, i = 1, 2, 3$ be pairwise distinct. There do not exist $f,\ g \in \mathcal{M}(\mathbb{K}) \setminus \mathbb{K}$, there do not exist $f,\ g \in \mathcal{A}_u(d(a, R^-))$, and there do not exist $f,\ g \in \mathcal{A}^c(D)$ such that $(g(x))^2 = (f(x) - a_1)(f(x) - a_2)(f(x) - a_3)$.*

Proof. Suppose two functions $f,\ g \in \mathcal{M}(\mathbb{K}) \setminus \mathbb{K}$ satisfy $(g(x))^2 = (f(x) - a_1)(f(x) - a_2)(f(x) - a_3)$. Then each zero of $f - a_i,\ i = 1,\ 2,\ 3$ must be of order at least 2. And each pole of $(f - a_1)(f - a_2)(f - a_3)$ is a pole of g^2, hence is of even order and hence each pole of f is at least of order 2. f admits 4 totally branched values: $a_1, a_2, a_3, and\ \infty$, what is impossible by Theorem C.7.8. □

Remark. We don't know whether there exists a function $f \in \mathcal{A}_u(d(a, R^-))$ admitting two perfectly branched values. The only case when we can improve Theorem C.7.11 is the case when $\mathbb{K}$ has residue characteristic 0.

In the proofs of Theorems C.7.8 and C.7.9, without loss of generality, we can obviously assume that all supposed perfectly branched values of the functions we consider are finite, what we will do for simplicity.

Proof. (Theorems C.7.8, C.7.9, and C.7.11) If f lies in $\mathcal{M}_u(d(a, R^-))$, we assume that $a = 0$. Suppose f has q perfectly branched values b_j with $j = 1, \dots, q$. For each j, let s_j be the number of simple zeros of $f - b_j$ and let $s = \sum_{j=1}^{q} s_j$. Applying Theorem C.4.24, we have

$$(1) \qquad (q-1)T(r, f) \leq \sum_{j=1}^{q} \overline{Z}(r, f - b_j) + \overline{N}(r, f) - \log r + O(1).$$

But since $f - b_j$ has s_j simple zeros, we have

$$\overline{Z}(r, f - b_j) \leq \frac{Z(r, f - b_j) + s_j \log r}{2} + O(1) \leq \frac{T(r, f) + s_j \log r}{2} + O(1) \ \forall j = 1, \dots, q,$$

hence

$$(2) \qquad (q-1)T(r, f) \leq \frac{qT(r, f)}{2} + T(r, f) + \left(\frac{s}{2} - 1\right) \log r + O(1).$$

By (2) clearly we have $q \leq 4$ in all cases, which shows the first statement of Theorem C.7.8 whenever $f \in \mathcal{M}(\mathbb{K})$ or $f \in \mathcal{M}_u(d(0, R^-))$.

Further, suppose that f lies in $\mathcal{M}(\mathbb{K})$ or in $\mathbb{K}(x)$ and that $b_1, \dots, b_q$ are totally branched values. Then $s = 0$, hence by (2) we have

$$(3) \qquad \left(\frac{q}{2} - 2\right) T(r, f) \leq -\log r + O(1).$$

In Theorem C.7.8, since f is transcendental, we have $\log r = o(T(r, f))$, hence $q \leq 3$.

Consider now the hypotheses of Theorems C.7.9 and C.7.11. Let t be the number of poles of f, taking multiplicity into account. We have $N(r, f) \leq t \log r + O(1)$, hence by (1) we obtain $(q-1)T(r, f) \leq \frac{q}{2} T(r, f) + O(\log r)$ and hence $q \leq 2$. Thus, Theorem C.7.11 is proved.

For the proof of Theorem C.7.9, without loss of generality, we may assume that these perfectly branched values are 0 and b. Suppose first that f has infinitely many zeros of order ≥ 3. Then $Z(r, f) - 2\overline{Z}(r, f)$ is a function $\psi(r)$ such that

$$(4) \qquad \lim_{r \to +\infty} \frac{\psi(r)}{\log r} = +\infty.$$

Therefore,

$$(5) \qquad \overline{Z}(r, f) \leq \frac{T(r, f) - \psi(r)}{2}.$$

On the other hand, by (1), with $q = 2$ we have

$$T(r, f) \leq T(r, f) - \frac{\psi(r)}{2} + O(\log r)$$

and then, by (4), we can see a contradiction proving that f cannot admit 0 and b as branched values.

Suppose now that all zeros of both f and $f-b$ are of order 2 except finitely many. So, there exists $S>0$ satisfying the following properties:

(i) all poles of f lie in $d(0,S)$,
(ii) $|f|(r)>|b|\ \forall r>S$, and
(iii) all zeros of f and of $f-b$ in $\mathbb{K}\setminus d(0,S)$ are of order 2 exactly.

We can then write f in the form $\frac{Pg^2}{V}$ with $P,V\in\mathbb{K}[x]$, $g\in\mathcal{A}(\mathbb{K})$ and $\deg(P)=k$. Similarly, $f-b$ is in the form $\frac{Qh^2}{V}$ with $Q\in\mathbb{K}[x]$, $h\in\mathcal{A}(\mathbb{K})$, where all zeros of P,Q,V lie in $d(0,S)$ and all zeros of g, h lie in $\mathbb{K}\setminus d(0,S)$ and are simple zeros. Set $\deg(V)=t$. We notice that g is transcendental.

By (ii), we have $|f|(r)=|f-b|(r)\ \forall r>S$. Consequently, by Lemma C.1.20, f and $f-b$ have the same number of zeros in $d(0,S)$ and hence $\deg(P)=\deg(Q)$. Let $P(x)=\prod_{i=1}^{k}(x-\alpha_i)$ and $Q(x)=\prod_{i=1}^{k}(x-\beta_i)$. Let $\Theta(x)=\prod_{i=1}^{k}\left(\frac{1-\frac{\beta_i}{x}}{1-\frac{\alpha_i}{x}}\right)$ and let $\Xi(x)=\frac{V(x)}{P(x)}$. Of course, $\lim_{r\to+\infty}\frac{|\Xi|(r)}{r^{t+1}}=0$.

Now, we have

$$g(x)^2=h(x)^2\Theta(x)+b\Xi(x).$$

By Lemma C.7.7, we can derive

$$\lim_{r\to+\infty}\frac{\left(|g^2-h^2\Theta|(r)\right)}{r^m}=+\infty\ \forall m\in\mathbb{N},$$

a contradiction to $\lim_{r\to+\infty}\frac{|\Xi|(r)}{r^{t+1}}=0$. This completes the proof of Theorem C.7.9. □

Remark: Corollary C.7.12 may suggest that Theorem B.19.4 is wrong. Indeed, by Theorem B.19.4, given sequences $(a_n)_{n\in\mathbb{N}}$, $(b_n)_{n\in\mathbb{N}}$, $(c_n)_{n\in\mathbb{N}}$ in $d(0,R^-)$ such that

$$\lim_{n\to+\infty}|a_n|=\lim_{n\to+\infty}|b_n|=\lim_{n\to+\infty}|c_n|=R,$$

together with $\prod_{n=0}^{\infty}\frac{R}{|a_n|}=+\infty$, there exists functions $f\in\mathcal{A}(d(0,R^-))$ admitting each a_n as a zero of order 2 and such that $f-1$ admits each b_n as a zero of order 2 and such that $f-2$ admits each c_n as a zero of order 2. Moreover, since $\prod_{n=0}^{\infty}\frac{R}{|a_n|}=+\infty$, we can check that $f\in\mathcal{A}_u(d(0,R^-))$.

The explanation is that, given such a function f, either f or $f-1$ or $f-2$ has infinitely many other zeros of order 1 and therefore, at least one of these three values (0, 1, 2) is not perfectly branched.

Theorem C.7.8 gives an easy proof of the impossibility to parametrize elliptic and hyperelliptic curves by meromorphic functions in all $\mathbb{K}$.

Corollary C.7.14: *Let $a_i \in \mathbb{K}$, $i = 1, 2, 3$ be pairwise distinct. There do not exist $f, g \in \mathcal{M}(\mathbb{K})$, there do not exist $f, g \in \mathcal{A}_b((d(a, R^-))$, and there do not exist $f, g \in \mathcal{A}^c(D)$ such that $(g(x))^2 = (f(x) - a_1)(f(x) - a_2)(f(x) - a_3)$.*

Proof. Suppose two functions $f, g\mathcal{M}(\mathbb{K}) \setminus \mathbb{K}$ satisfy $(g(x))^2 = (f(x) - a_1)(f(x) - a_2)(f(x) - a_3)$. Then each zero of $f - a_i, (i = 1, 2, 3)$ must be of order at least 2. And each pole of $(f - a_1)(f - a_2)(f - a_3)$ is a pole of g^2, hence is of even order and hence each pole of f is at least of order 2. Thus, f admits 4 totally branched values: $a_1, a_2, a_3, and\ \infty$, which is impossible by Theorem C.7.8. □

Remark: In general, it is proven that curves of genus $n \geq 1$ admit no parametrization by meromorphic functions $f \in \mathcal{M}(\mathbb{K})$ [39].

C.8. Exceptional values of functions and derivatives

The chapter is aimed at studying various properties of derivatives of meromorphic functions, particularly their sets of zeros. Many important results are due to Jean-Paul Bezivin.

We will first notice a general property concerning quasi-exceptional values of meromorphic functions and derivatives.

Definitions and notations: Let $f \in \mathcal{M}(\mathbb{K})$ (resp. $f \in \mathcal{M}(d(0, R^-))$) and let $\mathcal{T}$ be a property satisfied by f at certain points. Let $r \in]0, R[$. Assume that $f(0) \neq 0, \infty$. We denote by $Z(r, f \mid \mathcal{T})$ the counting function of zeros of f in $d(0, r)$ at the points where f satisfies $\mathcal{T}$, i.e., if (a_n) is the finite or infinite sequence of zeros of f in $d(0, R^-)$ with respective multiplicity order s_n, where $\mathcal{T}$ is satisfied, we put $Z(r, f) = \sum_{|a_n| \leq r, \mathcal{T}} s_n(\log r - \log |a_n|)$.

Given two meromomorphic functions $f,\ g \in \mathcal{M}(\mathbb{K})$ or $f,\ g \in \mathcal{M}(d(a, R^-))$ ($a \in \mathbb{K},\ R > 0$), we will denote by $W(f, g)$ the Wronskian of f and g: $f'g - fg'$.

Theorem C.8.1: *Let $f \in \mathcal{M}(\mathbb{K}) \setminus \mathbb{K}(x)$ (resp. Let $f \in \mathcal{M}_u(d(\alpha, R^-))$). If f admits a quasi-exceptional value, then f' has no quasi-exceptional value different from 0.*

Proof. Without loss of generality, we may assume $\alpha = 0$ and that f has no zero and no pole at 0. Let $b \in \mathbb{K}$ and suppose that b is a quasi-exceptional value of f. There exist $P \in \mathbb{K}[x]$ and $l \in \mathcal{A}(\mathbb{K}) \setminus \mathbb{K}[x]$ (resp. and $l \in \mathcal{A}_u(d(0, R^-)))$ without common zeros, such that $f = b + \frac{P}{l}$.

Let $c \in \mathbb{K}^*$. Remark that $f' - c = \frac{P'l - Pl' - cl^2}{l^2}$. Let $a \in \mathbb{K}$ (resp. let $a \in d(0, R^-)$). If a is a pole of f, it is a pole of $f' - c$ and we can check that

(1) $\omega_a(P'l - Pl' - cl^2) = \omega_a(l') = \omega_a(l) - 1,$

because a is not a zero of P.

Now suppose that a is not a pole of f. Then

(2) $\omega_a(f' - c) = \omega_a(P'l - Pl' - cl^2).$

Consequently, $Z(r, f' - c) = Z(r, (P'l - Pl' - cl^2) \mid l(x) \neq 0)$. But, by (1) we have

(3) $Z(r, (P'l - Pl' - cl^2) \mid l(x) = 0) < Z(r, l),$

and therefore by (2) and (3) we obtain

(4) $Z(r, f' - c) = Z(r, (P'l - Pl' - cl^2) \mid l(x) \neq 0) > Z(r, P'l - Pl' - cl^2) - Z(r, l).$

Now, let us examine $Z(r, P'l - Pl' - cl^2)$. Let $r \in]0, +\infty[$ (resp. let $r \in]0, R[$). Since $l \in \mathcal{A}(\mathbb{K})$ is transcendental (resp. since $l \in \mathcal{A}_u(d(0, R^-)))$, we can check that when r is big enough, we have $|Pl'|(r) < |c|\big(|l|(r)\big)^2$ and $|Pl|(r) < |c|\big(|l|(r)\big)^2$, hence clearly $|P'l - Pl'|(r) < |c|\big(|l|(r)\big)^2$ and hence $|P'l - Pl' - cl^2|(r) = |c|\big(|l|(r)\big)^2$. Consequently, when r is big enough, by Theorem C.4.2 we have $Z(r, P'l - Pl' - cl^2) = Z(r, l^2) + O(1)$. But $Z(r, l^2) = 2Z(r, l)$, hence $Z(r, P'l - Pl' - cl^2) = 2Z(r, l) + O(1)$ and therefore by (4) we check that when r is big enough,

(5) $Z(r, f' - c) > Z(r, l).$

Now, if $l \in \mathcal{A}(\mathbb{K})$, since l is transcendental, by (5), for every $q \in \mathbb{N}$, we have $Z(r, f' - c) > Z(r, l) > q \log r$, when r is big enough, hence $f' - c$ has infinitely many zeros in $\mathbb{K}$. And similarly if $l \in \mathcal{A}_u(d(0, R^-))$, then by (5), $Z(r, f' - c)$ is unbounded when r tends to R, hence $f' - c$ has infinitely many zeros in $d(0, R^-)$. □

We will now notice a property of differential equations of the form $y^{(n)} - \psi y = 0$ that is almost classical [52].

The problem of a constant Wronskian is involved in several questions.

Theorem C.8.2: *Let $h,\ l \in \mathcal{A}(\mathbb{K})$ (resp. $h,\ l \in \mathcal{A}(d(\alpha, R^-)))$ and satisfy $h'l - hl' = c \in \mathbb{K}$, with h non-affine. If $h,\ l$ belong to $\mathcal{A}(\mathbb{K})$, then $c = 0$ and $\frac{h}{l}$ is a constant. If $c \neq 0$ and if $h,\ l \in \mathcal{A}(d(\alpha, R^-))$, there exists $\phi \in \mathcal{A}(d(\alpha, R^-))$ such that $h'' = \phi h,\ l'' = \phi l$.*

Proof. Suppose $c \neq 0$. If $h(a) = 0$, then $l(a) \neq 0$. Next, h and l satisfy

$$(1) \quad \frac{h''}{h} = \frac{l''}{l}.$$

Remark first that since h is not affine, h'' is not identically zero. Next, every zero of h or l of order ≥ 2 is a trivial zero of $h'l - hl'$, which contradicts $c \neq 0$. So, we can assume that all zeros of h and l are of order 1.

Now suppose that a zero a of h is not a zero of h''. Since a is a zero of h of order 1, $\frac{h''}{h}$ has a pole of order 1 at a and so does $\frac{l''}{l}$, hence $l(a) = 0$, a contradiction. Consequently, each zero of h is a zero of order 1 of h and is a zero of h'' and hence, $\frac{h''}{h}$ is an element ϕ of $\mathcal{M}(\mathbb{K})$ (resp. of $\mathcal{M}(d(\alpha, R^-))))$ that has no pole in $\mathbb{K}$ (resp. in $d(\alpha, R^-))$. Therefore, ϕ lies in $\mathcal{A}(\mathbb{K})$ (resp. in $\mathcal{A}(d(\alpha, R^-)))$.

The same holds for l and so, l'' is of the form ψl with $\psi \in \mathcal{A}(\mathbb{K})$ (resp. in $\mathcal{A}(d(\alpha, R^-)))$. But since $\frac{h''}{h} = \frac{l''}{l}$, we have $\phi = \psi$.

Now, suppose h, l belong to $\mathcal{A}(\mathbb{K})$. Since h'' is of the form ϕh with $\phi \in \mathcal{A}(\mathbb{K})$, we have $|h''|(r) = |\phi|(r)|h|(r)$. But by Theorem C.2.10, we know that $|h''|(r) \leq \frac{1}{r^2}|h|(r)$, a contradiction when r tends to $+\infty$. Consequently, $c = 0$. But then $h'l - hl' = 0$ implies that the derivative of $\frac{h}{l}$ is identically zero, hence $\frac{h}{l}$ is constant. □

Corollary C.8.3: *Let h, $l \in \mathcal{A}(\mathbb{K})$ with coefficients in $\mathbb{Q}$, also be entire functions in $\mathbb{C}$, with h non-affine. If $h'l - hl'$ is a constant c, then $c = 0$.*

Theorem C.8.4: *Let $\psi \in \mathcal{M}(\mathbb{K})$ (resp. let $\psi \in \mathcal{M}_u(d(\alpha, R^-)))$ and let $(\mathcal{E})$ be the differential equations $y'' - \psi y = 0$. Let E be the sub-vector space of $\mathcal{A}(\mathbb{K})$ (resp. of $\mathcal{A}(d(\alpha, R^-)))$ of the solutions of $(\mathcal{E})$. Then, the dimension of E is 0 or 1.*

Proof. Suppose E is not $\{0\}$. Let $h, l \in E$ be non-identically zero. Then $h''l - hl'' = 0$ and therefore $h'l - hl'$ is a constant c. On the other hand, since h, l are not identically zero, neither are h'', l''. Therefore, h, l are not affine functions.

Suppose ψ belongs to $\mathcal{M}(\mathbb{K})$ and that h, l belong to $\mathcal{A}(\mathbb{K})$. By Theorem C.8.2, we have $c = 0$ and hence $\frac{h}{l}$ is a constant, which proves that E is of dimension 1.

Suppose now that ψ lies in $\mathcal{M}_u(d(\alpha, R^-))$ and that h, l belong to $\mathcal{A}(d(\alpha, R^-))$. If ψ lies in $\mathcal{A}(d(\alpha, R^-))$, then by Theorem C.8.1, $E = \{0\}$. Finally, suppose that ψ lies in $\mathcal{M}_u(d(\alpha, R^-)) \setminus \mathcal{A}(d(\alpha, R^-))$. If $c \neq 0$, by Theorem C.8.2, there exists $\phi \in \mathcal{A}(d(\alpha, R^-))$ such that $h'' = \phi h$, $l'' = \phi l$. Consequently, $\phi = \psi$, hence $\psi \in \mathcal{A}(\mathbb{K})$ and therefore $c = 0$. Hence, $h'l - hl' = 0$ again and hence $\frac{h}{l}$ is a constant. Thus, we see that E is at most of dimension 1. □

Remark: The hypothesis ψ *unbounded in* $d(\alpha, R^-)$ is indispensable to show that the space E is of dimension 0 or 1, as shown in the example given again by the p-adic hyperbolic functions $h(x) = \cosh(x)$ and $l(x) = \sinh(x)$. The radius of convergence of both h, l is $p^{\frac{-1}{p-1}}$ when $\mathbb{K}$ has residue characteristic p and is 1 when $\mathbb{K}$ has residue characteristic 0. Of course, both functions are solutions of $y'' - y = 0$ but they are bounded.

Theorem C.8.5 given in [8] is an improvement of Theorem C.8.2. It follows previous results [7].

Theorem C.8.5: *Let $f, g \in \mathcal{A}(\mathbb{K})$ be such that $W(f,g)$ is a non-identically zero polynomial. Then both f, g are polynomials.*

Proof. First, by Theorem C.8.2 we check that the claim is satisfied when $W(f,g)$ is a polynomial of degree 0. Now, suppose the claim holds when $W(f,g)$ is a polynomial of certain degree n. We will show it for $n+1$. Let $f, g \in \mathcal{A}(\mathbb{K})$ be such that $W(f,g)$ is a non-identically zero polynomial P of degree $n+1$

Thus, by hypothesis, we have $f'g - fg' = P$, hence $f''g - fg'' = P'$. We can extract g' and get $g' = \frac{(f'g-P)}{f}$. Now consider the function $Q = f''g' - f'g''$ and replace g' by what we just found: we can get $Q = f'(\frac{(f''g-fg'')}{f}) - \frac{Pf''}{f}$.

Now, we can replace $f''g - fg''$ by P' and obtain $Q = \frac{(f'P'-Pf'')}{f}$. Thus, in that expression of Q, we can write $|Q|(R) \leq \dfrac{|f|(R)|P|(R)}{R^2|f|(R)}$, hence $|Q|(R) \leq \frac{|P|(R)}{R^2}$ $\forall R > 0$. But by definition, Q belongs to $\mathcal{A}(\mathbb{K})$. Consequently, Q is a polynomial of degree $t \leq n-1$.

Now, suppose Q is not identically zero. Since $Q = W(f', g')$ and since $\deg(Q) < n$, by the induction hypothesis f' and g' are polynomials and so are f, g. Finally, suppose $Q = 0$. Then $P'f' - Pf'' = 0$ and therefore f', P are two solutions of the differential equation of order 1 for meromorphic functions in $\mathbb{K}$: $(\mathcal{E})$ $y' = \psi y$ with $\psi = \frac{P'}{P}$, whereas y belongs to $\mathcal{A}(\mathbb{K})$. By Theorem C.8.4, the space of solutions of $(\mathcal{E})$ is of dimension 0 or 1. Consequently, there exists $\lambda \in \mathbb{K}$ such that $f' = \lambda P$, hence f is a polynomial. The same holds for g. □

Here, we can find again the following result that is known and may be proved without ultrametric properties:

Let F be an algebraically closed field and let P, $Q \in F[x]$ be such that $PQ' - P'Q$ is a constant c, with $\deg(P) \geq 2$. *Then $c = 0$.*

Definitions and notations: Let $f \in \mathcal{A}(\mathbb{K})$. We can factorize f in the form $\overline{f}\widetilde{f}$ where the zeros of $\overline{f}$ are the distinct zeros of f each with order 1. Moreover, if $f(0) \neq 0$ we will take $\overline{f}(0) = 1$.

Lemma C.8.6: *Let $U, V \in \mathcal{A}(\mathbb{K})$ have no common zero and let $f = \dfrac{U}{V}$. If f' has finitely many zeros, there exists a polynomial $P \in \mathbb{K}[x]$ such that $U'V - UV' = P\widetilde{V}$.*

Proof. If V is a constant, the statement is obvious. So, we assume that V is not a constant. Now $\widetilde{V}$ divides V' and hence V' factorizes in the way $V' = \widetilde{V}Y$ with $Y \in \mathcal{A}(\mathbb{K})$. Then no zero of Y can be a zero of V. Consequently, we have

$$f'(x) = \frac{U'V - UV'}{V^2} = \frac{U'\overline{V} - UY}{\overline{V}^2\widetilde{V}}.$$

The two functions $U'\overline{V} - UY$ and $\overline{V}^2\widetilde{V}$ have no common zero since neither have U and V. So, the zeros of f' are those of $U'\overline{V} - UY$, which therefore has finitely many zeros and consequently is a polynomial. □

Theorem C.8.7: *Let $f \in \mathcal{M}(\mathbb{K})$ have finitely many multiple poles, such that for certain $b \in \mathbb{K}$, $f' - b$ has finitely many zeros. Then f belongs to $\mathbb{K}(x)$.*

Proof. Suppose first $b = 0$. Let us write $f = \frac{U}{V}$ with $U,\ V \in \mathcal{A}(\mathbb{K})$, having no common zeros. By Lemma C.8.6, there exists a polynomial $P \in \mathbb{K}[x]$ such that $U'V - UV' = P\widetilde{V}$. Since f has finitely many multiple poles, $\widetilde{V}$ is a polynomial, hence so is $U'V - UV'$. But then by Theorem C.8.5, both $U,\ V$ are polynomials, which ends the proof when $b = 0$. Consider now the general case. $f' - b$ is the derivative of $f - bx$ that satisfies the same hypothesis, so the conclusion is immediate. □

Definitions and notations: For each $n \in \mathbb{N}^*$, we set $\lambda_n = \max\{\frac{1}{|k|},\ 1 \leq k \leq n\}$. Given positive integers $n,\ q$, we denote by C_n^q the combination $\dfrac{n!}{q!(n-q)!}$.

For convenience, in this chapter Log is the Neperian logarithm and we denote by e the number such that $\text{Log}(e) = 1$ and Exp is the real exponential function.

Remark: For every $n \in \mathbb{N}^*$, we have $\lambda_n \leq n$ because $k|k| \geq 1\ \forall k \in \mathbb{N}$. The equality holds for all n of the form p^h.

Lemmas C.8.8 and C.8.9 are due to Jean-Paul Bézivin [13].

Lemma C.8.8: *Let $U,\ V \in \mathcal{A}(d(0, R^-))$. Then for all $r \in]0, R[$ and $n \geq 1$ we have*

$$|U^{(n)}V - UV^{(n)}|(r) \leq |n!|\lambda_n \frac{|U'V - UV'|(r)}{r^{n-1}}.$$

More generally, given $j,\ l \in \mathbb{N}$, we have

$$|U^{(j)}V^{(l)} - U^{(l)}V^{(j)}|(r) \leq |(j!)(l!)|\lambda_{j+l}\frac{|U'V - UV'|(r)}{r^{j+l-1}}.$$

Proof. Set $g = \frac{U}{V}$ and $f = g'$. Applying Theorem C.2.10 to f for $k-1$, we obtain

$$|g^{(k)}|(r) = |f^{(k-1)}|(r) \leq |(k-1)!|\frac{|f|(r)}{r^{k-1}} = |(k-1)!|\frac{|U'V - UV'|(r)}{|V^2|(r)r^{k-1}}.$$

As in the proof of Theorem C.2.10, we set $U = V\Big(\frac{U}{V}\Big)$. By Leibniz formula again, now we can obtain

$$U^{(n)} = \sum_{q=1}^{n} C_n^q V^{(n-q)}\Big(\frac{U}{V}\Big)^{(q)} + V^{(n)}\Big(\frac{U}{V}\Big),$$

hence

$$(1) \qquad U^{(n)} - V^{(n)}\Big(\frac{U}{V}\Big) = \sum_{q=1}^{n} C_n^q V^{(n-q)}\Big(\frac{U}{V}\Big)^{(q)}.$$

Now, we have

$$\Big|\Big(\frac{U}{V}\Big)^{(q)}\Big|(r) = |g^{(q)}|(r) \leq |(q-1)!|\frac{|U'V - UV'|(r)}{|V^2|(r)r^{q-1}}$$

and

$$|V^{(n-q)}|(r) \leq |(n-q)!|\frac{|V|(r)}{r^{n-q}}.$$

Consequently, the general term in (1) is upper bounded as

$$\begin{aligned}\Big|C_n^q V^{(n-q)}\Big(\frac{U}{V}\Big)^{(q)}\Big|(r) &\leq \frac{|(n!)((n-q)!)((q-1)!)|}{|(q!)((n-q)!)|}\frac{|U'V - UV'|(r)}{|V|(r)r^{n-1}}\\ &\leq \lambda_n\frac{|n!||U'V - UV'|(r)}{|V|(r)r^{n-1}}.\end{aligned}$$

Therefore, by (1) we obtain

$$\Big|U^{(n)} - V^{(n)}\Big(\frac{U}{V}\Big)\Big|(r) \leq |n!|\lambda_n\frac{|U'V - UV'|(r)}{|V|(r)r^{n-1}}$$

and finally

$$\Big|U^{(n)}V - V^{(n)}U\Big|(r) \leq |n!|\lambda_n\frac{|U'V - UV'|(r)}{r^{n-1}}.$$

We can now generalize the first statement. Set $P_j = U^{(j)}V - UV^{(j)}$. By induction, we can show the following equality that already holds for $l \leq j$.

$$U^{(j)}V^{(l)} - U^{(l)}V^{(j)} = \sum_{h=0}^{l} C_l^h(-1)^h P_{j+h}^{(l-h)}.$$

Then, the second statement gets just an application of the first. □

Lemma C.8.9: *Let $U, V \in \mathcal{A}(\mathbb{K})$ and let $r,\ R \in]0, +\infty[$ satisfy $r < R$. For all $x, y \in \mathbb{K}$ with $|x| \leq R$ and $|y| \leq r$, we have the inequality*

$$|U(x+y)V(x) - U(x)V(x+y)| \leq \frac{R|U'V - UV'|(R)}{e(\text{Log}R - \text{Log}r)}.$$

Proof. By Taylor's formula at the point x, we have

$$U(x+y)V(x) - U(x)V(x+y) = \sum_{n\geq 0} \frac{U^{(n)}(x)V(x) - U(x)V^{(n)}(x)}{n!} y^n.$$

Now, by Lemma C.8.8, we have

$$\begin{aligned}\left|\frac{U^{(n)}(x)V(x) - U(x)V^{(n)}(x)}{n!} y^n\right| &\leq \lambda_n \frac{|U'V - UV'|(R)}{R^{n-1}} r^n \\ &= \lambda_n R|U'V - UV'|(R) \left(\frac{r}{R}\right)^n.\end{aligned}$$

Consequently, $\lim_{n\to+\infty} \left|\frac{U^{(n)}(x)V(x) - U(x)V^{(n}(x)}{n!} y^n\right| = 0$, therefore we can define $B = \max_{n\geq 1}\left\{\lambda_n\left(\frac{r}{R}\right)^n\right\}R|U'V - UV'|(R)$ and we have $|U(x+y)V(x) - U(x)$ $V(x+y)| \leq B$. Now, as remarked above, we have $\lambda_n \leq n$. We can check that the function h defined in $]0, +\infty[$ as $h(t) = t\left(\frac{r}{R}\right)^t$ reaches it maximum at the point $u = \frac{1}{e(\text{Log}R - \text{Log}r)}$. Consequently, $B \leq \frac{1}{e(\text{Log}R - \text{Log}r)}$ and therefore

$$|U(x+y)V(x) - U(x)V(x+y)| \leq \frac{R|U'V - UV'|(R)}{e(\text{Log}R - \text{Log}r)}.$$

□

Definitions and notations: Let $f \in \mathcal{M}(d(0, R^-))$. For each $r \in]0, R[$, we denote by $\zeta(r, f)$ the number of zeros of f in $d(0, r)$, taking multiplicity into account and set $\xi(r, f) = \zeta(r, \frac{1}{f})$. Similarly, we denote by $\beta(r, f)$ the number of multiple zeros of f in $d(0, r)$, each counted with its multiplicity and we set $\gamma(r, f) = \beta(r, \frac{1}{f})$.

Theorem C.8.10: *Let $f \in \mathcal{M}(\mathbb{K})$ be such that for some $c, q \in]0, +\infty[$, $\gamma(r, f)$ satisfies $\gamma(r, f) \leq cr^q$ in $[1, +\infty[$. If f' has finitely many zeros, then $f \in \mathbb{K}(x)$.*

Proof. Suppose f' has finitely many zeros and set $f = \frac{U}{V}$. If V is a constant, the statement is immediate. So, we suppose V is not a constant and hence it admits at least one zero a. By Lemma C.8.6, there exists a polynomial $P \in \mathbb{K}[x]$ such that $U'V - UV' = P\widetilde{V}$. Next, we take $r, R \in [1, +\infty[$ such that $|a| < r < R$ and $x \in d(0, R)$, $y \in d(0, r)$. By Lemma C.8.9, we have

$$|U(x+y)V(x) - U(x)V(x+y)| \leq \frac{R|U'V - UV'|(R)}{e(\text{Log}R - \text{Log}r)}.$$

Notice that $U(a) \neq 0$ because U and V have no common zero. Now set $l = \max(1, |a|)$ and take $r \geq l$. Setting $c_1 = \frac{1}{e|U(a)|}$, we have

$$|V(a+y)| \leq c_1 \frac{R|P|(R)|\widetilde{V}|(R)}{\text{Log}R - \text{Log}r}.$$

Then taking the supremum of $|V(a+y)|$ inside the disk $d(0,r)$, we can derive

$$(1) \qquad |V|(r) \leq c_1 \frac{R|P|(R)|\widetilde{V}|(R)}{\text{Log}R - \text{Log}r}.$$

Let us apply Corollary B.13.30, by taking $R = r + \dfrac{1}{r^q}$, after noticing that the number of zeros of $\widetilde{V}(R)$ is bounded by $\beta(R,V)$. So, we have

$$(2) \qquad |\widetilde{V}|(R) \leq \Big(1 + \frac{1}{r^{q+1}}\Big)^{\beta((r+\frac{1}{r^q}),V)} |\widetilde{V}|(r).$$

Now, due to the hypothesis: $\beta(r,V) = \gamma(r,f) \leq cr^q$ in $[1,+\infty[$, we have

$$(3) \qquad \Big(1 + \frac{1}{r^{q+1}}\Big)^{\beta((r+\frac{1}{r^q}),V)} \leq \Big(1 + \frac{1}{r^{q+1}}\Big)^{[c(r+\frac{1}{r^q})^m]}$$

$$= \text{Exp}\Big[c\Big(r + \frac{1}{r^q}\Big)^q \text{Log}\Big(1 + \frac{1}{r^{q+1}}\Big)\Big].$$

The function $h(r) = c(r + \frac{1}{r^m})^m \text{Log}(1 + \frac{1}{r^{m+1}})$ is continuous on $]0,+\infty[$ and equivalent to $\dfrac{c}{r}$ when r tends to $+\infty$. Consequently, it is bounded on $[l,+\infty[$. Therefore, by (2) and (3) there exists a constant $M > 0$ such that, for all $r \in [l,+\infty[$ by (3), we obtain

$$(4) \qquad |\widetilde{V}|\Big(r + \frac{1}{r^q}\Big) \leq M|\widetilde{V}|(r).$$

On the other hand,

$$\text{Log}\Big(r + \frac{1}{r^q}\Big) - \text{Log}r = \text{Log}\Big(1 + \frac{1}{r^{q+1}}\Big)$$

clearly satisfies an inequality of the form

$$\text{Log}\Big(1 + \frac{1}{r^{q+1}}\Big) \geq \frac{c_2}{r^{q+1}}$$

in $[l,+\infty[$ with $c_2 > 0$. Moreover, we can obviously find positive constants c_3, c_4 such that

$$\Big(r + \frac{1}{r^q}\Big)|P|\Big(r + \frac{1}{r^q}\Big) \leq c_3 r^{c_4}.$$

Consequently, by (1) and (4) we can find positive constants c_5, c_6 such that $|V|(r) \leq c_5 r^{c_6}|\widetilde{V}|(r)$ $\forall r \in [l,+\infty[$. Thus, writing again $V = \overline{V}\widetilde{V}$, we have $|\overline{V}|(r)|\widetilde{V}|(r) \leq c_5 r^{c_6}|\widetilde{V}|(r)$ and hence

$$|\overline{V}|(r) \leq c_5 r^{c_6} \; \forall r \in [l,+\infty[.$$

Consequently, by Corollary B.13.31, $\overline{V}$ is a polynomial of degree $\leq c_6$ and hence it has finitely many zeros and so does V. But then, by Corollary C.8.7, f must be a rational function. □

Corollary C.8.11: *Let f be a meromorphic function on $\mathbb{K}$ such that, for some $c, q \in]0, +\infty[$, $\gamma(r, f)$ satisfies $\gamma(r, f) \leq cr^q$ in $[1, +\infty[$. If for some $b \in \mathbb{K}$ $f' - b$ has finitely many zeros, then f is a rational function.*

Proof. Suppose $f' - b$ has finitely many zeros. Then $f - bx$ satisfies the same hypothesis as f, hence it is a rational function and so is f. □

Corollary C.8.12: *Let $f \in \mathcal{M}(\mathbb{K}) \setminus \mathbb{K}(x)$ be such that $\xi(r, f) \leq cr^q$ in $[1, +\infty[$ for some $c, q \in]0, +\infty[$. Then for each $k \in \mathbb{N}^*$, $f^{(k)}$ has no quasi-exceptional value.*

Proof. Indeed, if $k = 1$, the statement just comes from Corollary C.8.11. Now suppose $k \geq 2$. Each pole a of order n of f is a pole of order $n + k$ of $f^{(k)}$ and $f^{(k)}$ has no other pole. Consequently, we have $\gamma(r, f^{k-1}) = \xi(r, f^{(k-1)}) \leq kcr^q$. So, we can apply Corollary C.8.11 to $f^{(k-1)}$ to show the claim. □

C.9. Small functions

Small functions with respect to a meromorphic functions are well known in the general theory of complex functions. Particularly, one knows the Nevanlinna theorem on 3 small functions. Here we will construct a similar theory.

Definitions and notations: Throughout the chapter, we set $a \in K$ and $R \in]0, +\infty[$ and we still denote by D the set $\mathbb{K} \setminus d(0, R^-)$. For each $f \in \mathcal{M}(\mathbb{K})$ (resp. $f \in \mathcal{M}(d(a, R^-))$, resp. $f \in \mathcal{M}(D)$) we denote by $\mathcal{M}_f(\mathbb{K})$, (resp. $\mathcal{M}_f(d(a, R^-))$, resp. $\mathcal{M}_f(D)$) the set of functions $h \in \mathcal{M}(\mathbb{K})$ (resp. $h \in \mathcal{M}(d(a, R^-))$, resp. $\mathcal{M}(D)$) such that $T(r, h) = o(T(r, f))$ when r tends to $+\infty$ (resp. when r tends to R, resp. when r tends to $+\infty$). Similarly, if $f \in \mathcal{A}(\mathbb{K})$ (resp. $f \in \mathcal{A}(d(a, R^-))$, $f \in \mathcal{A}(D)$) we shall denote by $\mathcal{A}_f(\mathbb{K})$ (resp. $\mathcal{A}_f(d(a, R^-))$, resp. $\mathcal{A}_f(D)$) the set $\mathcal{M}_f(\mathbb{K}) \cap \mathcal{A}(\mathbb{K})$ (resp. $\mathcal{M}_f(d(a, R^-)) \cap \mathcal{A}(d(a, R^-))$, resp. $\mathcal{M}_f(D) \cap \mathcal{A}(D)$).

The elements of $\mathcal{M}_f(\mathbb{K})$ (resp. $\mathcal{M}_f(d(a, R^-))$, resp. $\mathcal{M}_f(D)$) are called small meromorphic functions with respect to f, small functions in brief. Similarly, if $f \in \mathcal{A}(\mathbb{K})$ (resp. $f \in \mathcal{A}(d(a, R^-))$, resp. $f \in \mathcal{A}(D)$) the elements of $\mathcal{A}_f(\mathbb{K})$ (resp. $\mathcal{A}_f(d(a, R^-))$, resp. $\mathcal{A}_f(D)$) are called small analytic functions with respect to f small functions in brief.

Theorems C.9.1 and C.9.2 are immediate consequences of Theorem C.4.3.

Theorem C.9.1: *Let $a \in \mathbb{K}$ and $r > 0$. $\mathcal{A}_f(\mathbb{K})$ is a $\mathbb{K}$-subalgebra of $\mathcal{A}(\mathbb{K})$, $\mathcal{A}_f(d(a, R^-))$ is a $\mathbb{K}$-subalgebra of $\mathcal{A}(d(a, R^-))$, $\mathcal{A}_f(D)$ is a $\mathbb{K}$-subalgebra of $\mathcal{A}(D)$, $\mathcal{M}_f(\mathbb{K})$ is a subfield field of $\mathcal{M}(\mathbb{K})$, $\mathcal{M}_f(d(a, R^-))$ is a subfield of field of $\mathcal{M}(a, R^-))$, and $\mathcal{M}_f(D)$ is a subfield field of $\mathcal{M}(D)$. Moreover, $\mathcal{A}_b(d(a, R^-)$ is a sub-algebra of $\mathcal{A}_f(d(a, R^-)$ and $\mathcal{M}_b(d(a, R^-)$ is a subfield of $\mathcal{M}_f(d(a, R^-)$.*

Theorem C.9.2: *Let $f \in \mathcal{M}(\mathbb{K})$ (resp. $f \in \mathcal{M}(d(0,R^-))$, resp. $f \in \mathcal{M}(\mathbb{K})$) and let $g \in \mathcal{M}_f(\mathbb{K})$ (resp. $g \in \mathcal{M}_f(d(0,R^-))$, resp. $g \in \mathcal{M}_f(D)$). Then $T(r, fg) = T(r,f) + o(T(r,f))$ and $T\Big(r, \dfrac{f}{g}\Big) = T(r,f) + o(T(r,f))$ (resp. $T(r,fg) = T(r,f) + o(T(r,f))$ and $T\Big(r, \dfrac{f}{g}\Big) = T(r,f) + o(T(r,f))$, resp. $T_R(r,fg) = T_R(r,f) + o(T_R(r,f))$ and $T_R\Big(r, \dfrac{f}{g}\Big) = T_R(r,f) + o(T_R(r,f))$).*

Here we can mention some precisions to Theorem C.9.1 that will be useful later.

Theorem C.9.3: *Let $f \in \mathcal{A}(\mathbb{K})$ (resp. let $f \in \mathcal{A}_u(d(a,r^-))$, resp. let $f \in \mathcal{A}(D)$). Let $g,h \in \mathcal{A}_f(\mathbb{K})$ (resp. let $g,h \in \mathcal{A}_f(d(a,r^-)$, resp. let $g,h \in \mathcal{A}_f(D)$) with g and h not identically zero. If gh belongs to $\mathcal{A}_f(\mathbb{K})$ (resp. to $\mathcal{A}_f(d(a,r^-))$, resp. to $\mathcal{A}_f(D)$), then so do g and h.*

Proof. Concerning the claim on $f \in \mathcal{A}_u(d(a,r^-))$, we can obviously assume $a = 0$. By Theorems C.4.3 and C.5.7, we have $T(r, g.h) = T(r,g) + T(r,h) + O(1)$. Consequently, $T(r,g.h) = o(T(r,f))$ if and only if $T(r,g) = o(T(r,f))$ and $T(r,h) = o(T(r,f))$. □

Theorem C.9.4: *Let $f,g \in \mathcal{A}(\mathbb{K})$ (resp. let $f,g \in \mathcal{A}_u(d(0,r^-))$, resp. let $f,g \in \mathcal{A}(D)$) and let $q \in \mathbb{N}^*$. If $\dfrac{f}{g}$ is not a q-th root of 1, then $f^q - g^q$ does not belong to $\mathcal{A}_f(\mathbb{K})$ (resp. to $\mathcal{A}_f(d(0,r^-))$, resp. to $\mathcal{A}_f(D)$).*

Proof. Concerning the claim on $f \in \mathcal{A}_u(d(a,r^-))$, we can obviously assume $a = 0$. Let $h = \dfrac{f}{g}$. Since h is not a q-th root of 1, neither $f - g$ nor the function $F(x) = \sum_{j=0}^{q-1} f^j g^{q-1-j}$ is identically zero. Suppose that $f^q - g^q \in \mathcal{A}_f(\mathbb{K})$ (resp. $f^q - g^q \in \mathcal{A}_f(d(0,r^-)$, resp. $\mathcal{A}_f(D)$. So, by Theorem C.9.3, both $f - g$ and F belong to $\mathcal{A}_f(\mathbb{K})$ (resp. to $\mathcal{A}_f(d(0,r^-)$, resp to $\mathcal{A}_f(D)$. Let $w = f - g$, hence $g = f + w$. Then $F(x) = \sum_{j=0}^{q-1} f^j (f+w)^{q-1-j}$. Thus, we can check that $F(x)$ is of the form $f^{q-1} + P(f(x))$ with $P(Y)$ a polynomial in Y of degree at most $q-2$, with coefficients in $\mathcal{A}_f(\mathbb{K})$ (resp. in $\mathcal{A}_f(d(0,r^-)$, resp. in $\mathcal{A}_f(D)$). Consequently, by Theorems C.4.3 and C.5.7, $T(r, F(x))$ is of the form $(q-1)T(r,f) + o(T(r,f))$ because $T(r, P(f(x))) \leq (q-2)T(r,f) + o(T(r,f))$, which proves that F does not belong to $\mathcal{A}_f(\mathbb{K})$ (resp. to $\mathcal{A}_f(d(0,r^-)$, resp. to $\mathcal{A}_f(D)$). □

In the proof of Theorem C.9.5 and in the sequel, we will have to use the following notation:

Definitions and notations: Let $h \in \mathcal{M}(\mathbb{K})\backslash\mathbb{K}$ (resp. $h \in E(x)\backslash E$) and let $\Xi(h)$ be the set of zeros c of h' such that $h(c) \neq h(d)$ for every zero d of h' other than c. If $\Xi(h)$ is finite, we denote by $\Upsilon(h)$ its cardinal and if $\Xi(h)$ is not finite, we put $\Upsilon(h) = +\infty$. Let $f \in \mathcal{M}(\mathbb{K})$ (resp. let $f \in E(x)$, resp. let $f \in \mathcal{M}(d(0, R^-))$, let $f \in \mathcal{M}(D)$. We will denote by

$Z(r, f \mid "f(x)$ satisfying Property P$")$
(resp. $Z(r, f \mid "f(x)$ satisfying Property P$")$,
resp. $Z_R(r, f \mid "f(x)$ satisfying Property P$"))$
the counting function of zeros of f when Property P is satisfied.

Similarly, we will denote by $\overline{Z}(r, f \mid "f(x)$ satisfying Property P$")$

(resp. $\overline{Z}(r, f \mid "f(x)$ satisfying Property P$")$,
resp. $\overline{Z}_R(r, f \mid "f(x)$ satisfying Property P$"))$
the counting function of zeros of f without counting multiplicity, when Property P is satisfied.

We will denote by $N(r, f \mid "f(x)$ satisfying Property P$")$,

(resp. $N(r, f \mid "f(x)$ satisfying Property P$")$,
resp. $N_R(r, f \mid "f(x)$ satisfying Property P$"))$
the counting function of poles of f when Property P is satisfied.

And we will denote by $\overline{N}(r, f \mid "f(x)$ satisfying Property P$")$

(resp. $\overline{N}(r, f \mid "f(x)$ satisfying Property P$")$,
$\overline{N}_R(r, f \mid "f(x)$ satisfying Property P$"))$
the counting function of poles of f without counting multiplicity when Property P is satisfied.

Theorem C.9.5 is a wide generalization of Theorem C.4.8. It consists of the following claim: given a meromorphic function f and a rational function G of degree n whose coefficients are small functions with respect to f, then $T(r, G(f))$ is equivalent to $nT(r, f)$. The big difficulty consists of showing that $T(r, G(f))$ is not smaller than $nT(r, f)$. The proof, based on an elementary property of Bezout's Theorem, was given in $\mathbb{C}$ by F. Gackstatter and I. Laine [60] and was made in a field such as $\mathbb{K}$ by C.C. Yang and Peichu Hu [66].

Theorem C.9.5: *Let $f \in \mathcal{M}(\mathbb{K})$ (resp. $f \in \mathcal{M}(d(0, R^-))$, let $f \in \mathcal{M}(D)$). Let $G(Y) \in \mathcal{M}_f(\mathbb{K})(Y)$ (resp. $G \in \mathcal{M}_f(d(0, R^-))(Y)$, resp. $G(Y) \in \mathcal{M}_f(D)(Y)$), and let $n = \deg(G)$. Then $T(r, G(f)) = nT(r, f) + o(T(r, f))$ (resp. $T(r, G(f)) = nT(r, f) + o(T(r, f))$, resp. $T_R(r, G(f)) = nT_R(r, f) + o(T_R(r, f))$.*

Proof. Let $G = \dfrac{P}{Q}$ with P, Q relatively prime in the ring $\mathcal{M}_f(\mathbb{K})$ (resp. $\mathcal{M}_f(d(a, R^-))$, $\mathcal{M}_f(D)$). Suppose first $G(Y) \in \mathcal{M}_f(\mathbb{K})[Y]$ (resp. $G \in$

$\mathcal{M}_f(d(0,R^-))[Y]$, resp. $G(Y) \in \mathcal{M}_f(D)[Y])$), hence $G = P$. Let $P(X) = \sum_{j=0}^n b_j X^j$ with $c_j \in \mathcal{M}_f(\mathbb{K})$ (resp. $c_j \in \mathcal{M}_f(d(0,R^))$, $c_j \in \mathcal{M}_f(D)$). By Theorems C.4.8 and C.5.7 we have $T(r, P(f)) = T(r, b_n^{-1}P(f)) + o(T(r,f))$ (resp. $T(r,P(f)) = T(r,b_n^{-1}P(f)) + o(T(r,f))$, resp. $T_R(r,P(f)) = T_R(r,b_n^{-1}P(f)) + o(T_R(r,f))$). Consequently, without loss of generality, we may also assume that P is monic.

Let $\widehat{\mathbb{K}}$ be an algebraically closed spherically complete extension of $\mathbb{K}$. Given $a \in \mathbb{K}$, we denote by $\widehat{d}(a,R^-)$ the disk $\{x \in \widehat{\mathbb{K}} \mid |x-a| < R\}$.

Suppose first $f \in \mathcal{M}(\mathbb{K})$ or $f \in \mathcal{M}(D$. We can write $f = \frac{h}{l}$ with $h, l \in \mathcal{A}(\mathbb{K})$ having no common zero. Now suppose that $f \in \mathcal{M}(d(0,R^-))$. Since f has continuation to a function $\widehat{f}$ meromorphic in the disk $\widehat{d}(0,R^-)$ of the field $\widehat{\mathbb{K}}$, by Lemma C.4.1 we know that the Nevanlinna functions $T(r,\widehat{f})$ in $\widehat{\mathbb{K}}$ is exactly this of f in $\mathbb{K}$. Consequently, without loss of generality we may assume that $\mathbb{K}$ is spherically complete. Thus, we can write f in the form $f = \frac{h}{l}$ with $h, l \in \mathcal{A}(d(0,R^-))$ having no common zero. Then $P(f)$ is in the form $\frac{\sum_{j=0}^n b_j h^j l^{n-j}}{Bl^n}$ with B, $b_j \in \mathcal{A}_f(\mathbb{K})$ (resp. B, $b_j \in \mathcal{A}_f(d(0,R^-))$, resp. B, $b_j \in \mathcal{A}_f(D)$). Clearly, we have $Z(r, \frac{\sum_{j=0}^n b_j h^j l^{n-j}}{l^n}) \leq Z(r, \sum_{j=0}^n b_j h^j l^{n-j})$ and by Theorem C.4.8 (resp. by Theorem C.4.8, resp. by Theorem C.5.7)

$$Z\left(r, \sum_{j=0}^n b_j h^j l^{n-j}\right) \leq \max_{0 \leq j \leq n} Z(r, b_j h^j l^{n-j}) \leq nT(r,f) + o(T(r,f),$$

(resp.

$$Z\left(r, \sum_{j=0}^n b_j h^j l^{n-j}\right) \leq \max_{0 \leq j \leq n} Z(r, b_j h^j l^{n-j}) \leq nT(r,f) + o(T(r,f),$$

resp.

$$Z_R\left(r, \sum_{j=0}^n b_j h^j l^{n-j}\right) \leq \max_{0 \leq j \leq n} Z_R(r, b_j h^j l^{n-j}) \leq nT_R(r,f) + o(T_R(r,f)).$$

On the other hand, $N(r,P(f)) \leq nN(r,f) + o(T(r,f))$ (resp. $N(r,P(f)) \leq nN(r,f) + o(T(r,f))$, resp. $N_R(r,P(f)) \leq nN_R(r,f) + o(T_R(r,f))$), hence $T(r,P(f)) \leq nT(r,f) + o(T(r,f))$ (resp. $T(r,P(f)) \leq nT(r,f) + o(T(r,f))$, resp. $T_R(r,P(f)) \leq nT_R(r,f) + o(T_R(r,f))$).

So, now it remains us to prove the reverse inequality. Indeed, suppose that inequality does not hold. For simplicity, we will first suppose that f lies either in $\mathcal{M}(\mathbb{K})$ or in $\mathcal{M}(d(0,R^-))$.

Then there exists $\rho \in]0,1[$ and a sequence of intervals $]r'_m, r''_m[$ such that $\lim_{m\to+\infty} r'_m = +\infty$ (resp. $\lim_{m\to+\infty} r'_m = R$, resp. $\lim_{m\to+\infty} r'_m = +\infty$) and that

$T(r, P(f)) \leq \rho n T(r, f)\ \forall r \in \bigcup_{m=1}^{\infty}]r'_m, r''_m[$. Particularly, we have $N(r, P(f)) \leq \rho n T(r, f)\ \forall r \in \bigcup_{m=1}^{\infty}]r'_m, r''_m[$.

Let us write again $P(f)$ in the form $\sum_{j=0}^{n} c_j f^j$. We shall prove the following:

$$N(r, P(f) \geq nN(r, f) - n \sum_{j=0}^{n} N(r, c_j) = nN(r, f) + o(T(r, f)). \tag{1}$$

Indeed, let α be a pole of f and suppose it is not a pole of order $\geq -n\omega_\alpha(f)$ of $P(f)$. We can check that there must exist $j \in \{0, \dots, n-1\}$ such that $-\omega_\alpha(c_j) \geq -\omega_\alpha(f)$. Consequently, at α we have $-\omega_\alpha(P(f)) \geq -n\omega_\alpha(f) - \max_{j=0,\dots,n-1}(-n\omega_\alpha(c_j))$, hence

$$N\big(r, f \mid \omega_\alpha(P(f)) < n\omega_\alpha(f)\big) \leq n \sum_{j=0}^{n-1} N(r, c_j),$$

hence (1) follows clearly.

Consequently, there exists $q \in \mathbb{N}$ such that $nN(r, f) \leq \rho n T(r, f)\ \forall r \in \bigcup_{m=q}^{\infty}]r'_m, r''_m[$ and therefore there exists $s \in \mathbb{N}$ and $\lambda \in]0, 1[$ such that

$$N(r, Bl^n) < n\lambda T(r, f)\ \forall r \in \bigcup_{m=s}^{\infty}]r'_m, r''_m[. \tag{2}$$

In particular, there exists $t \in \mathbb{N}$ such that

$$T(r, b_j h^j l^{n-j}) < \lambda T(r, b_n h^n)\ \forall j = 0, \dots, n-1, \forall r \in \bigcup_{m=t}^{\infty}]r'_m, r''_m[.$$

Now, since $T(r, b_n h^n) > T(r, b_j h^j l^{n-j})\ \forall j = 0, \dots, n-1, \forall r \in \bigcup_{m=t}^{\infty}]r'_m, r''_m[$, we notice that $|\sum_{j=0}^{n} b_j h^j l^{n-j}|(r) = |b_n|(r)\ \forall r \in \bigcup_{m=t}^{\infty}]r'_m, r''_m[$, hence

$$T\left(r, \sum_{j=0}^{n} b_j h^j l^{n-j}\right) = T(r, b_n h^n)\ \forall r \in \bigcup_{m=t}^{\infty}]r'_m, r''_m[.$$

Consequently, there exists $\sigma \in]\rho, 1[$ and $u \in \mathbb{N}$ $(u \geq t)$ such that

$$T\left(r, \sum_{j=0}^{n} b_j h^j l^{n-j}\right) > \sigma n T(r, h)\ \forall r \in \bigcup_{m=u}^{\infty}]r'_m, r''_m[. \tag{3}$$

Now, by (2) we have $T(r, h) = T(r, f)\ \forall r \in \bigcup_{m=u}^{\infty}]r'_m, r''_m[$ and

$$T(r, P(f)) = T\left(r, \sum_{j=0}^{n} b_j h^j l^{n-j}\right)\ \forall r \in \bigcup_{m=u}^{\infty}]r'_m, r''_m[,$$

hence finally, by (3) we obtain $T(r, P(f)) > \sigma n T(r, f)\ \forall r \in \bigcup_{m=u}^{\infty}]r'_m, r''_m[$, a contradiction that proves the claim when G is a polynomial.

Similarly, replacing N by N_R, T by T_R, we can make the same reasoning when f belongs to $\mathcal{M}(D)$.

We now consider the general case $G(Y) \in \mathcal{M}_f(\mathbb{K})(Y)$ (resp. $G \in \mathcal{M}_f(d(0,R^-))(Y)$). Without loss of generality, we may assume $\deg(G) = \deg(P)$.

Since P and Q are relatively prime, by Bezout's theorem in a ring of polynomials on a field, we can find $A,\ B \in \mathcal{M}_f(\mathbb{K})[Y]$ (resp. $A,\ B \in \mathcal{M}_f(d(0,R^-))[Y]$) such that $AQ + PB = 1$. Since $\deg(Q) \le \deg(P)$ of course $\deg(B) \le \deg(A)$, hence $\deg\left(\frac{A}{B}\right) = \deg(A)$. Now,

$$T\left(r, \frac{B(f)}{A(f)} + \frac{Q(f)}{P(f)}\right) = T\left(r, \frac{1}{A(f)P(f)}\right) = T(r, A(f)P(f)) + O(1).$$

Consequently, by the theorem already proven when G is a polynomial, we have

$$T\left(r, \frac{B(f)}{A(f)} + \frac{Q(f)}{P(f)}\right) = (\deg(A) + \deg(P))T(r,f) + o(T(r,f))$$

and since $\deg(P) = \deg(G)$, actually we have

$$T\left(r, \frac{B(f)}{A(f)} + \frac{Q(f)}{P(f)}\right) = (\deg(A) + \deg(G))T(r,f) + o(T(r,f)). \tag{4}$$

Now, $T\left(r, \frac{B(f)}{A(f)} + \frac{Q(f)}{P(f)}\right) \le T\left(r, \frac{B(f)}{A(f)}\right) + T\left(r, \frac{Q(f)}{P(f)}\right) + o(T(r,f)$ and by the first inequality already proven above, we obtain

$$T\left(r, \frac{B(f)}{A(f)} + \frac{Q(f)}{P(f)}\right) \le \deg\left(\frac{B}{A}\right)T(r,f) + T\left(r, \frac{Q(f)}{P(f)}\right) + o(T(r,f).$$

But since $\deg(B) \le \deg(A)$, actually we have $T\left(r, \frac{B(f)}{A(f)} + \frac{Q(f)}{P(f)}\right) \le \deg(A)T(r,f) + T\left(r, \frac{Q(f)}{P(f)}\right) + o(T(r,f))$, i.e.,

$$T\left(r, \frac{B(f)}{A(f)} + \frac{Q(f)}{P(f)}\right) \le \deg(A)T(r,f) + T(r, G(f)) + o(T(r,f)). \tag{5}$$

Now by (4) and (5) we can see that $\deg(G)T(r,f) \le T(r, G(f)) + o(T(r,f))$. Similarly, when f belongs to $\mathcal{M}(D)$ and G belongs to $\mathcal{M}_f(D)(Y)$, we can make the same reasoning, as above. This completes the proof. □

Theorem C.9.6: *Let $a \in \mathbb{K}$ and $r > 0$. Let $f \in \mathcal{M}(\mathbb{K}) \setminus \mathbb{K}(x)$ (resp. $f \in \mathcal{M}_u(d(a,R^-))$, resp. $f \in \mathcal{M}^c(D)$). Then, f is transcendental over $\mathcal{M}_f(\mathbb{K})$ (resp. over $\mathcal{M}_f(d(a,R^-))$, resp. over $\mathcal{M}_f(D)$).*

Proof. Suppose there exists a polynomial $P(Y) = \sum_{j=0}^{n} a_j Y^j \in \mathcal{M}_f(\mathbb{K})[Y] \neq 0$ such that $P(f) = 0$. If f belongs to $\mathcal{M}_u(d(a,R^-))$, we may obviously suppose that $a = 0$. By Theorem C.9.5, we have $T(r, a_n f^n) = nT(r,f) + o(T(r,f))$ whenever f

belongs to $\mathcal{M}(\mathbb{K})\setminus\mathbb{K}(x)$ or $\mathcal{M}_f(d(0,R^-))$ and $T_R(r,a_nf^n)=nT_R(r,f)+o(T_R(r,f))$ whenever f belongs to $\mathcal{M}^c(\mathbb{K}$, whereas $T(r,\sum_{j=0}^{n-1}a_jf^j)=(n-1)T(r,f)+o(T(r,f))$, a contradiction. □

Corollary C.9.7: *Let $a\in\mathbb{K}$ and $r>0$. Let $f\in\mathcal{M}(\mathbb{K})\setminus\mathbb{K}(x)$ (resp. $f\in\mathcal{M}_u(d(a,R^-))$, resp. $f\in\mathcal{M}^c(D)$. Then, f is transcendental over $\mathbb{K}(x)$.*

A function $h\in\mathcal{M}_b(d(a,R^-))$ is obviously small with respect to any function $f\in\mathcal{M}_u(d(a,R^-))$. So, we have the following corollary:

Corollary C.9.8: *Let $a\in\mathbb{K}$ and $r>0$ and let $f\in\mathcal{M}_u(d(a,R^-))$. Then, f is transcendental over $\mathcal{M}_b(d(a,R^-))$.*

By Corollary C.1.23, we know that a meromorphic function $f\in\mathcal{M}(\mathbb{K})$ or $f\in\mathcal{M}(d(0,R^-))$ admits at most one quasi-exceptional value. Here, we will generalize that statement.

Theorem C.9.9: *Let $a\in\mathbb{K}$ and $r>0$. Let $f\in\mathcal{M}(\mathbb{K})\setminus\mathbb{K}(x)$(resp. $f\in\mathcal{M}_u(d(a,R^-))$, resp. $f\in\mathcal{M}^c(D)$). There exists at most one function $g\in\mathcal{M}_f(\mathbb{K})$ (resp. $g\in\mathcal{M}_f(d(a,R^-))$, resp. $g\in\mathcal{M}_f(D)$) such that $f-g$ have finitely many zeros. Moreover, if f belongs to $\mathcal{A}(\mathbb{K})\setminus\mathbb{K}[x]$ (resp. to $\mathcal{A}_u(d(a,R^-))$, resp. to $\mathcal{A}^c(\mathbb{K})$), then there exists no function $g\in\mathcal{M}_f(\mathbb{K})\setminus\mathbb{K}(x)$, (resp. $g\in\mathcal{M}_f(d(a,R^-))$) such that $f-g$ have finitely many zeros.*

Proof. Concerning claims on $\mathcal{M}_u(d(a,R^-))$, we can obviously assume $a=0$. Suppose that there exist two distinct functions $g_1,\ g_2\in\mathcal{M}_f(\mathbb{K})$ (resp. $g_1,\ g_2\in\mathcal{M}_f(d(0,R^-))$) such that $f-g_k$ has finitely many zeros. So, there exist $P_1,\ P_2\in\mathbb{K}[x]$ and $h_1,\ h_2\in\mathcal{A}(\mathbb{K})$ (resp. $h_1,\ h_2\in\mathcal{A}(d(0,R^-))$) such that $f-g_k=\dfrac{P_k}{h_k}$, $k=1,\ 2$ and hence we notice that

$$T(r,f)=T\left(r,\frac{P_k}{h_k}\right)+o(T(r,f))=T(r,h_k)+o(T(r,f))\ k=1,\ 2. \tag{1}$$

Consequently, putting $g=g_2-g_1$, we have

$$\frac{P_1}{h_1}=\frac{P_2}{h_2}+g$$

and by Theorem C.9.1, g belongs to $\mathcal{M}_f(\mathbb{K})$ (resp. to $\mathcal{M}_f(d(0,R^-))$). Therefore, $P_1h_2-P_2h_1=gh_1h_2$ and hence

$$T(r,P_1h_2-P_2h_1)=T(r,gh_1h_2). \tag{2}$$

Now, by Theorem C.4.3, we have

$$\begin{aligned}T(r,P_1h_2-P_2h_1)&\le\max(T(r,P_1h_2),T(r,P_2h_1))\\&\le\max(T(r,h_1),T(r,h_2))+o(T(r,f))\end{aligned}$$

and hence by (1), we obtain

$$T(r, P_1h_2 - P_2h_1) \leq T(r, f) + o(T(r, f)). \tag{3}$$

On the other hand, by Theorem C.9.2, we have

$$T(r, gh_1h_2) = T(r, h_1h_2) + o(T(r, h_1h_2)) = 2T(r, f) + o(T(r, f)),$$

a contradiction to (3).

Now, if f belongs to $\mathcal{M}^c(D)$ we can make the same reasoning with T_R instead of T.

Suppose now that f belongs to $\mathcal{A}(\mathbb{K}) \setminus \mathbb{K}[x]$ and that there exists a function $w \in \mathcal{M}_f(\mathbb{K})$ such that $f - w$ has finitely many zeros. Set $w = \frac{l}{t}$ where l and t belong to $\mathcal{A}_f(\mathbb{K})$ and have no common zeros. Thus, $f - w = \frac{tf - l}{t}$ and each zero of $tf - l$ cannot be a zero of t, hence is zero of $f - w$. Consequently, since $f - w$ has finitely many zeros, $tf - l$ has finitely many zeros and hence is a polynomial. But since l belongs to $\mathcal{A}_f(\mathbb{K})$, when r is big enough we have $|f|(r) > |l|(r)$ and hence $|tf|(r) > |l|(r)$, therefore $|tf - l|(r) = |tf|(r)$. And since f is transcendental, by Corollary B.13.23, for every fixed $q \in \mathbb{N}$, $|f|(r) > r^q$ when r is big enough. Similarly, $|tf - l|(r) > r^q$ when r is big enough. Consequently, by Corollary B.13.23, $tf - l$ is not a polynomial, which proves that w does not exist.

Suppose now that f belongs to $\mathcal{A}_u(d(0, R^-))$ and that there exists a function $w \in \mathcal{M}_f(d(0, R^-))$ such that $f - w$ has finitely many zeros. Without loss of generality, we can assume that the field $\mathbb{K}$ is spherically complete because both f and w have continuation to an algebraically closed spherically complete extension of $\mathbb{K}$ where their zeros are the same as in $\mathbb{K}$. Consequently, we can write $w = \frac{l}{t}$ where l and t have no common zeros. Now, the zeros of $f - w$ are those of $tf - l$, hence $tf - l$ has finitely many zeros and hence, is bounded in $d(0, R^-)$. But since w belongs to $\mathcal{M}_f(d(0, R^-))$, so does l and hence $|tf|(r) > |l|(r)$ when r tends to R. Consequently, $|tf - l|(r) = |tf|(r)$ is not bounded in $d(0, R^-)$, a contradiction proving again that w does not exist.

Suppose finally that f belongs to $\mathcal{A}^c(\mathbb{K})$. We can make the same reasoning as in $\mathcal{A}(\mathbb{K})$ by replacing T by T_R. □

Theorem C.9.10 is known as Second Main theorem on Three Small Functions in p-adic analysis [66]. It holds as well as in complex analysis, where it was shown first. Notice that this theorem was generalized to any finite set of small functions by K. Yamanoi in complex analysis, through methods that have no equivalent on a p-adic field [92]. However, Corollary C.4.22 provides us with a kind of Second Main theorem on q-bounded functions, inside a disk.

Theorem C.9.10: *Let $f \in \mathcal{M}(\mathbb{K})$ (resp. $f \in \mathcal{M}_u(d(0, R^-))$, resp. $f \in \mathcal{M}^c(D)$) and let w_1, $w_2, w_3 \in \mathcal{M}_f(\mathbb{K})$ (resp. $w_1, w_2, w_3 \in \mathcal{M}_f(d(0, R^-))$, resp. $w_1, w_2, w_3 \in \mathcal{M}_f(D)$) be pairwaise distinct. Then $T(r, f) \le \sum_{j=1}^{3} \overline{Z}(r, f - w_j) + o(T(r, f))$, resp $T(r, f) \le \sum_{j=1}^{3} \overline{Z}(r, f - w_j) + o(T(r, f))$, resp. $T_R(r, f) \le \sum_{j=1}^{3} \overline{Z}_R(r, f - w_j) + o(T(r, f))$.*

Proof. We will make the proof when f belongs to $\mathcal{M}(\mathbb{K})$ or to $\mathcal{M}_u(d(0, R^-))$. Let $\phi(x) = \dfrac{(f(x) - w_1(x))(w_2(x) - w_3(x))}{(f(x) - w_3(x))(w_2(x) - w_1(x))}$. By Theorem C.4.24, we have

(1) $T(r, \phi) \le \overline{Z}(r, \phi) + \overline{Z}(r, \phi - 1) + \overline{N}(r, \phi) + O(1)$.

On the other hand, we have $T(r, f) \le T(r, f - w_j) + T(r, w_j)$ $(j = 1, 2, 3)$, hence $T(r, f) \le T\Big(r, \dfrac{w_3 - w_1}{f - w_3}\Big) + o(T(r, f))$, thereby

$$T(r, f) \le T\left(r, \frac{w_3 - w_1}{f - w_3} + 1\right) + o(T(r, f)) = T\left(r, \frac{f - w_1}{f - w_3}\right) + o(T(r, f)).$$

Now, $T\Big(r, \dfrac{w_2 - w_1}{w_2 - w_3}\Big) = o(T(r, f)$. Consequently, by writing $\dfrac{f - w_1}{f - w_3} = \phi\Big(\dfrac{w_2 - w_1}{w_2 - w_3}\Big)$ we have $T\Big(r, \dfrac{f - w_1}{f - w_3}\Big) \le T(r, \phi) + T\Big(r, \dfrac{w_2 - w_1}{w_2 - w_3}\Big) \le T(r, \phi) + o(T(r, f))$ and finally $T(r, f) \le T(r, \phi) + o(T(r, f))$. Thus, by (1) we obtain

(2) $T(r, f) \le \overline{Z}(r, \phi) + \overline{Z}(r, \phi - 1) + \overline{N}(r, \phi) + o(T(r, f))$.

Now, we can check that $\overline{Z}(r, \phi) + \overline{Z}(r, \phi - 1) + \overline{N}(r, \phi) \le \sum_{j=1}^{3} \overline{Z}(r, f - w_j) + \sum_{1 \le j < k \le 3} \overline{Z}(r, w_k - w_j) \le \sum_{j=1}^{3} \overline{Z}(r, f - w_j) + o(T(r, f))$, which, by (2), completes the proof when f belongs to $\mathcal{M}(\mathbb{K})$ or $\mathcal{M}_u(d(0, R^-))$. When f belongs to $\mathcal{M}(D)$ we can make a similar proof just by replacing T by T_R and Z by Z_R. □

Theorem C.9.11: *Let $f \in \mathcal{M}(\mathbb{K})$ (resp. $f \in \mathcal{M}_u(d(0, R^-))$, resp. $f \in \mathcal{M}^{\rfloor}(D)$ and let w_1, $w_2 \in \mathcal{M}_f(\mathbb{K})$ (resp. $w_1, w_2 \in \mathcal{M}_f(d(0, R^-))$, resp. $w_1, w_2 \in \mathcal{M}_f(D)$) be distinct. Then $T(r, f) \le \overline{Z}(r, f - w_1) + \overline{Z}(r, f - w_2) + \overline{N}(r, f) + o(T(r, f))$, resp. $T(r, f) \le \overline{Z}(r, f - w_1) + \overline{Z}(r, f - w_2) + \overline{N}(r, f) + o(T(r, f))$, resp. $T_R(r, f) \le \overline{Z}_R(r, f - w_1) + \overline{Z}_R(r, f - w_2) + \overline{N}_R(r, f) + o(T_R(r, f)))$.*

Proof. Suppose first $f \in \mathcal{M}(\mathbb{K})$ or $f \in \mathcal{M}_u(d(0, R^-))$. Let $g = \dfrac{1}{f}$, $h_j = \dfrac{1}{w_j}$, $j = 1, 2$, $h_3 = 0$. Clearly,

$$T(r, g) = T(r, f) + O(1),\ T(r, h) = T(r, w_j),\ j = 1, 2,$$

so we can apply Theorem C.9.10 to g, h_1, h_2, h_3. Thus, we have $T(r, g) \le \overline{Z}(r, g - h_1) + \overline{Z}(r, g - h_2) + \overline{Z}(r, g) + o(T(r, g))$.

But we notice that $\overline{Z}(r, g-h_j) = \overline{Z}(r, f-w_j)$ for $j = 1, 2$ and $\overline{Z}(r, g) = \overline{N}(r, f)$. Moreover, we know that $o(T(r, g)) = o(T(r, f))$. Consequently, the claim is proved when $w_1 w_2$ is not identically zero.

Now, suppose that $w_1 = 0$. Let $\lambda \in \mathbb{K}^*$, let $l = f + \lambda$ and $\tau_j = u_j + \lambda$, $(j = 1, 2, 3)$. Thus, we have $T(r, l) = T(r, f) + O(1)$, $T(r, \tau_j) = T(r, w_j) + O(1)$, $(j = 1,\ 2)$, $\overline{N}(r, l) = \overline{N}(r, f)$. By the claim already proven whenever $w_1 w_2 \neq 0$ we may write $T(r, l) \leq \overline{Z}(r, l - \tau_1) + \overline{Z}(r, l - \tau_2) + \overline{N}(r, l) + o(T(r, l)))$, hence $T(r, f) \leq \overline{Z}(r, f - w_1) + \overline{Z}(r, f - w_2) + \overline{N}(r, l) + o(T(r, f)))$.

Suppose now $f \in \mathcal{M}(D)$. By replacing T by T_R, Z by Z_R, N by N_R, we can check that the same reasoning applies. □

Next, by setting $g = f - w_1$ and $w = w_1 + w_2$, we can write Corollary C.9.12.

Corollary C.9.12: *Let $g \in \mathcal{M}(\mathbb{K})$ (resp. $g \in \mathcal{M}_u(d(0, R^-))$, resp. $g \in \mathcal{M}^c(D)$) and let $w \in \mathcal{M}_g(\mathbb{K})$ (resp. $w \in \mathcal{M}_g(d(0, R^-))$, resp. $w \in \mathcal{M}_g(D)$). Then $T(r, g) \leq \overline{Z}(r, g) + \overline{Z}(r, g - w) + \overline{N}(r, g) + o(T(r, g))$, resp. $T(r, g) \leq \overline{Z}(r, g) + \overline{Z}(r, g - w) + \overline{N}(r, g) + o(T(r, g))$, resp. $T_R(r, g) \leq \overline{Z}_R(r, g) + \overline{Z}_R(r, g - w) + \overline{N}({}_R r, g) + o(T_R(r, g)))$.*

Corollary C.9.13: *Let $f \in \mathcal{A}(\mathbb{K})$ (resp. $f \in \mathcal{A}_u(d(0, R^-))$, resp. $f \in \mathcal{A}^c(D)$) and let $w_1,\ w_2 \in \mathcal{A}_f(\mathbb{K})$ (resp. $w_1, w_2 \in \mathcal{A}_f(d(0, R^-))$, resp. $w_1, w_2 \in \mathcal{A}_f(D)$) be distinct. Then $T(r, f) \leq \overline{Z}(r, f - w_1) + \overline{Z}(r, f - w_2) + o(T(r, f))$, resp. $T(r, f) \leq \overline{Z}(r, f - w_1) + \overline{Z}(r, f - w_2) + o(T(r, f)))$, resp. $T_R(r, f) \leq \overline{Z}_R(r, f - w_1) + \overline{Z}_R(r, f - w_2) + o(T_R(r, f)))$.*

And similarly to Corollary C.9.12, we get Corollary C.9.14.

Corollary C.9.14: *Let $f \in \mathcal{A}(\mathbb{K})$ (resp. $f \in \mathcal{A}_u(d(0, R^-))$, resp. $f \in \mathcal{A}^c(D)$) and let $w \in \mathcal{A}_f(\mathbb{K})$ (resp. $w \in \mathcal{A}_f(d(0, R^-))$, resp. $w \in \mathcal{A}_f(D)$). Then $T(r, f) \leq \overline{Z}(r, f) + \overline{Z}(r, f - w) + o(T(r, f))$ (resp. $T(r, f) \leq \overline{Z}(r, f) + \overline{Z}(r, f - w) + o(T(r, f))$, resp. $T_R(r, f) \leq \overline{Z}_R(r, f) + \overline{Z}_R(r, f - w) + o(T_R(r, f)))$.*

Here is now an application of that theory.

Theorem C.9.15: *Let $h, w \in \mathcal{A}_b(d(a, R^-))$ and let $m, n \in \mathbb{N}^*$ be such that* $\min(m, n) \geq 2$, $\max(m, n) \geq 3$. *Then the functional equation*

$$(\mathcal{E})\ (g(x))^n = h(x)(f(x))^m + w(x)$$

has no solution in $\mathcal{A}_u(d(a, R^-))$.

Proof. Without loss of generality, we can obviously assume $a = 0$. Let $F(x) = g(x)^n$. Thanks to Corollary C.9.14. We can write

$$T(r, F) \leq \overline{Z}(r, F) + \overline{Z}(r, F - w) + o(T(r, F)).$$

Now, it appears that $\overline{Z}(r,F) \leq \frac{1}{n}Z(r,F)$. Moreover, since h is bounded, $Z(r,h)$ is bounded, hence $\overline{Z}(r,hf^m) \leq Z(r,f) + Z(r,h) = Z(r,f) + O(1)$, hence

$$\overline{Z}(r,hf^m) \leq \frac{1}{m}Z(r,hf^m) + O(1) = \frac{1}{m}Z(r,F) + O(1). \tag{1}$$

On the other hand, $Z(r,F) = Z(r,F-w) + O(1) = T(r,F) + O(1)$. Consequently, by (1), we can derive

$$T(r,F) \leq \left(\frac{1}{m} + \frac{1}{n}\right)T(r,F) + o(T(r,F)).$$

Therefore, we have $\frac{1}{m} + \frac{1}{n} \geq 1$, a contradiction to the hypothesis, which implies $\frac{1}{m} + \frac{1}{n} \leq \frac{5}{6}$. □

Theorem C.9.16: *Let $f \in \mathcal{M}(\mathbb{K})$ be transcendental (resp. $f \in \mathcal{M}_u(d(0,R^-))$, resp. $f \in \mathcal{M}^c(D)$) and let $w_j \in \mathcal{M}_f(\mathbb{K})$ $(j = 1,\dots,q)$ (resp. $w_j \in \mathcal{M}_f(d(a,R^-))$, resp. $w_j \in \mathcal{M}_f(D)$) be q distinct small functions other than the constant ∞. Then*

$$qT(r,f) \leq 3\sum_{j=1}^{q}\overline{Z}(r,f-w_j) + o(T(r,f)),$$

(resp.

$$qT(r,f) \leq 3\sum_{j=1}^{q}\overline{Z}(r,f-w_j) + o(T(r,f)),$$

resp.

$$qT_R(r,f) \leq 3\sum_{j=1}^{q}\overline{Z}_R(r,f-w_j) + o(T_R(r,f))).$$

Moreover, if f has finitely many poles in $\mathbb{K}$ (resp. in $d(0,R^-)$, resp. in D), then

$$qT(r,f) \leq 2\sum_{j=1}^{q}\overline{Z}(r,f-w_j) + o(T(r,f)),$$

(resp.

$$qT(r,f) \leq 2\sum_{j=1}^{q}\overline{Z}(r,f-w_j) + o(T(r,f)),$$

resp.

$$qT_R(r,f) \leq 2\sum_{j=1}^{q}\overline{Z}_R(r,f-w_j) + o(T_R(r,f))).$$

Proof. Suppose first that f and w_j, $(j = 1,\dots,q)$ belong to $\mathcal{M}(\mathbb{K})$ or to $\mathcal{M}(d(0,R^-))$. By Theorem C.9.10, for every triplet (i,j,k) such that $1 \le i \le j \le k \le q$, we can write

$$T(r,f) \le \overline{Z}(r,f-w_i) + \overline{Z}(r,f-w_j) + \overline{Z}(r,f-w_k) + o(T(r,f)).$$

The number of such inequalities is C_q^3. Summing up, we obtain

(1)
$$C_q^3 T(r,f) \le \sum_{(i,j,k),\ 1\le i\le j\le k\le q} \overline{Z}(r,f-w_i)+\overline{Z}(r,f-w_j)+\overline{Z}(r,f-w_k)+o(T(r,f)).$$

In this sum, for each index i, the number of terms $\overline{Z}(r,f-w_i)$ is clearly C_{q-1}^2. Consequently, by (1) we obtain

$$C_q^3 T(r,f) \le C_{q-1}^2 \sum_{i=1}^{q} \overline{Z}(r,f-w_i) + o(T(r,f))$$

and hence

$$\frac{q}{3} T(r,f) \le \sum_{i=1}^{q} \overline{Z}(r,f-w_i) + o(T(r,f)).$$

Suppose now that f has finitely many poles. By Theorem C.9.11, for every pair (i,j) such that $1 \le i \le j \le q$, we have

$$T(r,f) \le \overline{Z}(r,f-w_i) + \overline{Z}(r,f-w_j) + o(T(r,f)).$$

The number of such inequalities is then C_q^2. Summing up, we now obtain

(2)
$$C_q^2 T(r,f) \le \sum_{(i,j,\ 1\le i\le j\le q} \overline{Z}(r,f-w_i) + \overline{Z}(r,f-w_j) + o(T(r,f)).$$

In this sum, for each index i, the number of terms $\overline{Z}(r,f-w_i)$ is clearly $C_{q-1}^1 = q-1$. Consequently, by (1) we obtain

$$C_q^2 T(r,f) \le (q-1) \sum_{i=1}^{q} \overline{Z}(r,f-w_i) + o(T(r,f))$$

and hence

$$\frac{q}{2} T(r,f) \le \sum_{i=1}^{q} \overline{Z}(r,f-w_i) + o(T(r,f)).$$

Now, if f and w_j, $(j = 1,\dots,q)$ belong to $\mathcal{M}(D)$, we can make the same reasoning with T_R instead of T and Z_R instead of Z. □

Definitions and notations: Let $f, g \in \mathcal{M}(\mathbb{K})$ (resp. $f,\ g \in \mathcal{M}_u(d(a, R^-))$, resp. $f,\ g \in \mathcal{M}^c(D)$). Then f and g will be to share a small function $w \in \mathcal{M}(\mathbb{K})$ (resp. $w \in \mathcal{M}(d(a, R^-))$ resp. $w \in \mathcal{M}(D)$) I.M. if $f(x) = w(x)$ implies $g(x) = w(x)$ and if $g(x) = w(x)$ implies $f(x) = w(x)$.

Theorem C.9.17: *Let $f, g \in \mathcal{M}(\mathbb{K})$ be transcendental (resp. $f, g \in \mathcal{M}_u(d(a, R^-))$, resp. $f, g \in \mathcal{M}^c(D)$) be distinct and share q distinct small functions I.M. $w_j \in \mathcal{M}_f(\mathbb{K}) \cap \mathcal{M}_g(\mathbb{K})$ $(j = 1, \dots, q)$ (resp. $w_j \in \mathcal{M}_f(d(a, R^-)) \cap \mathcal{M}_g(d(a, R^-))$ $(j = 1, \dots, q)$, (resp. $w_j \in \mathcal{M}_f(D) \cap \mathcal{M}_g(D)$ $(j = 1, \dots, q)$), other than the constant ∞. Then*

$$\sum_{j=1}^{q} \overline{Z}(r, f - w_j) \leq \overline{Z}(r, f - g) + o(T(r, f)) + o(T(r, g)).$$

Proof. Suppose that f and g belong to $\mathcal{M}(\mathbb{K})$, are distinct, and share q distinct small functions I.M. $w_j \in \mathcal{M}_f(\mathbb{K}) \cap \mathcal{M}_g(\mathbb{K})$ $(j = 1, \dots, q)$.

Lat b be a zero of $f - w_i$ for a certain index i. Then it is also a zero of $g - w_i$. Suppose that b is counted several times in the sum $\sum_{j=1}^{q} \overline{Z}(r, f - w_j)$, which means that it is a zero of another function $f - w_h$ for a certain index $h \neq i$. Then we have $w_i(b) = w_h(b)$ and hence b is a zero of the function $w_i - w_h$, which belongs to $\mathcal{M}_f(\mathbb{K})$. Now, put $\widetilde{Z}(r, f - w_1) = \overline{Z}(r, f - w_1)$ and for each $j > 1$, let $\widetilde{Z}(r, f - w_j)$ be the counting function of zeros of $f - w_j$ in the disk $d(0, r^-)$ ignoring multiplicity and avoiding the zeros already counted as zeros of $f - w_h$ for some $h < j$. Consider now the sum $\sum_{j=1}^{q} \widetilde{Z}(r, f - w_j)$. Since the functions $w_i - w_j$ belong to $\mathcal{M}_f(\mathbb{K})$, clearly, we have

$$\sum_{j=1}^{q} \overline{Z}(r, f - w_j) = \sum_{j=1}^{q} \widetilde{Z}(r, f w_j) = o(T(r, f)).$$

It is clear, from the assumption, that $f(x) - w_j(x) = 0$ implies $g(x) - w_j(x) = 0$ and hence $f(x) - g(x) = 0$. Since $f - g$ is not the identically zero function, it follows that

$$\sum_{j=1}^{q} \overline{Z}(r, f - w_j) \leq \overline{Z}(r, f - g).$$

Consequently,

$$\sum_{j=1}^{q} \overline{Z}(r, f - w_j) \leq \overline{Z}(r, f - g) + o(T(r, f)) + o(T(r, g)).$$

Now, if f and g belong to $\mathcal{M}(d(0, R^-))$ or to $\mathcal{M}(D)$, the proof is exactly the same. □

Theorem C.9.18: *Let $f,\ g\ \in\ \mathcal{M}(\mathbb{K})$ be transcendental (resp. $f,\ g\ \in \mathcal{M}_u(d(a,R^-))$, resp. $f,\ g \in \mathcal{M}^c(D)$ be distinct, and share 7 distinct small functions (other than the constant ∞) I.M. $w_j \in \mathcal{M}_f(\mathbb{K}) \cap \mathcal{M}_g(\mathbb{K})$ $(j = 1, \dots, 7)$ (resp. $w_j \in \mathcal{M}_f(d(a,R^-)) \cap \mathcal{M}_g(d(a,R^-))$, resp. $w_j \in \mathcal{M}_f(D) \cap \mathcal{M}_g(D)$ $(j = 1, \dots, 7),$). Then $f = g$.*

Moreover, if f and g have finitely many poles and share 3 distinct small functions (other than the constant ∞) I.M., then $f = g$.

Proof. We put $M(r) = \max(T(r,f), T(r,g))$. Suppose that f and g are distinct and share q small function I.M. w_j, $(1 \le j \le q)$. By Theorem C.9.16, we have

$$qT(r,f) \le 3\sum_{j=1}^{q} \overline{Z}(r, f - w_j) + o(T(r,f)).$$

But thanks to Theorem C.9.17, we can derive

$$qT(r,f) \le 3T(r, f - g) + o(T(r,f))$$

and similarly

$$qT(r,g) \le 3T(r, f - g) + o(T(r,g)),$$

hence

$$qM(r) \le 3T(r, f - g) + o(M(r)). \tag{1}$$

By Theorem C.4.8, we can derive that

$$qM(r) \le 3(T(r,f) + T(r,g)) + o(M(r)))$$

and hence $qM(r) \le 6M(r) + o(M(r))$. That applies to the situation when f and g belong to $\mathcal{M}(\mathbb{K})$ as well as when when f and g belong to $\mathcal{M}_u(d(0,R^-))$ and when f and g belong to $\mathcal{M}(D)$, after replacing T by T_R and Z by Z_R. We have a similar proof whenever f, g, and w_j belong to $\mathcal{M}(d(0,R^-))$ or $\mathcal{M}(D)$ after replacing T by T_R and Z by Z_R. Consequently, it is impossible if $q \ge 7$ and hence the first statement of Theorem C.9.18 is proved.

Suppose now that f and g have finitely many poles. By Theorem C.4.8, Relation (1) gives us

$$qM(r) \le 2M(r) + o(M(r)),$$

which is obviously absurd whenever $q \ge 3$ and proves that $f = g$ when f and g belong to $\mathcal{M}(\mathbb{K})$ as well as when f and g belong to $\mathcal{M}_u(d(0,R^-))$ or to $\mathcal{M}^c(D)$, after replacing T by T_R and Z by Z_R. □

Corollary C.9.19: *Let $f, g \in \mathcal{A}(\mathbb{K})$ be transcendental (resp. $f, g \in \mathcal{A}_u(d(a,R^-))$, resp. $f,\ g \in \mathcal{A}^c(D)$ be distinct and share 3 distinct small functions (other than the constant ∞) I.M. $w_j \in \mathcal{A}_f(\mathbb{K}) \cap \mathcal{A}_g(\mathbb{K})$ $(j = 1,2,3)$ (resp. $w_j \in \mathcal{A}_f(d(a,R^-)) \cap \mathcal{A}_g(d(a,R^-))$, $(j = 1,2,3)$, resp. $w_j \in \mathcal{A}_f(D) \cap \mathcal{A}_g(D)$ $(j = 1,2,3)$). Then $f = g$.*

C.10. The p-adic Hayman conjecture

In the 1950s, Walter Hayman asked the question whether, given a meromorphic function in $\mathbb{C}$, the function $g'g^n$ might admit a quasi-exceptional value $b \neq 0$ [65]. W. Hayman showed that $g'g^n$ has no quasi-exceptional value, whenever $n \geq 3$. Henceforth, the problem was solved for $n = 2$ by E. Mues in 1979 [76] and next for $n \geq 1$, in 1995 by W. Bergweiler and A. Eremenko [5] and separately by H. Chen and M. Fang [38]. The same problem is posed on the field $\mathbb{K}$, both in $\mathcal{M}(\mathbb{K})$ and in a field $\mathcal{M}(d(a, R^-))$ ($a \in \mathbb{K},\ R > 0$).

Lemma C.10.1 is immediate.

Lemma C.10.1: *Let $g \in \mathcal{M}(\mathbb{K})$ (resp. let $g \in \mathcal{M}(d(a, R^-))$, $a \in \mathbb{K},\ R > 0$), set $f = \dfrac{1}{g}$ and let $n \in \mathbb{N}^*$. Then $g'g^n$ admits a quasi-exceptional value $b \in \mathbb{K}^*$ if and only if $f' + bf^{n+2}$ has finitely many zeros that are not zeros of f.*

Remark: We can also consider the same problem when $n = -1$, i.e., the question whether $f' + bf$ has infinitely many zeros. We will examine this in p-adic analysis. When $n = 0$, in $\mathbb{C}$ the well-known counter-example furnished by the function $\tan x$ shows that $f' - f^2$ may have no zero. On the field $\mathbb{K}$, we will examine the cases $n = -1$ and $n = 0$.

Henceforth, we will examine that problem by considering the set of zeros of $f' + bf^m$, with $b \neq 0$. In the field $\mathbb{K}$, two theorems are specific to p-adic analysis. Both are based on Lemma C.10.2.

Lemma C.10.2: *Let $f \in \mathcal{M}(\mathbb{K})$ (resp. let $f \in \mathcal{M}(d(0, R^-))$, $R > 0$, suppose that f admits infinitely many zeros and suppose that there exists a sequence of intervals $[r'_n, r''_n]$ such that $\lim_{n\to+\infty} r'_n = +\infty$ (resp. $\lim_{n\to+\infty} r'_n = \lim_{n\to+\infty} r''_n = R$) and such that $|(f' + f^m)|(r) = |f^m|(r)\ \forall r \in \bigcup_{n\in\mathbb{N}} [r'_n, r''_n]$. Let $m \in \mathbb{N}^*$ be $\neq 2$. Then $f' + f^m$ has infinitely many zeros that are not zeros of f.*

Proof. Let $J = \bigcup_{n\in\mathbb{N}} [r'_n, r''_n]$. By Corollary B.13.6, we have

$$\nu^+(f' + f^m, \log r) = \nu^+(f^m, \log r),\ \ \nu^-(f' + f^m, \log r) = \nu^-(f^m, \log r)\ \forall r \in J.$$

Consequently, in each disk $d(0, r)$ with $r \in J$, f and $f' + f^m$ have the same difference between the number of zeros and poles. Now, if $m \geq 3$ the poles of $f' + f^m$ and f^m are the same taking multiplicity into account. And when $m = 1$, each pole of f is a pole of $f' + f$ with a strictly greater order. Consequently, for each $r \in J$, the number of zeros of $f' + f^m$ in $d(0, r)$ is superior or equal to this of f^m.

Now, for each $n \in \mathbb{N}$, let s_n be the number of distinct zeros of f in $d(0, r''_n)$. Since f has infinitely many zeros, the sequence s_n is increasing and tends to $+\infty$.

On the other hand, for each zero α of order u of f, either α is not a zero of $f' + f^m$ (when $u = 1$), or it is a zero of order $u - 1$. Consequently, the number of zeros of $f' + f^m$ in $d(0, r''_n)$, which are not zeros of f, is at least s_n. Thus, we have proved that $f' + f^m$ has infinitely many zeros that are not zeros of f. □

We will prove together Theorems C.10.3 and C.10.4.

Theorem C.10.3: *Let $f \in \mathcal{M}(\mathbb{K}) \setminus \mathbb{K}(x)$ satisfy $\limsup\limits_{r\to\infty} |f|(r) > 0$ and let $b \in \mathbb{K}^*$. Let $m \in \mathbb{N}^*$ be ≥ 3. Then $f' + bf^m$ has infinitely many zeros that are not zeros of f.*

Theorem C.10.4: *Let $f \in \mathcal{M}_u(d(a, R^-))$ satisfy $\limsup\limits_{r\to R} |f|(r) = +\infty$ and let $b \in \mathbb{K}^*$. Let $m \in \mathbb{N}^*$ be ≥ 3. Then $f' + bf^m$ has infinitely many zeros that are not zeros of f.*

Proof. (Theorems C.10.3 and C.10.4) Without loss of generality, we can assume $b = 1$ and when $f \in \mathcal{M}(d(a, R^-))$, we may assume $a = 0$. By hypotheses, there exists a sequence of intervals $[r'_n, r''_n]$ such that $\lim\limits_{n\to+\infty} r'_n = +\infty$ (resp. $\lim\limits_{n\to+\infty} r'_n = \lim\limits_{n\to+\infty} r''_n = R$) and such that, putting $J = \bigcup\limits_{n\in\mathbb{N}} [r'_n, r''_n]$, we have $\limsup\limits_{\substack{r\to\infty,\\ r\in J}} |f|(r) > 0$ (resp. $\lim\limits_{\substack{r\to R^-\\ r\in J}} |f|(r) = +\infty$).

Suppose first we assume the hypothesis of Theorem C.10.3. Let $M = \dfrac{\limsup_{r\to+\infty} |f|(r)}{2}$. We will prove that there exists $t > 0$ such that $|f' + f^m|(r) = |f^m|(r)\ \forall r \in J \cap [t, +\infty[$. By Theorem C.2.10, we have $|f'|(r) \leq \dfrac{|f|(r)}{r}$. Consequently, when r lies in J, there exists $s > 0$ such that $|f|(r) \geq M\ \forall r \in [s, +\infty[\cap J$.

$$\big(|f|(r)\big)^m \geq |f|(r) M^{m-1} \geq r|f'|(r) M^{m-1}.$$

Next, when r is big enough, rM^{m-1} is greater than 1, hence $(|f|(r))^m > |f'|(r)$. Thus there exists $t \geq s$ such that $(|f|(r))^m > |f'|(r)\ \forall r \in J \cap [t, +\infty[$. Let $J' = J \cap [t, +\infty[$. So we have $|f' + f^m|(r) = |f^m|(r)\ \forall r \in J'$.

Suppose now that we assume the hypothesis of Theorem C.10.4. We have $|f'|(r) \leq \dfrac{|f|(r)}{r} \leq \dfrac{|f|(r)}{R}$. Set $B = \dfrac{1}{R}$. Then we have

$$\big(|f|(r)\big)^m \geq B|f'|(r)(|f|(r))^{m-1}.$$

Now, when r is close enough to R, $r \in J$, $B|f(x)|^{m-1}$ is strictly greater than 1, hence $(|f|(r))^m > |f'|(r)$. Thus, there exists $t > 0$ such that $(|f|(r))^m > |f'|(r)\ \forall r \in [t, +\infty[\cap J$. We can set again $J' = J \cap [t, R[$ and then we have $|f' + f^m|(r) = |f^m|(r)\ \forall r \in J'$.

We can now conclude Theorems C.10.3 and C.10.4. For each $n \in \mathbb{N}$, let q_n be the number of zeros of f in $d(0, r''_n)$. Suppose the sequence $(q_n)_{n\in\mathbb{N}}$ is bounded. Then, f has finitely many zeros, hence it is of the form $\frac{P}{h}$ with $P \in \mathbb{K}[x]$ and $h \in \mathcal{A}(\mathbb{K})$ (resp. $h \in \mathcal{A}_u(d(0, R^-))$). Consequently, we have $\lim_{r\to+\infty} |f|(r) = 0$ (resp. $\lim_{r\to R^-} |f|(r) = 0$), a contradiction to the hypothesis in both theorems. Therefore, the sequence $(q_n)_{n\in\mathbb{N}}$ that is increasing by definition, tends to $+\infty$. Now, in each Theorems C.10.3 and C.10.4, we may apply Lemma C.10.2 showing that $f' + f^m$ has infinitely many zeros that are not zeros of f. □

In the case $m = 1$, we can have a better conclusion in $\mathcal{M}(\mathbb{K})$.

Theorem C.10.5: *Let $f \in \mathcal{M}(\mathbb{K}) \setminus \mathbb{K}(x)$. For each $b \in \mathbb{K}^*$, $f' + bf$ has infinitely many zeros that are not zeros of f.*

Proof. Without loss of generality, we can assume again $b = 1$. By Theorem C.2.10, we have $|f'|(r) < |f|(r)$ when r is big enough and hence $|f' + f|(r) = |f|(r)$ in an interval $I = [s, +\infty[$. Suppose first that f has infinitely many zeros. We can then apply Lemma C.10.2 and get the conclusion.

Suppose now that f has finitely many zeros. Then f has infinitely many poles c_n of respective order t_n. Since $\mathbb{K}$ has characteristic zero, f' admits each c_n as a pole of order $t_n + 1$ and similarly, $f' + f$ also admits each c_n as a pole of order $t_n + 1$. Thus, we have $N(r, f' + f) = N(r, f) + \overline{N}(r, f)$. But since $|f' + f|(r) = |f|(r)$ holds in I, we have $\nu(f' + f, \log r) = \nu(f, \log r)\ \forall r \in I$ and hence $Z(r, f' + f) - N(r, f' + f) = Z(r, f) - N(r, f)$, therefore $Z(r, f' + f) - (N(r, f) + \overline{N}(r, f)) = Z(r, f) - N(r, f)$ and hence $Z(r, f' + f) = Z(r, f) + \overline{N}(r, f)$. Since we have supposed that f has finitely many zeros and since f has infinitely many poles, $f' + f$ has infinitely many zeros and all but finitely many are not zeros of f. □

Concerning functions $f' + bf^2$, we can obtain a first conclusion when f is analytic.

Theorem C.10.6: *Let $f \in \mathcal{A}(\mathbb{K}) \setminus \mathbb{K}(x)$ (resp. let $a \in \mathbb{K}$, let $R \in]0, +\infty[$ and let $f \in \mathcal{A}_u(da(, R^-)))$. For each $b \in \mathbb{K}^*$, $f' + bf^2$ has infinitely many zeros that are not zeros of f.*

Proof. Without loss of generality, we can assume $b = 1$ and $a = 0$. Clearly, when r is big enough, in $]0, +\infty[$ (resp. in $]0, R[$), we have $|f' + f^2|(r) = |f^2|(r)$ therefore, by Corollary B.13.6, f^2 and $f' + f^2$ have the same number of zeros in $C(0, r)$ (taking multiplicity into account). Let $\alpha \in C(0, r)$ be a zero of f of order q. When r is big enough, it is a zero of order $2q$ for f^2 and it is a zero of order $q - 1$ for $f' + f^2$. Consequently, by Corollary B.13.6, $f' + f^2$ has at least $q + 1$ zero in $C(0, r)$ that are not zeros of f (taking multiplicity into account). This is true for every such zeros of f and hence $f' + f^2$ has infinitely many zeros that are not zeros of f. □

Corollary C.10.7: *Let $m \in \mathbb{N}^*$ be ≥ 1, let $f \in \mathcal{A}(\mathbb{K}) \setminus \mathbb{K}(x)$. For each $b \in \mathbb{K}^*$, $f' + bf^m$ has infinitely many zeros that are not zeros of f.*

Corollary C.10.8: *Let $m \in \mathbb{N}$ be ≥ 2, let $a \in \mathbb{K}$, let $R \in]0, +\infty[$, and let $f \in \mathcal{A}_u(d(a, R^-)))$. For each $b \in \mathbb{K}^*$, $f' + bf^m$ has infinitely many zeros that are not zeros of f. Theorem* C.10.9 *was published in* [23] *and partially in* [78].

Theorem C.10.9: *Let $f \in \mathcal{M}(\mathbb{K}) \setminus \mathbb{K}(x)$ (resp. let $a \in \mathbb{K}$ and $R \in \mathbb{R}_+^*$ and let $f \in \mathcal{M}_u(d(a, R^-)))$ and let $m \in \mathbb{N}$. If $m \geq 5$, then for each $b \in \mathbb{K}^*$, $f' + bf^m$ has infinitely many zeros that are not zeros of f. If $m = 4$, if $f \in \mathcal{M}(\mathbb{K}) \setminus \mathbb{K}(x)$, and if f admits at least s multiple zeros and at least t multiple poles, then $f' + bf^4$ admits a number of zeros that are not zeros of f (taken account of multiplicity), which is strictly superior to $\frac{s+t}{2}$.*

Proof. By Corollary B.13.2, the zeros of $f'+bf^m$ in $\mathbb{K}$ are the same as in a spherically complete algebraically closed extension $\widehat{\mathbb{K}}$ of $\mathbb{K}$. So, for simplicity, we can suppose that the field $\mathbb{K}$ is spherically complete without loss of generality. We can also suppose that $b = 1$. Then if $f \in \mathcal{M}(\mathbb{K}) \setminus \mathbb{K}(x)$, we can obviously write $f = \frac{h}{l}$ with $h, l \in \mathcal{A}(\mathbb{K})$, having no common zeros and if $f \in \mathcal{M}(d(a, R^-))$, since $\mathbb{K}$ is spherically complete, we can write $f = \frac{h}{l}$ with $h, l \in \mathcal{A}(d(a, R^-))$, having no common zeros again.

Let $g = \frac{1}{f}$ and let $n = m - 2$. So, by Lemma C.10.1, the problem is reduced to show that $g'g^n - 1$ has infinitely many zeros. Then, $g'g^n - 1 = \frac{(l'h - h'l)l^n - h^{n+2}}{h^{n+2}}$ and since h, l have no common zeros, this is of the form $\frac{P}{h^{n+2}}$ where P is a polynomial of degree q. Now, set $F = (l'h - h'l)l^n$. Applying Corollary C.9.13 to F, we have

(1) $T(r, F) = Z(r, F) + O(1) \leq \overline{Z}(r, F) + \overline{Z}(r, F - P) + T(r, P) + O(1).$

By (1) we derive $Z(r, l'h - h'l) + nZ(r, l) \leq \overline{Z}(r, l'h - h'l) + \overline{Z}(r, l) + \overline{Z}(r, F - P) + T(r, P) + O(1)$. Actually, $\overline{Z}(r, F - P) = \overline{Z}(r, h)$, hence $nZ(r, l) \leq \overline{Z}(r, l) + \overline{Z}(r, h) + T(r, P) + O(1)$ and hence $(n-1)Z(r, l) \leq Z(r, h) + T(P) + O(1)$. But since $T(r, P) = q \log r + O(1)$, we have

(2) $(n - 1)Z(r, l) \leq Z(r, h) + q \log r + O(1).$

Now, consider the hypothesis $f \in \mathcal{M}(\mathbb{K})$. By Theorem C.10.3, if $\liminf_{r \to +\infty} |f|(r) > 0$, i.e., if $\liminf_{r \to +\infty} Z(r, f) - N(r, f) > -\infty$ the claim is proved. Consequently, if the claim is not true, we can assume $\liminf_{r \to +\infty} Z(r, f) - N(r, f) = -\infty$, i.e.,

(3) $\liminf_{r \to +\infty} Z(r, l) - Z(r, h) = +\infty.$

Since f is transcendental, by (3) we notice that l is transcendental. Consequently, (2) is impossible whenever $n \geq 3$, i.e., $m \geq 5$.

Now, suppose $m = 4$, i.e., $n = 2$. More precisely, $\overline{Z}(r,l) \leq Z(r,l) - \frac{s\log r}{2}$ and $\overline{Z}(r,h) \leq Z(r,h) - \frac{t\log r}{2}$, so by Relation (1) we have

$$(4)\quad (n-1)Z(r,l) \leq Z(r,h) + \Big(q - \frac{s+t}{2}\Big)\log r + O(1).$$

Then Relation (3) implies $q - \frac{s+t}{2} > 0$ and hence $f'f^n$ admits a number of zeros strictly superior to $\frac{s+t}{2}$.

Now, suppose that $f \in \mathcal{M}_u(d(0,R^-))$. By Theorem C.10.4, if $\lim_{r\to R^-} |f|(r) = +\infty$, i.e., if $\liminf_{r\to R^-} Z(r,f) - N(r,f) = +\infty$, the claim is proved. Consequently, if the claim is not true, we can assume

$$(5)\quad \liminf_{r\to R^-} Z(r,f) - N(r,f) < +\infty.$$

But by (2), we see that (5) is impossible whenever $n \geq 3$, i.e., $m \geq 5$. □

Corollary C.10.10: *Let $a \in \mathbb{K}$, $R > 0$ and let $f \in \mathcal{M}_u((a,R^-))$. For every $n \in \mathbb{N}$, $n \geq 3$, for every $b \in \mathbb{K}^*$, $f'f^n - b$ has infinitely many zeros.*

Corollary C.10.11: *Let $f \in \mathcal{M}(\mathbb{K}) \setminus \mathbb{K}(x)$ have s multiple zeros and t multiple poles. Let $b \in \mathbb{K}^*$. Then if f has infinitely many multiple zeros or poles, then $f'+bf^4$ has infinitely many zeros that are not zeros of f.*

We will now thoroughly examine the situation when $m = 4$, i.e., $n = 2$, as made in [55]. This requires several basic lemmas.

Lemma C.10.12: *Let $f \in \mathcal{M}(\mathbb{K})$ be transcendental and such that f' has finitely many multiple zeros. Then $\frac{f''f}{(f')^2}$ has no quasi-exceptional value.*

Proof. Let $g = \frac{f}{f'}$. A pole of g is a zero of f', hence by hypothesis, g has finitely many multiple poles. Consequently, by Theorem C.8.7, g' has no quasi-exceptional value. And hence neither has $1-g'$. But $g' = \frac{(f')^2 - f''f}{(f')^2} = 1 - \frac{f''f}{(f')^2}$. Therefore, $\frac{f''f}{(f')^2}$ has no quasi-exceptional value. □

Lemma C.10.13: *Let $f \in \mathcal{M}(\mathbb{K})$ be transcendental and have finitely many multiple zeros. Then $f''f + 2(f')^2$ has infinitely many zeros that are not zeros of f.*

Proof. Suppose first that f' has infinitely many multiple zeros. Since f has finitely many multiple zeros, the zeros of f' are not zeros of f except at most finitely many. Hence, f' has infinitely many multiple zeros that are not zeros of f. And then, they are zeros of f'', hence of $f''f + 2(f')^2$, which proves the statement.

So we are now led to assume that f' has finitely many multiple zeros. By Lemma C.10.12, $\dfrac{f''f + 2(f')^2}{(f')^2}$ has infinitely many zeros. Let $c \in \mathbb{K}$ be a pole of order q of f. Without loss of generality, we can suppose $c = 0$. The beginning of the Laurent development of f at 0 is of the form $\dfrac{a_{-q}}{x^q} + \dfrac{\varphi(x)}{x^{q-1}}$, whereas $\varphi \in \mathcal{M}(\mathbb{K})$ has no pole at 0. Consequently, $\dfrac{f''f + 2(f')^2}{(f')^2}$ is of the form

$$\frac{(a_{-q})^2(3q^2 + q) + x\phi(x)}{(a_{-q})^2(q^2) + x\psi(x)},$$

whereas ϕ, $\psi \in \mathcal{M}(\mathbb{K})$ have no pole at 0. So, the function $\dfrac{f''f + 2(f')^2}{(f')^2}$ has no zero at 0. Therefore, each zero of $\dfrac{f''f + 2(f')^2}{(f')^2}$ is a zero of $f''f + 2(f')^2$ and hence $f''f + 2(f')^2$ has infinitely many zeros.

Now, let us show that the zeros of $f''f + 2(f')^2$ are not zeros of f, except maybe finitely many. Let c be a zero of $f''f + 2(f')^2$ and suppose that c is a zero of f. Then, it is a zero of f' and hence it is a multiple zero of f. But by hypotheses, f has finitely many multiple zeros, hence the zeros of $f''f + 2(f')^2$ are not zeros of f, except at most finitely many. That finishes proving the claim. □

Lemma C.10.14: *Let $f \in \mathcal{M}(\mathbb{K})$ be transcendental and let $b \in \mathbb{K}^*$ be such that $f^2 f' - b$ has finitely many zeros. Then, $N(r, f) \le Z(r, f) + O(1)$.*

Proof. Let $F = f^2 f'$. Since $F - b$ is transcendental and has finitely many zeros, it is of the form $\dfrac{P(x)}{h(x)}$ with $h \in \mathcal{A}(\mathbb{K}) \setminus \mathbb{K}[x]$. Consequently, $|F|(r)$ is a constant when r is big enough and therefore, by Theorem C.4.2 we have $Z(r, F) = N(r, F) + O(1)$ when r is big enough. Now, $Z(r, F) = 2Z(r, f) + Z(r, f')$ and, by Theorem C.4.14, $Z(r, f') \le Z(r, f) + \overline{N}(r, f) - \log r + O(1)$. On the other hand, by Theorem C.4.14 again, we have $N(r, F) = 3N(r, f) + \overline{N}(r, f)$. Consequently, $3N(r, f) + \overline{N}(r, f) \le 3Z(r, f) + \overline{N}(r, f) - \log r + O(1)$, which proves the claim. □

Theorem C.10.15 was published in [55].

Theorem C.10.15: *Let $f \in \mathcal{M}(\mathbb{K}) \setminus \mathbb{K}(x)$. Then for each $b \in \mathbb{K}^*$, $f'f^2 - b$ has infinitely many zeros.*

Proof. Let $b \in \mathbb{K}^*$ and suppose that the claim is wrong, i.e., $f^2 f' - b$ has s zeros, taking multiplicity into account. By Theorem C.10.9, we may assume that f has finitely many multiple zeros and finitely multiple poles. Set $F = f^2 f'$. Then $F' = f(f''f+2(f')^2)$. By Lemma C.10.13, $f''f+2(f')^2$ has infinitely many zeros that are not zeros of f. Consequently, F' admits for zeros: the zeros of f and the zeros of $f''f+2(f')^2$. And by Lemma C.10.13, there exists a sequence of zeros of $f''f+2(f')^2$ that are not zeros of f.

Let $S = \{0, b\}$ and let $Z_0^S(r, F')$ be the counting function of zeros of F' when $F(x)$ is different from 0 and b. Since $F - b$ has finitely many zeros, the zeros c of F', which are not zeros of f, cannot satisfy $F(c) = b$ except at most finitely many. Consequently, there are infinitely many zeros of F' counted by the counting function $Z_0^S(r, F')$ and hence for every fixed integer $t \in \mathbb{N}$, we have

$$Z_0^S(r, F') \geq t \log r + O(1). \tag{1}$$

Let us apply Theorem C.4.24 to F. We have

$$T(r, F) \leq \overline{Z}(r, F) + \overline{Z}(r, F - b) + \overline{N}(r, F) - Z_0^S(r, F') - \log(r) + O(1). \tag{2}$$

Now, we have

$$\overline{Z}(r, F) \leq Z(r, f) + Z(r, f') \tag{3}$$

$$\overline{N}(r, F) = \overline{N}(r, f) \tag{4}$$

and since the number of zeros of $F - b$ is s, taking multiplicity into account,

$$\overline{Z}(r, F - b) \leq s \log r + O(1). \tag{5}$$

Consequently, by (2), (3), (4), and (5) we obtain

$$T(r, F) \leq Z(r, f) + Z(r, f') + \overline{N}(r, f) - Z_0^S(r, F') + (q - 1) \log r + O(1). \tag{6}$$

On the other hand, by construction, $T(r, F) \geq Z(r, F) = 2Z(r, f) + Z(r, f')$, hence by (6) we obtain (7).

$$Z(r, f) \leq \overline{N}(r, f) - Z_0^S(r, F') + (s - 1) \log r + O(1). \tag{7}$$

Now, by Lemma C.10.14, we have $N(r, f) \leq Z(r, f) + O(1)$, hence by (7) we obtain $0 \leq (s-1)\log r - Z_0^S(r, F') + O(1)$ and hence by (1), fixing $t > s - 1$ we can derive $0 \leq (s - 1) \log r - t \log r + O(1)$, a contradiction. That finishes the proof of Theorem C.10.15. □

By Lemma C.10.1 and Theorems C.10.9 and C.10.15, we can now state the general result on the p-adic Hayman conjecture.

Corollary C.10.16: *Let $f \in \mathcal{M}(\mathbb{K})$ be transcendental and let $b \in \mathbb{K}^*$. Then for every $n \geq 2$, $f'f^n - b$ has infinitely many zeros. For every $m \geq 4$, $f' + bf^4$ has infinitely many zeros that are not zeros of f.*

Concerning the case $m = 3$, i.e., $n = 1$, which remains unsolved, thanks to Theorem C.8.10. Corollary C.10.16 has an immediate application to the conjecture with additional hypotheses.

Corollary C.10.17: *Let $f \in \mathcal{M}(\mathbb{K})$. Suppose that there exists $c, q \in]0, +\infty[$, such that $\xi(r, f) \leq cr^q \ \forall r \in [1, +\infty[$. If $f'f^n - b$ has finitely many zeros for some $b \in \mathbb{K}^*$, with $n \in \mathbb{N}$, then $f \in \mathbb{K}(x)$.*

Proof. Suppose f is transcendental. By hypothesis, f^{n+1} satisfies $\zeta(r, \frac{1}{f^{n+1}}) = \xi(r, f^{n+1}) \leq c(n+1)r^q \ \forall r \in [1, +\infty[$. Hence, by Corollary C.10.16 and Theorem C.8.10, $f'f^n$ has no quasi-exceptional value different from 0. □

Corollary C.10.17 may be written in another way.

Corollary C.10.18: *Let $f \in \mathcal{M}(\mathbb{K}) \setminus \mathbb{K}(x)$. Suppose that there exists $c, q \in]0, +\infty[$, such that $\zeta(r, f) \leq cr^q \ \forall r \in [1, +\infty[$. Then for all $m \in \mathbb{N}$, $m \geq 3$ and for all $b \in \mathbb{K}$, $f' - bf^m$ admits infinitely many zeros that are not zeros of f.*

Proof. We set $g = \dfrac{1}{f}$. Then by Corollary C.10.17 $g'g^{m-2}$ has no quasi-exceptional value. Consequently, given $b \in \mathbb{K}^*$, $g'g^{m-2} + b$ has infinitely many zeros and hence $f' - bf^m$ has infinitely many zeros that are not zeros of f. Next, if $b = 0$, by Theorem C.8.10, f' has infinitely many zeros. □

Consider now the case $m = 3$, i.e., $n = 1$.

Theorem C.10.19: *Let $f \in \mathcal{M}(\mathbb{K})$. Suppose that there exists $c, q \in]0, +\infty[$, such that $\beta(r, f) \leq cr^q \ \forall r \in [1, +\infty[$. Then, for all $b \in \mathbb{K}$, $\dfrac{f'}{f^2} - b$ has infinitely many zeros.*

Proof. Set $g = \dfrac{1}{f}$ again. Since the poles of g are the zeros of f, we have $\gamma(r, g) \leq cr^q$. Consequently, by Corollary C.8.11, g' has no quasi-exceptional value. □

Remark: Using Theorem C.10.19 to study the zeros of $f' - bf^2$ that are not zeros of f is not so immediate, as we will see below because of residues of f at poles of order 1. Of course, if $\frac{1}{f}$ is an affine function, $f' + f^2$ has no zeros, except if it is identically zero. And if it is not identically zero, the residue at the pole is not 1 in the general case.

Lemma C.10.20: *Let $f = \dfrac{h}{l} \in \mathcal{M}(\mathbb{K})$ with $h, l \in \mathcal{A}(\mathbb{K})$ having no common zero, let $b \in \mathbb{K}^*$, and let $a \in \mathbb{K}$ be a zero of $h'l - hl' + bh^2$ that is not a zero of $f' + bf^2$. Then a is a pole of order 1 of f and $\mathrm{res}(f, a) = \dfrac{1}{b}$.*

Proof. Clearly, if $l(a) \neq 0$, a is a zero of $f' + bf^2$. Hence, a zero a of $h'l - hl' + bh^2$ that is not a zero of $f' + bf^2$ is a pole of f. Now, when $l(a) = 0$, we have $h(a) \neq 0$, hence $l'(a) = bh(a) \neq 0$ and therefore a is a pole of order 1 of f such that $\dfrac{h(a)}{l'(a)} = \dfrac{1}{b}$. But since a is a pole of order 1, we have $\text{res}(f, a) = \dfrac{h(a)}{l'(a)}$, which ends the proof. □

Theorem C.10.21 is not a result specific to p-adic analysis but it will be useful in Theorem C.10.23.

Theorem C.10.21: *Let $f \in \mathcal{M}(\mathbb{K})$ (resp. let $a \in \mathbb{K}$, let $f \in \mathcal{M}(d(a, R^-))$), let $b \in \mathbb{K}^*$, and let $\alpha \in \mathbb{K}$ (resp. let $\alpha \in d(a, R^-)$) be a point that is not a zero of f and such that the residue of f at α is different from $\dfrac{1}{b}$. Then α is a zero of $f' + bf^2$ if and only if it is a zero of $\dfrac{f'}{f^2} + b$. Moreover, if it is a zero of both functions, it has the same multiplicity with both.*

Proof. Suppose first α is a zero of $f' + bf^2$. If α is not a pole of f, of course it is a zero of $\dfrac{f'}{f^2} + b$ with same multiplicity. Suppose now that α is a pole of f. Since it is not a pole of $f' + bf^2$, it must be a pole of order 1 of f. Without loss of generality, we may assume that $\alpha = 0$ (resp. $a = \alpha = 0$). Consider the Laurent series of f at 0: $f(x) = \dfrac{a_{-1}}{x} + a_0 + a_1 x + x^2\phi(x)$ with $\phi \in \mathcal{M}(\mathbb{K})$ (resp. $\phi \in \mathcal{M}(d(0, R^-))$ and $\phi(0) \neq \infty$. Then $f' + bf^2$ is of the form

$$f'(x) + bf(x)^2 = \frac{a_{-1}(-1 + ba_{-1})}{x^2} + \frac{2ba_0a_1}{x} + a_1 + b(a_0^2 + 2a_1a_{-1}) + x\eta(x)$$

with $\eta \in \mathcal{M}(\mathbb{K})$ (resp. $\eta \in \mathcal{M}(d(0, R^-))$ and $\eta(0) \neq \infty$ and hence, we have $a_{-1}(-1 + ba_{-1}) = 0$, $a_0a_{-1} = 0$, $a_0^2 + 2a_1a_{-1} = 0$. Since by hypothesis $\text{res}(f, \alpha) \neq -\frac{1}{b,}$ we have $(1 + ba_{-1}) \neq 0$, hence $a_{-1} = 0$, a contradiction. Consequently, every zero of $f' + bf^2$ that is not a zero of f is a zero of $\dfrac{f'}{f^2} + b$ with same multiplicity.

Conversely, suppose now that α is a zero of $\dfrac{f'}{f^2} + b$. If α is not a pole of f, it is a zero of $f' + bf^2$, with the same multiplicity, because by hypothesis it is not a zero of f. Now suppose that α is a zero of $\dfrac{f'}{f^2} + b$ and is a pole of f. Clearly, it is a pole of order 1 and again, we may assume that $\alpha = 0$.

Consider again the Laurent series of f at 0: $f(x) = \dfrac{a_{-1}}{x} + a_0 + a_1 x + x^2\phi(x)$ with $\phi \in \mathcal{M}(\mathbb{K})$ and $\phi(0) \neq \infty$. Then

$$\frac{f'}{f^2} = \frac{\frac{-a_{-1}}{x^2} + a_1 + x\psi(x)}{\frac{(a_{-1})^2}{x^2} + \frac{2a_0a_1}{x} + a_0^2 + 2a_1a_{-1} + x\eta(x)},$$

where both ψ, $\eta \in \mathcal{M}(\mathbb{K})$ have no pole at 0. Clearly, $\frac{f'}{f^2}$ is analytic at 0 and its value is $\frac{-1}{a_{-1}}$. But since 0 is a zero of $\frac{f'}{f^2} + b$, we have $a_{-1} = \frac{1}{b}$, what is excluded by hypothesis. Thus, we have proved that every zero of $\frac{f'}{f^2} + b$ is a zero of $f' + bf^2$ (that is not a zero of f) with the same multiplicity and this ends the proof of Theorem C.10.21. □

Theorem C.10.22: *Let $b \in \mathbb{K}^*$ and let $f \in \mathcal{M}(\mathbb{K})$ have finitely many zeros and finitely many residues at its simple poles equal to $\frac{1}{b}$ and be such that $f' + bf^2$ has finitely many zeros. Then f belongs to $\mathbb{K}(x)$.*

Proof. Let $f = \frac{P}{l}$ with $P \in \mathbb{K}[x]$, $l \in \mathcal{A}(\mathbb{K})$ having no common zero with P. Then $f' + bf^2 = \frac{P'l - l'P + bP^2}{l^2}$. By hypothesis, this function has finitely many zeros. Moreover, if a is a zero of $P'l - l'P + bP^2$ but is not a zero of $f' + bf^2$, then by Lemma C.10.20 it is a pole of order 1 of f such that $\text{res}(f, a) = \frac{1}{b}$. Consequently, $P'l - l'P + bP^2$ has finitely many zeros and hence, we can write $\frac{P'l - l'P + bP^2}{l^2} = \frac{Q}{l^2}$ with $Q \in \mathbb{K}[x]$, hence $P'l - l'P = -bP^2 + Q$. But then, by Theorem C.8.5, l is a polynomial, which ends the proof. □

Remark: If $f(x) = \frac{1}{x}$, the function $f' + bf^2$ has no zero whenever $b \neq 1$.

Theorem C.10.23: *Let $f \in \mathcal{M}(\mathbb{K})$ be transcendental and have finitely many zeros of order ≥ 2 and let $b \in \mathbb{K}$. Then $\frac{f'}{f^2} + b$ has infinitely many zeros. Moreover, if $b \neq 0$, every zero α of $\frac{f'}{f^2} + b$ that is not a zero of $f' + bf^2$ is a pole of f of order* 1 *such that the residue of f at α is equal to $\frac{1}{b}$.*

Proof. Let $g = \frac{f'}{f^2} + b$. Since all zeros of f are of order 1 except maybe finitely many, g has finitely many poles of order ≥ 3, hence a primitive G of g has finitely many poles of order ≥ 2. Consequently, by Theorem C.8.7, g has infinitely many zeros.

Now, suppose $b \neq 0$. Let α be a zero of g. If α is not a pole of f, it is a zero of $f' + bf^2$ and we can see that it is not a zero of f.

Finally, suppose that α is a pole of f. Then it must be a pole of order 1 and then, by Lemma C.10.20, the residue of f at α is $\frac{1}{b}$. □

Corollary C.10.24: *Let $f \in \mathcal{M}(\mathbb{K}) \setminus \mathbb{K}(x)$ have finitely many zeros of order ≥ 2 and finitely many poles of order* 1 *and let $b \in \mathbb{K}^*$. Then $f' + bf^2$ has infinitely many zeros that are not zeros of f.*

Remark: On the field $\mathbb{K}$, we find the same we don't know whether a meromorphic function f similar to the function tan is such that $f' + bf^2$ have finitely many zeros.

C.11. Bezout algebras of analytic functions

Let us recall that the ring of analytic functions on a region of the complex number field is well known to be a Bezout ring. The two fundamental theorems necessary for a proof are the Weierstrass factorization theorem and the Mittag–Leffler theorem.

According to the results of [72], it appears that in several hypotheses, rings of analytic functions on complete ultrametric algebraically closed fields are Bezout rings. However, that interesting property is not stated. Moreover, it derives from a Mittag–Leffler theorem referred in general topology whose justification is not relevant. Here, we plan to give proofs of all these properties, using results on quasi-invertible analytic elements and on a Mittag–Leffler theorem for meromorphic functions similar to this of complex analysis, but is quite different from Krasner's Mittag–Leffler theorem for analytic elements on an infraconnected subset of $\mathbb{K}$ [42].

Definitions and notations: Given a pole b of order q of a meromorphic function $f \in \mathcal{M}(\mathbb{K})$ (resp. $f \in \mathcal{M}(d(a, R^-))$), there exists a unique rational function g of the form $\frac{Q(x)}{(x-b)^q}$ with $Q \in \mathbb{K}[x]$, $\deg(Q) \leq q - 1$, called *the singular part of f at the pole b*.

As explained in Chapter B.18, the $\mathbb{K}$-algebra $\mathcal{A}(\mathbb{K})$ (resp. $\mathcal{A}(d(a, R^-)))$ is provided with the following topology of $\mathbb{K}$-algebra: given $\in \mathcal{A}(\mathbb{K})$ (resp. $f \in \mathcal{A}(d(a, R^-)))$, the neighborhoods of f are the sets $\mathcal{W}(f, r, \epsilon) = \{h \in \{\mathcal{A}(\mathbb{K}) \mid |f - h|(r) \leq \epsilon\}$ (resp. $\mathcal{W}(f, r, \epsilon) = \{h \in \{\mathcal{A}(d(a, R^-)) \mid |f - h|(r) \leq \epsilon\}$, with $0 < r < R$, $\epsilon > 0$). The topology defined on $\mathcal{A}(\mathbb{K})$ (resp. $\mathcal{A}(d(a, R^-))$ has expansion to $\mathcal{M}(\mathbb{K})$ (resp. to $\mathcal{M}(d(a, R^-))$. The neighborhoods of a function $f \in \mathcal{M}(\mathbb{K})$ (resp. $\mathcal{M}(d(a, R^-))$ are the sets $\mathcal{W}(f, r, \epsilon) = \{h \in \{\mathcal{M}(\mathbb{K}) \mid |f - h|(r) \leq \epsilon\}$, with $r > 0$ (resp. $\mathcal{W}(f, r, \epsilon) = \{h \in \{\mathcal{M}(d(a, R^-)) \mid |f - h|(r) \leq \epsilon\}$, with $r > 0$), which implies that f and h have the same singular part at each pole in $d(0, r)$.

As explained in Chapter B.18, given a divisor T on $\mathbb{K}$, there is no problem to construct an entire function whose divisor is T. But given a divisor T on a disk $d(a, R^-)$, if $\mathbb{K}$ is not spherically complete, it is not always possible to find an analytic function (in that disk) whose divisor is T due to Lazard's problem.

The set of divisors of $\mathbb{K}$ (resp. of $d(a, R^-)$) is provided with a natural multiplicative law that makes it a semi-group. It is also provided with a natural order relation: given two divisors T and T', we can set $T \leq T'$ when $T(\alpha) \leq T'(\alpha)$ $\forall \alpha \in d(a, R^-)$. Moreover, if T, T' are two divisors such that $T(\alpha) \geq T'(\alpha)$ $\forall \alpha \in d(0, R^-)$, we can define the divisor $\frac{T}{T'}$.

Given $f \in \mathcal{A}(\mathbb{K})$ (resp. $f \in \mathcal{A}(d(a, R^-))$), we have defined *the divisor of* f, denoted by $\mathcal{D}(f)$ on $\mathbb{K}$ (resp. of $d(a, R^-)$) as $\mathcal{D}(f)(\alpha) = 0$ whenever $f(\alpha) \neq 0$ and $\mathcal{D}(f)(\alpha) = s$ when f has a zero of order s at α.

Similarly, given an ideal I of $\mathcal{A}(\mathbb{K})$ (resp. of $\mathcal{A}(d(a, R^-))$), we will denote by $\mathcal{D}(I)$ the lower bound of the the $\mathcal{D}(f)$ $f \in I$ and $\mathcal{D}(I)$ will be called *the divisor of* I.

Remark: Let $f \in \mathcal{A}(d(a, R^-))$ and let $(a_n, q_n)_{n\in\mathbb{N}} = \mathcal{D}(f)$. Then $\omega_{a_n}(f) = q_n$ $\forall n \in \mathbb{N}$ and $\omega_\alpha(f) = 0$ $\forall \alpha \in d(a, R^-) \setminus \{a_n \mid n \in \mathbb{N}\}$.

Mittag–Leffler for meromorphic functions is similar to the classical Mittag–Leffler theorem for complex meromorphic functions.

Theorem C.11.1 (Mittag–Leffler Theorem for Meromorphic Functions): *Let* $(a_m, q_m)_{m\in\mathbb{N}}$ *be a divisor of* $\mathbb{K}$ *(resp. of* $d(a, R^-)$*),* $a \in \mathbb{K}$*,* $R > 0$*) and for every* $n \in \mathbb{N}$*, let* $Q_m \in \mathbb{K}[x]$ *be of degree* $< q_m$*. There exists* $f \in \mathcal{M}(\mathbb{K})$ *(resp.* $f \in \mathcal{M}(d(a, R^-))$*) admitting for poles each* a_m *of order* q_m *and no other pole and such that its singular part at* a_m *is* $\dfrac{Q_m}{(x - a_m)^{q_m}}$.

Proof. The proof is similar to that in the complex case. Without loss of generality, we can suppose that $|a_m| \leq |a_{m+1}|$. Let $(t(n))_{n\in\mathbb{N}}$ be the strictly increasing sequence such that $|a_{t(n)}| < |a_{t(n+1)}|$ and let $r_n = |a_{t(n)}|$ $n \in \mathbb{N}$.

For each $m \in \mathbb{N}$, set $S_m(x) = \dfrac{Q_m(x)}{(x - a_m)^{q_m}}$. Now, for each $n \in \mathbb{N}$, we can set $f_n = \displaystyle\sum_{m\in L_n} S_m(x)$. So, by construction, f_n belongs to $H(d(0, r_{n-1}))$, hence there exists $P_n \in \mathbb{K}[x]$ such that

$$\|f_n - P_n\|_{d(0;r_{n-1})} \leq \left(\frac{1}{2}\right)^n.$$

Consequently, the sequence $(f_n - P_n)_{n\in\mathbb{N}}$ converges to 0 with respect to the topology of $\mathcal{M}(\mathbb{K})$ (resp. of $\mathcal{M}(d(a, R^-))$). Set $f(x) = f_1(x) + \displaystyle\sum_{n=2}^{\infty}(f_n(x) - P_n(x))$. By construction, f belongs to $\mathcal{M}(d(0, r^-))$ $\forall r > 0$, hence f belongs to $\mathcal{M}(\mathbb{K})$ (resp. to $\mathcal{M}(d(a, R^-))$). Moreover, the poles of f are the points a_m, $m \in \mathbb{N}$. Let us take $q \geq 1$ and $\rho > 0$ such that $|a_m - a_q| \geq \rho$ $\forall m \neq q$. Then $f - S_q$ belongs to $H(d(a, \rho))$, so S_q is the singular part of f at S_q. $\square$

Theorem C.11.2: *Let* $a \in \mathbb{K}$*,* $R > 0$*. Let* f*,* $g \in \mathcal{A}(\mathbb{K})$ *(resp.* f*,* $g \in \mathcal{A}(d(a, R^-))$*) be such that* $\mathcal{D}(f) \geq \mathcal{D}(g)$*. Then there exists* $h \in \mathcal{A}(\mathbb{K})$ *(resp.* $h \in \mathcal{A}(d(a, R^-))$*) such that* $f = gh$.

Proof. Let $T = \mathcal{D}(g) = (a_n, q_n)_{n\in\mathbb{N}}$. Let us fix $r > 0$ (resp. $r \in]0, R[$), let $s \in \mathbb{N}$ be such that $|a_n| \leq r$ $\forall n \leq s$ and $|a_n| > r$ $\forall n > s$. Let $P_r(x) = \displaystyle\prod_{n=0}^{s}\left(1 - \frac{x}{a_n}\right)^{q_n}$. We can factorize f in the form $P_r\widehat{f}$ and similarly, we can factorize g in the form $P_r\widehat{g}$, hence $\frac{f}{g} = \frac{\hat{f}}{\widehat{g}}$. Since $\hat{g}$ has no zero in $d(0,r)$ it is invertible in $H(d(0,r))$, hence $\frac{f}{g}$ belongs to $H(d(0,r))$. This is true for all $r > 0$ (resp. for all $r \in]0, R[$) and hence $\frac{f}{g}$ belongs to $\mathcal{A}(\mathbb{K})$ (resp. to $\mathcal{A}(d(a, R^-))$). □

Corollary C.11.3: *Let $a \in \mathbb{K}$, $R > 0$. Let I be an ideal of $\mathcal{A}(\mathbb{K})$ (resp. an ideal of $\mathcal{A}(d(a, R^-))$) and suppose that there exists $g \in I$ such that $\mathcal{D}(g) = \mathcal{D}(I)$. Then $I = g\mathcal{A}(\mathbb{K})$ (resp. $I = g\mathcal{A}(d(a, R^-))$).*

Definitions and notations: Given a divisor E of $\mathbb{K}$, we will denote by $\mathcal{T}(E)$ the ideal of the $f \in \mathcal{A}(\mathbb{K})$ such that $\mathcal{D}(f) \geq E$. Similarly, given $a \in \mathbb{K}$ and $R > 0$ and a divisor E of $d(a, R^-)$, we will denote by $\mathcal{T}_{a,R}(E)$ the ideal of the $f \in \mathcal{A}(d(a, R^-))$ such that $\mathcal{D}(f) \geq E$.

Theorem C.11.4: *For every divisor E of $\mathbb{K}$, $\mathcal{T}(E)$ is a closed ideal of $\mathcal{A}(\mathbb{K})$. Moreover, $\mathcal{T}$ is a bijection from the set of divisors of $\mathbb{K}$ onto the set of closed ideals of $\mathcal{A}(\mathbb{K})$. Further, given a closed ideal I of $\mathcal{A}(\mathbb{K})$, then $I = \mathcal{T}(\mathcal{D}(I))$.*

Corollary C.11.5: *Every closed ideal of $\mathcal{A}(\mathbb{K})$ is principal.*

Proof. Indeed, consider a closed ideal I and let $E = \mathcal{D}(I)$. By Theorem C.11.4, I is of the form $\mathcal{T}(E)$ with $E = \mathcal{D}(I)$. By Corollary B.18.5, there exists $g \in \mathcal{A}(\mathbb{K})$ such that $\mathcal{D}(g) = E$ and of course, g belongs to I. Hence, $g\mathcal{A}(\mathbb{K}) \subset I$. Now, let $f \in I$. Then $\mathcal{D}(f) \geq E$, hence by Corollary B.18.6 f factorizes in the form gh with $h \in \mathcal{A}(K)$, hence $I = g\mathcal{A}(\mathbb{K})$. □

Similarly, we have Theorem C.11.6.

Theorem C.11.6: *Let $a \in \mathbb{K}$ and take $R > 0$. For every divisor E of $d(a, R^-)$, $\mathcal{T}_{a,R}(E)$ is a closed ideal of $\mathcal{A}(d(a, R^-))$. Moreover $\mathcal{T}_{a,R}$ is a bijection from the set of divisors of $d(a, R^-)$ onto the set of closed ideals of $\mathcal{A}(d(a, R^-))$. Further, given a closed ideal I of $\mathcal{A}(d(a, R^-)))$, then $I = \mathcal{T}_{a,R}(\mathcal{D}(I))$.*

Corollary C.11.7: *Suppose $\mathbb{K}$ is spherically complete. Then all closed ideals of $\mathcal{A}(d(a, R^-))$ are principal.*

Proof. Let I be a closed ideal of $\mathcal{A}(d(a, R^-))$ and let $E = \mathcal{D}(I)$. By Theorem C.11.6, we have $I = \mathcal{T}_{a,R}(E)$. Now, by Corollary B.18.5 there exists $g \in \mathcal{A}(\mathbb{K})$ such that $\mathcal{D}(g) = E$ and of course g belongs to $\mathcal{T}_{a,R}(E)$, hence to I. Consequently, $g\mathcal{A}(d(a, R^-)) \subset I$. Conversely, by Corollary C.11.3, we have $I = g\mathcal{A}(d(a, R^-))$. □

Proof. (Theorems C.11.4 and C.11.6) Let E be a divisor of $\mathbb{K}$ (resp. of $d(a, R^-)$). First, let us check whether $\mathcal{T}(E)$ (resp. $\mathcal{T}_{a,R}(E)$) is a closed ideal of $\mathcal{A}(\mathbb{K})$ (resp. of $\mathcal{A}(d(a, R^-))$). Let $E = (a_n, q_n)_{n\in\mathbb{N}}$ and let $(f_m)_{m\in\mathbb{N}}$ be a sequence of elements of $\mathcal{T}(E)$ (resp. of $\mathcal{T}_{a,R}(E)$) converging to a limit f in $\mathcal{A}(\mathbb{K})$ (resp. in $\mathcal{A}(d(a, R^-))$). Let us fix $n \in \mathbb{N}$ and take $r > 0$ such that $|a_n - a_s| > r \ \forall s \neq n$. The convergence in $\mathcal{A}(\mathbb{K})$ (resp. in $\mathcal{A}(d(a, R^-))$) implies the convergence in $H(d(a_n, r))$ with respect to the norm $\| \ . \ \|_{d(a_n,r)}$. Next, for every $n \in \mathbb{N}$, each f_m admits a_n as a zero of order at least q_n, hence by Theorem B.15.3, so does f. Consequently, f belongs to $\mathcal{T}(E)$ (resp. f belongs to $\mathcal{T}_{a,R}(E)$).

Now, let us show that $\mathcal{T}$ (resp. $\mathcal{T}_{a,R}$) is injective. Let E, F be two distinct divisors of $\mathbb{K}$ (resp. of $d(a, R^-)$). Without loss of generality, we can suppose that E admits a pair (b, s) with $s > 0$ and that F either does not admit any pair (b, m) or admits a pair (b, m) with $m < s$. Let $f \in \mathcal{T}(F)$ (resp. let $f \in \mathcal{T}_{a,R}(F)$) and suppose that $\omega_b(f) \geq s$. Then by Theorem B.15.2, f factorizes in the form $(x - b)^{s-m}g$ with $g \in \mathcal{A}(\mathbb{K})$ (resp. $g \in \mathcal{A}(d(a, R^-))$) and of course g belongs to $\mathcal{T}(F)$ (resp. to $\mathcal{T}_{a,R}(F)$). But by construction, g does not belong to $\mathcal{T}(E)$ (resp. to $\mathcal{T}_{a,R}(E)$) because $\omega_b(g) < s$. Therefore, $\mathcal{T}(E) \neq \mathcal{T}(F)$ (resp. $\mathcal{T}_{a,R}(E) \neq \mathcal{T}_{a,R}(F)$). So, $\mathcal{T}$ (resp. $\mathcal{T}_{a,R}$) is injective.

Let us show that it is also surjective. Let I be a closed ideal of $\mathcal{A}(\mathbb{K})$ (resp. of $\mathcal{A}(d(a, R^-))$) and let $E = \mathcal{D}(I)$. Then E is of the form $(a_n, q_n)_{n\in\mathbb{N}}$ with $|a_n| \leq |a_{n+1}|$ and $\lim\limits_{n\to+\infty} |a_n| = +\infty$ (resp. $\lim\limits_{n\to+\infty} |a_n| = R$), hence there is a unique $s \in \mathbb{N}$ such that $a_n \in d(0, r) \ \forall n \leq s$ and $a_n \notin d(0, r) \ \forall n > s$.

Let $J = \mathcal{T}(E)$ (resp. $J = \mathcal{T}_{a,R}(E)$). Then of course, $I \subset J$. Let us show that $J \subset I$. Let $f \in J$ and take $r > 0$. Denoting by P_r the polynomial $\prod\limits_{i=0}^{s}(X - a_i)^{q_i}$, by Theorem B.15.2, we have $I \cap H(d(0, r)) = P_r(x)H(d(0, r))$. But now all functions $g \in J \cap H(d(0, r))$ also are of the form $P_r(x)h(x)$ with $h \in H(d(0, r))$. Consequently, in $H(d(0, r))$ we can write f in the form $f = \sum\limits_{j=1}^{m} g_j h_j$ with $g_j \in I$ and $h_j \in H(d(0, r))$. Let $\epsilon > 0$ be fixed.

For each $j = 1, \ldots, m$, narrowing each h_j by a polynomial ℓ_j in $H(d(0, r))$, we can find $\ell_j \in \mathbb{K}[x]$ such that $|g_j(h_j - \ell_j)|(r) \leq \epsilon$. Now, let $\phi_r = \sum\limits_{j=1}^{m} g_j \ell_j$. Then ϕ_r belongs to I and satisfies $|\phi_r - f|(r) \leq \epsilon$. This is true for each $r > 0$ and for every $\epsilon > 0$. Consequently, since I is closed, f belongs to I. This finishes proving that $\mathcal{T}$ (resp. $\mathcal{T}_{a,R}$) is surjective. Further, we have proven that $I = \mathcal{T}(\mathcal{D}(I))$ (resp. $I = \mathcal{T}_{a,R}(\mathcal{D}(I))$). □

Lemma C.11.8: *Let E be a divisor of $\mathbb{K}$ (resp. a divisor of $d(a, R^-)$, $a \in \mathbb{K}$, $R > 0$) and for each $r > 0$ (resp. $r \in]0, R[$), let $g_r \in H(C(0, r))$. There exists $g \in \mathcal{A}(\mathbb{K})$ (resp. $g \in \mathcal{A}(d(a, R^-))$), not depending on r, such that $\mathcal{D}(g - g_r) \geq E_r$.*

Proof. Let $f \in \mathcal{A}(\mathbb{K})$ (resp. $f \in \mathcal{A}(d(a, R^-))$) be such that $\mathcal{D}(f) \geq E$. By Theorem C.11.1, there exists $F \in \mathcal{M}(\mathbb{K})$ whose principal parts at the poles located in $C(0, r)$ are respectively the same as those of $g_r f^{-1}$ for each $r > 0$. Then fF belongs to $\mathcal{A}(\mathbb{K})$ (resp. to $\mathcal{A}(d(a, R^-))$). Putting $g = fF$, we can see that $\mathcal{D}(g - g_r) \geq E_r$, which ends the proof. □

Theorem C.11.9: *Every ideal of finite type of $\mathcal{A}(\mathbb{K})$ (resp. of $\mathcal{A}(d(a, R^-))$, $a \in \mathbb{K}$, $R > 0$) is closed and is of the form $\mathcal{T}(E)$ (resp. $\mathcal{T}_{a,R}(E)$) with E a divisor of $\mathbb{K}$ (resp. of $d(a, R^-)$).*

Proof. Let I be an ideal of finite type of $\mathcal{A}(\mathbb{K})$ (resp. of $\mathcal{A}(d(a, R^-))$) generated by $f_1, \ldots, f_q$ and let $E = \mathcal{D}(I)$. By Theorem C.11.4, the closure J of I is $\mathcal{T}(E)$ (resp. by Theorem C.11.6 the closure J of I is $\mathcal{T}_{a,R}(E)$). Consequently, we can see that $E = \min(\mathcal{D}(f_1), ..., \mathcal{D}(f_q))$. Let us fix $r > 0$. In $H(C(0, r))$, there exists $g_{1,r}, \ldots, g_{q,r} \in H(C(0, r))$ such that $g = \sum_{j=1}^{q} g_{j,r} f_j$. For each $j = 2, \ldots, q$, let $f_{j,r}$ be the polynomial of the zeros of f_j in $C(0, r)$. By Lemma C.11.8, there exists $g_j \in \mathcal{A}(\mathbb{K})$ (resp. $g_j \in \mathcal{A}(d(a, R^-))$, not depending on r, such that $g_{j,r} - g_j$ be divisible in $H(C(0, r))$ by $\mathcal{D}(f_{1,r})$. Now, set $h = g - \sum_{j=2}^{q} g_j f_j$. We have $\mathcal{D}(h) \geq \mathcal{D}(f_1)$, hence h factorizes in the form $g_1 f_1$ with $g_1 \in \mathcal{A}(\mathbb{K})$ (resp. $g_1 \in \mathcal{A}(d(a, R^-))$) and then

$$g = h + \sum_{j=2}^{q} g_j f_j = \sum_{j=1}^{q} g_j f_j.$$

□

Corollary C.11.10: *$\mathcal{A}(\mathbb{K})$ is a Bezout ring.*

Proof. Indeed, consider an ideal of finite type I. By Theorem C.11.9, it is closed and hence, by Corollary C.11.5 it is principal. □

And by Corollary C.11.7, we have Corollary C.11.11.

Corollary C.11.11: *Let $a \in \mathbb{K}$ and let $R > 0$. If $\mathbb{K}$ is spherically complete, $\mathcal{A}(d(a, R^-))$ is a Bezout ring.*

Proof. Indeed, consider an ideal of finite type I. By Theorem C.11.9, it is of the form $\mathcal{T}_{a,R}(E)$ with E a divisor of $d(a, R^-)$ and hence, by Theorem C.11.6, it is closed. But then, by Corollary C.11.7, it is principal. □

Bibliography

1. Adams, W.W. and Straus, E.G. *Non-Archimedean analytic functions taking the same values at the same points*, Illinois J. Math., 15, 418–424 (1971).
2. Amice, Y. *Les nombres p-adiques*, P.U.F. (1975).
3. Amice, Y. *Dual d'un espace $H(D)$ et transformation de Fourier*, Groupe d'étude d'Analyse Ultramétrique (IHP), Paris (1973–1974).
4. Bartels, S. *Meromorphic functions sharing a set with 17 elements, ignoring multiplicities*, Complex Variable Theory Appl., 39, no. 1, 85–92 (1999).
5. Bergweiler, W. and Eremenko, A. *On the singularities of the inverse to a meromorphic function of finite order*, Rev. Mat. Iberoam., 11, 355–373 (1995).
6. Berkovich, V. *Spectral Theory and Analytic Geometry over Non-archimedean Fields*, Issue 33 of AMS Surveys and Monographs, American Mathematical Society (1990).
7. Bezivin, J.-P. *Wronskien et equations differentielles p-adiques*, Acta Arith., 158, no. 1, 61–78 (2013).
8. Bezivin, J.-P., Boussaf, K. and Escassut, A. *Zeros of the derivative of a p-adic meromorphic function,* Bull. Sci. Math., 136, no. 8, 839–847 (2012).
9. Bezivin, J.-P., Boussaf, K. and Escassut, A. *Some old and new results on zeros of the derivative of a p-adic mermorphic function*, Contem. Math., 596, 23–30 (2013).
10. Boussaf, K. *Motzkin factorization in algebras of analytic elements*, Ann. Math. Blaise Pascal, 2, no. 1, 73–91 (1995).
11. Boussaf, K., Hemdaoui, M. and Maïnetti, N. *Tree structure on the set of multiplicative semi-norms of Krasner algebras $H(D)$*, Rev. Mat. Complut., XIII, no. 1, 85–109 (2000).
12. Boussaf, K. *Strictly analytic functions on p-adic analytic open sets*, Publ. Mat., 43, no. 1, 127–162 (1999).
13. Boussaf, K. *Shilov boundary for algebras $H(D)$*, Ital. J. Pure Appl. Math., 8, 75–82 (2000).
14. Boussaf, K. *Image of circular filters*, Int. J. Math. Game Theory Algebra, 10, no. 5, 365–372 (2003).
15. Boussaf, K., Escasssut, A. and Maïnetti, N. *Mappings in the tree $Mult(K[x])$*, Bull. Belg. Math. Soc. Simon Stevin, 9, 25–47 (2002).
16. Boussaf, K., Boutabaa, A. and Escassut, A. *p-adic sets of range uniqueness*, Proc. Edingburgh Math. Soc., 50, no. 2, 263–276 (2007).
17. Boussaf, K. *p-adic sets of range uniqueness*, Rend. Circ. Mat. Palermo, 56, no. 3, 381–390 (2007).
18. Boussaf, K. *Identity sequences,* Bull. Sci. Math., 134, no. 1, 44–53 (2010).
19. Boussaf, K. *Picard values of p-adic meromorphic functions*, p-Adic Numbers Ultrametric Anal. Appl., 2, no. 4, 285–292 (2010).

20. Boussaf, K. Boutabaa, A. and Escassut, A. *Growth of p-adic entire functions and applications*, Houston J. Math., 40, no. 3, 715–736 (2014).
21. Boussaf, K. and Ojeda, J. *Value distribution of p-adic meromorphic functions*, Bull. Belg. Math. Soc. Simon Stevin, 18, no. 4, 667–678 (2011).
22. Boussaf, K., Ojeda, J. and Escassut, A. *Primitives of p-adic meromorphic functions*, Contemp. Math., 551, 51–56 (2011).
23. Boussaf, K., Ojeda, J. and Escassut, A. *Zeros of the derivative of a p-adic meromorphic function and applications,* Bull. Belg. Math. Soc. Simon Stevin, 19, no. 2, 367–372 (2012).
24. Boussaf, K., Ojeda, J. and Escassut, A. *p-adic meromorphic functions* $f'P'(f)$, $g'P'(g)$ *sharing a small function*, Bull. Sci. Math., 136, no. 2, 172–200 (2012).
25. Boussaf, K., Ojeda, J. and Escassut, A. *Complex meromorphic functions* $f'P'(f)$, $g'P'(g)$ *sharing a small function*, Indagationes (N.S.), 24, no. 1, 15–41 (2013).
26. Boussaf, K., Boutabaa, A. and Escassut, A. *Growth of p-adic entire functions and applications*, Houston J. Math., 40, no. 3, 715–736 (2014).
27. Boussaf, K., Escassut, A. and Ojeda J. *New results on applications of Nevanlinna methods to value sharing problems*, P-Adic Numbers, Ultrametric Anal. Appl., 5, no. 4, 278–301 (2013).
28. Boussaf, K., Escassut, A. and Ojeda J. *Growth of complex and p-adic meromorphic functions and branched small functions*, Preprint.
29. Boutabaa, A. *Théorie de Nevanlinna p-adique*, Manuscripta Math., 67, 251–269 (1990).
30. Boutabaa, A., Escassut, A. and Haddad, L. *On uniqueness of p-adic entire functions*, Indag. Math., 8, 145–155 (1997).
31. Boutabaa, A. and Escassut, A. *URS and URSIMS for p-adic meromorphic functions inside a disk*, Proc. Edinburgh Math. Soc., 44, 485–504 (2001).
32. Boutabaa, A. and Escassut, A. *Applications of the p-adic Nevanlinna theory to functional equations,* Ann. l'Institut Fourier, 50, no. 3, 751–766 (2000).
33. Boutabaa, A. and Escassut, A. *Nevanlinna theory in characteristic p and applications*, Ital. J. Pure Appl. Math., 23, 45–66 (2008).
34. Boutabaa, A., Cherry, W. and Escassut, A. *Unique range sets in positive characteristic*, Acta Arith., 103, no. 2, 169–189 (2002).
35. Boutabaa, A. *About the p-adic Yosida equation inside a disk,* Indag. Math. (N.S.), 20, no. 3, 397–413 (2009).
36. Carleson, L. *Interpolation by bounded analytic functions and the corona problem,* Ann. Math., 76, 547–559 (1962).
37. Charak, K.S. *Value distribution theory of meromorphic functions*, Math. Newslett., 18, no. 4, 1–35 (2009).
38. Chen, H. and Fang, M. *On the value distribution of* $f^n f'$, Sci. China, 38, no. 7, 789–798 (1995).
39. Cherry, W. and Yang, C.C. *Uniqueness of non-Archimedean entire functions sharing sets of values counting multiplicities*, Proc. Am. Math. Soc., 127, 967–971 (1998).
40. Diarra, B. *Ultraproduits ultramétriques de corps valués*, Ann. Sci. l'Université de Clermont II, Série Math. Fasc., 22, 1–37 (1984).
41. Diarra, B. *The continuous coalgebra endomorphism of* $C(\mathbb{Z}_p, K)$, Bull. Belg. Math. Soc. Simon Stevin, 9, 63–79 (2002).
42. Diarra, B. and Escassut, A. *Survey on Bezout rings of p-Adic analytic functions*, Southeast Asian Bull. Math., 39, 1–8 (2015).

43. Escassut, A. *Algèbres d'éléments analytiques en analyse non archimedienne*, Indagat. Math., 36, no. 4, 339–351 (1974).
44. Escassut, A. *Algèbres de Krasner-Tate et algèbres de Banach ultramétriques*, Astérisque, 10, 1–107 (1973).
45. Escassut, A. *Transcendence order over* $\mathbb{Q}_p$ *in* $\mathbb{C}_p$, J. Number Theory, 16, no. 3, 395–402 (1982).
46. Escassut, A. and Sarmant, M.C. *Sufficient conditions for injectivity of analytic elements,* Bull. Sci. Math., 118, 29–46 (1994).
47. Escassut A. and Sarmant, M.C. *Mittag-Leffler series and Motzkin product for invertible analytic elements*, Rivista di Matematica Pura ed Applicata, 2, 61–73 (1992).
48. Escassut A. and Sarmant, M.C. *Injectivity, Mittag-Leffler series and Motzkin products,* Ann. Sci. Math. Quebec, 16, 155–173 (1992).
49. Escassut, A. *Analytic Elements in p-adic Analysis*, World Scientific Publishing Co. Pte. Ltd., Singapore (1995).
50. Escassut, A. *Ultrametric Banach Algebras*, World Scientific Publishing Co. Pte. Ltd., Singapore (2003).
51. Escassut, A. *p-adic value distribution*, Some Topics on Value Distribution and Differentability in Complex and P-adic Analysis, Mathematics Monograph, Series 11, Science Press, Beijing, pp. 42–138 (2008).
52. Escassut, A. and Ojeda, J. *Exceptional values of p-adic analytic functions and derivatives*, Complex Var. Elliptic Equations, 56, no. 1–4, 263–269 (2011).
53. Escassut, A., Ojeda, J. and Yang, C.C. *Functional equations in a p-adic context*, J. Math. Anal. Appl., 351, no. 1, 350–359 (2009).
54. Escassut, A. and Ojeda, J. *Branched values and quasi-exceptional values for p-adic mermorphic functions*, Houston J. Math., 39, no. 3, 781–795 (2013).
55. Escassut, A. and Ojeda, J. *The p-adic Hayman conjecture when* $n = 2$, Complex Var. Elliptic Eqn., 59, no. 10, 1451–1455 (2014).
56. Escassut, A. and Riquelme, J.L. *Applications of branched values to p-adic functional equations on analytic functions*, p-Adic Numbers, Ultrametric Anal. Appl., 6, no. 3, 188–194 (2014).
57. Escassut, A. *Survey and additional properties on the transcendence order over* $\mathbb{Q}_p$ *in* $\mathbb{C}_p$, p-Adic Numbers, Ultrametric Anal. Appl., no. 1, 17–23 (2015).
58. Escassut, A. *Value Distribution in p-adic Analysis*, World Scientific Publishing Co. Pte. Ltd., Singapore (2015).
59. Frank, G and Reinders, M. *A unique range set for meromorphic functions with 11 elements*, Complex Var. Theory Appl., 37, 185–193 (1998).
60. Gackstatter, F. and Laine, I. *Zur Theorie der gewhnlichen differentialgleichungen im Komplexen,* Ann. Polon. Math., 38, no. 3, 259–287. (1980).
61. Garandel, G. *Les semi-normes multiplicatives sur les algèbres d'éléments analytiques au sens de Krasner*, Indag. Math., 37, no. 4, 327–341 (1975).
62. Guennebaud, B. *Sur une notion de spectre pour les algèbres normées ultramétriques*, thèse Université de Poitiers, Poitiers, France (1973).
63. Ha, H.K. *On p-adic meromorphic functions*, Duke Math. J., 50, 695–711 (1983).
64. Hayman, W.K. *Meromorphic Functions,* Oxford University Press (1975).
65. Hayman, W.K. *Picard values of meromorphic functions and their derivatives*, Ann. Math., 70, 9–42 (1959).
66. Hu, P.C. and Yang, C.C. *Meromorphic Functions over non-Archimedean Fields*, Kluwer Academic Publishers (2000).

67. Kaplansky, I *Fields and rings. Reprint of the second (1972) edition*, Chicago Lectures in Mathematics. University of Chicago Press, Chicago, IL (1995).
68. Krasner, M. *Prolongement analytique uniforme et multiforme dans les corps valués complets. Les tendances géométriques en algèbre et théorie des nombres*, Centre National de la Recherche Scientifique, Clermont-Ferrand, France, pp. 94–141 (1964). (Colloques internationaux de C.N.R.S. Paris, 143).
69. Krasner, M. *Nombre d'extensions d'un degré donn'e d'un corps p-adique. Les tendances géométriques en algèbre et théorie des nombres*, Centre National de la Recherche Scientifique, Clermont-Ferrand, France, pp. 143–169 (1964). (Colloques internationaux de C.N.R.S. Paris, 143).
70. Lang, S. *Algebra*, Addison Wesley, Reading, MA-London, England-Don Mills, Canada (1965).
71. Lang, S. *Introduction to Transcendental Numbers*, Addison-Wesley Publishing Co., Reading, MA-London, England-Don Mills, Canada (1966).
72. Lazard, M. *Les zéros des fonctions analytiques sur un corps valué complet*, IHES, Publ. Math., 14, 47–75 (1962).
73. Malher, K. *Ein Beweis der Transzendenz der P-adischen Exponential funktion*, J. Reine Angew. Math., 169, 61–66 (1932).
74. Motzkin, E. and Robba, Ph. *Prolongement analytique en analyse p-adique*, Séminaire de theorie des nombres, année 1968-1969, Faculté des Sciences de Bordeaux.
75. Motzkin, E. *La décomposition d'un élément analytique en facteurs singuliers*, Ann. Inst. Fourier, 27, no. 1, 67–82 (1977).
76. Mues, E. and Reinders, M. *On a question of C. C. Yang*, Complex Var. Theory Appl., 34, no. 1-2, 171–179 (1997).
77. Nevanlinna, R. *Le théorème de Picard-Borel et la théorie des fonctions méromorphes*, Gauthiers-Villars, Paris, France (1929).
78. Ojeda, J. *On Hayman's Conjecture over a p-adic field*, Taiwanese J. Math. 12, no. 9, 2295–2313 (2008).
79. Ojeda, J. *Applications of the p-adic Nevanlinna theory to problems of uniqueness, Advances in p-adic and Non-Archimedean analysis*, Contemp. Math., 508, 161–179 (2010).
80. Ojeda, J. *Zeros of ultrametric meromorphic functions* $f'f^n(f-a)^k - \alpha$, Asian-Eur. J. Math., 1, no. 3, 415–429 (2008).
81. Ojeda, J. *Uniqueness for ultrametric analytic functions*, Bull. Math. Sci. Math. Roumanie, 54, no. 2, 153–165 (2011).
82. Picard, E. *Mémoire sur les fonctions entières, Annales de l'Ecole Normale Supérieure* (1880).
83. Robba, P. *Fonctions analytiques sur les corps valués ultramétriques complets. Prolongement analytique et algèbres de Banach ultramétriques*, Astérisque, no. 10, 109–220 (1973).
84. Robert, A. *A Course in P-Adic Analysis*, Graduate texts, Springer (2000).
85. Ru, M. *A note on p-adic Nevanlinna theory*, Proc. Amer. Math. Soc., 129, 1263–1269 (2001).
86. Rubel, L.A. *Entire and Meromorphic Functions*, Springer-Verlag, New York, NY (1996).
87. Sarmant, M.-C. *Factorisation en Produit Méromorphe d'un élément semi-inversible*, Bull. Sci. Math., 115, 379–394 (1991).
88. Shamseddine, K. *A brief survey of the study of power series and analytic functions on the Levi-Civita fields*, Contemp. Math., 596, 269–279, (2013).

89. Schikhof, W.H. *Ultrametric Calculus. An Introduction to p-adic Analysis,* Cambridge University Press (1984).
90. Serre, J.-P. *Dépendance d'exponentielles p-adiques*, Séminaire Delange-Pisot-Poitou. Thé orie des nombres, 7, no. 2 (1965–1966), Exp. No. 15.
91. Waldschmidt, M. *Nombres transcendants*, Lecture Notes in Mathematics 402, Springer-Verlag (1974).
92. Yamanoi, K. *The second main theorem for small functions and related problems,* Acta Math., 192, 225–294 (2004).

Definitions

Chapter A.1.

closure, adherence,
interior, opening,
value group, valuation group,
absolute value,
valuation ring, valuation ideal,
residue class, residue class field, residue characteristic,
dense valuation, discrete valuation, trivial valuation,
dense absolute value, discrete absolute value, trivial absolute value,
hole of a set,
infraconnected set, infraconnected component,
affinoid set,
empty annulus of a set.

Chapter A.2.

filter thinner than another one,
sequence thinner than a filter,
filter secant with a set, with another filter,
increasing distances sequence, decreasing distances sequence,
monotonous distances sequence,
equal distances sequence,
increasing filter of center a and diameter r,
decreasing filter of center a and diameter r,
decreasing filter with no center of diameter r, of canonical basis (D_n),
monotonous filter,
spherically complete field,
pierced monotonous filter,
circular filter of center a and diameter r,
peripheral of a bounded set,
circular filter with no center, of diameter r, of canonical basis (D_n),
large circular filter, punctual circular filter,
$\mathcal{F}$-affinoid.

Chapter A.3.
Gauss norm.

Chapter A.4.
quasi-monic polynomial.

Chapter A.5.
p-adic absolute value.

Chapter A.6.
Eisenstein polynomial,
uniformizer of an extension of $\mathbb{Q}_p$,
ramification index of an extension of $\mathbb{Q}_p$.

Chapter A.7.
principal ultrafilter,
incomplete ultrafilter,
immediate extension of an ultrametric field.

Chapter A.8.
transcendence order $\leq \tau$ over $\mathbb{Q}_p$,
denominator of an algebraic number,
transcendence type $\leq \tau$ in $\mathbb{C}_p$,
infinite transcendence type.

Chapter B.2.
analytic element on D,
invertible element of a space $H(D)$,
pole of order q of an element $f \in H(D)$,
polynomial of poles of an element of $H(D)$ in $\overline{D} \setminus D$,
residue of an element $f \in H(D)$ at a pole a.

Chapter B.3.
bianalytic element from A onto B.

Chapter B.4.
punctual semi-norm.

Chapter B.5.
radius of convergence,
entire function,
power series, Laurent series,
zero of multiplicity order q.

Chapter B.6.

f-hole, Mittag–Leffler term of f associated to a hole T_n,
principal term of f,
Mittag–Leffler series of f on the infraconnected set D,
specific circular filter,
residue of an analytic element on a hole.

Chapter B.7.

polynomial of zeros of an analytic element on a subset A,
semi-invertible analytic element,
quasi-invertible analytic elements,
quasi-minorated analytic element.

Chapter B.8.

analytic element vanishing along a filter $\mathcal{F}$,
analytic element properly vanishing along a filter $\mathcal{F}$.

Chapter B.9.

piercing of a subset,
well-pierced subset.

Chapter B.11.

analytic element strictly vanishing along a monotonous filter $\mathcal{F}$,
analytic element collapsing along a monotonous filter $\mathcal{F}$.

Chapter B.12.

analytic set.

Chapter B.18.

divisor on K, divisor on a disk $d(a, r^-)$,
bounded divisor in a disk,
divisor of a function,
Euclidean division by a polynomial.

Chapter B.20.

index of an analytic element,
pure factor associated to a hole,
Motzkin factor in a hole,
Motzkin index,
f-supersequence,
Motzkin factorization,
principal factor.

Chapter B.21.

order of growth of an entire function,
cotype of growth of an entire function.

Chapter B.22.

type of growth of an entire function,
a function satisfies Hypothesis L.

Chapter B.24.

order of growth of a function in an open disk,
cotype of growth of a function in an open disk,
type of growth of a function in an open disk.

Chapter C.1.

meromorphic function in K, in a disk $d(a, R^-)$,
pole of order s of a meromorphic function,
divisor of a meromorphic function in K, in a disk $d(a, R^-)$,
divisor of the poles of f,
exceptional value,
pseudo-exceptional value,
quasi-exceptional value.

Chapter C.2.

residue of a meromorphic function at a point a,
analytic element meromorphic in a hole,
singular part of f at b,
polynomial of the poles in a hole.

Chapter C.3.

meromorphic unction out of a hole,
exceptional value, for a meromorrphic function out of a hole,
pseudo-exceptional value,
quasi-exceptional value.

Chapter C.4.

counting function of zeros, counting multiplicity,
counting function of zeros, ignoring multiplicity,
counting function of poles, counting multiplicity,
counting function of polcs, ignoring multiplicity,
characteristic function Nevanlinna function of a meromorphic function.

Chapter C.5.

counting function of zeros, counting multiplicity in $\mathcal{M}(D)$,
counting function of zeros, ignoring multiplicity n $\mathcal{M}(D)$,
counting function of poles, counting multiplicity n $\mathcal{M}(D)$,
counting function of poles, ignoring multiplicity, n $\mathcal{M}(D)$,
characteristic function Nevanlinna function of a meromorphic function n $\mathcal{M}(D)$.

Chapter C.6.

function of uniqueness,
function of strong uniqueness.

Chapter C.7.

perfectly branched value,
totally branched value.

Chapter C.9.

small meromorphic function,
small analytic function.

Notations

Chapter A.1.

$\mathbb{N}$, $\mathbb{Z}$, $\mathbb{Q}, \mathbb{R}\mathbb{C}$
$\overline{S}$, $\overset{\circ}{S}$
$|x|$, $(x \in E)$
$|t|_\infty$, $(t \in \mathbb{R})$
$\log$, θ, v, Ψ
$|E|$
U_E, M_E, $\mathcal{L}$
$d(a, r)$
$d(a, r^-)$
$C(a, r)$
$\Delta(a, r', r'')$
$\Gamma(a, r', r'')$
$\delta(a, D)$
$\delta(D, E)$
$\mathrm{diam}(D)$
$\mathrm{codiam}(D)$
$\widehat{L}$
$\widehat{d}(a, r)$
$\widehat{d}(a, r^-))$
$\widehat{C}(a, r))$
$\widehat{D}$
$\widetilde{D}$
$\mathcal{J}(\Lambda)$
$\mathcal{E}(\Lambda)$
$\chi(D)$

Chapter A.2.

$\mathcal{C}_D(\mathcal{F})$
$\mathcal{B}_D(\mathcal{F})$
$\mathrm{diam}(\mathcal{F})$
$\widehat{G}$

Chapter A.3.

Chapter A.4.

Chapter A.5.

Chapter A.6.

Chapter A.7.

Chapter A.8.

Chapter B.1.

$\mathcal{U}_D$

$R(D),\ R_b(D),\ R_0(D)$

$\|\ .\ \|_D$

$Mult(A)$

$Mult(A, \|\ .\ \|),\ Mult_m(A, \|\ .\ \|),\ Mult_1(A, \|\ .\ \|)$

Chapter B.2.

$H(D),\ H_b(D),\ H_0(D)$

Alg

$\mathrm{res}(f, a)$

Chapter B.3.

$\Xi(D', D)$

Chapter B.4.

$Mult(H(D), \mathcal{U}_D)$

${}_D\varphi_{\mathcal{F}},\ {}_D\varphi_{a,r},\ \varphi_a,\ {}_D\varphi_D,\ {}_D\varphi_D,\ {}_D\varphi_\infty$

$\mathcal{J}(\mathcal{F}),\ \mathcal{J}_0(\mathcal{F}),\ \mathcal{J}(a)$

Chapter B.5.

$\mathcal{A}(\mathbb{K}),\ \mathcal{A}(d(a, r^-)),\ \mathcal{A}_b(d(a, r^-)),\ \mathcal{A}_u(d(a, r^-)),\ \mathcal{A}(\mathbb{K} \setminus d(a, r)),$

$\mathcal{A}_b(\mathbb{K} \setminus d(a, r)),\ \mathcal{A}_u(\mathbb{K} \setminus d(a, r))\ \mathcal{A}(\Gamma(a, r', r'')),\ \mathcal{A}_b(\Gamma(a, r', r''))$

$|f|(r)$

$\nu^+(f, \mu),\ \nu^-(f, \mu),\ \nu(f, \mu)$

Chapter B.6.

$\overline{\overline{f_0}},\ \overline{\overline{f_T}}$

$\widehat{H}(\widehat{D})$

$\mathrm{res}(f, T)$

$E^{\circledast}$

Chapter B.7.

$\omega_\alpha(f)$

Chapter B.8.

Alg

$f^*, \overline{f}$

$\overline{f}$

$\mathcal{J}(a)$

Condition A, Condition B

Chapter B.10.

$\lambda(a)$

$\Psi(x),\ \Psi(f,\mu),\ \Psi_a(f,\mu)$

Chapter B.13.

$f^{<n>}$

Chapter B.15.

$\mathcal{Q}_n(D),\ \mathcal{Q}(D)$

Chapter B.16.

r_q

Log, exp

Chapter B.18.

$\mathcal{D}(f)$

$\mathcal{T},\ |T|(r)$

Chapter B.19.

$\zeta(q,s,r),\ \tau(q,s,n)$

$\mathcal{T}_R$

Chapter B.20.

$m(h,T)$

f^T

$\sqrt[q]{1+u}$

Chapter B.21.

$\rho(f)$ for an entire function

$\zeta(r,f),\ \xi(r,f)$

$\psi(f)$ for an entire function

Chapter B.22.

$\sigma(f),\ \widetilde{\sigma}(f)$ for an entire function

Chapter B.24.

$\rho(f)$ for an analytic function inside a disk

$\zeta(r,f),\ \xi(r,f)$

$\psi(f)$ for an analytic function inside a disk

$\zeta(r,f)$ and $\theta(f)$ for an analytic function inside a disk

$\mathcal{M}(\mathbb{K})$

$\mathcal{M}(d(a,r^-)),\ \mathcal{M}_b(d(a,r^-)),\ \mathcal{M}_u(d(a,r^-))$

$\mathcal{D}(f)$

$\omega_\alpha(f)$ (f meromorphic function)

Chapter C.2.

res(f, α) (f meromorphic function)

Chapter C.3.

$\mathcal{M}^c(D)$

Chapter C.4.

$Z(r,f), \overline{Z}(r,f)$

$N(r,f), \overline{N}(r,f)$

$T(r,f)$

Chapter C.5.

$Z_R(r,f), \overline{Z}_R(r,f)$

$N_R(r,f), \overline{N}_R(r,f)$

$T_R(r,f)$

Chapter C.6.

$g.c.d.(m,n)$

Chapter C.9.

$\mathcal{M}_f(\mathbb{K})$, $\mathcal{M}_f(d(0,R^-)))$

$\mathcal{A}_f(\mathbb{K})$, $\mathcal{A}_f(d(0,R^-))$

$\Xi(h)$, $\Upsilon(h)$

$Z(r, f \mid$ "$f(x)$ satisfying Property P")

$\overline{Z}(r, f \mid$ "$f(x)$ satisfying Property P")

$N(r, f \mid$ "$f(x)$ satisfying Property P")

$\overline{N}(r, f \mid$ "$f(x)$ satisfying Property P")

Index

Printed in the United States
by Baker & Taylor Publisher Services